HANDBOOK OF ELECTRICAL INSTALLATION PRACTICE

HANDBOOK OF ELECTRICAL INSTALLATION PRACTICE

THIRD EDITION

Edited by

ALAN SMITH
IEng, Hon FIEIE

b

Blackwell
Science

© Third edition Blackwell Science Ltd.
1996
© First and second editions E. A. Reeves
1983, 1990

Blackwell Science Ltd
Editorial Offices:
Osney Mead, Oxford OX2 0EL
25 John Street, London WC1N 2BL
23 Ainslie Place, Edinburgh EH3 6AJ
350 Main Street, Malden
  MA 02148 5018, USA
54 University Street, Carlton
  Victoria 3053, Australia
10, rue Casimir Delavigne
  75006 Paris, France

Other Editorial Offices:

Blackwell Wissenschafts-Verlag GmbH
Kurfürstendamm 57
10707 Berlin, Germany

Blackwell Science KK
MG Kodenmacho Building
7–10 Kodenmacho Nihombashi
Chuo-ku, Tokyo 104, Japan

The right of the Author to be
identified as the Author of this Work
has been asserted in accordance
with the Copyright, Designs and
Patents Act 1988.

First edition published by Granada
Publishing Ltd –
  Technical Books Division 1983
Second edition published by Blackwell
  Scientific Publications 1990
Reprinted 1993
Reprinted by Blackwell Science Ltd 1994
Third edition 1996
Reprinted 1997, 1999

Set by Setrite Typesetters Ltd
Printed and bound in the United Kingdom
at the University Press, Cambridge

DISTRIBUTORS
  Marston Book Services Ltd
  PO Box 269
  Abingdon, Oxon OX14 4YN
  (*Orders*: Tel: 01235 465500
          Fax: 01235 465555)

USA
  Blackwell Science, Inc.
  Commerce Place
  350 Main Street
  Malden, MA 02148 5018
  (*Orders*: Tel: 800 759 6102
              781 388 8250
          Fax: 781 388 8255)

Canada
  Login Brothers Book Company
  324 Saulteaux Crescent
  Winnipeg, Manitoba R3J 3T2
  (*Orders*: Tel: 204 837 2987
          Fax: 204 837 3116)

Australia
  Blackwell Science Pty Ltd
  54 University Street
  Carlton, Victoria 3053
  (*Orders*: Tel: 3 9347 0300
          Fax: 3 9347 5001)

A catalogue record for this title
is available from the British Library

ISBN 0–632–03882–9

Library of Congress
Cataloging-in-Publication Data

Handbook of electrical installation practice/
  edited by Alan Smith. – 3rd ed.
      p.   cm.
  Includes index.
    ISBN 0-632-03882-9 (alk. paper)
    1. Electric wiring, Interior –
Handbooks, manuals, etc.
2. Electric apparatus and appliances –
Installation – Great Britain –
Handbooks, manuals, etc.   I. Smith,
Alan.
TK3271.H28   1996
621. 319′24 – dc20   95–30821
                                          CIP

For further information on
Blackwell Science, visit our website:
www.blackwell-science.com

ref 621.31924 Smi

25638

Contents

Preface

My first words must be to thank Eric Reeves for his work in the production of the two previous editions of this book. I trust that he will approve of this third edition.

Undoubtedly times are changing in all aspects of the electrical installation industry.

- The revolution in electronic microtechnology has introduced complex controls into both industrial and domestic electrical systems.
- Rationalization and European harmonization have brought about changes in standards and the Wiring Regulations.
- On the political and financial front, the supply industry has been privatized.

The effects of these changes have meant major updating in the content of this comprehensive handbook. It is difficult to highlight significant chapter revisions. Readers of earlier editions will be advised to study all of the chapters and to take note of the consequences of both technical and commercial aspects.

Sections dealing with the safety of electrical installations, most particularly Chapter 11, now take into account Electricity at Work Regulations. To a large extent this statutory document complements the new BS 7671, known previously as the IEE Wiring Regulations.

All chapters required some revision of British Standards. Chapter 12 has been completely rewritten to explain the intricacies of European harmonization and international standardization. To some extent this subject is a moving target but every effort has been made to bring references into line with current documentation.

There has been a change in emphasis on some topics. Despite efforts by the industry, the universal introduction of building automation has not taken off. Probably growth has been inhibited by financial strictures. The previous chapter on this subject has therefore been removed and replaced with a new Chapter 4 dealing with cable management systems.

Although developers may not be taking full advantage of systems for building management, there is no doubt that data processing equipment has come to the forefront in offices, factories and even the domestic environment. The sensitivity of computer software and hardware to interference and damage by voltage surges is emphasized in the expansion of Chapter 9 dealing with lightning protection.

Lighting design has been revolutionized by changes in light sources. Extra-low voltage luminaires are now used extensively for display and feature illumination, and the installation of security lighting has almost become an industry in itself. Chapter 18 takes these factors into account.

It is quite impossible to forecast the direction of national or international engineering developments as we approach the millenium. Changes happen quite rapidly. All that can be said is that everyone involved with electrical installations in building will need to keep abreast of new technology.

Together with the team of distinguished chapter authors, I trust that this third edition will be helpful to devotees of earlier versions of the handbook and to new readers.

Alan Smith

Addendum

Even while this book has been in production there have been significant changes in the industry that cannot be handled within the text of chapters.

- The declared low voltage supply has been harmonized across Europe to 400/230 V with tolerances of +10% and −6%. The traditional UK supply of 415/240 V is within these tolerances but all references or calculations of installation characteristics should now use the 400/230 V figures.
- The consequences of privatization of the supply industry are only now being fully appreciated. Companies are merging or being taken over and the management of the industry is in a continual state of flux. This will ultimately affect the tariffs and supply conditions that are given in some of the chapters of this book.

Alan Smith
October 1995

CHAPTER 1

Power Supplies in the UK

D.C. Murch, CEng, FIEE

Revised by G.S. Finlay, BSc, CEng, MIEE
(Electricity Association)

In Great Britain the electricity supply industry was fully nationalized on 1 April 1948. Changes were made in 1957 which created a UK structure consisting of the Central Electricity Generating Board (CEGB) responsible for major generation and transmission throughout England and Wales, twelve Area Boards responsible for distribution within their areas in England and Wales, two Scottish boards and a Northern Ireland board responsible for both generation and distribution within their areas, and the Electricity Council which had a co-ordinating role and performed special functions in relation to research, industrial relations and finance.

Privatization of the supply industry was implemented in England, Wales and Scotland under the 1989 Electricity Act and in Northern Ireland by the Electricity (Northern Ireland) Orders 1992. A company which holds licences under the Act and Orders is designated a public electricity supplier (PES). Most of the industry is now owned by private investors following flotation on the stock market and returns on capital investment are now made to shareholders instead of the Government. The Electricity Council was abolished by the Act but some of its engineering functions are now carried out by two new companies. EA Technology Ltd now offers a research and development service to the electricity industry (EI), whilst the Electricity Association (EA) operates as a trade association on behalf of all its member companies, most of whom are mentioned in the following text.

The organization of a PES is usually set out in detail in the local telephone directory. For those who need the full organization this is set out in the *Electricity Supply Handbook* published annually by the *Electrical Times*. Parts of the following text are based on a document published by the Electricity Association as *The British Electricity System*, which describes and appraises the structure and operation of the privatized electricity industry in greater depth. The structure of the UK electricity industry is shown in Fig. 1.1.

Most electricity production at present in England and Wales comes from the three generating companies created out of the CEGB; they are National Power and PowerGen who are non-nuclear generators and Nuclear Electric which remains in Government ownership. The main pumped storage facilities at Dinorwig and Festiniog and the d.c. cross channel link are operated by a subsidiary of the

1

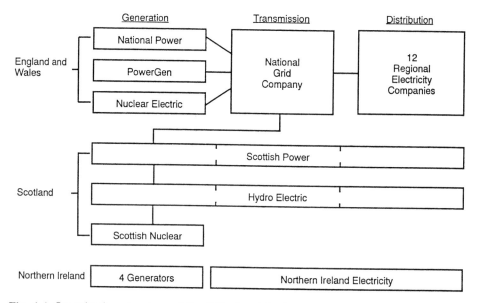

Fig. 1.1 Organization structure of the UK electricity industry.

National Grid Company (NGC). The twelve regional electricity companies (RECs), who are the successors of the Area Boards and are also PESs, jointly own the NGC which operates the transmission system and enables the bulk transmission of electricity between generators and suppliers.

Each REC supplies electricity to a franchise market in its region, although customers with a demand of over 100 kW may seek to purchase their electricity from outside the franchise market. The franchise market will be removed in 1998, allowing all customers to contract supply from competing companies.

In Scotland there are two companies operating franchise markets: Scottish Power and Hydro Electric, and Scottish Nuclear which remains state-owned.

In Northern Ireland there are four private companies operating the four power stations and one company, Northern Ireland Electricity, operating the franchise market for supply, transmission and distribution. The franchise areas are shown in Fig. 1.2.

The 1988 Electricity Supply Regulations and amendments contain the statutory technical requirements for Public Electricity Suppliers. There are other technical requirements of a statutory nature in the Health and Safety at Work Act 1989. It is the Electricity Act which created the regulatory system for the privatized industry headed by a Director-General of Electricity Supply who is answerable to the Secretaries of State for Trade and Industry and for Scotland. The role of Director-General is to oversee implementation of the Electricity Act particularly with regard to price regulation, protection of the interests of users (customers) by setting standards of service and the promotion of competition for electricity generation and supply. His department, which is the Office of Electricity Regulation is known as OFFER.

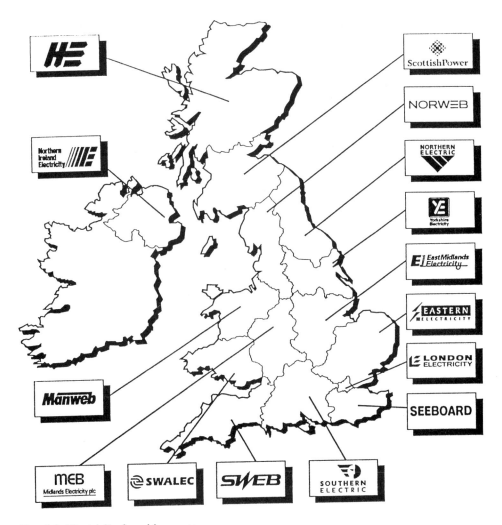

Fig. 1.2 Electricity franchise areas.

NGC was obliged to produce a Grid Code approved by the Director-General of Electricity Supply which defines the technical aspects of the working relationship between NGC and all those connected to its transmission system. Some generation embedded in PES networks is subject to the provisions of the Grid Code. NGC provides open access to the grid under terms which are non-discriminatory. However, all generators and distributors seeking connection must meet the appropriate standards to ensure that technical difficulties are not caused for others connected to the system.

The Grid Code covers items such as planning requirements, connection conditions, demand forecasting, coordination and testing requirements, together with the registration of appropriate data.

Before companies were issued with PES Licences to enable them to supply customers within their franchise area they had to produce Distribution Codes which set down the basic technical terms of connection for users. They listed the technical planning standards which form the basis of their system design and enable them to meet their obligations under the Grid Code.

Each company is required to have a Distribution Code although in the case of the RECs one common document is published as they all use the same pre-privatization Engineering Recommendations of the Electricity Council as technical references. In practice the supply systems are managed to the same standards as before privatization.

The engineering publications referred to in the RECs Distribution Code are now published by the Electricity Association and are as follows:

(1) Engineering Recommendation G.5/3, *Limits for Harmonics in the United Kingdom Electricity Supply System*.
(2) Engineering Recommendation G.12/2, *Application of Protective Multiple Earthing to Low Voltage Networks*.
(3) Engineering Recommendation G.59/1, *Recommendations for the Connection of Embedded Generating Plant to Public Electricity Suppliers' Distribution Systems*.
(4) Engineering Recommendation P2/5, *Security of Supply*.
(5) Engineering Recommendation P.25, *The Short Circuit Characteristics of Electricity Boards' Low Voltage Distribution Networks and the Coordination of Overcurrent Protective Devices on 240 V Single Phase Supplies up to 100 A*.
(6) Engineering Recommendation P.26, *The Estimation of the Maximum Prospective Short Circuit Current for Three Phase 415 V Supplies*.
(7) Engineering Recommendation P.28, *Planning Limits for Voltage Fluctuations Caused by Industrial, Commercial and Domestic Equipment in the United Kingdom*.
(8) Engineering Recommendation S.5/1, *Earthing Installations in Substations*.
(9) Engineering Recommendation S.34, *A Guide for Assessing the Rise of Earth Potential at Substation Sites*.
(10) Engineering Recommendation P.29, *Planning Limits for Voltage Unbalance in the United Kingdom*.

All the documents listed above have been exempted from the requirements of the Restrictive Trade Practices Act 1976.

Distribution Code Review Panels comprised of representatives of PES, generators and users as well as OFFER meet regularly in order to agree amendments to the Distribution Codes and their reference documents.

The Scottish companies operate under common distribution and transmission codes.

Licences are also granted under Section 6(2)a of the Electricity Act 1989 to second tier electricity suppliers authorizing them to sell electricity to customers with whom they have no physical connection.

Second tier licences allow suppliers to sell to any customer with a peak demand of more than 100 kW. Up to 1994, this freedom of choice in supply was restricted to 1 MW customers, a potential market of 5000. The 100 kW market is much larger, with some 50 000 potential customers.

Competition to supply customers in the non-franchise market has been very fierce. In the 1 MW market, nearly 40% of customers now take their supply from a company other than their local distributor. In the much larger 100 kW market, opened to competition in April 1994, over 11 000 customers changed their supplier in the first two months of operation. From 1998, freedom of choice will be extended to all customers when the franchise disappears altogether. This totally free market will allow any customer to buy electricity from any seller.

Second tier licences are held by all the regional electricity companies, National Power, PowerGen, Nuclear Electric and the Scottish companies. All these licence holders are competing for supplies, with the result that prices have been subjected to significant downward pressure. Among the most active players are the two main generators, who are seeking to supply the major industrial customers.

There is a price formula, policed by OFFER, which requires that the average use of system charge per unit distributed should not increase year on year by more than $RPI - x$, after adjusting for an incentive to reduce electrical losses. The initial values for x were agreed by the Government and the electricity companies, and range from 0 to -2.5 according to the company. In fact, increases in use of system charges have generally been kept well below the permitted limit.

The biggest change in the EI, resulting from privatization and Government incentives, has been the reduction of coal fired generation, the increase in gas fired generation, particularly combined cycle gas turbines (CCGTs), and the increase of generation embedded in PES networks.

CCGTs are particularly attractive as they offer major environmental advantages over coal fired plant; they consume 27% less fuel, emit 58% less CO_2 and 80% less NO_x for each kWh of electricity generated. Moreover they emit no SO_2 and so they do not contribute towards acid rain and global warming.

Also increasing are the numbers and size of wind farms connected to PES networks. Schemes of 10 MW asynchronous generation are becoming common and development of photo-voltaic generation systems has just commenced. These non-fossil fuel systems are encouraged by Government financial incentives as they are environment friendly. In 1995 the completion of Nuclear Electric's Sizewell B station will add another 1250 MW to the system and a similar station is proposed for the same site.

Finally, mention should be made of the special arrangements made by Government for nuclear decommissioning and renewable generation. Renewables are not, at present, competitive with either new or existing generating plant. To encourage investment in renewable energy sources in England and Wales, a subsidy is granted by the Director-General of Electricity Supply to the development of certain projects. This development is financed by a levy on electricity produced from fossil fuel, and currently amounts to 10% of the final cost of electricity. The levy is set to decline, until it is phased out in 1998. The bulk of the levy is allocated to the nuclear power producers to meet the historic CEGB liabilities for station decommissioning. These liabilities arose before the industry was privatized. The subsidy to renewable sources is set to continue beyond 1998. Although subsidies are undesirable in a market system, the levy does allow for greater transparency of costs.

In Scotland, the costs associated with nuclear decommissioning are incorporated in the price which Scottish Power and Hydro-Electric pay for the output of Scottish Nuclear.

THE ELECTRICITY POOL

Whatever the generating method, all the major generating companies having a generation output greater than 50 MW are required to sell all the electricity they produce into an open commodity known as 'the Pool'. The Pool is a rather complex trading mechanism.

Essentially, each generating unit has to declare a day in advance its availability to the market, together with the price at which it is prepared to generate, for each and every half hour of the day. The units are then called to generate by the National Grid Company in ascending order of price. The most expensive unit used sets the system marginal energy price which all others receive for that half hour. In addition, there is a separate pricing mechanism for capacity made available to the Pool.

This form of virtual real-time pricing will inevitably tend to produce volatility in prices which neither purchaser nor seller necessarily welcomes. To overcome this, the Pool has been overlaid with contracts, both short and long term, to make capacity and energy prices more predictable for customers and generators. About 90% of the electricity sold by the major generating companies is now subject to contracts, both with the PESs and large industrial users. These contracts are essentially financial instruments, the primary purpose of which is to hedge risk. About 1000 customers currently buy electricity 'direct' from the Pool.

In practical terms, the majority of generating stations are connected to the National Grid although some are 'embedded' in the systems of PESs. Metered values of supplies generated are summated and aggregated as to each generating company. Supplies taken by PESs are likewise summated and referenced to grid supply points (GSPs) from which they are taken. The Pool then carries out a reconciliation between generators and suppliers, on a day by day basis, for each half hour of that day. Figure 1.3 illustrates the concept of the metering arrangements. An important principle arises in the case of GSPs shared by two PESs (A and B). Total supplies to the GSP and supplies to PES B are metered, the 'take' of PES A then being the difference. This principle of 'difference metering' is also applied to embedded generation, inter-PES supplies and second tier customers.

VOLTAGE AND FREQUENCY

The supply system in Great Britain was standardized at 240 V single-phase and 415 V three-phase with a frequency of 50 Hz following nationalization in 1947 but was recently redeclared. The supply voltage for R.V. consumers is now 400 V + 10%, − 6%, for three-phase supplies, and 230 V + 10%, − 6%, for single-phase

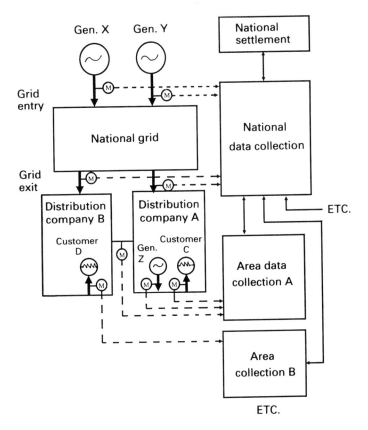

Fig. 1.3 Pool metering arrangements.

supplies at a frequency of $50\,Hz \pm 1\%$. In the year 2003 it is expected that all voltages in Europe will be redeclared at $400/230\,V \pm 10\%$ in line with the Cenelec Harmonised Document, HD472, S. Distribution network design and voltage control will ensure that existing consumers' supplies will be unaffected by the new declaration and that their apparatus will continue to function satisfactorily. The greater part of electrical energy is sold at $230\,V$ and thus is used directly by most apparatus. Major generation is connected to the supergrid system which operates at $400\,kV$ although some sections are at $275\,kV$. In England and Wales this network is owned and operated by the NGC. The original $132\,kV$ grid system which was started in 1926 was transferred from the CEGB to the area boards in the 1970s and is now the higher strata of PES distribution systems.

The major primary distribution voltage is $11\,kV$ although a small proportion still operates at $6.6\,kV$. For most of the system there is an intermediate stage of $33\,kV$, but direct transformation between $132\,kV$ and $11\,kV$ is becoming common policy in city areas, where over $150\,MW$ can be economically distributed at $11\,kV$ from one site. The frequency is maintained by the NGC which regulates the input power to the generators to match the instantaneous load on the system and thus maintain their speed and thereby the frequency. This is manually controlled to very close limits, thus enabling synchronous clocks, time-switches and other motors to be used.

The voltage supplied to the consumers is mainly regulated by on-load tap-change gear on the transformers which supply the 11 kV system. Fixed tappings are used on the 11 kV to 433/250 V transformers as it would not be economic to put on-load tap-changing on these. For large machines or equipment supply is at a higher voltage, usually 11 kV or transformed to 3.3 kV.

Apart from the size of the largest item, the need to bring an h.v. supply on to a consumer's premises is determined by the relative strength of the local network in relation to the total load. In order to determine what supply can be made available at any particular location, the local office of the appropriate PES should be consulted at an early stage. While generators produce a near perfect waveform and the on-load tap-change gear maintains voltage at the 11 kV source within fine limits, all load on the system creates some distortion. The extent to which the various types of consumer load can be connected to the system depends upon the distortion they are likely to create and the nuisance this disturbance causes to other consumers. This subject is dealt with more fully in a later section.

SYSTEM IMPEDANCE AND SHORT-CIRCUIT LEVELS

From the consumer's point of view another important parameter of the supply system is its impedance as viewed from his terminals. On the one hand, the lower the impedance the greater will be the stress on his switchgear and protective devices, but on the other hand, the higher the impedance the greater will be the risk of annoyance due to distortion caused by either the consumer's own load or by that of a nearby consumer. High network impedances are troublesome to installation designers because they result in low values of fault current, which severely limit the number of series graded protection devices and cause an increase in the I^2t energy let through of inverse characteristic devices such as fuses. The 16th edition of the IEE Wiring Regulations, BS 7671, requires installation designers to have a knowledge of the limits of system impedance to which the supply will be kept in order that they may install the necessary protective devices to an appropriate rating and to operate within the required time. All supply systems are dynamic and many PES staff are continually employed laying cables and moving and installing plant in order to ensure that the system configuration meets the demands of the customers. For this reason it is not possible to give an exact impedance figure for any one location, but the appropriate local area office should be able to give installation designers the maximum and minimum likely to be encountered for a particular location. A maximum earth loop impedance figure of 0.35 Ω is quoted nationally for l.v., single-phase, PME system supplies of 100 A or less. An appropriate maximum prospective short-circuit current of 16 kA is quoted for many urban supplies; further information may be gained from Engineering Recommendation P25.

For many years it has been common practice to express the energy available on short-circuit in terms of 'short-circuit MVA'. This is simply $\sqrt{3}\ VA \times 10^{-6}$ where V is the normal system voltage between phases and A the symmetrical component of the short-circuit current.

In 1971 the International Electrical Commission (IEC) introduced a standard for switchgear ratings (IEC 56) which specified that the working voltage rating should be expressed in terms of the system maximum, for example 12 kV for an 11 kV system, and that short-circuit ratings should be expressed in terms of the maximum symmetrical fault current. A range of ratings was specified, for example 12.5, 16, and 25 kA for 12 kV gear.

For any three-phase system voltage the short-circuit level and the system impedance are inverse functions of each other, $kA = kV/\sqrt{3}\ Z$.

On l.v. systems the cables will generally be the major contributors to the system impedance as the h.v./l.v. transformers are of low impedance. The governing factor is thus the distance of the consumer from the nearest substation.

On PES h.v. systems the short-circuit ratings of the switchgear have a considerable economic significance and, therefore, system designers aim at keeping these to as low a figure as practicable. A common method of achieving this is to employ high impedance 33/11 kV or 132/11 kV transformers. The high impedance is achieved by judicious spacing of the windings and does not increase the transformer losses or costs to any appreciable extent. The high impedance does not affect the voltage output as the tap-changer regulates accordingly. At 11 kV the impedance of the cables is generally much less significant.

Until the publication of IEC 56 many 11 kV systems were designed for a maximum level of 250 MVA which is equal to 13.1 kA. The new rating method therefore poses a problem where new or additional switchgear is required on an existing system, since the 16 kA switchgear is more expensive. In general, however, British manufacturers can supply switchgear tested to 13.1 kA at similar to 12.5 kA prices. With the low growth of demand, these 250 MVA systems will remain for many years.

LOADING EFFECTS ON THE SYSTEM

Any normal load causes a voltage drop throughout the system. This is allowed for in the design and the cost associated with the losses incurred are recovered in the related unit sales.

Unbalanced loads

Unequal loading between the phases of the network causes an unequal displacement of the voltages. Extreme inequality causes motors and other polyphase equipment to take unequal current and perhaps become overloaded on one phase. For this reason PESs impose limits on the extent to which they accept unbalanced loads at any particular location in order to ensure that other consumers are not adversely affected. Installation designers need to ensure that the same problem does not arise due to an unbalanced voltage drop within the consumer's installation itself.

While most voltage unbalance is caused by single-phase loading, the effect on a three-phase motor can best be assessed in terms of the negative phase sequence component of the voltage thereby created. Providing that this is less than 2% the inequality of current between phases should not be more than the motor has been designed to withstand. Engineering Recommendation P29 aims to limit continuous levels of voltage unbalance to 1%.

Power factor

Many types of apparatus such as motors and fluorescent lighting also require reactive power and thereby take a higher current than is necessary to supply the true power alone. This extra current is not recorded by the kWh meter but nevertheless has to be carried by the distribution system and uses up its capacity thereby. It also increases the losses on the system. A power factor of 0.7 means that the current is $1/0.7 = 1.43$ times as great as absolutely necessary and thus doubles the losses ($I^2 R$). If all the loads in the UK were permitted to have as low a power factor as this, the additional cost of the losses (if the system could stand the burden) would be in the order of £200 million per annum.

It will readily be seen, therefore, why PESs are keen to ensure that their consumer's power factors are near to unity. Where practical, some penalty for poor power factor is built into the tariff or supply agreement.

Power factor correction is therefore an important aspect in the design of installations, although too often forgotten at the outset, see Chapter 23. The most simple and satisfactory method is to have each equipment individually corrected as this saves special switching and reduces the loading on such circuits. Bulk correction of an installation is, however, quite commonly used, particularly where it is installed as an afterthought. There are then problems of switching appropriate blocks of capacitance to match the load, as overcorrection again increases losses and, in addition, creates voltage control problems at light load.

Switching transients

One of the primary uses of electricity is for general lighting and the local PES must ensure that its supply is suitable for this purpose. Repeated sudden changes in voltage of a few per cent are noticeable and are liable to cause annoyance. The local PES must ensure that these sudden variations are kept within acceptable levels and this means placing limits on consumers' apparatus which demand surges of current large enough to cause lighting to flicker.

In order to evaluate flicker in measurable terms, two levels have been selected, first, the threshold of visibility and secondly, the threshold of annoyance. Both are functions of frequency of occurrence as well as voltage change. Since both these thresholds are subjective it has been necessary to carry out experiments with various forms of lighting and panels of observers to ascertain consensus relationships between frequency of occurrence and percentage voltage change for the two thresholds. The PESs have used this information in setting the planning levels for flicker, contained in Engineering Recommendation P28, which govern motor starting currents, etc.

The network impedance from the source to the point of common coupling between the lighting and the offending load is of paramount importance and thus the local office of the PES should be consulted in cases where the possibility of creating an annoyance arises.

Intermittently loaded or frequently started motors, such as those on lifts, car crushers, etc., together with instantaneous waterheaters, arc welders and furnaces, are all potential sources of trouble. Large electric furnaces present a particular problem and it is frequently necessary to connect them to a higher voltage system than is necessary to meet their load in order to achieve a lower source impedance.

Fluctuations occurring about ten times a second exhibit the maximum annoyance to most people, but even those as intermittent as one or two an hour will annoy if the step change is of sufficient magnitude.

The UNIPEDE publication entitled *Characteristics of Low Voltage Supplies* gives very useful information on supply variations and contains the frequency/voltage change characteristic appropriate to the manufacture of equipment to CENELEC Standard EN61000-3-3, BS 5406 Part 3, limits voltage fluctuation emissions from equipment rated less than 16 A.

Harmonics

Harmonics on the supply system are becoming a greater problem due to the increasing use of fluorescent lighting and semiconductor equipment. Cases have been known where large balanced loads of fluorescent lighting have resulted in almost as much current in the neutral as in the phases. This current is almost entirely third harmonic. The use of controlled rectifiers and inverters for variable speed drives, as used on many continuous process lines, can also be a problem and PESs have frequently to insist that 6- or 12-phase rectification be used. It is another case for local consultation to determine what the system will stand. Even a multiplicity of small equipment can summate to create a big problem, for example television set rectification, and for this reason international standards have been agreed. Engineering Recommendation G5/3 sets out the limits of harmonic voltage distortion which are acceptable to the supply industry for the connection of load at various voltage levels. CENELEC Standard EN61000-3-2 limits harmonic emissions from equipment rated less than 16 A.

SUPERIMPOSED SIGNALS

For many years load control systems which superimposed signals on the normal supply system have been in use. The earlier ones were usually at a higher frequency and were referred to as ripple control, though one system used a d.c. bias. Tuned relays respond to the appropriate signal and switch public lighting, change various tariffs according to time of day or system load, switch water or space heating according to tariff availability or weather conditions, and perform various other functions, even to calling emergency staff.

Although these did and still do perform satisfactorily, they did not come into universal use, probably because the full expenditure on signal generators was necessary to cover the system even though initially there were only a few receivers and this caused cash flow problems.

A decade or so ago equipment came on the market which could communicate from one premises to another over the supply system. The EI predicted severe interference problems if bandwidths were not allocated for specific types of usage. Self regulation has evolved under the guidance of TACMA and the EI has produced its own working document Engineering Recommendation G22. The British Standards Institution has published BS 6839 on the subject.

With the advent of the thyristor and the transistor, supply companies started experimenting with them on their network as a communication media, using pulse techniques both for one-way and two-way communication. Practical installations have been developed and a number of one-way systems installed to perform similar functions to those carried out by the earlier ripple control method.

One such system developed by London Electricity was a range of codes transmitted by short pulses injected at the zero point of the 50 Hz wave, Fig. 1.4. Injection at this point requires less power and thus enables smaller transmitters to be used. This method has been taken up by a major UK manufacturer and several one-way systems called Cyclocontrol are now controlling over 150 MW of off-peak load in London. The two-way communication objective is to read meters remotely and a workable system has been in use for several years, but only as a pilot scheme because it was not economically viable using discrete components.

The advent of the microprocessor and its relatively low cost in quantity production has widened the development field and made many schemes attractive. Solid state metering with sophisticated control and tariff regulating functions will definitely be in use in the near future. Deciding what these 'black boxes' should be made to do is the big problem facing the EI, remembering that there are some 20 million energy meters in the UK alone and the task of changing these is a mammoth one. If the changes were carried out as part of the normal recertification programme it would take 15–20 years to complete. Consumers who had to wait this long for an attractive tariff or control function to be available to them would be far from happy!

There is another side to the picture. Sophisticated control systems which are already avaiable for use within a consumer's premises enable a master programme controller to send signals over the main conductors to slave controllers on appropriate apparatus. The master controller being flexibly programmed via a keyboard and display is even capable of responding to signals sent via the telephone.

Interference, both accidental and wilful is a problem which needs careful consideration and rejection circuits may be needed where such systems are in use on both sides of the supply terminals.

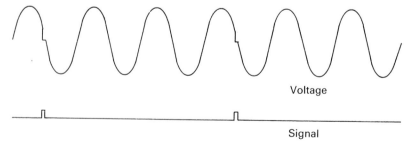

Voltage

Signal

Fig. 1.4 Cyclocontrol signals.

While the use of the network itself, as a control medium, is attractive to a PES, it has its drawbacks and other methods using superimposed signals over the radio or telephone are being used. The radio is attractive for one-way communication as the transmitter cost is minimal and full coverage is available. The telephone is attractive for two-way communication as the addressability exists within the telephone system.

From the electrical contractor's point of view this expanding field of superimposed control is likely to extend the range of his work and so add to the complexity of some installations.

RADIO TELESWITCHING

The electricity industry (EI) in collaboration with the BBC investigated the possibility of superimposing digital data on BBC long wave transmissions for the purposes of sending information and control signals from its offices to its customers' premises. Suitable technology to enable this was developed and the radio tele-switching (RT) system was approved by the relevant authorities. An important requirement was that the superimposed signals should not impair the reception of normal audio (Radio 4) programmes transmitted on long wave band. Fully operational facilities became available in 1984.

The EI holds a patent on the development and issues licences to approved manufacturers of RT devices. The RT device consists of three essential components – a radio receiver capable of satisfactory performance under low signal strength and high noise level areas, a microprocessor which decodes the received signals and includes a time clock and memory, and a set of contacts or switches which are controlled by the microprocessor.

Companies supporting the RT system use it for one-way communication with load and metering apparatus in their customers' premises. The system does not need routing or rerouting of signal paths. Nationwide broadcast coverage and the use of unique codes enables a user company to send signals to its customers no matter where they are located in the UK and prevents one company's data affecting customers of another company. A company can store an entire week's programme for all its groups of RT devices in the central computer of the RT system or, through the same computer system, it can send immediate command control signals addressed to any of its groups. This central teleswitch control unit (CTCU) is managed by the Electricity Association which also manages the development and operational arrangements of the whole RT system. The CTCU converts the programme information into the required codes, minimizes the signals required and schedules and despatches them for transmission at the appropriate times. Off-air monitors are installed for each transmitter area, and the signals from these are fed back to the CTCU as a check that all despatched messages have been broadcast and therefore should have been received by all RT devices.

The most common type of RT device used by companies has a 24 hour clock and a memory which stores programme messages for a day. The RT device then controls customer load and tariff according to the received and stored 24 hour programme. The problem of inaccuracy and drifting of load and tariff switching times is eliminated by the broadcasting of time signals every minute. A stored programme can be removed and replaced by an entirely different 24 hour programme at any time, or it can be gradually changed to the following day's programme on a rolling basis. A programme may also be suspended and replaced by an immediate command broadcast at any time. The devices have a standard fallback programme which they follow under certain conditions. To keep new devices updated and help ensure reception, broadcast messages are repeated as a routine. If a device misses a programmed broadcast and its repeats it will follow the last programme it received.

By being first in establishing radio teleswitching as a fully operational cost effective option for use in energy management on a national scale the EI has given the UK a world lead in this field. Further developments of the RT system are intended to increase its data carrying capability, enable weather and cost related information to be broadcast as energy management parameters and allow the system to be integrated with other developing technologies, such as smart cards, so that the full potential of the system can be realized.

SYSTEM AND INSTALLATION EARTHING

In the UK the Electricity Supply Regulations govern the way systems are connected to earth.

Low voltage systems

For many years the Regulations required that each l.v. system should be solidly connected to earth at only one point, that being the neutral of the source transformer. Special permission was necessary to earth at more than one point. The Regulations also required that cables buried in the highway must have a metallic sheath. Systems earthed at only one point require the neutral conductor to be electrically separate and are now known as SNE (separate neutral and earth).

It was, and still is, the responsibility of each consumer to provide the earth connection for his own installation. This was commonly achieved by connection to a metallic pipe water main. The growing use of PVC water mains makes this impossible for new installations and causes problems with existing ones when water mains are replaced. Gradually, supply companies developed a practice of providing consumers with an earth terminal connected to the sheath of their service cable. This is, of course, a very satisfactory arrangement but it is not universally practical as many cables laid in the 1920s or earlier are still in use and many of these are not bonded across at joints. The arrangement is not practical on most overhead systems.

In Germany and elsewhere in Europe an earthing system known as 'nulling' grew up. This employed the principle of earthing the neutral at as many points as possible. It simplified the problem of earthing in high resistance areas and by combining the sheath with the neutral conductor permitted a cheaper cable construction. These benefits were attractive and during the 1960s the official attitude in the UK gradually changed to permit and then encourage a similar system known as PME (protective multiple earthing). Blanket approvals for the use of this system and setting out the required conditions to be met were finally given to all area boards in 1974. These approvals have now been incorporated into the Electricity Supply Regulations 1988. In the 16th edition of the IEE Wiring Regulations this system is classified as TN-C-S.

Providing the consumer with an earth terminal which is connected to the neutral conductor ensures that there is a low impedance path for the return of fault currents, but without additional safeguards there are possibilities of dangerous situations arising under certain circumstances.

If the neutral conductor becomes disconnected from the source of supply then the earthed metalwork in the consumer's premises would be connected via any load to the live conductor and thus present a shock hazard from any metalwork not bonded to it, but which has some connection with earth. In order to eliminate this rare potential hazard the Secretary of State, in his official Regulations, requires that all accessible metalwork should be bonded together as specified in the IEE Ring Regulations and so render the consumer's premises a 'Faraday cage'. This is the reason for the more stringent bonding regulations associated with PME.

Under the extremely rare circumstances of a broken service neutral and intact live conductor, there may be a danger of shock on the perimeter of the 'cage' to someone using an earthed metal appliance in a garden, even though the appliance may be protected by an RCD (residual current device) in accordance with the IEE Wiring Regulations. For the same reason metal external meter cabinets are undesirable.

In order to eliminate as far as possible the chance of a completely separated neutral, a number of precautions are taken. First, all cables must be of an approved type with a concentric neutral, either solid or stranded, of sufficient current carrying capacity. Secondly the neutral conductor of a spur end on the system is connected to an earth electrode if more than four consumers' installations are connected to the spur, or if the length of the spur connection from the furthest connected consumer to the distributing main exceeds 40 metres. Where reasonably practicable cable neutrals are joined together to form duplicate earth connections. A faulty or broken neutral will give an indication of its presence by causing supply voltages to fluctuate, which, of course, should be reported to the local PES as soon as possible. All these measures contribute to a system which is as safe as practicable and self monitoring.

It is the declared intention of the EI in the UK to provide earth terminals wherever required and practicable within the foreseeable future. The local PES should be contacted regarding their requirements for the use of PME earth terminals for TN-C-S systems.

High voltage system earthing

The Electricity Supply Regulations 1988 specify that every h.v. system shall be connected with earth at or as near as reasonably practicable to a source of voltage in the system. They also specify that conductors, which respectively connect a supply neutral conductor with earth and any apparatus used in a h.v. system with earth, shall not be interconnected unless the combined resistance to earth is 1 Ω or less; the conductors shall not be connected to separate earth electrodes unless any overlap between the resistance areas of those electrodes is sufficiently small as not to cause danger.

There are various methods used to earth an h.v. system and practices between the PESs in the UK vary considerably. Most of the 11 kV systems are derived from 33/11 kV delta star transformers and thus there is a neutral point available. In quite a number of cases these are solidly earthed and the resultant potential earth fault current can be slightly higher than on a phase-to-phase fault. It is also quite common to restrict the current by the insertion of a resistance or reactance in the neutral to earth connection to limit the fault current to about the full load of the supply transformer.

Where the h.v. supply is designed directly for 132 or 66 kV, it is usual to have a star wound primary to minimize the transformer cost. In order to get the correct phase relationship, it is necessary to use a delta secondary winding in some cases and a star winding in others. When delta secondaries are used it is necessary to employ a separate earthing transformer to create a star point and this transformer limits the earth fault current, though frequently a resistor is also used. Where the secondary winding is star connected, this is either directly connected to earth or through a resistor. Usually the resistors for this purpose are water-filled but in some cases a reactor is employed.

Where an h.v. supply is afforded, the installation designer needs to know the range of the protective earth fault current in order to provide the correct protection.

Where a consumer transforms an h.v. supply it is necessary to earth the secondary system. Some PESs allow the use of a common earthing system, which means that the sheaths of their cables are probably providing the major return path for any earth fault current. Where there is a requirement for the segregation of earthing systems for technical or security reasons, such as the provision of 'clean' earths to computer systems, special care must be taken with the earthing system design. All electrical and other metal installations within an associated structure must be considered when assessing the overall design to ensure that earthing systems extended by equipotential bonding are not close enough to cause safety problems related to indirect contact. The earthing problem takes on an extra dimension where generating plant is embedded in PES's distribution networks. Engineering Recommendation G59/1 and Engineering Technical Report ETR113 (1994) give the requirements and guidance, respectively, for the connection of embedded generating plant up to 5 MW to PES distribution systems less than 20 kV.

PROTECTION

The design of the protective arrangements on a supply authority's system is influenced by the need to provide as economic a supply as possible commensurable with a reasonable degree of reliability. Because of the high cost of circuit-breakers this means using as few as possible. Experience also indicates that the simplest system is often the most reliable as it is less prone to human error.

Most PES systems employ the simple principle of time and current grading up to the 33 kV level. Besides the benefits of simplicity and economy it can cater for the ever-changing topography of the network to meet the changing load pattern, and each protective stage also provides back-up to the one nearer the load if that fails to operate.

On rural overhead systems a fairly high proportion of faults are transient and immediately clear as soon as the supply is disconnected and it is therefore common practice to install automatically reclosing circuit-breakers on the h.v. system. These eventually lock out if the fault persists.

Throughout the life of the industry there has been a continuous evolution of electromagnetic relays for protection and control purposes and these now perform satisfactorily for the majority of applications. The development of modern semi-conductors makes the use of analogue circuits a practical alternative with the benefit of no moving parts. Their adoption, apart from one or two specialized roles, has been slow, probably because there were a number of early failures and their benefits in terms of increased reliability or cost reduction were insufficiently attractive.

With the advent of microprocessors entirely new concepts of protection are offered. Their speed, flexibility and memory capacity, coupled with their low power requirement, have put the design of the power system back in the melting pot. It is generally appreciated that a lot of trial and error experience will need to be undertaken before they come into general use on old or new systems, but their ultimate take-over seems to be inevitable.

The Electricity Supply Regulations require that each consumer is fed through a suitable fusible cut-out or automatic switching device, the purpose of which is to protect the supplier's works on the consumer's premises which are not under the control of the consumer. It is the general practice to install the maximum size of cut-out fuse which matches the incoming service capacity, in order to give the widest discrimination with the consumer's first protective device.

A typical urban system

Urban l.v. cables are typically protected at the h.v./l.v. substation by 400–600 A fuses, Fig. 1.5. Using the simple proven ratio of 2 to 1 for correct discrimination of series connected fuses, means that 200–300 A fuses may be installed for individual supplies to offices, factories, shops, flats, etc. Larger loads frequently need to be fed directly from an h.v./l.v. substation. The transformers in these substations are protected by h.v. fuses as these are less expensive than a circuit-breaker and operate faster under short-circuit conditions.

The next stage of protection is at the h.v. substation where each outgoing h.v. circuit is protected by inverse definite minimum time lag (idmtl) relays, two elements connected to detect and operate on overcurrents and the third detecting and operating on earth faults.

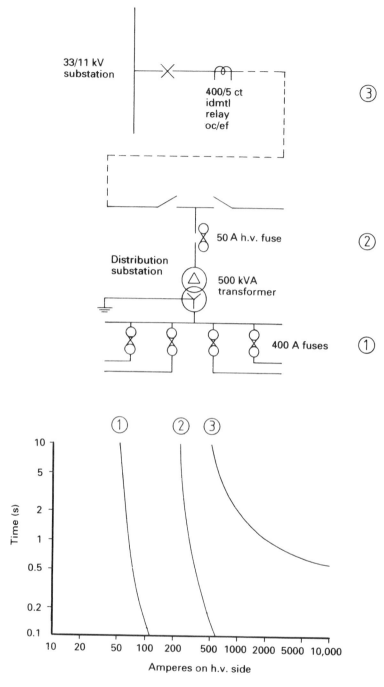

Fig. 1.5 Typical urban system protection.

RELIABILITY

Electricity plays a vital part in modern society and a loss of supply causes much consternation, inconvenience, financial loss and some danger to life. By putting in surplus generation, alternative circuits and high standards of construction it is possible to achieve a very high standard of supply. It is not possible to achieve absolute security and the question of how much money should be spent on security to obtain a reasonable level of reliability is one which is under continuous review and sophisticated techniques have been developed by the industry in order to obtain the most cost effective arrangements.

The Electricity Association operates a comprehensive computer-based system, the National Fault and Interruption Reporting Scheme (NAFIRS) on behalf of its member companies which builds up statistics to highlight weaknesses, etc., and assists in improving techniques. The average UK consumer loses supply about once every two years for about three hours. Rural supplies, by virtue of their sparseness, are less reliable than those in urban areas.

Having developed systems and equipment which can provide a high degree of reliability, the supply industry finds that third-party damage is a major hazard and in conjunction with the Health and Safety Executive it is trying to instil into those who work in the public highway the need to know the location of supply mains in the vicinity of the work. New sites are particularly vulnerable in this respect.

Where a consumer requires a more reliable supply than normally afforded by the system to which he is or will be connected, then it may be possible to provide him with an alternative source from another substation, the cost being chargeable. It would, of course, be vulnerable to a mutual failure including industrial action, but might, nevertheless, be worthwhile as the alternative of standby generators is not 100% reliable.

EMBEDDED GENERATION

The Energy Act 1983 was the statutory instrument which placed an obligation on the EI to publish tariffs for the purchase of energy from companies operating generating plant and to support and adopt combined heat and power (CHP) schemes. These tariffs also include the charges appropriate to the provision of standby capacity to a supplier and the charges associated with using the EI power systems between a generator or supplier and their dedicated customers. On occasions PESs have been unable to permit the connection of embedded generating plant because either the resulting fault levels would have been in excess of plant ratings or the devices used to limit fault current have not been approved by the EI approvals panel as being safe.

Engineering Recommendation G59/1 sets out the EI's basic requirements with regard to the connection of generating plant installations up to 5 MW to systems up to 20 kV. Engineering Technical Report ETR113 (1994) gives guidance on the connection of embedded generating plant to PES networks.

A consumer may install his own generators because he feels he may be able to generate more cheaply or because he hopes it will give him a more reliable supply.

There are many possible modes of operation:

(1) Full generation of electricity only, without any incoming public supply. This is unlikely to be profitable, particularly if hidden costs are evaluated.
(2) Full generation in association with process steam.
   (a) With public supply as standby. Providing the steam and electricity demands are well matched throughout the working period this can be a viable arrangement.
   (b) With public supply meeting some of the load which is switched to suit the generation capability at the time. This could be a viable arrangement subject to a satisfactory supply agreement. It has the advantage of avoiding parallel operation, which imposes technical problems.
   (c) Generating in parallel with the public supply system. This can permit any surplus or deficiency of generation to be exported or imported within agreed levels and controlled by the metering and protection. Separate metering is used for import and export as the price paid for the excess units will be less than for those from the public supply. Because excess units are not of prime value at times of light load the price offered often appears mean to those who are not aware of the wide range of incremental value with time and committed availability. PESs are generally keen to make this arrangement work as it can be financially attractive to both sides, as well as giving a high degree of reliability.

   One major problem is the increase in short-circuit level created by the generators, as each power system is likely to have a source short-circuit level near that of the switchgear rating. Switchable reactors may be required to overcome this problem, one such arrangement being shown in Fig. 1.6. There is also a problem of earthing the generators should they become separated from the power system and its source earth point.

   This method of supply seems a good one for the application of microprocessors. On the metering side they could calculate the value of the units on an incremental worth basis to suit a time/load/power factor agreement. On the control side they could monitor the number of generators and reactors to contain the short-circuit level within limits. They could be used to control voltage and reactive power generated, and also for protection.
(3) Standby generation. Load shedding due to shortage of fuel or other industrial action caused an upsurge in the installation of small generators by those consumers whose business could not stand the risk of loss of supply, ranging from just a few kilowatts at low voltage to several thousand kilowatts switched at high voltage. These cause some anxiety on safety grounds as cases have occurred where the generator has been started up and connected to the installation without the latter being disconnected from the incoming supply.

   It is easy to imagine a small l.v. generator being plugged in to supply a milking machine on a farm at the end of a line and a linesman repairing the fault receiving a shock and being thrown to the ground. Problems have also

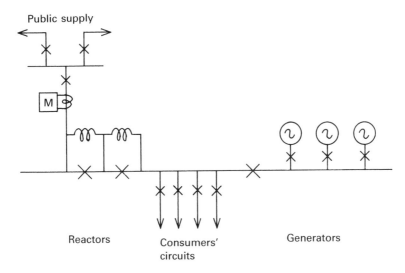

Fig. 1.6 High voltage consumer with switchable reactors.

occurred on large sets due to incorrect changeover arrangements, for example, in one case a loss of supply on one phase only brought in the standby set without the normal feed being disconnected.

The local PES is generally pleased to give advice on appropriate changeover arrangements.

SUPPLY ARRANGEMENTS

Electricity supply tariffs are normally specified according to the type of consumer, i.e. domestic, commercial or industrial, and it is convenient to use the same categories to discuss the supply arrangements.

Domestic supplies

Houses are now provided with a 100 A single-phase supply. Flats without off-peak heating are provided with a 60 A supply. Service installation is an expensive labour-intensive activity and it is therefore important that the capacity and location are such as to ensure that very few installations need early alteration. On the other hand, the number that need installing every year means that even the most modest economy is worthwhile.

Domestic meter reading is a heavy recurring expense and ease of access is therefore an important factor in the siting of the meter. External meter cabinets are thus preferred wherever practical but their use is not without problems. It has taken quite a long time to get their general acceptance architecturally as they do little to enhance the appearance of premises.

From the structural point of view there have also been problems, because the hole they occupy in the wall causes a reduction in its fire resistance. The entry of the supply cable also presents problems. It is important to arrange for the builder to install a pipe which has a suitable diameter and radius, and that if any cavity insulation is to be used then it should be omitted adjacent to the pipe as otherwise it will derate the cable. The pipe has to be sealed at the bottom end in order to prevent the possible entry of gas into the premises should a leakage occur outside the house, Fig. 1.7.

An external meter cabinet is outside the Faraday cage and therefore it is preferable to use one of insulated construction.

The gas companies are also keen on external meter reading and there are pressures for a common cabinet to be used, although clearly there are problems of bonding and ensuring that there is no danger of ignition if a gas leak occurs.

Not all domestic development is in the form of individual housing; quite a large proportion is made up of maisonettes and flats, either low- or high-rise, although the latter have declined considerably in popularity. Where more than one dwelling is to be fed from one service it is necessary to extend the mains from the

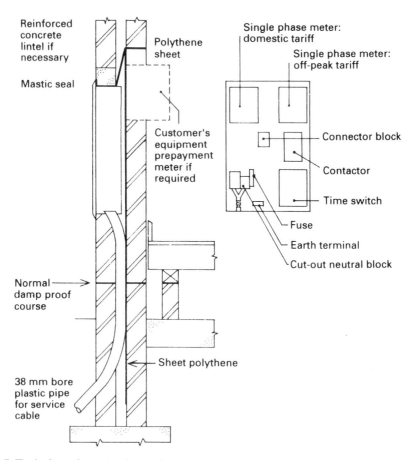

Fig. 1.7 Typical service entry into a house.

termination of the incoming cable to the remote dwelling units. For this purpose
individual rising or lateral connections will need to be installed. High-rise buildings
may also require a common main, Fig. 1.8.

The ownership and installation of these mains are dependent upon agreement
between the developer and the local PES. In many cases the PES will pull their
cables into pipes, ducts or chases provided by the developer. Such installations
must resist interference and the spread of fire. Early agreement and architectural
provision is essential as some modern methods of construction can present acute
problems of routeing. If the local PES is to be the owner then a PME system may
be employed using CNE cables.

The question which arises is the siting of the meters; the possibilities are for
them to be in a common intake room, in meter cabinets outside each flat or inside
each flat itself. None of these is ideal as each fails to meet one of the following
requirements: capable of being read by the consumer, capable of being read by a
meter reader without entry to premises, not observable by persons with malicious
intent (no movement of the meter might give an 'all clear' to a break-in), allow a

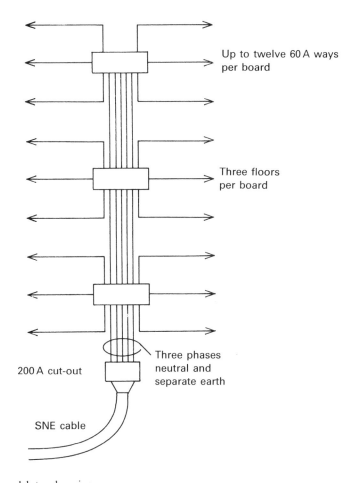

Fig. 1.8 Rising and lateral mains.

prepayment meter to be used. This latter possibility was not an attractive one from the local PES' point of view as the collection of the money was a big problem. Fortunately, the development of token meters has solved the problem of theft. Trial schemes have been so successful that existing money prepayment meters are rapidly being replaced by token meters; 25 000 were installed in London alone in 1988. Each token meter has a unique key identified with a consumer, which is recharged at streetside vending machines; hence the keys have no intrinsic value. Load and address data stored in a key's memory chip can be recovered by a vending machine and used for load analysis purposes.

In each case the meter position is decided on which factors are important and which disadvantages can be overcome.

In high-rise buildings electric heating can simplify the construction and be an economic solution but the interdependence of the heating gives problems of individual control and payment. In many cases it has been necessary for the landlords to be responsible and make an appropriate charge with the rent. If designed for at the outset this may be done by a completely separate installation. Where the heating is taken over at a later stage, control can be by the use of superimposed signals.

Commercial supplies

Commercial development can be either purpose or speculatively built. If the occupier and the intended use of the building is known before it is built the electrical installation is designed to meet the consumer's requirements. If it is to be a branch of a departmental chainstore or something similar then the consumer may well have his own electrical engineer who is able to give an accurate estimate of the loading requirement based on the experience of other branches.

In other cases it may be that consultants are used and the supply engineer needs to be on guard against being led to install capacity which will not be fully utilized due to a liberal load estimate. Providing that space is earmarked for the contingency of putting in extra or larger plant, the local PES may on its own experience discount the estimate and put in smaller plant, being prepared to be wrong on the odd occasion.

Despite the fact that many new office developments remain empty for long periods speculation still goes on. The design of the electrical installation poses many problems, some of which also concern the local PES. Is there a possibility of multiple occupancy and, if so, can the installation be split up to permit individual metering?

In a tower block this can sometimes be achieved by using a number of cable rising mains each feeding, say, three or five floors with facilities to swing the intermediate floors either way. Cases have been known where a developer has had to change his intention of letting to a single tenant. With a single heavy busbar riser this means that sub-metering had to be used which is not a very satisfactory arrangement, particularly with regard to maximum demand. In other cases this has not been acceptable and the rising main system has had to be replaced.

The increasing use of office automation and air conditioning makes the problem of estimating the loading more difficult and thereby the designing of the integral substation. No local PES is happy about installing plant which may not be used for several years but which, nevertheless, has to be kept live and maintained. It may therefore not only want a full contribution to cover the initial cost but also some provision for the maintenance.

Industrial supplies

While most large companies have their factories purpose built, light industry frequently rents its premises which may well be part of a factory estate. It is now quite common for small factory estates to be set up speculatively with the purpose of encouraging local industry. Because in most cases the use of these buildings is not known, the problem of providing suitable services is a difficult one. The developers want to be able to offer adequate availability of electricity but the local PES does not want to put in plant to meet any eventuality with little chance that the full capacity will be used.

The range of consumption and demand per square metre for industrial use is very wide – it can range from a storage warehouse to a heavy power consuming process. Some quite small units may require an h.v. supply in order to withstand load variations without undue voltage fluctuations. Even a continuously running motor can cause problems if it is driving a cutting or stamping machine or a similar machine which makes sudden demands on power and draws heavy currents.

The local PES may well compromise by putting down a skeleton network, earmarking possible substation sites and laying down spare ducts to cater for the possible future h.v. cables to feed them. Without any guaranteed load or customers the local PES almost certainly requires a full expenditure contribution.

INTAKE ARRANGEMENTS

Installation designers and contractors may find it helpful to understand some of the factors which influence local PES supply arrangements. Prior consultation and co-operation is necessary to ensure a neat and efficient combined arrangement.

Low voltage

Except for supplies to street lighting and furniture, the smallest service size normally used is 60 A single-phase. In most locations now, 100 A single-phase services are provided. Above this size the load must be spread over three phases and the typical range of sizes is 60, 100, 200, 300, 400, 600, 800 and 1000 A. Those up to 200 A may come from the network if there is sufficient capacity. The larger sizes normally come from an adjacent substation. Those below 600 A are generally provided by multicore cables. The larger sizes need single-core cables to provide the necessary capacity. Where the service is fed from a normal distributor it is terminated in a sealed chamber containing the appropriate fuse. Where the service is fed direct from a substation in the consumer's premises the fuses may be in the l.v. board in the substation.

As there is a need to insert and withdraw main fuses with one side live they are arranged in carriers. Round tagless fuses are used for the 60 and 100 A sizes where the fuse-carriers rely on spring contacts. For all sizes above these the contacts are tightened by screwed wedges which compress the double tags on the fuses on to the contact fingers. It is preferable to terminate the service as soon as possible on the consumer's premises as the small section service cables are not fully protected by the large fuses which protect the distributors.

The common practice of using metal-clad enclosures for the fuses on three-phase services is no longer favoured by PESs as some flashovers have occurred when the PES' staff have been inserting or withdrawing fuses. Those installed need to remain for many years and safe working practices have been developed. New designs are of all-insulated construction with separate covers on each phase so that only one needs to be exposed at any one time. These cut-outs are sealed to prevent access by other than the local PES' authorized staff and should not be tampered with even by highly skilled contracting staff; besides being illegal there have been some very serious accidents.

Where a PME terminal is provided this is sealed to prevent its use until compliance with the appropriate regulations has been checked by the local PES.

High voltage

Every h.v. supply has to be tailored to suit the needs of the consumer and the capability of the local PES network.

The Electricity Supply Regulations require provision to be made to guard against excess energy. The most simple arrangement is an h.v. switch-fuse and this is a satisfactory arrangement for a single transformer.

The most common arrangement for the provision of an h.v. supply is to loop in an h.v. circuit from a ring main network. These circuits are usually operated with a switch open at the mid-point, to provide a radial system. The open point can be altered by switching to permit any part of the ring to be maintained or altered, Fig. 1.9. By looping in an h.v. consumer the supply is maintained while such work is carried out. With this simple arrangement supply is lost if the circuit develops a fault but can be restored in an hour or so by simple switching operations. In some cases it is possible to arrange for the open point to be at the consumer's premises and this enables the supply to be restored by automatic changeover equipment or by the consumer carrying out the switching. A system of interlocks is used to prevent the circuits being paralleled.

In densely loaded areas it is often possible to tee on to two circuits to provide a normally available alternative. In London a variation of this has been developed and there are now a considerable number of installations which have proved successful in maintaining supply, Fig. 1.10. There are two incoming circuit-breakers, both are normally closed. One set of relays provides overcurrent and earth leakage protection, having a definite minimum time multiplier setting of 0.15 which operates in about half a second, and another faster operating directional relay. The operation of the overcurrent or earth fault relays opens one of the two circuit-breakers. The directional relay, by operating faster, decides which is the faulty feeder and switches the tripping circuit to the appropriate circuit-breaker.

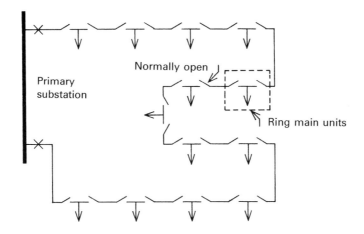

Fig. 1.9 High-voltage ring main.

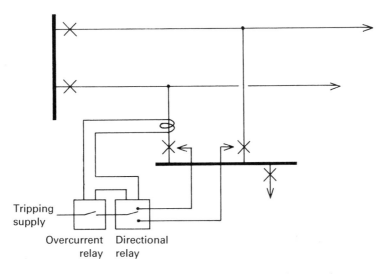

Fig. 1.10 High-voltage supply to a consumer using directional protection.

The protection on the outgoing circuit-breakers at the source substation has a longer time setting and thus the feeders are separated and only the faulty feeder loses its supply. This simple arrangement can only be used $n-1$ times between n feeders forming a group, otherwise parallel paths are formed for the flow of fault current and discrimination between DOC circuit-breakers and source substation circuit-breakers is jeopardized.

In areas where there is likely to be a considerable number of h.v. consumers in close proximity it becomes a practical proposition to employ a unit protected closed ring main, Fig. 1.11. Here each cable section is fitted at both ends with similar current transformers, summators and relays which compare outputs via a

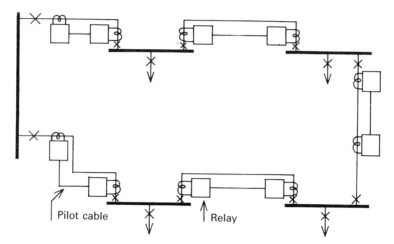

Fig. 1.11 Unit protected ring.

pilot cable and the circuit-breakers at both ends are tripped if there is a certain level and proportion of difference.

Two typical systems in use are Solkor and Translay, (see Chapter 22 on Protective Systems). Both names have existed for many years and each have passed through a number of revisions to increase speed, improve stability, reduce size and manufacturing costs. These systems by their very nature do not provide back-up for each other and it is thus prudent for a few strategically placed slower operating overcurrent relays to be installed to provide back-up protection.

This back-up protection also operates on a busbar fault in one of the substations. Such faults are, however, very rare with modern equipment and the risk acceptable. There are several ways that have been used to protect the busbars, for example, one is to overlap the unit protection schemes, but such arrangements complicate the issue and increase the risk of human error and cost and thus the more simple arrangement is now preferred.

A hybrid DOC/unit protection ring system has been developed in London in order to afford firm supplies to a group of h.v. consumers without having to install pilot cables some distance back to a source substation. The busbars of two DOC-protected consumers have been coupled together by unit protected cable links which contain several h.v. supplies. Overcurrent protection is fitted at one end of the unit protected circuits to act as a ring-buster and limit supply failure in the event of a busbar fault.

CONSUMERS' SUBSTATIONS

In order to control and transform an h.v. supply, the consumer needs one or more substations. They have to be designed to meet the regulations which apply for the particular location. Regulations cover such matters as fire protection, safety, means of escape, etc. To ensure that these are complied with it is advisable to employ an experienced consultant. The work of installing and maintaining h.v. equipment is a specialist activity and PESs are generally prepared to quote for such work.

The main consumer switchboard is fed by the local PES through its service circuit-breaker with associated metering. At first sight it sounds attractive to use a composite PES/consumers' switchboard, but taking all things into consideration it is probably better to have the local PES' switchgear in a separate switchroom. The single switchboard poses problems of meeting both specifications and of agreeing the supplier. There are problems of authorization of entry and of keys and security locking. By providing an intervening brick wall and a short through-connection the problem of co-ordination is much reduced and the cost is recovered in terms of reduced overheads and a faster completion.

The consumer needs to have some of his staff authorized to operate the h.v. gear and for this they require training. Depending upon the equipment and layout it may be appropriate for this to be done by the plant manufacturers but it may also be appropriate for some training in safety procedures to be given by the local PES. Many have training establishments for their own staff and they may be prepared to give the required training for an appropriate fee.

H.V. or L.V. SUPPLY?

The local PES generally charge slightly less for supplies at h.v. than at l.v., the difference being determined by what they estimate to be the extra cost of transforming and distributing the l.v. Providing all the supply is to be transformed to l.v. at a point where the local PES could equally well do it with its standard range of equipment, then it probably pays to take a supply at l.v., since the local PES has the advantage of economy of scale in its activities and can exchange plant to match the load most economically. Where economies can be made by deploying transformers around the site or by using equipment at a higher voltage then this will tip the balance in favour of the use of h.v.

METERING

Metering of 1 MW customers

The principle of 'metering of differences', described in the last paragraph of the section on the electricity pool, is applied to all second tier customers within a PES area. In addition the data recorded is 'tagged' to its nearest GSP to allow the local PES to allocate its use of system charges and NGC to allocate its use of grid charges.

Figure 1.12 indicates the system used for data collection and the contractual links between the contracted supplier, the local PES and the customer. The contractual supplier is deemed registrant of the meter and is responsible for appointing a meter operator, who had to be the local PES, to install and maintain equipment to agreed standards.

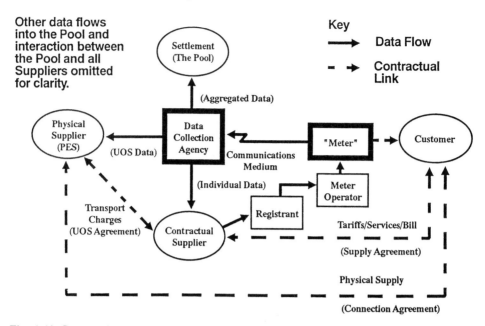

Fig. 1.12 Communication systems for competitive supply – parties involved for 1 MW customers.

Code of Practice G – metering of second tier supplies of 1 MW and above specifies the following metering equipment:

Class 2 active energy metering for customers up to 10 MW and Class 0.5 or 0.5 S above this;
CTs Class 0.5/0.2 and VTs Class 1.0/0.5 (<10 MW/>10 MW).
For each circuit measured, main kWh, check kWh, lagging kVArh and leading kVArh.

The Pooling and Settlement Agreement – Metering Clause 56 and Agreed Procedures, stipulates 'Agreed Procedures' for the collection, processing, checking, correcting and security of data collected. The National Data Collection Agency has taken over the franchise, in an attempt to reduce costs.

Metering of 100 kW customers

The system for 100 kW customers is basically the same as for 1 MW customers as shown in Fig. 1.12 except that as from 1994 the meter operator is appointed by the customer (this retrospectively applies also to 1 MW customers) and that the equipment is subject to Code of Practice 5 – Metering of Energy Transfers with a Maximum Demand of up to and including 1 MW, covered by the Pooling and Settlement Agreement. This code specifies the following:

Active energy meters of Class 2.

CTs of Class 0.5, VTs of Class 1.

For each circuit measured, import kWh, import kVArh and if appropriate export kWh.

Local communications access via an IEC 1107 port.

Remote communications outstation to display cumulative kWh and kVArh and time/date and to indicate reverse running. Options to indicate kW/kVA maximum demand, cumulative MD, number of MD resets and a rate sequence.

Monitoring of phase failure, errors/faults, battery condition and access to communications port.

Provision to the customer of kWh pulses and demand reset pulse.

Metering of less than 100 kW customers

In preparation for the removal of all franchised markets in 1998, OFFER issued a 'Consultation Paper on Metering, January 1992' which called for the installation of a two-way communication infrastructure by each PES to support new technology metering at customer premises. The practicalities and commercial benefits of such an infrastructure are still under consideration.

In the meantime supplies to domestic customers continue as before.

In the UK a domestic consumer is responsible for providing his own metering but in practice this is supplied by the local PES which recovers the cost either generally through the appropriate tariff or by rental in some cases.

Except where specifically agreed in writing, it is necessary to have all meters certified by a meter examiner who is under the control of the Department of Trade and Industry. This certification is generally carried out at authorized testing stations belonging to the local PES and every batch is checked by a meter examiner and appropriate records kept. Periodically every meter and, if applicable, its associated current (CT) and voltage (VT) transformers must be re-certified. The effective period is dependent upon the type of meter and its statistical performance.

At present the limit of accuracy in the UK is $+2\frac{1}{2}\%$ fast and $-3\frac{1}{2}\%$ slow. This is under review as the procedures and limits in the EU are different. There the manufacturers are also permitted to certify meters and in the interests of harmonization the UK must come into line.

For currents up to 100 A the full line current is passed through the coils of the meter whether single- or three-phase. For currents above this it is necessary to use CTs to produce 5 A at full load. When this is done it is also necessary to install fuses for the potential loads which, of course, need to be sealed to prevent removal.

One convenient arrangement for the CTs is to have all three cast in a single block of epoxy resin. This simplifies the arrangement and reduces the number of connections to be made at installation time as well as reducing the possibility of a reverse connection of one transformer. Where two or more services are provided in close proximity it is possible to summate the CTs and use one meter. Sometimes two 2.5 A secondaries are used and in other cases a special summation transformer. Where practical a more satisfactory arrangement is to put all the services at one end of a substation busbar and install the CTs on a removable section of the busbar.

Most commercial and industrial load is supplied on a maximum demand tariff. There are two methods, one based on kVA and the other on kW. The kVA charge is based on the argument that it covers the capacity of the supply taken. Its disadvantage, however, is that it is not a quantity that is directly measurable. Expensive meters generally of the 'Trivector' type are necessary.

The measurement of kW is more simple as it only requires a half-hour summation mechanism on the kWh meter. To cover the capacity argument it is usual to have a clause in the agreement specifying a minimum acceptable average lagging power factor. This is easy to check as a kWh meter with a 90 degrees connection will measure kVArh and a directional brake will eliminate the leading component. The average power factor is thus determined by

$$\text{p.f.} = kW/\sqrt{(kW^2 + kVAr^2)}.$$

High voltage metering requires the use of VTs as well as CTs. These usually have an output of 110 V between phases with one phase used as an earth connection; there is no neutral. Exceptions to this latter arrangement can occur in unusual cases such as h.v. electrode boilers where it is necessary to simulate the voltages to earth and a five limb VT with a neutral connection is necessary.

Where a consumer transforms all his supply to l.v. it may be possible to eliminate the cost of the VT by the use of loss-compensated metering. This is connected to the secondary side of the power transformer and its losses are simulated by injecting a current proportional to voltage into the current coils and adding a voltage proportional to current to the voltage coils. These cover the 'iron' and 'copper' losses respectively.

CHAPTER 2

Substations and Control Rooms

D.M. Barr, BSc, CEng, FIEE

(Consultant, Balfour Beatty Engineering Ltd)

INTRODUCTION

Health and safety

In recent years there have been a number of developments in the industry, particularly in relation to health and safety issues. Dangerous practices must be avoided and equipment must be suitable for its purpose.

Substations are potentially lethal. During their training, virtually all electrical engineers have witnessed the enormous power available in even a low voltage system. Careless mistakes made with a test set or battery illustrate the potential consequences of a high voltage accident.

Privatization

Privatization has drawn attention to the fact that electricity is a commodity to be sold to the consumer at a profit which depends upon the time of day or season when it is produced. Tariffs are constructed by calculating the cost of energy production at the time that it is required.

Costs are made up of:

(1) *Capital invested* This was probably borrowed at a commercial rate of interest.
(2) *Cost of upkeep* A combination of labour and material costs.
(3) *Cost of fuel* In the UK this is a mixture of nuclear, coal, oil, gas with a contribution from other forms such as hydro, waste disposal and wind power.

The fuel cost is the main basis for the variable kWh element in a tariff. Transmission and distribution costs are virtually all for capital and maintenance but do include an element to cover energy losses. This provides an incentive to make the network as efficient as possible.

The tariff has to take into account the huge cost of maintaining standby capacity to cater for outages of a large generator, or generating station or of an interconnecting line. Pumped storage stations will assist, if they are pumped up, but their overall efficiency is about 70%. This is the first additional cost that must be recovered for power at peak load times. Others include the additional capital, maintenance and fuel costs involved in providing spare capacity plant on standby.

33

Reactive power component

Expenditure is incurred for the provision of reactive power. Alternating current systems have to receive excitation current continuously. The effect is that there are large reactive currents being transmitted across the network causing real energy losses and voltage drop (regulation). The result produces additional capital and running costs. These costs are reflected in tariffs by use of appropriate metering which records the various quantities and time of day and date.

At peak load times, efforts are made to reduce the kW and kVAr maximum demand. This maximum demand charge can be onerous and difficult to reduce.

For many years there have been recommendations to install equipment to improve the power factor and to manage the energy demand. The long term financial recovery from such outlay has not been sufficiently attractive to those managements who are anxious to reduce forced outages.

New technology

The introduction of digital, data handling systems with programmable devices and transmission capability has provided a powerful new tool for control and monitoring of the supply. Protection and interlocking have been improved with the introduction of information technology equipment in substations.

A typical, classical substation secondary system would perhaps have 40 cores per unit marshalled at the switchgear for transmission, via the alarm and station control cubicle, to the network control centre.

The equivalent digital control system has single parallel connections between each of the switchgear units and their protection panels. There is a single serial connection from each unit to the local control panel, plus a further serial connection to each of the computer input/output modules and the network control centre.

Procedural planning and design

Remote monitoring and control, together with a degree of automation can be introduced into existing substations at a reasonable cost and with suitable checking facilities. This ensures that wrong procedures leading to unsafe conditions are not possible. Illustrations of modern equipment are shown in Figs 2.1 and 2.2.

The downside is that operational sequences based on logic diagrams must be produced with the co-operation and approval of the client's engineer. Comprehensive checks are needed, preferably physically on a mock up on the supplier's premises, to ensure that the procedures are safe. It should not be possible to defeat interlocks either deliberately or accidentally.

It is important that the installation contractor understands the purpose of sequences and is able to test the installed hardware connected to the various items of plant. A thorough knowledge of the software and the required settings is essential. The designer must be able to supply this information well in advance of the final commissioning procedures.

Electricity boards provide a single point of supply to a site which can be at different levels of security depending on the importance of the electrical loads

Fig. 2.1 11 kV switchboard, Jersey (*Siemens plc*).

Fig. 2.2 Digital control panel, Jersey (*Siemens plc*).

served. If the load demand is of sufficient magnitude the consumer will find it practical and economic to take the supply at high voltage (h.v.). For the purpose of this handbook the upper limit is 11 kV, although for very large complexes the intake might be at 33 kV or even higher. It is the consumer's responsibility to control and distribute the electricity around his site or premises to suit his own requirements.

The main substation serving a network incorporates the supply intake point, but there may be other substations located at strategic areas in a large complex and connected to the main substation by a cable network. The manner in which this interlinking occurs depends on a number of factors and is fully discussed in Chapter 3.

If the supply is taken at 11 kV the main substation will include 11 kV switchgear which may serve other substations at 11 kV throughout the site. Stepdown transformers at the site substations provide power at the consumer voltage. For very small installations the intake may be a single cable providing a three-phase 415/240 V supply to a switch and fuse board. In between are a variety of arrangements.

Depending upon the importance of the load served there may be a standby generator(s) linked to the system and arranged to supply essential loads. Some manufacturing complexes may generate their own power, particularly where the generation of process steam is related to the electrical requirements. Generally speaking it is uneconomic to rely on self-generation totally because of the cost of standby generators. Such sites would normally have a standby supply from the mains network. In that case synchronizing is required in the main substation to allow paralleling of the two systems. Alternatively, the two systems are interlocked to prevent accidental paralleling. Particular attention would be required to limit the fault power, the effects of reverse power and for discrimination under fault conditions. There would also be a problem in sharing responsibility. The magnitude and the importance of the load dictate the nature of the supply and distribution network and hence the design of the substations and control facilities.

Two points should be noted in respect of the demarcation between the supply authority and the consumer's equipment. The first is that there may be annual rates levied on any transformer between the supply point and the consumer's distribution system. Secondly, while the supply authority may provide, install, operate and own all of the equipment up to the point of supply it may require the consumer to make a contribution to the capital cost of making this supply available and will in many cases ask for a site on the consumer's premises to house it. It is common but not invariable for the supply authority to insist on a separate room or locked-off area for its apparatus, see Chapter 1.

There is therefore a wide variety of possible arrangements. The client's contractor is normally responsible for ensuring that the whole scheme is installed in a safe and secure manner with due regard to all health and safety regulations in accordance with the Health and Safety at Work Act and in compliance with the client's specifications. The site engineer therefore has a heavy responsibility. If a design drawing shows an unsafe arrangement or one which conflicts with good practice he must draw this to the attention of his employer so that it may be remedied. Electrical power systems can cause death and injury if not properly installed. The local authority and the fire authority inspect any new properties to check that they are safe. If they are not the remedial work could be expensive, especially if it involves civil works. The supply authority will normally require information about any new installation but apart from checking the busbars immediately after their own protection they do not have a responsibility for the client's installation.

As stated, the arrangement of supply to consumers' premises depends primarily on the nature of the load. For loads from say 100 to 300 kVA, a 415/240 V three-phase supply from a local authority substation is normally satisfactory. A load of between say 200 and 500 kVA may have its own 'local authority' transformer. Above 500 kVA one might expect an 11 kV supply and above 1000 kVA it would normally be split into two or more 11 kV circuits. Where a duplicate supply is provided the incomers are generally arranged to have manual changeover facilities or to be fully automatic with feeder protection depending upon whether or not a firm supply is required. Firm supplies are normally provided for all loads above 5 MVA.

Internal to the factory or other industrial complexes single radial feeders are generally acceptable provided that there is no safety hazard or process which is unduly sensitive to the loss of supply. A smelter or furnace could be badly damaged by the solidification of molten metal caused by a supply failure. In such circumstances a ring main may be installed or duplication of connections to the sensitive load. It is also important to remember that electrical equipment including busbars, transformers and circuit-breakers needs maintenance and therefore bus-section switches, duplicate transformers and interconnecting cabling may be inescapable for a continuous process.

An alternative supply may be better obtained from an in-house standby generator than a duplicate 'supply authority' feed. This is particularly so in remote areas where it is recognized that there is only a single overhead line or in overseas territories with unreliable power supply systems.

It is simpler and there is less risk of damage to equipment if the standby supply is not synchronized with the public supply. This is generally satisfactory since a small diesel generator can pick up full load in about a minute. Where there are installations such as computers for which a continuous supply is essential then a battery-operated inverter may be used. Emergency lighting is traditionally energized by direct current (d.c.) and it is wise to consider whether circuit-breaker closing and tripping supplies should be d.c. together with supplies to contactors and electrically operated valves.

SUBSTATIONS

Early consultation with the local supply authority is essential for agreement upon a mutually approved substation to act as the intake point for a particular site. This consultation is usually before detailed knowledge of the plant or project is known but it is essential that a fairly accurate load requirement be determined. Plant manufacturers must be approached to provide information relative to the equipment they are supplying and this, together with experience, enables a reasonably reliable load demand to be ascertained. Having this knowledge enables the rating of the power transformers and of associated switchgear to be decided.

All substations ought to be designed to be capable of extension unless it is obvious that this facility will not be needed. The extent of such provisions must be weighed against the monetary outlay and be agreed as a viable proposition. Depending on the nature of the system the substations may be required to accommodate h.v. and/or l.v. switchgear, transformer(s) and protective and control facilities. They can be wholly or partially outdoor or indoor and can take many forms.

Standard equipment should be selected as far as possible to keep cost and delivery to reasonable levels. This equipment should comply with an appropriate standard; in the UK this would be a British standard.

One can conveniently divide the subject into two sections, the various network supply arrangements possible which determine the nature of the equipment utilized and how this equipment is laid out and protected, i.e. the type of enclosure provided.

High voltage substations

There are three common ways that an 11 kV supply may be provided to a site. The first is by a ring main. The second is by duplicate feeders. The third is by a radial feeder or by a single spur from a radial feeder.

The duplicate supply may be provided with either automatic or manual change-over facilities.

Ring-main unit

A ring-main unit consists of two manually operated incoming isolators and a tee-off circuit to an 11 kV/415 V/240 V transformer. The tee-off may be controlled by a circuit-breaker or a fuse-switch, both providing a measure of protection to the transformer and acting as a back-up to the l.v. consumer's network. This is a very common arrangement.

If the site contains three-phase motors at various points remote from the substation, the designer may provide a motor control centre at the substation either integral with the main distribution board or adjacent to it. Alternatively, a number of circuits may be established to supply motor starters at sites closer to the motors. The main distribution board may also supply local general services such as lighting and small power directly in addition to circuits for more remote sub-distribution boards.

It is preferable to use motor control centres for process plant such as food or chemical manufacture where the production is automatically controlled. The cross connections between the monitoring equipment and the drive controller may be more readily effected. Duplicate supplies may also be provided economically to the motor control panel.

Where the supplies are only for small power and lighting the main l.v. switchboard may be less elaborate with fuse-switches, mccbs (or mcbs if the short-circuit level is not too high) and fuse boards. Where splitter boards are used a 415 V isolating switch is needed to enable fuses to be changed in safety. This should be so interlocked that access to the fuses is not possible until the switch is open. A typical ring-main unit is shown in Fig. 2.3 where the tee-off is controlled by a circuit-breaker and Fig. 2.4 where a fuse-switch is employed for this purpose.

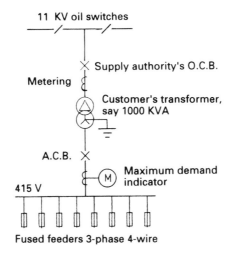

Fig. 2.3 A typical ring-main unit incorporating a circuit-breaker in the tee-off circuit.

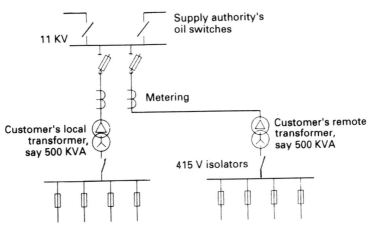

Fig. 2.4 A typical ring-main unit incorporating a fuse-switch in the tee-off circuit.

Duplicate supply substation
A more elaborate and expensive design of substation is shown in Fig. 2.5. This
would provide firm power to a site where security of supply is important. Generally
the bus-section switch is kept open and the two halves of the board are supplied
by their respective cable feeds. If there is a failure of one of the incoming cables
the bus-section switch is closed and the faulty cable isolated by its circuit-breaker.
Interlocking ensures that only two out of the three circuit-breakers (i.e. controlling
the supply cables and the bus-section switch) are closed at any one time. With this
system each intake must be capable of providing the full load of the network.

Single-supply substation
Where a consumer is able to accept an interruption in supply there is no necessity
to go to the expense of a duplicate feed, either from a ring-main unit or two

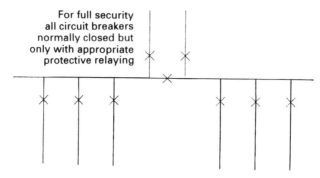

Fig. 2.5 Duplicate 11 kV supply substation.

separate cables. A single cable supplying an 11 kV switchboard can be used as shown in Fig. 2.6, and this could well be the responsibility of the supply authority.

Low voltage substations

For loads up to about 300 kVA the power is usually provided from the local supply authority's network at 415 V. As for h.v. substations, either single or duplicate feeds may be provided to the consumer's main switchboard. Quite often these boards are divided into two sections one supplying non-essential plant and the other essential equipment. In the event of a mains failure, the essential supplies can either be provided by a duplicate mains cable or more likely from a standby generator set.

Generally, the main switchboard is a factory-built assembly often of composite

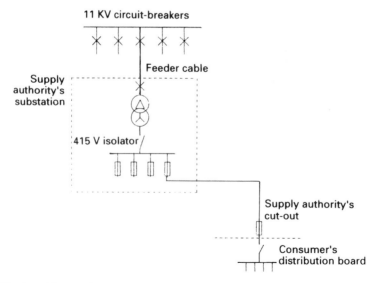

Fig. 2.6 Single-cable supply to an 11 kV substation.

design, incorporating circuit-breakers, fuse-switches and mcb panels. Motor starters may also be included or may form a separate mcc board.

The rating of circuits is discussed in detail elsewhere, in Chapter 3 'Site distribution systems', but there are a number of general points which are often overlooked in respect of the substation design which make it easier to understand the ratings which are commonly utilized.

A widespread site such as a dockyard or a large petrochemical plant may have a total load measured in megawatts. It is necessary therefore to have a number of substations.

Final sub-circuits should not have a voltage drop exceeding 4.0% of the nominal voltage at the design current. However, it may be necessary to use a conductor larger than that required for the voltage drop to satisfy the motor starting conditions. In addition the cables and protective gear must be designed to match the prospective fault current.

To size a cable therefore requires consideration of:

(1) full load continuous rating;
(2) voltage drop under full load conditions;
(3) motor starting voltage drop;
(4) prospective fault current short time rating.

Each of these conditions is subject to additional constraints. For example the full load rating must take into account the effect of a low voltage and a low power factor. These conditions will also apply to the voltage drop. In addition, continuous full load rating must be available despite proximity to other cables and high ambient temperature. The motor starting situation will also be made more difficult by low mains voltage and power factor. The fault current is obviously a function of the supply system. It is wise to allow for some strengthening of the supply system as time passes.

Fault clearance

The circuit-breakers and fusegear must be able to clear faults before cables are overheated. They must also themselves be capable of accepting the mechanical, thermal and electrical stresses imposed by faults. Transformers, busbars, cable boxes and insulators must also be suitable for the fault level.

To assist in the correct selection of fuses, manufacturers offer a variety of fuse characteristics. These cover variations in current, voltage, time/current, Joule integral, cut-offs, power dissipation and frequency including direct current.

The contractor must ensure that he is installing fuses which are appropriate to the duty. A fuse failure may result in an explosion or the emission of flame. Comprehensive guides are available from such organizations as ERA (Report No. 87–0186R) and the manufacturers on the selection of fuses. It is particularly important to ensure that fuses in substations (which are subject to the highest short-circuit levels on the system) have adequate breaking capacity and are fitted with the correct fuse link to protect the outgoing cables.

ENCLOSURES

Substations may be either outdoor or indoor types or a combination of both. Site substations generally are no different from main substations having the same equipment and layout but usually on a smaller scale. All the same precautions have to be taken with respect to safety, access arrangements, protection, etc., as with main substations.

Outdoor substations

Where all the equipment is mounted in the open the enclosures must be of weatherproof design; this generally relates to h.v. substations. Transformers are automatically suitable for outdoor mounting but if liquid-filled designs are employed they need to be provided with drainage facilities as discussed later.

Ring-main units are often used outdoors and designs are available that are suitable for this. Distribution cabinets are needed for the l.v. distribution feeder cables. Railings or anti-vandal fencing is provided to enclose the equipment to form the substation, Fig. 2.7.

There are also packaged h.v./l.v. substations that utilize standard indoor equipment mounted inside a weatherproof enclosure. The transformer may be outside the housing which contains the h.v. and l.v. switchboards separated by a corridor. Room for rear access to the switchboards may have to be provided. A claimed advantage for this type of substation is that it allows the foundation to be prepared well in advance and is ready to accept the assembled equipment direct from the manufacturer. It is not so popular today because of economic considerations but has the advantage that maintenance on the switchgear is possible in all weathers.

An economic arrangement of outdoor substations is the so-called integrated design which has switchgear mounted on the transformer cable boxes, permitting a naturally cooled transformer to be employed, Fig. 2.8.

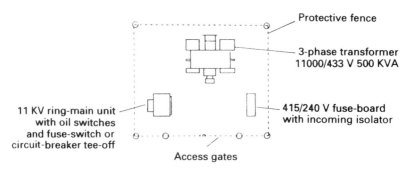

Fig. 2.7 Typical outdoor substation layout.

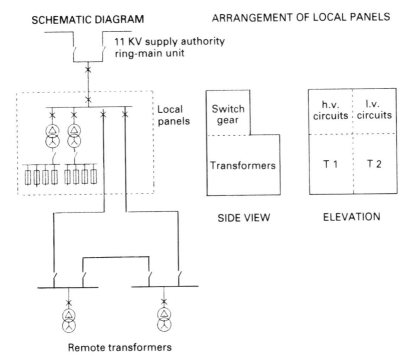

SCHEMATIC DIAGRAM   ARRANGEMENT OF LOCAL PANELS

Fig. 2.8 Typical multi-transformer substation.

Outdoor/indoor substations

The more conventional arrangement of h.v. substations is to have outdoor transformers with indoor 11 kV and 415 V switchboards. Transformers are housed in
an annexe to the switchroom and there may be separate chambers for 11 kV and
415 V switchboards, battery and charger and controls. It is often convenient to
have a separate control room for installations which include more than one
substation, standby generation, combined heat and power systems and process
plant requiring sequencing and interlocks.

This arrangement is preferable for town or urban substations because of the
convenience of using standard indoor switchgear rather than the more expensive
weatherproof designs. The difference in cost between indoor and outdoor transformers is not so great, and because of the need to provide adequate ventilation
the trend is to install them outside with protection against vandalism.

A common practice for multi-transformer substations is to have weatherproof
chambers for the h.v. and l.v. switchgear and metering and to mount the transformers in open pens as shown in Fig. 2.9.

Where the h.v. switchgear is under different operational management from the
l.v. side, the more conventional design of substation is probably the best. It is also
much more flexible to cater for special protection, extensions, non-standard
ratings or bulk-buying of components. However, these advantages must be weighed

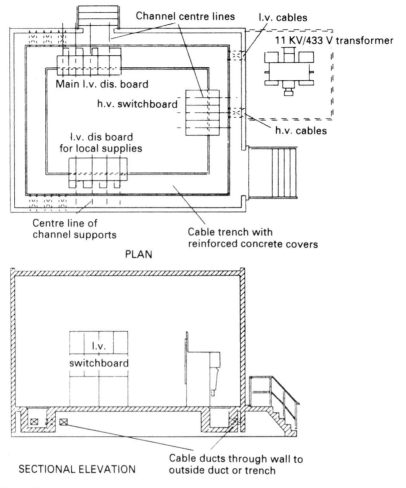

Fig. 2.9 Cabling of substation.

against the undoubted counter-advantages of packaged designs requiring less liaison at site between civil and electrical engineering, protection of sensitive equipment from weather and theft during transport and at site, lower design and installation costs, short completion times, and previous testing as a unit in the factory. The chance of a mistake in connecting up is also reduced.

It ought to be noted that a number of materials such as steel, glass fibre and concrete are available to house the switchgear, metering and distribution equipment. Extensions are also available from some manufacturers.

Indoor substations

Indoor substations require non-flammable equipment to make them safe or precautions to limit the fire risk. Basically there are two types of indoor substation. The first comprises separate components such as switchgear, control and metering panels installed in a room(s); the disposition depends on the space available, size of equipment and importance of segregation. The second, and becoming increasingly popular, is the packaged substation. This should not be confused with the outdoor package design which is quite different.

The indoor packaged substation is a single transformer assembly with integral l.v. switchgear housed in a suitably designed indoor enclosure which can be shipped as a complete unit or split up into sections for transport. The h.v. switchgear may be integral with the unit but it is more likely that an 11 kV cable will be brought from a nearby substation into the packaged substation.

A particular advantage of this form of h.v./l.v. substation is that it can be located at the load centre of the area it is serving, even on the factory floor. It is extremely compact and, like the outdoor equivalent, requires the minimum of site work.

SUBSTATION CABLING

Substation cabling may be run in trenches or ducts for outdoor substations or from overhead trays and cleats indoors. An attractive alternative for the large indoor installation is a cable basement with trays supporting cables running directly below the equipment. Such an arrangement facilitates changes and additions at a later date if it is known that the factory or process may be subject to alteration. It is also convenient if there is a separate control room because the control cabling then becomes significant so that it may be marshalled into reasonably large multicores for distribution to the associated controls, relays, interlocks and process plant.

For modest sized plants a cable trench is satisfactory. It is always true that where cables enter a building there are problems of preventing access of water and vermin. The best arrangement is to have the cable trench in the building at a higher level than outside. This is illustrated in Fig. 2.9. However there are many buildings where the cables are taken from a below-ground-level trench outside to a below-ground-level trench inside the building without letting water into the building. A set of salt glazed ducts installed at a level of say 200 mm above the floor of the trench prevents normal rain-water entering the building, particularly if they are sealed with a weak concrete mix of waterproof cement or some proprietary sealant such as Densoplast which is elastic enough to accommodate a modest relative movement of the cables.

In general, the more extensive substations, such as those illustrated in Figs 2.5 and 2.6, are made up as special combinations of standard parts. The choice of equipment is described below in general terms and in more detail in the chapters on each type of equipment. It is not proposed to describe all the combinations of circuit-breaker, isolator, fuse-switch, contactor, fusegear, transformer and control gear which are possible. One important feature which both designer and installer should strive for is standardization. The installer should be able to instigate site procedures which ensure that all the systems when completed are in accordance with good practice.

Transformers

In general 11 kV transformers are rated between 300 kVA and 2 MVA. If they are installed outdoors they are usually mineral oil filled, naturally cooled units. These transformers can be used indoors where the fire risk is not high but they must be provided with a bund and oil drainage facilities.

Where it is important to minimize the fire risk, i.e. for office blocks or commercial premises, one can use a transformer filled with a fire resistant liquid or a dry type Class C design. Until quite recently askarel liquids (composed of polychlorinated biphenyls, or pcbs) were used to replace mineral oil to provide a good cooling and insulating medium with fire resistant characteristics. Following accidents where pcbs have been spilled and have come in contact with human beings has come the realization that the liquid is toxic and is to be avoided.

Where transformers presently installed are filled with an askarel, care should be exercised and steps taken if necessary to retrofill them with a safer fire resistant liquid like silicone fluid. Askarels are marketed under a number of trade names such as Pyroclor, Aroclor, Pyranol, Phenocolor, Clophen and Fenchlor.

Silicone liquid is a satisfactory product to replace mineral oil as a cooling and insulating medium where this is desirable. It is much more economic than changing the transformer.

Dry type Class C transformers are an alternative fire resistant design but are not normally economic above 50 kVA, although where the situation demands it or the user prefers it, designs are available up to about 5 MVA. Not only is the cost much higher than liquid-filled designs but dimensions and weight may also be significantly increased. This may be critical if substation space is restricted or the foundations are not able to support the increased weight.

High voltage switchgear

There is a wide choice of 11 kV switchgear to choose from including oil break, air break, vacuum or sulphur hexafluoride (SF_6). The switchboards are either metal-clad or metal-enclosed designs with withdrawable circuit-breakers.

The bulk oil breaker board is probably cheapest but suffers from the risk of fire, high maintenance costs and a limited number of operations between overhaul. Vacuum gear is attractive where a large number of operations is envisaged but it has a relatively low fault breaking capacity and may impose a switching overvoltage on connected equipment. Air-break gear has a good fault rating, has low maintenance costs and has no fire risk but is very bulky and is expensive. Sulphur hexafluoride gear is relatively untried but promises to be attractive from the point of view of space, maintenance and safety.

The form of enclosure varies from company to company, some favouring metalclad designs with segregated busbars and separate current transformer (CT) chambers, cable entry chambers, voltage transformer (VT) chamber or tank, control and instrumentation and protective relay panels. Other companies provide a cheaper but less secure system without the earthed metal barriers between circuit chamber and busbars. There is a risk with such an arrangement that circuit faults will develop into busbar faults thus shutting down a whole network.

There are a number of variations to the main types of supply switchboard to suit particular circumstances, for example: double-bus switchboards where the circuit-breakers may be connected to one of two busbars; vertical or horizontal isolation; full interlocking to prevent operation of the circuit-breaker in the wrong position or withdrawal when closed; integral earthing where the circuit-breaker can earth either the circuit or the busbars when selected to do so; secondary isolating contacts which can be arranged to allow testing without danger to the commissioning staff; bus-section circuit-breakers; extensibility on either side of the switchboard; h.v. cable test facilities; and current injection test facilities.

There are also a wide variety of operating mechanisms to suit site conditions such as: independent manual closing; hand-charged spring-operated power closing mechanism with or without an electrical release; motor-charged spring-operated power closing mechanism; and solenoid closing mechanism.

There are also a number of 11 kV units which have been specially designed for ring-main supplies as discussed earlier. These usually consist of oil switches on the two connections to the ring and either an oil fuse-switch or exceptionally a circuit-breaker on the transformer tee connection. The oil switches are normally able to close on to a fault and to open full load current safely. This is an economic way of providing a secure supply and is widely used in the UK.

The ring-main unit can conveniently be provided as part of a package substation (discussed later) including the transformer and 415 V distribution fused pillar. Such a unit can be completely enclosed in a weatherproof cubicle and be transported to site in one piece. The cubicle also serves to keep out of danger members of the public or staff who are not concerned with its operation or maintenance. Figure 2.10 shows a typical 12 kV switchboard.

Control panels

In respect of control there may be separate panels or control may be entirely from the switchboard. This is often most satisfactory as a permanent 11 kV supply need only be switched off in an emergency. Such an occasion may never occur and the extra expense of separate controls is not worthwhile. However, where two or more sources of supply are involved, or if synchronizing of a private generator is necessary, or if there is a great deal of switching owing to load management or fluctuations of loads or supply systems, then a separate control panel is best. This is equipped with instruments indicating the normal electrical parameters such as voltage and current with perhaps, where necessary, power, vars, power factor and frequency; it could also have switchgear position indicator lamps together with a remote control switch for each circuit-breaker. The normally accepted colour code for indicator lamps is given in BS 4099: Part 1. There may be a requirement for additional facilities such as auto-trip alarm, overload alarm, earth-leakage alarms, trip circuit supervision and remote/local changeover switches.

Protective relays, alarm facias, mimic diagrams, secondary circuit fusegear and test blocks may all be mounted on the control panels which must therefore be of substantial construction and vermin proof with adequate space to mount the equipment, run the wiring and terminate the incoming cables. Control equipment may be on the front panel of a vertical cubicle or on the horizontal or sloping face

Fig. 2.10 Typical 12 kV metal-clad switchboard using SF$_6$ circuit-breakers (*Yorkshire Switchgear and Engineering Co. Ltd*).

of a desk-cubicle. The cubicles may have bottom cable entry with gland plates near the floor to leave room for making off, alternatively, they can be terminated in marshalling boxes below the floor with wiring run into the cubicle in conduits. Another design allows for the termination of the cables in terminal boxes mounted on the wall behind the cubicles with wiring running overhead in ducts or trunking to enter the cubicle from above.

Since a control board or desk is a man–machine interface its design and features are to some extent subjective. However, the indications and controls which are extended from the primary equipment to the control room need careful selection to ensure that the operator has sufficient information and control facilities without cluttering the panel with irrelevant signals or unnecessary switches. Data which are only required for post-fault analysis or long term records should not appear on the operator's panel.

There is also the possibility of bringing together individual alarms where a detailed knowledge of each item of plant which has failed will not assist in deciding the immediate course of action. In this case the cause of the alarm must be ascertained by going to the initiating device.

In the larger installations there may be scope for visual display units using small computers to receive, process, store and have available for display by instruction from the operator, the data needed to decide on the correct sequence of operations. Such systems are very powerful tools which can also provide facilities such as data recording and analysis, alarm limit setting and display, high-speed pre- and post-fault recording, sequencing control and monitoring, equipment status as well as loading information and the selection of sections of the network to be illustrated on the screen while the status is being changed.

On the more conventional installations, however, it is common to rely on a mimic diagram, possibly using a mosaic type construction for ease of alteration, and including either indicating lamps for important features such as circuit-breakers and isolating switches or discrepancy-type semaphores. The semaphore may also include the operating switch with a system of illumination to show whether the primary plant has altered position in accordance with that taken up by the control switch. Miniature ammeters, voltmeters, etc. may also be incorporated but as these become illegible on the larger panels unless the operators approach them closely, it may be preferable to mount them separately from the mimic panel on a desk immediately in front of a seated operator.

On the smaller installations a combined mimic and operator's panel is of course quite satisfactory and this would be the case in the majority of industrial premises. Such a panel is commonly placed in the same room as panels controlling factory boilers, compressed air systems, and processes of various kinds. Associated with them are energy management systems which are becoming more and more an economic necessity. It is important to have a reliable internal communication system associated with the control room both for the normal and emergency operation. Such a system enables the control room to be kept informed when switching operations are required or when emergencies arise.

The arrangement of lighting and of the panels themselves is an aesthetic question as well as one of ergonomics. The site supervision engineer will be wise to ensure that the designers have taken care to ensure that there is adequate access for the sections of panels being shipped and that there are clear working areas behind the panels for connection and testing. He should also check that the lighting provides adequate illumination for operation and maintenance without glare or reflections. The equipment should also be handled carefully as it can be badly scratched and chipped by carelessness during installation. Instruments and relays must be checked on arrival to ensure that they have not suffered damage during transit.

The layout of the equipment on the desks and panels should be such that the more important switches are nearest to hand and they are accompanied by the corresponding indications. Instruments should therefore be at eye level and switches at hand level. Switches should not be placed in such a position that they can be inadvertently operated by passing personnel or by carelessly pressing on them with books or with the hands when leaning. Instruments should not be in a position to be broken by vacuum cleaners or washing down mops and brushes. Fuses and links should also be out of harm's way and should not be so mounted that they can work loose by vibration.

All control panels and desks must be properly labelled so that there can be no ambiguity about the purpose of the indication or switch or the circuit being controlled. Many serious accidents have resulted from mistaken panel identities.

The mimic, which is a simplified circuit diagram, permanently secured to the panel, may itself be illuminated to indicate the status of the various sections of the distribution network, but it is more usually painted in bright colours using a different colour for each voltage. It often illustrates only the busbars of the main switchboards, with the minimum of instrumentation. Designers should ensure that the order of the circuits from left to right on the control panel corresponds with that on the switchboard. Synchronizing equipment must be visible to operators along the whole length of the control panel.

One of the important aspects from the site engineer's point of view is the illumination of the cubicles and the method of switching the lighting supply on. Accidents have occurred from inadvertent contact by electricians with the lighting supplies and it is essential to ensure that all live wiring is properly shrouded. A control board is shown in Fig. 2.11.

Low voltage switchgear

The variety of l.v. switchgear is also very wide. Typically a switchboard includes circuit-breakers for incoming supplies, a bus-section(s), important feeders and interconnectors, some or all of which may require interlocks. The individual supplies to equipment such as local lighting and small power, battery chargers, trace heaters and domestic supplies in general tend to be controlled by switch-fuse

Fig. 2.11 Final adjustments being made on a control and instrumentation panel for Equador. Note the semaphores on the panel front (*Arcontrol Ltd*).

units or fuse-switches. The difference between a switch-fuse and fuse-switch is defined in Chapter 16.

Where there is a substantial incoming power supply to an industrial complex then an air-break circuit-breaker is often utilized. This can be a single cubicle unit with cables both in and out or it can be the incoming unit on a main switchboard or be connected direct to the switchboard busbar system. Such a unit may be fitted with CT-operated overcurrent trips with inverse time characteristics, an automatic operating mechanism and other features such as on/off indication, fuses for metering supplies, direct acting earth faults trip, auxiliary switches, shunt trip release and undervoltage release.

Typically these circuit-breakers have thermal ratings up to 4000 A, a short time rating of 40–50 kA for one second and a short-circuit breaking capacity of a similar level. One point which is worth noting is that the rated insulation voltage should be about 660 V. As the normal insulation level of l.v. cable is 1000 V it is possible to consider the use of 600 V motors and control gear as a means of reducing costs. Increasingly popular as alternatives to fuses for fault protection are moulded case circuit-breakers. These are now offered in a wide variety with ratings in excess of 2500 A and breaking capacities up to 150 kA at 380/415 V. They are offered with either thermal/magnetic or electronic protection. Ratings up to 3000 A and 660 V a.c. are also available. However, the fault breaking capacity of these may be limited to 60 or 65 kA.

Open type power circuit-breakers in compact ranges are available up to 8000 A. These compact breakers are available in modular form with a wide range of accessories. They can be plugged in with terminal shields. They can be locally or remotely controlled and they have arrangements for earth leakage, shunt trip or undervoltage release. Auxiliary and alarm switches are fitted and a visible indication is available confirming that in the 'off' position all contacts are separated by the required distance.

The main outgoing cables can therefore be selected taking into account the most onerous of the circuit duties.

Mechanical strength and protection is also important. For example vibration may be significant if the substation is adjacent to a road or heavy machinery. Fire precautions are also necessary if the cables are being run in a confined space where personnel may be in danger from heat or smoke. Corrosion may also be a hazard. Single-core cables may cause overheating of adjacent metalwork and can also produce high induced voltages under fault conditions. They are also a source of interference unless suitable precautions are taken to segregate them from sensitive equipment.

The alternative to a circuit-breaker is an isolating switch with or without back-up fuses. The isolating switch should preferably have a making capacity of about ten times and a breaking capacity of about eight times normal full load current at say 0.3 p.f. This is to deal with the stalled rotor current or starting current of any motor connected to the board. Alternatively on-load switches can be obtained which provide a capability of making onto a fault and breaking a current of about three times normal rating.

On the incomer an alternative may be to use a fuse-switch. It is important in this case that the moving parts are connected to the busbars and not to the incoming cable, otherwise the fuses are still alive when the switch is open. On those boards which are protected by a circuit-breaker or fuse at the sending end of the incoming cable no fusegear is required on the incomer itself. This can then be an isolating switch using perhaps a fuse-switch with fuses replaced by links of adequate through-fault rating.

For outgoing units the same range of equipment is available, such as air circuit-breakers, mccbs, fuse-switches and isolators, but with the addition of switch-fuses, i.e. an isolator associated with separate fuses, distribution fuseboards and motor starters for the control of rotating equipment. All such gear is connected through busbars of appropriate rating to the incoming supply. It is imperative that the busbars are adequately insulated and protected from vermin, insects, dust and damp but most important that they should not be adversely affected by a fault on an outgoing circuit.

There are many ways of cabling a switchboard but the majority have rear access from the bottom. However, a modern trend is to have front access to the cable compartments allowing the board to be set against the rear wall. Additional space is required for aluminium cables and shrouded terminals are a useful safeguard.

Other variations and accessories available are bus-section switches, packaged substations where a Class C transformer is directly connected to the switchboard, power factor correction equipment including the capacitors, relays, contactors and fuse-switches for automatic multi-stage operation, together with instrumentation, metering, protective relaying and control switches. A 415 V switchboard is shown in Fig. 2.12.

Fuses

The various types of fuses available cannot be easily summarized, but in general high voltage fuses at substations are current limiting. This means that they will clear the fault before the prospective current reaches its first peak and thus limit both the mechanical and thermal stresses on the protected equipment.

Low voltage fuses in substations may be selected either to protect the a.c. electricity supply network or, where appropriate, industrial equipment. Special types are required for semiconductor devices.

Motor starters

Integral with the switchboard or in a separate panel may be motor starters. The majority of motor starters are direct-on-line designs for cage induction motors comprising contactors which are electrically held-in when energized and incorpor-ating simple no-volt release and a thermal overload protection: reversing contactors are a straightforward variation. Star-delta starters giving a reduced voltage across the star connections of machines are also standard arrangements. The starters are normally available as combinations with a disconnecting switch and short-circuit protection in the form of fuses either mounted on the contactor or separately.

These are compact and yet have good access for maintenance, more room to accommodate aluminium cables, starters which can be easily removed, boards which can be easily extended and high rating busbars both in terms of normal thermal rating and also short-circuit withstand. One important feature available in

Fig. 2.12 A 415 V power station distribution switchboard incorporating air circuit-breakers (*Electro-Mechanical Manufacturing Co. Ltd*).

some designs is a cable box for each starter, each with its own gland plate and metal cover segregating it completely from other circuits. Another feature which is useful to the client and his installer is the provision of vertical and horizontal cable trunking to enable interconnecting wiring to be run from starter to starter. This means that modifications to sequencing wiring or connections to relays controlling the starters can easily be made either at commissioning or subsequently. Of course the necessary contacts need to be available in the starter. A starter panel is shown in Fig. 2.13.

Auxiliary control equipment

For the control of the contactors and starters there is a wide range of equipment available including push buttons, rotary switches, solid-state control modules with input and output interfaces, logic elements and memories, timers, power supplies,

Fig. 2.13 The swing-out design on this l.v. motor control centre gives easy access to the components (*Arcontrol Ltd*).

amplifiers, decimal displays and monitoring lights. The system may be enhanced by providing adaptability. This is the basis of the programmable logic controller which enables the user to modify the control sequence for most standard system applications. The program need not be finalized before starting the control system design and assembly. Complete system change can be undertaken by re-employing the controller on a new installation if required.

Programming is simple and can be undertaken from a small portable 'memory loader' or other form of programming equipment such as a hand-held keyboard and display unit or a more sophisticated keyboard and VDU. Programming equipment can be provided also for on-line or off-line use and with hard copy of the logic statement and can be accompanied by data logging and management reporting. Such equipment is of course only required for supplies to process plant, automatic manufacturing, conveying or sorting and other relatively large or sophisticated installations where the financial consequences of faulty sequencing justify' the costs involved. A programmable logic controller is shown in Fig. 2.14.

Supplies to electronic devices

With the rise in the number of electronic and digital devices, filters may become necessary to isolate equipment from transient overvoltages and harmonics. It is particularly important to ensure that electronic protection does not inadvertently trip essential services for no good reason.

Sequencing equipment and energy management systems installed at central control points, such as substations, should also be isolated from interference by suitable enclosures and by the use of uninterruptible power supply units (UPS).

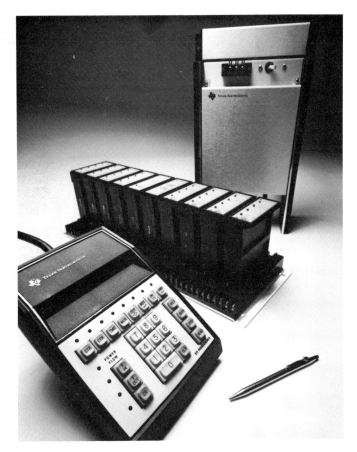

Fig. 2.14 A programmable logic control system (*Texas Instruments Ltd*).

These can be provided with powers up to several MVA and with output frequencies such as 50, 60 and 400 Hz. They may also have static and manual bypass switches for maintenance or overload. The basic principle is to rectify the incoming supply, to provide thereby a direct current for charging a battery and to operate an inverter to the desired frequency of output. The battery acts as a standby source but also provides an excellent filter and smoothing circuit. If necessary additional units can be combined to provide redundancy for important computer installations. The units would also normally be supplied with a status monitor which could energize a remote control unit.

INSTALLATION

Basic design

Consider the installation of a typical 11 kV/415 V substation feeding a main 415 V switchboard. The main switchhouse may also contain an mcc and a control board for energy management. The local supply authority provides power at 11 kV through its own switchgear but the contractor is responsible for installing the stepdown transformer, cabling the 415 V switchboard, starters and control panels, together with the auxiliary systems. The premises are not in a hazardous area and while the loss of supply is financially serious there is no danger inherent in a failure. The designers have provided a simple single storey building with a shallow cellar and with separate rooms for a switchboard associated with the transformer, l.v. switchgear, the mcc and energy management centre, and the battery.

Cabling is run in trenches inside the building and salt glazed ductwork outside. Earth conditions are reasonable and therefore the design of an earthing system of low resistance is obtained by sinking four copper rods, one at each corner of the building. No special provision against lightning is considered necessary. Small power, lighting, heating and ventilation, fire protection and a telephone system are provided. A typical layout of the substation is shown in Fig. 2.9.

Programme

The critical path network system shows the best order for the erection and cabling of electrical equipment which cannot commence until buildings are more or less complete and the machinery has been installed. On the other hand, once machinery has been installed the purchaser is anxious to have it operational. The site engineer however will be restricted to a reasonably small team of electricians because only one or two men can work on a switchboard at one time and on large sites there are accommodation and messing restrictions.

The programme indicates which items must be received on site first and where access can be made available by the civil constructors. The resident engineer must ensure that there is adequate storage for those items which arrive early and for the consumables which are needed to keep the work progressing in an efficient way. This includes checking that there are adequate lifting facilities and local transport aids with rollers or trollies for the items which are to be manhandled into place. The commonest problem of all is to find that the doors are too small for the cubicles. It is easy to forget that one needs some clearance for the transporter and the protuberances such as relays and operating handles.

Foundations

The foundations for h.v. switchgear must be capable of withstanding whatever short-circuit forces can be imposed on them together with the vibration of operating mechanisms. It is also necessary to have a smooth flat floor to prevent distortion of the cubicles and easy movement of truck-gear. If inset steel rails are called for they must be checked level to the limits stipulated by the manufacturer and any holding-down bolts grouted in well before tightening up becomes necessary.

Cable trenches and ducts should be examined to see that they line up with the terminating points and that there is adequate room for bending cables and laying them in an orderly manner. All supporting steelwork must be properly fixed to walls and ceilings by the recommended size and number of insets. All floors, walls and ceilings should be properly finished before commencing the erection of permanent electrical fittings including lighting fittings. Any chases for cable conduit, boxes and ducting should be completed.

Good practice also dictates that the buildings should have observation panels in the doors of control rooms, emergency escape doors should be fitted with power bolts and latches and external doors should open outwards. In tropical or sub-tropical locations, fly-screens should be provided to all opening windows and exterior doors. Switchroom ceiling heights should be not less than 3 m with windows fitted at a high level to provide maximum inside wall space for equipment. Windows should not be located above outside transformers. Tubular heaters should be installed if there is no air conditioning, to avoid moisture deposition on equipment, also air bricks or ventilators should be provided to maintain two or three air changes per hour. Adjacent transformers should be separated by a fire wall or spaced from each other or substation buildings by at least 3 m. All transformers should be fenced in. Special arrangements are required to prevent the spread of oil spillage from transformers and oil circuit-breakers.

Cable ducts or conduits between a building and a transformer bay must have means to prevent ingress of oil. This can be done by ensuring that the transformer end is raised above the level of the ground or pebble infill and is placed directly below the cable glands.

The floor and foundations must not only be capable of supporting the weight of any equipment placed upon it but must also be hard enough not to be damaged by roll-out switchgear and the handling of equipment. This may be overcome by using steel channels in the floors.

Location of equipment

It is common to place h.v. switchgear in one chamber with access given only to a person sanctioned by the supply authority. The switchgear under the control of the customer is placed in a second chamber. However, this is not always the case.

Where transformers are housed indoors, ventilation must be adequate to dissipate their heat losses. Fire-fighting equipment should be provided wherever a number of transformers and switchgear units are grouped together. Where liquid-filled transformers are installed, adequate drainage to soakaway pits is needed with a medium such as granite chips or pebbles to help quench any fire arising from an electrical fault.

Cables require protection from each other to reduce damage from an electrical fault and limit the degree of interference in the event of a fire. Complete separation of the lower and higher voltage cables is desirable. All pipes and ducts should be effectively sealed to prevent the ingress of vermin. Where the installation of a substation is indoors, the problems of ventilation and transformer cooling conflict with the reduction of transformer noise to an acceptable level. Some form of compromise may have to be made. For instance sound attenuators placed in the air inlet and outlet vents reduce the efficiency of the cooling system and must be allowed for in the design. A vertical duct provides a chimney effect and may assist air flow across the transformers. However, on other sites forced cooling with fans may be the only solution. Transformers should preferably be mounted on anti-vibration pads and any pipework arranged so that a minimum of vibration is transmitted to the building structure.

Fire fighting

The layout of the plant and the design of the building play a major part in reducing the spread of fire and the effect of explosions. For example, equipment and buildings should be arranged to have vents which rupture rather than allowing an explosion to damage the main fabric. Site supervisors should ensure that these vents are never obstructed. In the prevention of fire, cleanliness and tidiness are very important, as is the careful maintenance of tools. Most fires are caused either by carelessness or faulty equipment.

The choice of fire-fighting equipment is dependent on its suitability for electrical fires but also on cost and the importance of the electrical supplies at the point in question. Portable manual types are as follows: halon gas of various kinds, carbon dioxide chemical foam and powder.

Fixed systems use water sprinklers, carbon dioxide and halon gas. Both halon gas and carbon dioxide can suffocate personnel trapped in the discharge area. Strict precautions must therefore be taken to lock-off the equipment when staff are present. There is also the use of sand, blankets and fire hoses. Fire doors are a very important means of limiting the spread of fire and ventilating systems should also be provided with automatic shut down if not with automatic dampers in the event of fire. Fire drill is also important and should not be neglected on a building site.

Cabling may also be a cause of serious fires with risks of extensive damage to the installation and danger to personnel. Low smoke and fume (LSF) cables are now available in a number of forms, most of which will reduce the flammability as well as causing less poisonous gas to be released when they are heated.

The d.c. supplies are a particularly important and vulnerable part of any installation. They are generally derived from stationary batteries which give off flammable and toxic gases. Batteries should be in a separate room with an acid-resistant floor, special lighting fittings, a suitable sink and adequate water supplies. It is wise to have an acid-resistant drainage system. The room must be properly ventilated but sunlight must not be allowed to shine directly on to the cells.

ERECTION PROCEDURES

Once the equipment is on site there is a need to follow a well-defined procedure for inspection, erection, testing and commissioning.

High-voltage switchgear

Before commencing erection of h.v. switchgear check that all connections are tight on busbars and on main and auxiliary circuits. Examine the insulation carefully, because there have been cases of burned out connections due to failure to observe this precaution. Obvious faults such as distorted panel work, broken meter glass and damaged packing cases should be noted and immediate steps taken to rectify the damage or to return the equipment to the supplier. After erection, steps must be taken to prevent deterioration of the equipment due to damp, dust or casual damage. Substations should be cleared out and locked fast and should not be used as a site office, store or workshop after the equipment has been installed.

When all erection and jointing work is complete the equipment must be inspected and thoroughly cleaned to remove cuttings of cable, spare nuts, washers, accumulations of dust or copper filings and tools which may have been left behind. However, it is dangerous to blow out equipment because dust, filings, light metal scraps, rags and sawdust can be blown into positions where they become inaccessible and may pass unnoticed. It is better to clean manually and vacuum the plant. An important point to remember is the removal of all packing materials particularly in moving parts and of course make sure that oil circuit-breakers receive their first filling of water-free oil.

In the erection of the switchgear the following points should be emphasized. Where steel channels are used as foundations, ensure that they are correctly laid to an accuracy of about ± 1 mm in 3 m. Use a masking strip to prevent fouling by the floor materials. Read and apply the manufacturer's erection instructions and use the correct tools such as torque spanners to obtain the appropriate tightness of nuts. Make sure equipment is properly aligned. This means vertically and horizontally in both longitudinal and lateral directions. Incorrect alignment causes problems during operation of mechanical linkages and connections, difficulty in removing and replacing circuit-breakers and other withdrawable parts and puts stress on interpanel connections such as busbars and earth bars. When the floor is being laid and there are no steel inset channels it should be level both front to back and should not vary by more than a millimetre or so between cubicle centres. When the floor is being laid it is common practice to form pockets at the foundation hole positions in order to avoid having to cut them out of the solid floor at installation time. The floor should be marked out in accordance with the switchgear assembly drawings usually provided by the manufacturer to ensure that the switchgear is correctly located with respect to cable trenches, building walls and other equipment in the room.

A datum line requires to be established usually along the rear foundation bolts and using normal geometric methods the switchboard and foundation bolt holes can be located. The method of positioning the equipment is dependent upon a number of factors such as equipment size, building location, site accessibility and lifting tackle available. Lifting eyes are often either incorporated into switchboards or can be screwed into pre-machined holes, but generally slinging is necessary and this should be done strictly in accordance with the manufacturer's recommendations. Without the use of cranes, however, the traditional manual methods utilizing jacks and roller bars are effective. Care must be taken not to exert pressure on weak parts such as control handles during this manhandling. Positioning of the cubicles should start near to the centre of the switchboard, installing as early as possible the enclosures associated with any special chambers or trunking. The first enclosure should be positioned and checked to ensure the side sheets are plumb and that any runner rails are level in both planes. When the first enclosure panel is correctly set the remaining enclosures should be positioned successively on alternate sides of this panel to make the front form an unbroken line.

Adjacent enclosures should be bolted together after they have been correctly aligned using whatever shimming proves necessary to make sure that the cubicles are vertically and horizontally true. The fixing bolts should be positioned in the foundation holes and cemented in leaving adequate time for the cement to set before tightening up.

The equipment may now be connected up and cable joints made. In particular one must confirm that the earthing of the equipment has been carried out using the recommended cross section of material and satisfactory terminations. The l.v. system must then be proved for continuity, preferably by using a hand generator or other portable device and with a current of about 1.5 times the design current. A further examination of all incoming and outgoing circuit and auxiliary cables, including a test of the correctness of the connections at the remote ends, should be done; this should include measuring the insulation resistance and continuity of all cables and wiring including internal and auxiliary connections. Where appropriate, phase rotation checks must be made before three-phase drives are energized. All moving parts must be inspected to ensure they are all operative. Dashpots must be filled to the correct level with the right grade of fluid, and the operation and accuracy of meters and relays by secondary or primary injection tests should be checked. All settings should be agreed with the client's engineers.

All cable boxes must be properly topped up, compound filling spouts capped off. All insulators and spouts must be clean and dry and cover plates securely bolted up with all screws in position and tightened and breather vents clear of obstruction. Ensure that the top of the equipment is free from all dirt and rubbish.

A final check on all incoming and outgoing cable connections to terminals ensures that they are tight and have adequate clearance. High-voltage testing can then be carried out to the test figures laid down in the appropriate BS switchgear specification or as specified by the client's engineers who should witness the tests and sign the test results.

After testing, precautions must be taken to discharge any static and remove the test connections before bolting up any covers removed for testing. Before energizing, the operation of all circuit-breakers and relays should be confirmed manually and electrically to ensure that no sticking or malfunction is present, particular attention being given to manual trip and close operations and to the operation of overcurrent relays and residual current devices. When the equipment is ready to make alive all circuits not in service shall be locked off at each end and safety operation procedures adopted. All switching operations should be carried out by a competent person.

The substation's entry and emergency exist doors must be operative and kept clear and free from obstruction. All substations should be kept locked when the equipment is live, and access restricted to authorized personnel only. Danger, safety, shock cards and any statutory notices must be prominently displayed. Tools required for operating switchgear should be stored adjacent to the equipment in proper racks or cabinets. Circuit and interlock keys should also be contained in special cabinets under the control of authorized personnel and no spare keys must be allowed to abort the safety of the system.

Batteries should be examined to ensure that they have received their first charge and the electrolyte is at the correct level and of appropriate specific gravity. The fire-fighting requirement should be checked and if it is CO_2 it must be confirmed that the safety lock-off procedure is understood by the personnel authorized to enter the substation. External warning notices must be fixed to protect any strangers from inadvertently suffocating.

Transformers

Chapter 13 deals with transformers and their installation so we restrict ourselves here to the procedures associated with their commissioning. Transformers should be inspected for internal or external damage, particularly if they have been dropped or tipped over. This should include such items as drain valves, selector switches, conservator tanks, Buchholz relays and winding temperature indicators. All transformers must be tested for winding insulation resistance and the readings confirmed as acceptable. However h.v. d.c. tests on a cable connected transformer cannot be done because the windings of the transformer short out the d.c. test set.

If between-core tests are required on transformer feeder cables after installation then a link box must be provided to disconnect the cable from the transformer windings. If transformer covers are taken off to achieve internal disconnection the tools used must be clean and secured externally by white tape so that they may be recovered if inadvertently dropped. All nuts and washers must be accounted for and all operatives should be asked to empty their pockets while working over the open tank. Waterproof covers should be provided during the period when the transformer tank is open. If it is not possible to disconnect the cables after jointing they must be tested beforehand. This means that the jointing and testing programme must be carefully planned to avoid leaving cable ends unsealed for long periods.

Transformer diagrams should be inspected and the phasing diagram confirmed as correct. Also before energizing, the voltage selector must be set on the appropriate tapping having regard to the voltage level of the system. Transformers which are to operate in parallel must be set on the same tapping and they should be checked as having the same impedance. Voltage selectors should be locked in their set position and if they are of the 'off-circuit' type they must not be adjusted without the supply being first switched off. Earthing arrangements for the tank and the neutral or other system earthing must be confirmed and completed before testing and commissioning. Where special tests for losses, ratio, phase angle or winding resistances are specified the assistance of the manufacturer should be sought.

The following points should be checked on the particular type of transformer as appropriate.

Oil immersed naturally cooled (ON)
Ensure that the oil level is adequate and that breather tubes are clear. Commission silica gel units by removing the air-tight seals from the cannisters and filling the oil

sealing-well to the correct level with transformer oil. The colour of the silica gel must be checked and the filling changed if it shows dampness (red for wet, blue for dry).

Dry-type transformers

As these are more susceptible to external damage they must be carefully handled and stored on site. They must also be kept in a dry, warm atmosphere until they are put in service, to prevent ingress of moisture. Satisfactory insulation tests are imperative before commissioning. Because they are wholly dependent on surface radiation and air convection for cooling they must be checked for any accumulation of dust or dirt which can block the air ducts and reduce the flow of air. Cleanliness is essential, particularly where the connection leads leave the windings and at the terminal supports. Damp dust leads to tracking and causes expensive damage. It is particularly important to check such transformers for dust in package substations which may escape the notice of commissioning staff.

Meters, relays and protection equipment

Because these are precision items they must be treated with care and any broken glass fronts replaced immediately before dust and damp damage the mechanism permanently. If the operation of a relay or instrument is unsatisfactory it should not be adjusted without the proper tools and sound knowledge of its function and construction. It is often possible to improve the operation at one particular setting but render it completely ineffective over other important parts of its range. A relay setting study should have been done by the design engineers to provide the setting instructions for site commissioning staff.

Arbitrary settings are unlikely to be more than a stop gap and can create a dangerous situation which should not be allowed to persist. Fuller details of protective gear operation are given in Chapter 22.

Motor starters

Before energizing motor control equipment ensure that both motor and starter are correctly earthed. Check that all packing equipment has been removed and that equipment is free to operate. The starter rating should correspond to that of the motor. Set the overload protection devices to the correct point for the operation of the motor and its protection. Check that all dashpots, tanks and oil immersed resistor banks are filled with the correct grade of oil or other filling medium. Confirm settings of overloads by injection testing for all large or critical drives. Confirm the insulation resistance readings of starter, main and auxiliary wiring and motor. Confirm the rating of short-circuit protective fuses against the type of starter and rating of the motor. Ensure resistor banks are adequately ventilated and not obstructed either by other equipment or rubbish. Check that the motor drives are free to operate and that there is no danger to other personnel before starting up the motors. Check that remote start/stop buttons or emergency stop buttons are operative before energizing the plant, and that the motors rotate in the correct direction.

Distribution boards and switch-fuses

After testing all insulation, phasing and earth continuity ensure that the ratings and types of all fuses are suitable for the circuits before they are energized. Make sure that the fuses are the right type for the fuse carriers provided, and that the ratings are correct for the size of the cable or for the load supplied, whichever is the smaller. This will also confirm that the circuit wiring is correct for the load to be supplied. HBC fuse carriers must not be fitted with wire elements because the enclosure is not suitable.

In distribution boards wiring disposition should be examined and all wires dressed clear of live metal work.

Terminal connections on incoming and outgoing cables must be tested for tightness, correct stripping of insulation and clearances. Where a number of neutrals enter a distribution board they should be sensibly disposed along the neutral bar in the same order as that of the live conductors.

Earthing arrangements

The earthing of l.v. networks in the UK is largely determined by the l.v. supplies. However, if the incoming supplies are at 11 kV and the transformers are in the ownership of the user, the l.v. supplies may be earthed in a less conventional way using a high impedance. This arrangement is not allowed for public supplies. However, it is a useful system when it is more important to maintain supplies than it is to clear the first earth fault. For example, an emergency lighting scheme for the evacuation of personnel from a hazardous area could use a high impedance system if it were considered less dangerous to maintain supplies after a first earth fault than to disconnect the light completely. The Channel Tunnel could be such a case. Even in these circumstances the original earth fault should be corrected as quickly as possible.

The more conventional earthing arrangements are TN-C where the earth and neutral are combined (PEN) and TN-S where they are separated (5 wire) or TN-C-S where the supply is TN-C and the installations are TN-S. The latter is very common as it allows the single-phase loads to be supplied by live and neutral with a completely separate earth system connecting together all the exposed conductive parts before connecting them to the PEN conductor via a main earthing terminal which is also connected to the neutral terminal.

For protective conductors of the same material as the phase conductor the cross-sectional area shall be the same size as the phase conductor up to 16 mm^2. When the phase conductor is above 16 mm^2 then the protective conductor may remain at 16 mm^2 until the phase conductor is 35 mm^2, after which the protective conductor should be half the size of the phase conductor. For conductors which are not of the same material the cross-sectional area shall be adjusted in the ratios of the factor *k* from Table 43A in the Regulations. The *k* factor takes into account the resistivity, temperature coefficient, and heat capacity of the conductor materials and of the initial and final temperatures.

Lastly there is the TT system which uses mother earth as the earth return. The neutral and the earthed parts are only connected together via an electrode system back to the source earth (and neutral). To check that conventional systems are satisfactory, i.e. that the protection operates on the occurrence of an earth fault, it is necessary to calculate the earth fault loop impedance (Z_s) and ensure that the fault current through it will cause the protection to operate.

This is quite a tedious process, involving as it does the calculation of the impedances afforded not only by the earth return but also by

(1) the live conductor,
(2) supply transformer,
(3) supply network, and
(4) any neutral impedance.

This information must be requested early. The supply authority should be able to give the fault level or the equivalent impedance of the supply network and the manufacturer can provide the appropriate impedances for the transformer. However, time will be required to obtain the answers so enquiries should be made at the commencement of the project.

The substation will house the circuit-breakers or fuses for the main cable connections to the sub-distribution boards and motor control centres. These protective devices must discriminate with those further down the line nearer the ultimate loads. A system study must therefore establish the correct ratings of the substation equipment to discriminate with the distribution network. This subject is described in Chapters 3 and 4.

Earthing of equipment should be electrically complete and confirmed mechanically sound and tight. Earthing conductors (previously termed earth leads) must be checked for compliance with the IEE Regulations, i.e. they must not be aluminium and they must be not less than 25 mm² for copper and 50 mm² for steel, unless they are protected against corrosion. These conductors are for connection to the earth electrodes.

The protective conductors previously known as earth continuity conductors must also comply with the IEE Regulations and in general for phase conductors of less than 16 mm²; this means the protective conductors must be the same size as the phase conductors. When the phase conductor is above 16 mm² then the protective conductor remains at 16 mm² until the phase conductor is 35 mm², after which the protective conductor should be half the cross-sectional area of the phase conductor.

Another important point to bring out is that the earthing conductor to the earth electrode must be clearly and permanently labelled 'SAFETY ELECTRICAL CONNECTION – DO NOT REMOVE' and this should be placed at the connection of conductor to the electrode. Fuse ratings should also be checked in relation to other fuse ratings in the supply circuit or against the settings of protective relays to assure correct sequence of operation and discrimination. Circuit charts for distribution boards should be completed and designation labels fitted to ensure safe operation of switches and isolators. All tests should be carried out as required in the IEE Wiring Regulations, Part 7, and a completion certificate given by the contractor to the person ordering the work. The completion certificate should be supported by an inspection schedule and include recommendations for periodic reinspection.

Many installations now incorporate rcds and fault-current operated protective devices. These also must be tested using appropriate test equipment, full details of which can be found in the 16th edition of the IEE Wiring Regulations or for more elaborate apparatus in the Code of Practice CP 1013 and Guidance Notes which are published separately and amplify the requirements in the British Standard. It should also be noted that the first amendment published in 1994 includes changes to align with the technical requirements of CENELEC harmonization documents:

- extra low voltage systems
- recognition of the change in the nominal voltage from
  240 V ± 6% to 230 V + 10% and −6%
  415 V ± 6% to 400 V + 10% and −6%
  Appendix 5 List of external influences.

Record drawings

Site staff should prepare 'As fitted record drawings' as a matter of course. The contractor's resident engineer should ascertain who in the client's organization is responsible for approving and accepting the completed 'As fitted' records. He should then make sure of the form, size, scales, title block and other information required before commencing the drawing work. Normally the 'As fitted' records are marked up on a set of prints of the contract drawings and sent back to the drawing office at the contractor's headquarters for preparation of the master transparencies. Frequently it is worthwhile obtaining blank transparencies of building outlines so that prints can be obtained for marking up with different 'As fitted' records such as lighting, power, earthing, telephones, trenches, cabling equipment arrangements and so on.

In addition to layouts and arrangements the final schematic and wiring diagrams will probably be required to assist in fault-finding or subsequent extensions.

Another common requirement is to provide cable routeing and apparatus location records for all special systems such as fire alarms, clocks, bells, telephones, smoke detection, l.v. systems, public address and music speakers, temperature/humidity control, instrumentation, emergency lighting, underfloor or wall trunking systems and underfloor heating systems including the positions of all repair joints and joints to cold tails. Also required are circuit charts for each distribution board showing the correct fuse ratings for each circuit and the names of the outgoing circuits, lighting fittings schedules, switchgear arrangements, wiring diagrams, connection diagrams and details of instruments relays and protective gear settings.

For transformers the client requires general arrangements, fittings, cable box connections, vector diagram and rating plate details, including voltage ratios and tapping switch range, connections to earth (neutral and tank) temperature gauge and earth leakage devices. The tap setting on which the tapping switch is locked should be recorded.

Earthing and lightning protection system records, see Chapters 7 and 9, should show the location and layout of individual electrodes, the position of test links, a record of earth resistance readings as installed, connections to water pipes, building structures, items of plant and other incoming services such as gas or oil and the position of lightning protection filials and down conductors. Incidentally, lightning conductors must never be run inside steel pipes or conduits as the impedance of the path is very high to typical lightning surges.

Underground cables should have the following information recorded in addition to the route. Tie-in dimensions of all trenches to fixed points on the site layout (these points should preferably be themselves cross-referred to identifiable permanent features), positions of all marker posts, i.e. joint and route markers, positions of underground ducts through roads into buildings, positions of entries or manholes into ducts, core connections at through and terminal joints, positions of all underground joints with tie-in dimensions, schedules of joints giving date and name of jointer, depth of laying, depth of ducts and of joints, whether cables are protected by tiles or covers and of what size and type, and positions of cables in trays and ducts to assist in identification. Maintenance manuals must be supplied for all specialized equipment.

It is important that the correct procedure for obtaining the client's approval is followed and that his representative should be given the master 'As fitted' drawing to be signed only after it has been thoroughly checked both by the site engineer and the contract engineer.

CHAPTER 3

Site Distribution Systems

M. Twitchett, IEng, FIEIE, MIPlantE
(Consultant, Electrical and Plant Engineer)

The purpose of a site distribution system is to provide a means of economically and reliably distributing power from one, or occasionally more than one, main location to a number of geographically dispersed load centres within a defined site boundary. The source of power is usually an intake from the local regional electricity company (REC), but may also be from on-site generating plant. The economics of the system will be largely dependent on the voltage at which it is decided to operate it. The ability of the system to supply power reliably to where it is needed, when it is needed, will depend initially on the quality of the equipment and the installation workmanship. In the long-term it will depend also on the flexibility available to continue to supply that power to where it is needed, despite the need to maintain parts of the system and possibly despite the occurrence of faults on the system. While much that will be discussed may also apply to distribution systems within buildings, (especially buildings with a large electrical demand), such installations are dealt with in Chapter 4.

If the site is other than very compact or has only low demand, it is likely to prove economic to distribute power at high voltage rather than at 415 V or the harmonized EC voltage of 400 V. Most REC local distribution networks operate at 11 kV, or in some localities at 6.6 kV, and in most cases it is usually sensible to distribute power around the site at this 'intake voltage'.

Electricity can usually be purchased at lower cost when taken at high voltage. The level of demand which makes this economically viable must be assessed in the early stages of design. This handbook is confined to systems up to 11 kV but it should be borne in mind that on some very large, high demand sites it becomes economic to take a supply at 33 kV or even 132 kV. It is unusual, however, for the consumer to operate the site network at more than 11 kV.

The capital cost implications of operating at say 11 kV become clearer if one considers as an example the need to transmit 500 kVA over a distance of 500 m. At 415 V this would require a four-core cable (assuming a requirement for a neutral), of 300 mm² or possibly 400 mm². The cost of 500 m of such cable would be in the region of £20 000. Whilst, in theory, the same power could be conveyed at 11 kV by a three-core, 6 mm² cable, in practice the smallest available cable for this voltage is likely to be three-core, 35 mm². Nevertheless the cost of 500 m of this cable would still only be in the region of £8000. A transformer would be needed at the remote end and a 500 kVA oil-filled transformer would cost approximately £4100. The 11 kV system would thus show an approximate saving of £7900 over the 415 V system for what is quite a modest power distribution requirement. (These calculations are based on 1994 prices.)

INTAKE ARRANGEMENTS

Whilst low voltage intake arrangements are generally simple single feeders there are many possible configurations for the interface between the consumer and the REC on high voltage supplies. Some of the more common arrangements are described below.

The exact point of demarcation between the consumer and the REC should be the subject of formal agreement. It becomes particularly important, and sometimes contentious, in the event of faults and the allocation of repair costs, and may sometimes be pin-pointed down to the cable lugs and the bolts securing them inside a cable box. The simplest arrangement is for a single transformer to be owned by the consumer with the h.v. switchgear supplying it owned and operated by the REC as shown in Fig. 3.1. This arrangement relieves the consumer of any high voltage switching duties but makes him completely dependent on the REC for making live the transformer or isolating it for maintenance. (Facilities are usually provided to enable the consumer to trip the REC's switch or circuit breaker in an emergency but not to reclose it.) Figure 3.2 depicts a more flexible and more common arrangement for a single feeder supply from a REC.

Both the REC's and the consumer's circuit-breakers or switches are effectively connected to one set of busbars but responsibility for the busbars changes at the declared point of demarcation. With this arrangement the consumer has control over his own network 'downstream' of his circuit-breakers or switches but there is mutual dependence for isolation of the busbars.

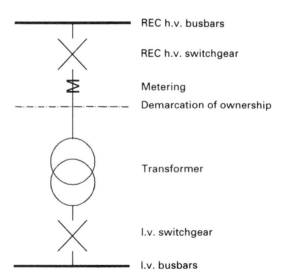

REC h.v. busbars

REC h.v. switchgear

Metering
Demarcation of ownership

Transformer

l.v. switchgear

l.v. busbars

Fig. 3.1 A single high voltage intake with h.v. switchgear under control of the REC.

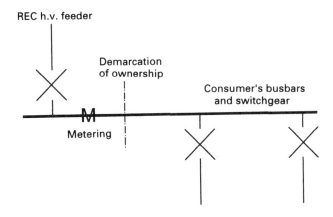

Fig. 3.2 Single feeder h.v. REC supply with consumer's h.v. switchgear.

Where the security of duplicate REC feeders is required the most common arrangement is as shown in Fig. 3.3. In this arrangement the REC's section of the busbars often forms a part of a ring-main. It may be equipped with either circuit-breakers or fault-making switches. The consumer's circuit-breaker (sometimes referred to as the metering circuit breaker), is actually configured as a bus-section switch and is normally the responsibility of the REC. There is mutual dependence for isolation of the consumer's section of the busbars, although the RECs are still able to keep their section of the busbars in service. In some cases the entire switchboard will be within one room but quite commonly the REC section will be partitioned off or in a separate room from the consumer's gear. In such cases either a busbar extension chamber or a bus-zone cable is used to carry the busbars through the dividing wall. In critical cases where a greater degree of supply security and flexibility is required, an arrangement like that shown in Fig. 3.4 may be necessary. Here the REC is usually responsible for the centre section of the switchboard and the bus-section circuit-breaker as shown.

The factors that must be carefully assessed before deciding on the intake configuration are as follows:

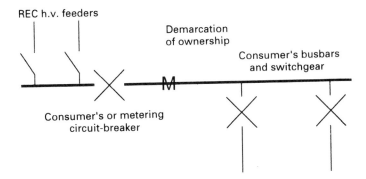

Fig. 3.3 Typical duplicate feeder REC h.v. intake arrangement.

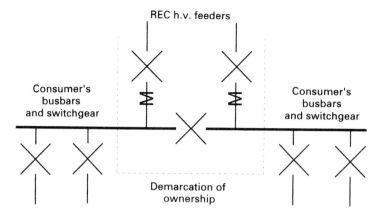

Fig. 3.4 Duplicate REC feeders supplying sectioned busbars.

(1) effects on the business of partial and total losses of supply;
(2) required availability;
(3) likelihood of faults on either REC's or consumer's equipment;
(4) maintenance requirements;
(5) possibility of and economics of generator usage to cover supply outages;
(6) cost.

Consultation with the local REC is required at an early stage in order to agree on a mutually acceptable design of intake substation and on the division of ownership and responsibilities. Accommodation for the REC's equipment will be required and they will require easy access to it at all times. All these aspects should be the subject of a formal agreement signed by both parties. Provision for metering must also be made and in this connection it should be remembered that either initially or at some later stage the consumer may wish to purchase electricity from someone other than the local REC.

SITE DISTRIBUTION NETWORKS

As with the configuration of the intake arrangements, so similar decisions need to be made regarding the configuration of the distribution network and its cost. Local supplies may be required at other than 415 V if large motor drives operating at high voltage (usually 3.3 kV or 6.6 kV but occasionally at 11 kV) are to be installed. The simplest arrangement is to supply single transformer substations via single radial feeders emanating from the intake substation as shown in Fig. 3.5. The major disadvantage of this system is that supply to any particular load centre will be lost if any outage due to a fault or for maintenance occurs on any item on the single supply route. Increased security of supply can be achieved by providing duplicate radial feeders and transformers as shown in Fig. 3.6. This is, however, likely to be an expensive option.

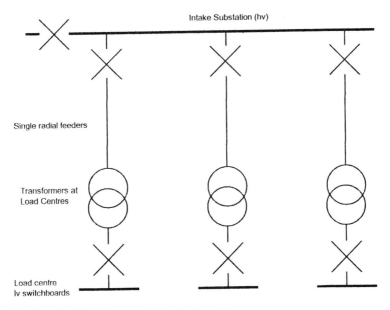

Fig. 3.5 Single radical feeders supplying load centres.

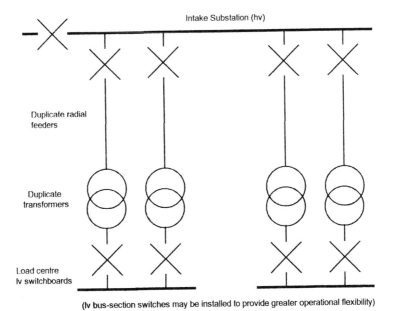

(lv bus-section switches may be installed to provide greater operational flexibility)

Fig. 3.6 Duplicate feeders and duplicate transformers.

Probably the most cost effective and common means of providing duplicate feeders to all the load centre substations is to employ one or more ring-mains. Figure 3.7 shows a typical ring-main supplying a number of both single- and two-transformer substations of various configurations. Such a system enables all sub-stations to be supplied in the event of an outage of any one leg of cable on the ring.

For economic reasons most ring-mains incorporate substations equipped with fault making switches rather than circuit-breakers on the ring-main cable feeders. Such switches do not provide automatic means of opening in the event of faults. It is therefore good practice in normal circumstances to operate the ring-main with one of the switches open at some strategically suitable point, so that a fault anywhere on the system causes a loss of supply to only the substations on that side of the network. If the fault is on a cable (which is the most likely event], once it has been located switching can be carried out to isolate the defective leg and restore supplies to all the substations. Of course, if a fault occurs on the busbars

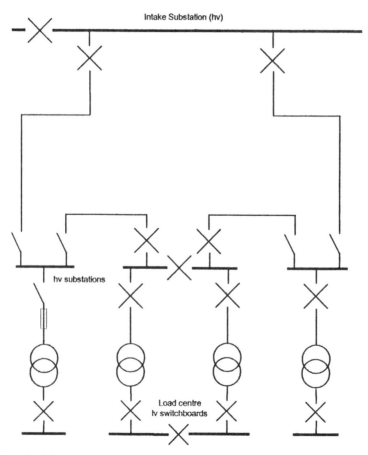

Fig. 3.7 A typical h.v. ring-main system.

of one of the substations (a rare but not unknown occurrence) then that substation (or section of busbars if provided with a bus-section switch) must be repaired before its supply can be restored. The location of faults is greatly facilitated by the provision of earth fault indicators at each cable terminating box around the system. These make unnecessary the practice of repeatedly closing onto a fault in order to locate its whereabouts or the time consuming technique of isolating each section in turn in order to conduct insulations tests.

Where continuity of supply is crucial and even short unplanned outages due to cable faults cannot be tolerated, it is necessary to provide circuit-breakers throughout the ring-main and to operate the system as a closed ring in normal circumstances. In this mode of operation conventional current and/or time graded protection will not provide proper discrimination because the direction of power flow around the ring-main is likely to vary under different load conditions. In order to provide continuity of supply to all substations in the event of a cable fault it is therefore necessary to employ differential protection schemes for each leg of the ring-main or to install a suitably graded scheme utilizing directional overcurrent and earth fault relays.

ON-SITE GENERATION

Generators may be installed on site for one or more of the following reasons:

(1) standby in the event of loss of REC supply;
(2) total stand alone system, possibly incorporating combined heat and power (CHP);
(3) peak lopping;
(4) utilization of available process steam.

Site generating plant may be operated in a number of different modes:

(1) 'island mode', i.e. disconnected from any REC supply;
(2) in parallel with the REC under normal conditions;
(3) in parallel with the REC supply only briefly to permit continuity during changeovers.

Consultation and formal agreement with the REC is required if paralleling with their supply, even briefly, is envisaged. Where generating plant is intended to have a 'standby' role it is essential that the risks that it is intended to cover are fully assessed and that it is decided exactly in what circumstances it is required in order to be able to deploy it. For example, in Fig. 3.8 the 11 kV generator provides an alternative in the event of a loss of the REC supplies but cannot be deployed if the intake substation busbars have to be isolated for maintenance.

Generators may be provided at strategic low voltage switchboards in order to maintain supplies to essential loads in the event of either loss of the high voltage REC supply or outages on parts of the site high voltage network. In some cases such generators are used to 'back-feed' the site h.v. network by using the local

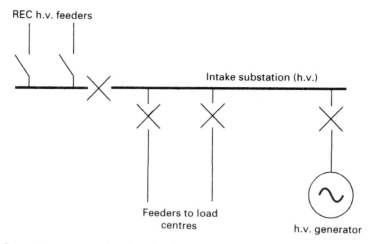

Fig. 3.8 Generator connected to h.v. intake substation.

transformer in a 'step-up' mode in order to provide emergency supplies to the entire site as shown in Fig. 3.9. (Precautions must of course be taken to avoid backfeeding into the REC system.) This practice is quite legitimate, but in the majority of cases all the transformers on the h.v. network will have delta connected h.v. windings, usually vector group Dyn11. This means that, when disconnected

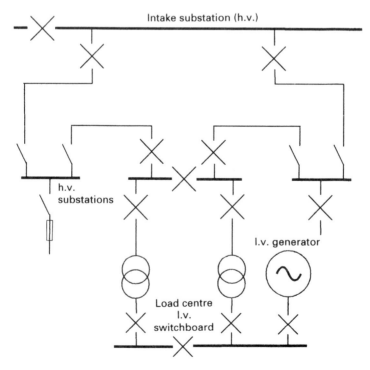

Fig. 3.9 Generator connected to l.v. substation.

from the REC supply, the site h.v. network is operating without any earthed reference point and that normal earth fault protection schemes will be ineffective. If an earth fault were to occur it would therefore go undetected. Beside the obvious undesirability of the system continuing to operate with an earth fault, there is the possibility of damage to any voltage graded insulation in machine windings. In order to overcome this problem either a star-connected 'earthing transformer' should be connected to the system or, perhaps less expensively, the system should be protected by a neutral point displacement relay supplied from a special voltage transformer.

SWITCHGEAR

Whilst switching devices with a wide range of capabilities are possible the devices that generally have practical application on h.v. site distribution networks are:

(1) *the circuit-breaker* This is a device capable of making and interrupting short-circuit current as well as operating on load current;
(2) *the fault making switch (FMS)* This is capable of operating on load current and of making short-circuit current but cannot interrupt short-circuit current;
(3) *the switch-fuse* This is a fault making switch with series fuses. (Such units are often provided with 'striker pin' fuses arranged to open the switch in the event of any fuse operating.)

All three of these devices have in the past been used in both fixed and withdrawable patterns but current practice for site distribution is to employ FMS's and switch-fuses in the form of integrated ring-main units (RMU). Such units incorporate two FMSs for connection to the ring-main cables and a switch-fuse for connecting the local transformer. A recent development has been the incorporation of a small circuit-breaker (instead of a switch-fuse) and two FMSs into compact SF, RMUs.

Circuit-breakers provide the advantages of being capable of automatic and remote opening and closing and are generally employed where rapid operation to restore supplies following a fault is necessary. Other than the small circuit-breakers incorporated into RMUs, it is generally preferable to employ circuit-breakers of the withdrawable pattern. These may be bodily isolated from the switchboard to facilitate maintenance.

Until recently, the vast majority of circuit-breakers used in h.v. site distribution systems were oil-filled with a small proportion being of the air-break type. The current choice will probably lay between vacuum and SF_6 types. When specifying h.v. switchgear it is important to ensure that provision is made for adequate means of earthing busbars, cables, transformers, and any other connected apparatus on which work will at some time need to be carried out.

Low voltage switchgear is generally of the air-break type and in the rare case of it forming part of a site distribution system is unlikely to need to be essentially different from its normal application at a l.v. load centre.

CABLES

For h.v. distribution, services cables with polymeric insulation, usually cross-linked polyethelene (XLPE) are finding favour over PILC types owing mainly to the simplicity with which they may be terminated and jointed. The most usual construction is XLPE/SWA/PVC. For l.v. systems the ubiquitous PVC/SWA/PVC is generally satisfactory. Where cables are routed through buildings or in cable tunnels safety requirements may in some cases dictate the use of low smoke/low fume cable types.

The choice between copper or aluminium conductors and, in the case of aluminium, solid or stranded construction, is influenced by cost and the practicalities of installation. It is also worth remembering that aluminium cables, where they are known to be such, are much less likely to be the subject of theft whilst awaiting, or during, installation.

As in any installation, conductor sizes must be chosen to meet the requirements of both current-carrying capacity and acceptable voltage drop. With regard to current-carrying capacity, decisions should be based on the usual factors and need not be discussed here. The question of cable size in relation to voltage drop is less clear. For low voltage systems the requirements of BS 7671: 1992 (IEE 16th edition) are deemed to be satisfied if the voltage drop between the origin of the installation and the fixed current-using equipment does not exceed 4% of the nominal voltage of the supply. This can be construed as being applicable to a 'site distribution system' operating at low voltage but there is at present no equivalent standard for high voltage systems. Where a h.v. system is supplying power to be utilized entirely at l.v., the voltage drop on the h.v. system may not be as important and can usually be compensated for by tap selection on the transformers at the load centre substations. (Overvoltage under light load conditions must not then exceed a value that might be damaging to the connected equipment.) Where loads such as large motors are supplied directly at the h.v. system voltage, cable sizes should be selected to limit the voltage drop to a value acceptable to the manufacturers of the equipment in question.

CABLE INSTALLATION

This chapter deals with site distribution. It is not intended to describe here cable installations within buildings although it should be recognized that some parts of the site distribution network may well be within buildings either in ducts, cable trenches or on cable tray or ladder systems. The major proportion of route cable length is, however, likely to be laid underground. Where the ground may be easily opened up for subsequent access to the cables they may be buried directly in the soil. In locations where cables are to pass beneath roads, railways or large areas of concrete or tarmac, the cost and inconvenience of subsequent excavation (and reinstatement) to gain access to the cables makes it desirable to run them in suitable ducts, at least beneath these areas.

Where ducts are employed it is necessary to provide draw-pits at all bends and at approximately every 50 m on straight and level runs. Closer intervals may be required in adverse circumstances. Figure 3.10 illustrates a typical draw-pit. Careful design of the drawn-pit is necessary to ensure that cables can be pulled into it and then drawn back down into the outgoing duct for the next leg of the run. Any joints that are required should be positioned in a draw-pit and its dimensions should allow for this. Suitable access covers should be provided and must be capable of withstanding the maximum traffic loading to which they will be subjected.

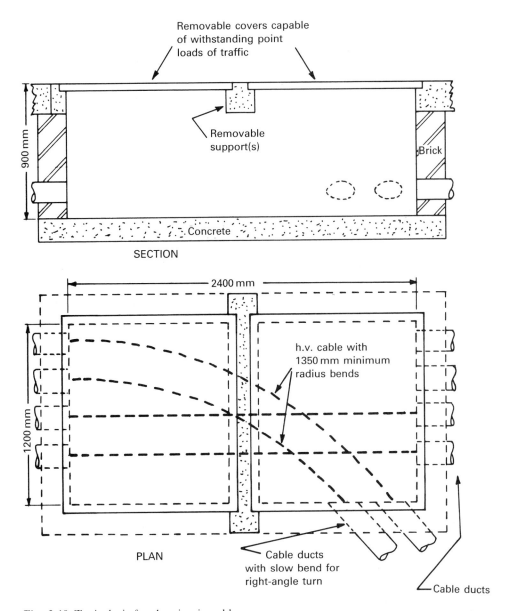

Fig. 3.10 Typical pit for drawing in cables.

Where cables are buried directly in the ground they should be approximately 750 mm deep for l.v. cables and preferably 1000 mm deep in the case of h.v. The presence of underground cables, even in ducts, is best indicated by laying proprietary plastic marker tape above the cable and about 150 mm below the final surface level. Such tape is usually yellow in colour and bears a repeated inscription such as 'ELECTRICITY CABLES BELOW'. These tapes are reasonably likely to be observed at an early stage by people digging on the route of the cable, whereas since the use of mechanical excavators became widespread, the employment of concrete cable-tiles has been found in practice to offer little protection. Figure 3.11 illustrates a cross-section through a suitable completed cable trench.

PROVISION FOR MAINTENANCE

Modern transformers and switchgear of reputable manufacture are generally very reliable and maintenance requirements have been steadily reduced over the years. Nevertheless all such apparatus, at some time, however infrequently, will need to be isolated for inspection, testing, maintenance or repair. The configuration of the system and the type of equipment installed will determine how much operational disruption will be caused whilst such work is undertaken. For example, a conventional non-extensible RMU has a relatively low first cost but in order to carry out maintenance involving any of its h.v. components the whole RMU must be isolated from all sources of supply so that the entire substation is out of service. On the other hand, a withdrawable circuit-breaker can be isolated and maintained whilst both its busbar and circuit spouts remain live. It may even be replaced by a

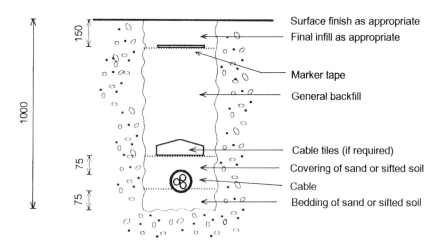

Approximate dimensions shown in millimetres

Fig. 3.11 Cross-section through typical cable trench (not to scale; approximate dimensions shown in millimetres).

spare circuit-breaker whilst maintenance is carried out. It does however have a high first cost.

Increased security of supply means that increased first cost and the degree of system sophistication to ensure continuity of supply during maintenance must be carefully weighed against operational requirements and the subsequent operational cost implications. Such computations are difficult enough when it is known what equipment is to be maintained and when it is to be maintained, but become virtually impossible if maintenance is left to random forces such as breakdowns.

SYSTEM OPERATION

There is often a period, in some cases quite protracted, when the installer is in charge of a high voltage system that is partially complete and partly live. This responsibility may also continue after completion into warranty periods. As installation continues the installer will need to isolate and gain access to various parts in order to connect the equipment in the next phase of the work. Properly thought out and well documented 'permit-to-work' procedures are absolutely essential for safety. The personnel carrying out the procedures must have adequate training, knowledge and experience of the equipment and the system to be able to avoid danger to themselves and to those who are to carry out the installation work. They should be formally appointed and their duties clearly defined.

Once the system is handed over to the end user he assumes the same responsibilities and must ensure that similar procedures are adopted and that suitably qualified persons are appointed to carry them out.

IDENTIFICATION OF SUBSTATIONS AND SWITCHGEAR

Confusion regarding the identity and function of any electrical equipment is dangerous and never more so than where a complex and geographically dispersed site h.v. distribution network is involved. It is therefore essential that all equipment is clearly and unambiguously identified both on the plant itself and on all relevant drawings. Unfortunately, at the design stage, several contractors may be involved and each may identify their particular parts of the overall job by various configurations of contract names and numbers, etc. It is not even unknown for items of plant to be identified by their height above Ordnance Datum (Newlyn), e.g., 'The 37.2 metre O.D. Switchboard'! Whilst such systems of identification may be adequate for the purposes of design and installation they are totally unsuited to the needs of those who are required to operate the system in a safe manner. Discussion should take place with the end user to decide upon a logical scheme for identifying both the substations, the switchboards and the individual switchgear and circuits.

With regard to the identification of substations, titles based on geographic or plant references, e.g., 'East Substation', 'Loading Bay Substation', etc., are much more likely to convey meaningful information than cryptic numbering systems. A switchboard within a particular substation is probably best identified simply by appending the voltage level at which it operates, e.g., 'Loading Bay 415 V Switchboard'. If there is more than one switchboard operating at the same voltage at a particular location it is usually quite easy to decide on a title based on, for example, its function.

When considering the identification of individual items of switchgear the simplest and most sensible practice is to identify a switch or circuit-breaker by what its 'circuit' side is connected to, e.g., 'Transformer T1', 'Administration Block', 'Pulp Pump 3', etc. Figure 3.12 illustrates a typical example.

It is undesirable to include what, for this purpose, is superfluous information on the identity plate. A typical gross bad example would be: 'Circuit-Breaker feeding 1000 kVA, 11 kV/433 V, Star-Delta, 4.75% Impedance Transformer Number 2'. In order to ensure safe operation this title should appear, in full, in every entry on switching schedules, permits-to-work and log books. When the identification is so protracted, the people operating the equipment will be likely to abbreviate the title and, worse still, each operator is likely to abbreviate it differently. The circuit-breaker in the above example need only bear a plate with the inscription 'Transformer T2'. All the other information should, of course, be available to the operator but it does not need to be on the identity plate. Expressions such as 'Feeder to ....', 'Incomer from ....', etc., are in general unnecessary and may easily create an impression that the circuit can only ever be made live from one direction which can be a dangerous assumption!

Bus-section switches and bus-couplers should simply be identified as such unless there is more than one on the switchboard, in which case it will be necessary to identify each in some unambiguous manner, e.g., 'West Bus-Section Switch' and 'East Bus-Section Switch'. In this connection there appears to be a very widespread

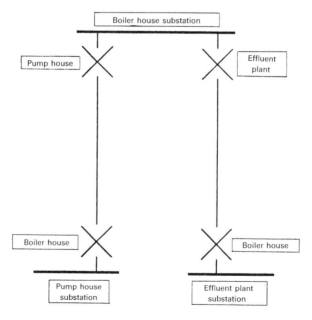

Fig. 3.12 Identification of switchgear.

misunderstanding regarding the difference between a 'bus-section switch' and a 'bus-coupler'. A 'bus-section switch' is connected between two sections of the same set of busbars, whereas a 'bus-coupler' is connected between two sections of two different sets of busbars and can therefore only exist on a duplicate busbar switchboard. Figure 3.13 depicts an example of the correct usage of the titles. It may be that the extreme rarity of duplicate busbar l.v. switchboards has led to the widespread misuse of the titles in the l.v. field. It is particularly galling, and very common, to see a 'bus-section switch' correctly identified as such on an 11 kV switchboard and the corresponding 'bus-section switch' on the relevant l.v. switchboard erroneously identified as a 'bus-coupler'. Further confusion is liable to occur when North American companies are involved because their equivalent for a 'bus-section switch' is a 'bus-tie'.

FAULTS LEVEL

In the event of a fault occurring on the system, all equipment that is subjected to the resulting fault current must be able to carry that current without damage and without causing danger until such time as the current is cut off by the operation of protective devices and must therefore be rated accordingly. The circuit-breakers or fuses that are called upon to interrupt the fault current must be able to do so safely. This requires that their rated breaking capacity is at least equal to and preferably greater than the maximum fault current that can flow at that point in the system. Generally, a symmetrical fault imposes the most arduous duty on the interrupting device and it is therefore the symmetrical breaking capacity that is normally quoted when specifying the switchgear.

Assessment of 'prospective short-circuit current' in l.v. systems is adequately dealt with in another chapter and also in many explanatory publications relating to BS 7671: 1992. In the case of h.v. systems the term 'fault level' is usually

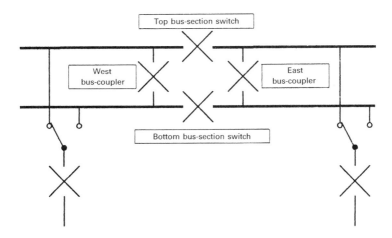

Fig. 3.13 Bus-section switches and bus-couplers.

employed and is expressed in MVA. This has the advantage of not being related to any particular voltage level. The fault current at a particular voltage level may be determined as follows:

$$I_{fault} = \frac{fault\ level\ (MVA)}{1.732 \times V_{line}}$$

A very simple but 'pessimistic' method of calculating fault level is to consider only the percentage impedance of the transformer supplying the system and to assume infinite fault level upstream of the transformer and neglible cable impedance downstream:

$$fault\ level\ (MVA) = \frac{trans\ rating\ (MVA) \times 100}{\%Z}$$

For example, consider a 1.0 MVA transformer having a percentage impedance of 4.75%. The worst possible fault level downstream of the transformer would be:

$$fault\ level\ (MVA) = \frac{1.0 \times 100}{4.75} = 21.05\ MVA$$

On an 11 kV system the fault current would be:

$$I_{fault} = \frac{21.05}{1.732 \times 11} = 1.105\ kA$$

(which is low compared with the breaking capacity of even the smallest h.v. switchgear). However on a 415 V system the fault current would be:

$$I_{fault} = \frac{21.05}{1.732 \times 0.415} = 29.29\ kA$$

(which is considerable).

Caution must be exercised if using this method to select switchgear because it might in some cases result in specifying gear of unnecessarily high rating. The fault level at the intake should be obtained from the REC, but as they must be careful never to quote a figure that is even marginally lower than the true value, or a figure that could, following reinforcement of their system be exceeded, they will tend to quote a value that may be considerably higher than the current true value. The present trend seems to be for the REC to quote a value equal to the rating of their immediate upstream switchgear.

The attenuating effect of cable resistance is significant at low voltage but diminishes as the system voltage increases. This is because the percentage impedance is a function of the reciprocal of the square of the line voltage. Cable resistance should therefore always be taken into account on l.v. systems but on an 11 kV system its effect will be small and the fault level can in most cases be assumed to be the same as the intake value throughout the network. Where cable resistance is to be taken into account it can only be added to the predominately reactive impedance of the rest of the system vectorially.

A thorough treatment of the subject of fault calculations is beyond the scope of this chapter but can be found in many textbooks on power engineering theory.

TESTING AND COMMISSIONING

All the normal inspections, compliance checks and tests that would be carried out on other electrical equipment and installations are required in the case of site distribution systems. Protection schemes installed on site networks are often sophisticated and this must be reflected in the calibre of the personnel entrusted to test and commission them. Where necessary, specialists should be employed for such work. Similarly, insulation testing on h.v. systems involves the use of potentially lethal voltages and should only be carried out by adequately qualified people under formal 'sanction-for-test' procedures.

Correct phasing of conductors around ring-mains is important. Cases have occurred of phases being 'crossed' during installation, this having gone unnoticed for several years because the 'open point' remained unchanged and all motors were connected for correct direction of rotation on the 'normal' supply. Such an error would of course become apparent if the 'open point' were changed, dramatically so if the normally legitimate practice of closing the ring first were employed. Another example of an installation error that should have been discovered and corrected during inspection but which went unobserved for several years was a case of an 11 kV RMU cabled up so that the switch-fuse was in the ring and the transformer was devoid of protection. Where two or more transformers are connected to a switchboard, even if interlocking is provided to prevent them being paralleled, it is good practice to ensure 'phase correctness', i.e. red to red, yellow to yellow, etc., as distinct from mere correct phase rotation.

In addition to the customary electrical tests the complexities of site distribution systems often entail the need to ascertain the correct functioning of such features as 'figure lock' and other interlock systems designed to prevent certain unacceptable operations such as parallelling unsynchronized supplies, parallelling transformers where this would entail excessive fault level, etc. It is not sufficient merely to check that such interlocks operate correctly in the 'normal' mode, their correct operation in preventing *all* the unacceptable modes must be positively ascertained.

CHAPTER 4

Cable Management Systems

M.J. Dyer
(Consultant Building Services)

INTRODUCTION

The increasing quantity of cables required in buildings has led to a relatively greater importance in the design and selection of cable management systems. Whereas there used to be considerable national preferences there is now a much more uniform international approach to the question of managing cable installations. Cables form a very large physical part of any electrical installation, but are not required to be visible or accessible except for maintenance and alteration purposes. With the increase in data and electronic systems in offices and in factory buildings, the volume of the cables installed has already increased considerably, and the need to 'manage' them has become proportionately more important.

A number of parameters require the attention of the designer when addressing cable management. It is to be assumed that the electrical requirements have already been determined, and that an electrical engineering design therefore exists. The cable types will probably have been selected and the points of utilization will be known. Certain cable types, such as steel wired armoured (SWA) cables and mineral insulated copper covered (MICC) cables mitigate against the use of most systems of cable management (except for tray and ladders in very large installations) as they are designed for direct fixing to structural surfaces, but all other cable types will require to be physically 'managed' in some way. The building structure will already have been designed, or may already exist, and this will affect the system selected, as there may or may not be specific provision for risers and/or cable access areas provided by the architect. Whatever the physical arrangement of the building, the designer must provide an efficient (and that also means a cost efficient) method of controlling the cable installation work involved, and throughout the length of the cables they must all be secured and segregated as may be required. The method selected may be continuous through the whole of the installation, or may vary from place to place within the area served by the installation, but in any case it must above all be suitable for the requirements of both the user and the environment in which it is located.

Cable management implies the enclosure, segregation, marshalling and mechanical protection of the cables within the building, and hence one must identify every cable and cable type which is required to implement the electrical design, and then allow for expansion of the installation in the future. Although some forms of cable installation do not involve the enclosure of the cables (see above), in general when the term 'cable management' is considered within buildings, enclosure (or at least part enclosure combined with support) is implied in the term. The segregation of various classifications of circuits and/or cable types is increasingly a major consideration in the selection of such systems, and more will be said of this later. Marshalling and mechanical protection tend to have very similar methods of attainment but the robustness of the materials required will depend as much on their anticipated environment as upon the properties of the cables to be so protected.

DECISION MAKING

The decision as to which cable management system to adopt for any particular installation needs to be an intensely practical decision. The cables to be accommodated may comprise low voltage (l.v.) cables supplying building services and points of utilization within the building itself, or passing through it, communications cabling (including telephones, data, fibre optic networks, etc.), and security wiring (including fire alarms, emergency lighting and other services at both mains and extra low voltage (ELV) levels). Thus consideration has to be given at an early stage to the need to segregate the circuit types, as well as to the need for derating power cables if they are to be enclosed and hence grouped for the purpose of the regulations. The routes along which each of the types of cables will principally run must be established, and a layout of the distribution services thus built up. This will allow a decision to be taken regarding the type of system to be specified for the main cable runs, based upon the number and physical size of the cables in each 'leg' of such a network. If the runs for the various types of electrical services are similar, then a single cable system with segregation for the various categories is indicated, whereas if the communication cables (say) take widely differing routes from those of the (say) power cables, then completely separate cable management systems for the various categories may be a more efficient solution.

Having come to some conclusions on the main distribution of the cabling, the designer must turn his attention to the final distribution cables at the points of utilization, whether that be at desks (work stations) in an office or at machines in a factory (or indeed at any other position where electrical lighting or power is required, or data is to be processed or accessed). Very often this final cable management system will be highly visible and if care is not taken it may become intrusive in the built environment; thus a more aesthetically influenced decision can be called for here than is the case with distribution cabling. In the following sections there is no attempt to differentiate between 'main' and 'final' distribution of cables, but the reader is cautioned to define these terms in his own mind before contemplating each variation of cable management which is available.

The category of use of the building itself is also an important fact to consider. Most large buildings fall into the categories of industrial, commercial or retail/hotel/conference buildings, and this will tend to dictate the mechanical protection required as well as the appearance criteria and the desirability of accessibility for the cabling (or the desirability of preventing unauthorized access to it). Whichever category the building or installation may be placed in, the importance of data cabling cannot be over-emphasized in the modern environment; the required volume of computer, voice and specialized communications cabling requires thought to be given at the design stage not just to accommodating the cables and the future additional cables, but to the requirements of electromagnetic compatibility (EMC) for such cables. The method of construction of the building may also influence the decision as to which method of cable management is most appropriate; the need for fire resisting floors and walls, the creation of fire zones and the provision (or otherwise), of risers, suspended floors and or ceilings, etc., all affect the designer's choice.

Where a design is being created for a particular client's requirements, the relative priority to be given to these aspects of the system are usually well defined. However in speculative jobs, or in the case of buildings where the usage may well change several times within the expected life and probably within the expected ownership of the building, things are less clear. It is necessary to make a fundamental decision regarding the priorities, before design selection can occur. For example, the physical protection aspect may be vital, and in a factory the mechanical robustness of the cable enclosures, or the placing of such enclosures out of the range of mechanical impact (from, say, fork lift trucks in a warehouse), may take precedence. In a commercial building with considerable multi-purpose office accommodation, the availability of all the various electrical services at each work station will probably be the overriding requirement, and so a system capable of this multi-compartment operation will be required. In every case the possibility of adding cables to the installation after it is theoretically completed must be considered.

Although rewiring is something which has always occupied the minds of those responsible for the electrical installation of buildings, it is now the extension of installations which is more important. With improved materials used in the construction of cables offering longer and longer service life, and the increased sophistication of electrical protection devices precluding the catastrophic failure of cables, rewiring is required very infrequently. However the changes to requirements brought about by new technologies, increased opportunities for communications and data exchange, and by differing commercial practices due to these and other factors, do mean that in many cases extra cables are required. Any cable management system must be suitable for cables to be added to the installation with relative ease, and systems should be designed with such potential expansion in mind. This might mean providing apparently oversized trunkings, etc., at the outset, or it might be interpreted as leaving provision (particularly in a modular system) for the addition of new cable ways in the future. These can often be attached to originally installed trunkings or trays, provided that space has been left at the design stage to accommodate such growth, and that the mechanical considerations of the original installation were appropriate for such increased loadings. Problems may occur at distribution boards and points, and these are more likely to be critical at DP distribution centres than at electrical power boards. This is because of the modular nature of contemporary designs of power distribution boards and the physical limitations of DP distribution equipment caused by greatly increased electronic capacity not requiring fundamentally more space to enclose the electronics: hence cable access problems at times of additions to the system.

TYPES OF SYSTEM

The principal systems available fall into one of three categories. Many installations will use a combination of such methods at various points throughout the building, but the systems selected need to be compatible with each other as well as suitable for the client's requirements. The principal categories are conduit and trunking, underfloor systems, and cable tray or basket systems. Each of these has its sub-divisions, and within each of the sub-divisions the individual manufacturers in the field offer an array of solutions to each and every circumstance, but the principles are the same and are restricted to the main categories. This chapter is not concerned with detailing the use of cable tray and ladder for large and heavy industrial cable installations, as the practice is well established and is strongly indicated where large numbers of already mechanically protected cables require to be fixed to a surface. Such methods are cable fixing rather than cable management techniques, and as such properly belong in the trade skills area of installation technology. Likewise the direct fixing (or clipping) of cables to a surface is a perfectly satisfactory and adequate method of installation provided that the necessary criteria of the electrical installation are complied with, but details are not felt to be needed in a work of this nature.

CONDUIT AND TRUNKING

Conduit and trunking is a very traditional British system, and has developed from metallic systems used in the early days in installation practice when wooden cable enclosures were found to be insufficiently fire resisting to provide a guarantee of safety for all installations and when rewirability was of paramount importance. Conduits may now be metallic (usually steel, or stainless steel if environmental cleanliness is critical) or plastic, and both materials have been developed into full systems providing fittings and accessories for most eventualities. Conduit is still an ideal way of marshalling a large number of small cables to a point of utilization or distribution, and is surprisingly cost effective in many instances. Non-metallic conduits are not only available in circular sections, but in ovals for use in restricted depth internal wall plaster or dry lining installation. When using plastics the reduced temperature tolerance of the material must be allowed for in designing the supports for luminaires and other items; with steel systems galvanized is to be preferred unless the cost limitations and a non-threatening atmosphere suggest that black enamelled will suffice. Stainless steel will be required only when the application is within a specific industry, such as food or medical areas. Conduit is the most easily rewirable system if it is designed correctly, and provides good mechanical protection; metallic systems give EMC advantages. It is good practice not to rely on the metallic conduit as a circuit protective conductor, but to draw into the conduit the appropriate (green/yellow insulated) conductor(s) for this purpose.

Trunking is available in the same materials as conduits, and in its traditional form is referred to as 'industrial trunking'. This is the (normally) rectangular enclosure for cables which is designed for surface mounting, and may be sub-divided for circuit segregation purposes. As it is unsightly it is kept out of sight except in industrial environments. However a large range of more aesthetically pleasing trunkings are available for use in exposed locations, particularly at skirting or dado levels. These provide one or more separate enclosures throughout their length to feed all categories of circuits to points of use, e.g., desks, hotel/hospital bed-heads, etc. The use of these systems in refurbishment work has led to their adoption in new construction because of their ease of rewirability and their pleasant appearance. They are, however, a costly way of running cables and are at their most effective when used also to mount socket outlets, control equipment and telephones, etc.

Somewhere between these extreme examples of trunking systems is the range of so-called 'mini-trunking' which accounts for some 50% of all trunking installed at the time of writing. This is the plastics extrusion with simple clip-on lid for enclosing a small number of cables on the surface. It may be as large as the smaller sizes of industrial trunking but has well designed clean lines and simple accessories. It is easy to decorate and blends into interior decorations in a way which no other low cost system is able to do. It is widely used for individual circuits and additions to installations. Conduit and trunking are normally combined, in order to utilize the larger sizes available in trunking to carry cables away from a distribution board or data processing centre, with individual small diameter conduits

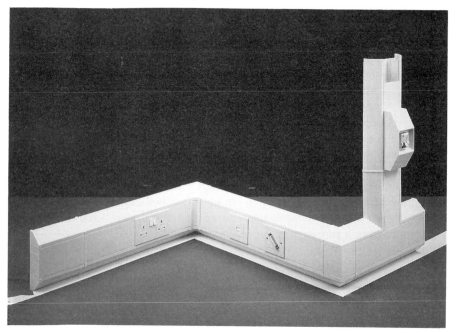

Fig. 4.1 Example of surface trunking for an office.

glanded off from the trunking to feed separate points. Conduit tends to be rigid in such applications, but flexible and semi-flexible materials are available in both insulating and conductive materials and have equivalent properties. The extra cost and difficulty of ensuring a neat and workman-like installation associated with the non-rigid conduit materials is offset by their ability to be routed through congested areas of construction with relative ease. In new construction, the casting of flexible conduits into screeds or even into structural slabs is an effective way of achieving outlets for lighting and power at fixed positions; dimensional specification is critical in such techniques.

UNDERFLOOR SYSTEMS

Underfloor systems fall into two main types: those designed for incorporation into the floor structure itself, such as trunking systems for casting into the screed, etc., and those designed to be installed below suspended (or so called 'computer') floors in commercial buildings. The principles of the two systems are the same, but the strength of the construction is different, as is the detailing of the designs. Systems for building into the floor structure itself are customarily of metallic construction, often galvanized steel, and need to be of high strength to withstand the rigours of building site conditions at the first fix stage. The spine of such systems is usually flat trunking with a top flange intended eventually to become flush with the floor screed and to which thin but robust covers are attached. A family of flush socket boxes is necessary to accommodate the various outlets required, and fittings such as tees and elbows are needed. As the trunking now invariably needs to be multi-compartment, considerable care is needed to arrange the compartments at junctions and crossovers. Utilization points can be taken above floor level by the use of 'power poles' (cf.) and conduit can be teed from the trunking if provision is made prior to screeding the floor. Suitable for traditional new build installations, this system needs careful planning in the design stage, particularly with respect to the termination of the trunking at distribution boards, etc.

The more recent approach to underfloor wiring is that adopted where the functional floor is raised above the structural floor (so called 'suspended flooring' or 'computer flooring'), and a void is formed for the cables between the two floors. Such flooring was originally restricted to computer suites and technical control rooms but the cost of this flooring is now justified throughout many commercial buildings as it provides flexibility of a higher order for developing the cable systems throughout the life of a structure. Cable management systems for such applications are numerous, and consist of trunking and conduit similar to conventional products but with design features specific to the environment proposed. Plastics material is very popular but metallic systems are available if mechanical or EMC protection is important in a specific application. Usually a number of parallel trunkings are run side by side to provide the segregated circuits, so that crossovers and junctions become straightforward to understand, and conduits (of the flexible type) can be glanded off to outlets, etc. The trunking is usually fixed to the structural floor and is accessed by lifting the tiles of the

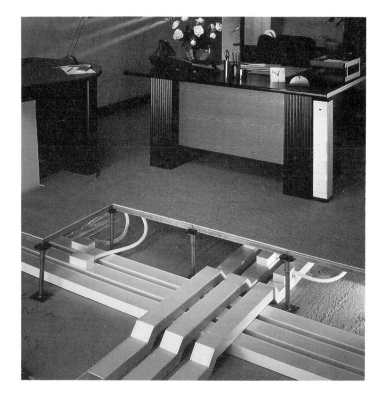

Fig. 4.2 Installed underfloor trunking. Note the segregated circuits.

suspended floor with specialist tools provide by the flooring contractor. Socket boxes in the floor tiles may be from the flooring specialist or the electrical manufacturer, but as they are connected only by flexible conduit compatibility is not usually a problem. Underfloor voids are also useful spaces for the installation of cable tray or basket for larger cables which are passing through the area as part of the distribution system in the building. All horizontal cable routes can be accommodated in such spaces with great savings to the electrical contractor, once the flooring system has been agreed upon. Figure 4.2 shows installed underfloor trunking.

The advent of communication cables in great numbers has lead in some instances to the loose installation of data and telephone cables below computer floors in office buildings. This technique, where the cables are loosely bunched by means of cable ties and then laid upon the structural floor, has its origin in the type of wiring used in television and associated control rooms in the broadcast industry. The adoption of the technique in the electrical installation industry has come about as a result of specialist contractors from the broadcasting business taking work in the construction area because of their knowledge of specific types of cabling. However the system is not satisfactory and would not meet the requirements of regulations (and therefore of insurance companies) regarding a 'suitable and workman-like' installation. Whereas in broadcasting the control and maintenance of such cables can confidently be left in the hands of engineering personnel, in the commercial world access will be made by unskilled persons and the integrity of such cables will be invalidated; the proximity of power distribution cabling then introduces a degree of risk which is not acceptable in the electrical industry.

CABLE TRAY AND CABLE BASKET

Mention has already been made of cable tray, which is the third main type of cable management system. Cable tray is particularly appropriate for situations where large numbers of cables have to run for considerable lengths and has traditionally been thought of as an industrial technique, but new designs have made it highly suitable for commercial installations as well. The metallic (often slotted) construction of heavy duty cable tray makes it appropriate for the support of heavy cables such as SWA mains and/or large numbers of MICC circuits in an industrial site. This method, combined with ladders for vertical runs, allows excellent mechanical support with accessibility and little problem from grouping factors due to bunching of cables. The same trays used for the main cables often form a convenient support structure for local circuit cables, or secondary tray may be installed alongside the main tray runs to accommodate circuits which need to be segregated or easily identifiable. Plastic coated steel tray is very effective for use in hostile and outdoor applications and is now as widely used as the hot dipped galvanized tray, which was previously the standard specification item for such areas.

Inside industrial buildings, if there are not a great number of heavy cables to accommodate, some less expensive system of support is called for, and cables are often seen directly clipped to structural surfaces for no other reason than to avoid the cost and size of a traditional tray installation. In many applications the use of cable basket provides the correct solution to these difficulties, at a cost effective price.

Cable basket is a system of cable support materials which includes all the shapes and accessories expected in a tray or trunking system, but which is constructed from a wire framework (rather like supermarket shopping baskets, hence the name), which makes it considerably lighter and easier to work. Indeed the reduction in labour costs when using basket rather than traditional tray is enormous, and this is one of the reasons for its recent introduction to the market and its popularity. Figure 4.3 shows the economic use of cable basket in industry. The skills required in the installation operatives are not as specialized as those for the fitting of galvanized tray, and generally skilled electricians are able to fit and install cable basket at a fast rate. The system requires less factory provided fittings and accessories, as the basic 'basket' component can be bent and adapted on site (owing to its wire construction) with considerable ease. Gland plates may be fitted to the side walls to accommodate conduit or trunking spurs to outlets, etc., and this is of particular use if the designer wishes to continue the installation below raised floors (cf.) for servicing floor outlets in offices. There is no reason why cable basket systems themselves should not be continued below raised flooring

Fig. 4.3 Economic use of cable basket in industry.

systems, provided that the access to them is satisfactory for the replacement of cables as and when required.

The cost saving possible from such installation techniques may well make them worthwhile, although consideration of the use of a low smoke and fume (LSF) cable sheathing material may offset some of the cost saving. In cases where intrinsically safe cables have been a requirement anyway, there will be no cost penalty in moving towards a less enclosed and cheaper to install system of this nature.

SEGREGATION

We have mentioned several times the need to segregate circuits, and this requirement must be a serious consideration in the selection of the cable management system to be used for any particular job. Designers must consult BS 7671 for details of safety requirements in as far as circuits of differing voltages and categories are concerned, and must also be aware of any special requirements regarding the physical separation required for data or control cables within the building. It may be necessary to consult the designers and/or suppliers of local area networks (LANs) or other data highway cable systems which need to be installed. It is worthwhile remembering that the building management system (BMS) is likely to be a DP system rather than a simple electro-mechanical system, and the BMS itself may have specific requirements for the segregation and protection of its cables. As the design of the BMS is likely to be the responsibility of the electrical engineer concerned with the cable management system, there should not be a conflict of interests as long as the requirements are clearly identified at the start of the design process. The need to protect data cables from the effects of electro-magnetic interference, and to prevent power cables radiating such potential interference, is set out in the EMC Directive from the EC. The requirements of this, and the data produced by manufacturers and research bodies on the properties of materials used in cable management products, will influence the selection of such products. The compliance with EMC requirements is to become a legal duty of those persons constructing electrical installations, and due importance must be given to these requirements.

General requirements to avoid problems of EMC are set out by the respective trade advisory bodies, and in the guidelines published by the IEE (originally in 1987), and these criteria should be regarded as a minimum for every installation. The particular needs of specialist services, or the specific need to contain the potential interference caused by certain types of electrical engineering equipment (thyristor drives and lift motor equipment amongst others) are extra conditions which must be considered. Not all of these are matters affecting the cable management system, but they might be, and evaluation is certainly necessary. In the routing of cables, some potential problems can be avoided by not exposing sensitive circuits to possible radiation – for example, the running of data cables down lift shafts is unlikely to lead to a good installation and may well have other problems of its own, not least those of maintaining the integrity of fire breaks.

SPECIAL CONSIDERATIONS

In any building environment, there will be a final connection from the cable distribution system to the point of utilization, which may well require different techniques from those used for the 'fixed' building wiring. In factories the use of flexible conduits to enclose cables to machines and work stations is acceptable, but in commercial interiors a more concealed method is required. One way is to provide 'power poles' where outlets of various types are mounted on vertical 'poles' with the cables passing to the floor or ceiling void. Such poles can be provided adjacent to each desk in an office or may be part of a furniture manufacturer's range of products. An alternative solution is the use of flat under-carpet cables which contain power and communication cores, and which are constructed to provide safe working in normal conditions of office use. Several such cable systems are available and those that comply with the appropriate BS can be regarded as satisfactory provided that the rated voltages are correct for the application. (Some imported cables are rated for 110 V centre tapped supplies and cannot be specified for 240 V phase to earth circuits.) Appropriate floor boxes must be used to terminate flat cables, and in the medium term there should be a maintenance and inspection programme for such flat cables as they are prone to damage from unauthorized tampering.

It is not unusual for the greatest cable density in a building to be the communication cables, and there is a trend which will make communications cables undoubtedly the largest part of any non-domestic installation in the future. The functions of such cables which are most often encountered are those for the transmission of voice, digital electrical data, and optically-encoded data. The first type includes telephone cables and comprises conventional cable construction with a large number of discrete pairs of wires inside a suitable sheath. The specification of 500% spare pairs is not unusual, and these cables can become quite large in new speculative office developments. Electrical digital data (which has almost replaced analogue data transmission) requires much more specialized cables, and the supplier of the data equipment will specify the cable types in detail. In many local area networks (LANs) within buildings, there will be a range of cable types, and there may well be EMC shielding requirements specified to protect the cables from invisible but potentially disastrous problems. The future trend is to work with combined data and voice systems, providing work stations with common purpose outlets to which the user may connect the required facility; again the cables are relatively basic, but their integrity must be respected.

The wide use of fibre-optic cables for optically-encoded data presents new requirements for the management of the cables, particularly with respect to the provision of outlet boxes. Fibre optics need outlet boxes with the facility to make off the core(s) as well as to retain the cable itself and to radius it to the required orientation. Such boxes are readily available, but must be specified and installed in conjunction with the fibre-optic cable supplier. The use of 'blown' optical fibre installations (where cables are not drawn into conduits but fibres are transported down them on a cushion of air) have removed some of the more onerous requirements of enclosing discrete fibre-optic cables, and this technique is likely to increase its penetration of the electrical installation industry in the near future. A feature of large data cable installations is that the cables will have their own termination areas in the building, (telephone frame room, computer suite, etc.), which are not likely to be at the same location as the mains intakes, etc. This means that the cable distribution system will have a different basic layout for the power cables to the data cables; hence multi-compartment trunkings (etc.) will not be appropriate, and separate techniques will be required for each type of cable.

The value of a cable management system to the installer and beneficial end user of the installation will be judged by the maintainability and rewirability of the cabling, including the need to increase the number of cables employed. However the designer must first take a responsible decision regarding the intrinsic safety of the designs, particularly in terms of electrical requirements, and also in terms of the need to preserve fire barriers in buildings and to respect the fire zones required. This may mean that certain methods of cable management cannot be employed, as effective fire barriers can only be installed in trunking type systems, unless structural materials are fused to surround the cables directly. The electrical system must also respect the building structure, in that there must be no reduction of the structural integrity of the building during the installation of the electrical distribution system.

In considering the question of cables crossing fire zone barriers, it is first necessary for the designer to have a complete understanding of the fire prevention architecture of the building in question. Large structures will be divided into zones, with fire barrier times specified between each such zone. At the very least each floor in a multi-storey building will form a zone, and in any building of size, the floors will be further divided into zones themselves. According to the nature of the activity planned in each zone, a fire barrier time will be specified, for which time the progress of the fire must be delayed by the building structure at that point. Where hazardous activities or materials are anticipated (for example, the storage of motor vehicle or petroleum spirit), the requirements will become more stringent, but even in straightforward commercial developments 30 minute fire barriers are required at numerous internal points in the building. Wherever cable distribution systems, or even individual cables, cross these zone boundaries, suitable firebreaks must be installed in the conduits, trunkings or whatever system is employed. Where the architect has provided ducts to transport services and electrical cables from floor to floor or transversely across the structure of a building, the maintenance of the integrity of the fire breaks is a consideration for the building services engineer. Proprietary products exist for creating such fire breaks, and their specification and use is essential if the installation is to comply with the Building Regulations and BS 7671.

Multi-storey buildings will also contain smoke lobbies which are areas protected against the ingress of smoke from a fire in the zone in which they are located; stairwells are a typical example of such lobbies. Cable management systems passing through such areas must be designed to prevent the passing of smoke along the cable ways, and in fact avoidance of large numbers of cables in such areas is the better practice. Cables themselves can produce toxic products of combustion when exposed to flames or high temperatures, and it may be necessary to use cables with low smoke and fume (LSF) emission in circumstances where cables could contribute to the hazard caused by fire or explosion in a building. If circuits are expected to function normally under abnormal conditions, then a further consideration of the cable performance is that of its reliability under such environmental disasters. This can relate particularly to communications cables and alarm circuits, and the full treatment of this topic is covered elsewhere.

Although the selection of the cable management system may be dictated by these apparently mundane requirements, the overall aesthetic effect can still be of the highest quality given the range of materials available to today's designers.

CHAPTER 5

Electricity on Construction Sites

G. Parvin

Revised by R.A. Hardy
(Electrical Design Engineer)

Electrical installations for constructional purposes are not in many instances afforded the time and attention warranted. Relative to the main contract the value of the constructional electrics is low; but if sufficient attention is given to this discipline prior to the commencement of site works, considerable expense will be spared.

The term 'temporary' is often used for electrical installations for constructional purposes and this term conjures up visions of a length of twin and earth cable connected into a 30 A single-phase and neutral switch-fuse, trailing across rough ground eventually disappearing into a 13 A metal-clad socket outlet mounted on a pattress. The 'installation' is carefully engineered to fulfil all the site electrical requirements for a modest price!

Fortunately, due to the efforts of the Health and Safety at Work Executive, equipment manufacturers and consulting engineers have greatly reduced the numbers of serious accidents resulting from improper use of equipment or the use of inferior quality products, thereby ensuring that the methods we employ in the design and application of products enable the UK to be proud of the 'Safety at Work' situation.

It is the responsibility of the main contractor to ensure that the construction programme is adhered to, and he will require the assurance that the electrical system is suitable to provide reliable power distribution, whether the contract period is over six months or six years, and involving a 2 kVA supply or a 2 MW supply.

The equipment and designs discussed in this chapter are for use on low voltage (l.v.) systems only, although it is realized that occasionally it is necessary to accept supplies at higher voltages, such as 11 kV and 33 kV. Such intake voltages involve the use of h.v. switchgear and transformers which are outside the scope of the British Standards and Codes of Practice mentioned herein.

This chapter has been prepared to assist the electrical designer and contractor to apply regulations and codes of practice in the compilation of a comprehensive electrical distribution system which will provide power for all of his site requirements.

EQUIPMENT DESIGN AND MANUFACTURE

British Standard 4363, *Specification for distribution units and electricity supplies for construction and building sites* was published in 1968 and re-written in 1991 and provides an invaluable base against which to manufacture construction site distribution equipment. The maximum current rating reffered to in BS 4363 is 315 A three-phase; but with the increased use of electricity on sites it is now necessary to manufacture units of a greater capacity but still maintain the features recommended in the document. It is quite common for supply authorities in the UK to be asked to provide 500 kVA at l.v. which entails the installation of 800 A switchgear at the supply intake point.

There are a number of important aspects involved in the general design of distribution units which are discussed below.

Mechanical

Robust construction is necessary to enable the equipment to fulfil its on-site requirements, as exposure to rough handling, accidental vehicular nudging and unit repositioning are common occurrences. In addition to this, the environmental conditions experienced on construction sites require careful consideration. Equipment must be capable of continuous operation in coastal locations without any additional protection; similarly, tropical climates must be catered for with the humidity and sandstorms that accompany them. Additional measures are necessary for very hostile environments such as the use of filters, anti-condensation heaters, etc., but the basic design of the unit remains unaltered.

Electrical

Prevention of direct contact with live parts is achieved by segregation of circuits or adequate shielding enabling certain operations to be effected with ease and safety while maintaining versatility and compactness. Protection against indirect contact is effected by efficient earthing and careful selection of protective devices.

Ruggedness, reliability, versatility and safety of operation are all combined to provide an assembly which will operate satisfactorily for long periods in onerous environments.

RANGE OF EQUIPMENT

The range of equipment likely to be found on a typical construction complex is detailed below.

Incoming supply assembly (ISA)

The purpose of the ISA on a construction site is primarily to accept the supply authority's main cable, afford metering equipment space and consumer's main protection. The ISA comprises two compartments each with its own means of access to authorized personnel. The metering compartment houses the supply authority's equipment and can be sealed. The consumer's main switch compartment comprises the main isolator (lockable in the OFF position) and protective equipment. Sufficient facilities are afforded for the maximum cable size likely to be terminated.

If the metering equipment is housed within a substation or the authority's own enclosure, then the ISA provides the main site protection and control only. Figure 5.1 shows typical forms of ISA.

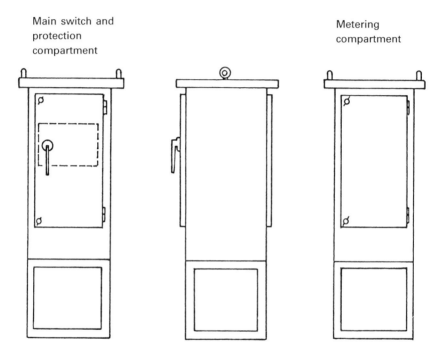

Main switch and protection compartment

Metering compartment

Fig. 5.1 Incoming supply assembly.

Incoming supply and distribution assembly (ISDA)

The ISDA comprises a compartment for the incoming feeder cable and metering equipment; a main/isolation compartment for main control; an outgoing compartment housing sub-circuit control and protection devices.

Incoming feeder compartment

The incoming feeder compartment provides facilities for the incoming cable termination and the supply authority's cut-out. When called for, the facility must exist for incorporation of current transformers (CTs) and metering equipment, as well as potential fuses. All the equipment contained in this section belongs to the supply authority and access is restricted to their personnel.

Main switch compartment

The main switch compartment contains a moulded case circuit breaker (mccb), air circuit-breaker (acb), or a switch-fuse. It is normal practice to interlock the compartment door with the main switch to prevent access to live equipment, similarly a lockable dolly is recommended for securing it while in the OFF position. Termination facilities require particular attention and must be capable of accepting cables of capacity commensurate with the rating of the ISDA and in line with BS 5372.

Outgoing feeder compartment

The number of outgoing circuits depends on the particular application, but a minimum quantity of twelve should be provided for. Loads from this section may vary from perimeter lighting to large tower cranes or sub-main distribution units. Another extremely important consideration is the prospective fault level at the incoming terminals of the ISDA and this information must be known at an early stage of the contract. It is conceivable that the ISDA will be positioned adjacent to the substation and therefore the protective devices incorporated should reflect the fault level associated with the main supply transformer.

Multi-section removable gland plates assist when making provision for outgoing circuits, and the termination of protective and control devices must be provided behind a removable cover plate to enable safe connection of equipment to be made while adjacent circuits are energized. Figure 5.2 shows a typical ISDA.

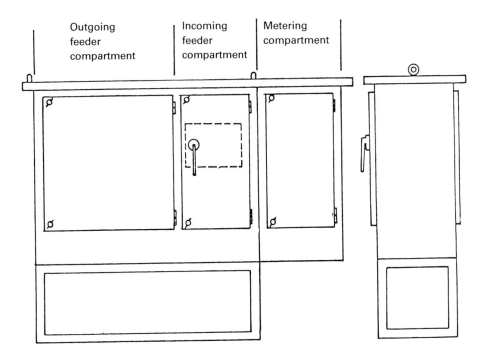

Fig. 5.2 Incoming supply and distribution assembly.

Mains distribution assembly (MDA)

When metering equipment is installed elsewhere, the ISDA then becomes an
MDA, see Fig. 5.3. The assembly consists of two sections, incoming and outgoing. It
is not economically viable to utilize a single size for MDAs, and therefore they
are normally embraced in the following ratings: 100–300 A, 400–600 A, 600–
800 A, 800–1250 A and above 1250 A. (IEC recommendations may adjust the
foregoing capacities to 125 A, 315 A, 400 A, 630 A, 800 A and 1250 A.)

The positioning of an MDA within a distribution network generally depends on
its size and application; it may serve as the major distribution device or be
supplied from a ISDA or larger MDA. To this end versatility is of paramount
importance while maintaining safety of operation.

Incoming feeder compartment

Accessibility and ease of termination are prime factors of the MDA incoming
feeder compartment. Distance between gland plate and cable termination point
requires special attention, for, unlike the ISDA, it is more likely that the supply
cable will be PVCSWAPVC and ease of glanding and terminating will need the
stated space of BS 5372.

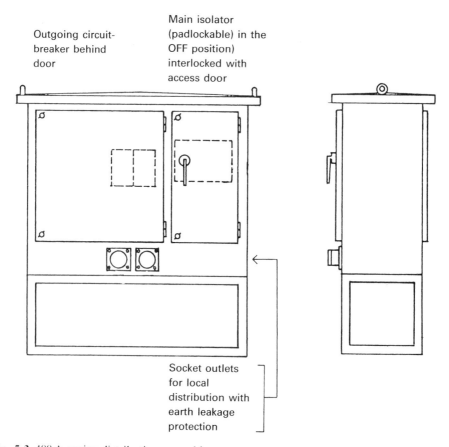

Fig. 5.3 400 A mains distribution assembly.

If the incoming feeder compartment door is lockable, the handle of the main isolator should be operable externally to the enclosure, and preferably interlocked with the access panel or door. As with the ISDA the main isolator handle should be lockable in the OFF position.

Outgoing feeder compartment
It is recommended that provision is made in the outgoing feeder compartment of an MDA to provide facilities to include at least twelve circuit-breakers, and with the variety of loads which can be introduced on to a construction site this quantity is certainly not over-specified. To reduce overall length it is normal practice to fit mcbs or mccbs in both the back and front of an MDA, and additional breakers can be incorporated with ease at a later date if space is left for this purpose. A low-level neutral bar assists when connecting to outgoing circuit-breakers allowing them to be installed safely and quickly. When residual current devices (rcds) are fitted as additional protection it is important to ensure that this neutral conductor passes through the device to an individual neutral connector. Outgoing supplies utilize armoured or sheathed cables although the use of socket outlets should not be discounted and to this end the enterprising manufacturer designs and builds an MDA which suits both methods of distribution. Figure 5.3 includes two socket outlets.

The purpose of the standards relating to construction site electrical distribution equipment is to provide a system which can be used on another site. A certain MDA may be suitable for application where fault levels are low, but one must guard against the possibility of the unit, after it has completed its term of application, being transported to a larger site with possibly a prospective fault current of 26 kA. The check list in Table 5.1 highlights the information required to assess the suitability of an MDA for a particular application.

Table 5.1 Check list/questionnaire for site distribution equipment.

---

(1) Location/atmospheric conditions.
(2) Trip rating of devices.
(3) Polarity of devices.
(4) Bolted or switched neutral.
(5) Main supply details, voltage and frequency, system protection, etc.
(6) Full or half size neutral.
(7) If a fault level is not specified try to establish:
    (a) transformer rating,
    (b) whether the transformer feeds the distribution unit direct,
    (c) if the answer to (b) is no, what other switching device is interposed between?
    (d) size and type of cable(s) between transformer and distribution unit,
    (e) distance between transformer and distribution unit.
(8) Size and type of cables – emphasize if aluminium cable is being used.
(9) Front or rear access.
(10) Enclosure protection, environmental.
(11) Are glands supplied by cable contractor?
(12) Instrumentation requirements.
(13) Any restrictions on dimensions.
(14) Label details (if known).

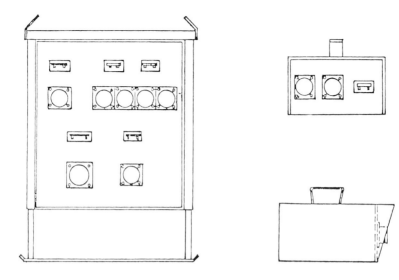

10 kVa 415/110 V CTN power
transformer fitted with 110 V
BS 4343 socket-outlets and mcbs to
BS 3871: Pt 1

2 kVA 240/110 V CTE tool
transformer fitted with 110 V
BS 4343 socket-outlets and mcbs to
BS 3871: Pt 1

Fig. 5.4 Typical transformer assemblies.

Transformer assembly (TA/1/3)

The MDAs distribute the mains voltage direct to the larger electrical loads or to the TAs which are located at load centres wherever possible. They operate from a 415 V three-phase supply or a 240 V single-phase supply and step down the voltage to 110 V. The standard design of a TA comprises a double-wound transformer housed in a weatherproof enclosure and fitted with socket outlets and protective devices. The star point of the secondary side of the three-phase unit is earthed and the single-phase unit secondary is centre-tapped to earth, reducing the potential to earth to 63.5 V and 55 V respectively. The widespread use of transformers providing power at 110 V has resulted in a considerable reduction in injuries due to electric shock over the last 25 years.

There are two main types of TAs, 1 kVA to 25 kVA continuously rated, and 0.5 kVA to 3 kVA power tool rated. It is usual to associate power transformers with final distribution on a construction complex, and the power tool rated portable transformers with smaller projects, such as service industries and house building. Figure 5.4 shows the two types of TAs.

The same basic parameters apply to each type of transformer, these being 110 V centre-tapped or star-point earthed. Secondary distribution is provided by 110 V socket outlets to BS 4343/IEC309−1. The quantity of socket outlets available on a TA can vary slightly, depending on company standards. Table 5.2 identifies the

Table 5.2 Number of socket outlets related to transformer rating with normal protective arrangements.

Rating (kVA)	Socket outlets number and rating (A)	Protection	
		primary	secondary
0.5–3 (PTR)*	2 × 16		mcb or fuse
3	2 × 16		mcb or fuse
5 (1-phase)	4 × 16; 1 × 32	mcb	mcb[†]
5 (3-phase)	3 × 16 (1−ph) 1 × 32 (3−ph)	mcb	mcb[†]
10 (1-phase)	6 × 16; 2 × 32	mcb	mcb[†]
10 (3-phase)	5 × 16 (1-ph) 1 × 16 (3−ph) 1 × 32 (1−ph)	mcb	mcb[†]
25+ (3-phase)	According to specific requirements[‡]	mcb	mcb[†]

* Power tool rated.
[†] All mcbs are to BS 3871: Part 1. It is permissible to supply two 16 A single-phase socket outlets from one double-pole circuit breaker of 20 A rating, provided that the cable attached to the associated plug top is suitably sized.
[‡] The wide variety of socket outlets possible prevents a typical arrangement being specified.

more common arrangements currently in use. Full flexibility is important and the availability of 110 V imperative. When recommending the number of socket outlets it is better to err on the side of too many, as long as adequate protection is provided.

Transformer supply cabling
It is normal practice to use PVC insulated armoured cables to connect an MDA to TAs: however for flexibility some TAs may be fitted with an appliance inlet and socket outlet for connection and loop on of a flexible armoured supply cable. This arrangement enables speedy re-siting of TAs with minimum inconvenience.

110 V distribution equipment

A particularly important aspect of the sub-distribution within the concentrated working areas of a building under construction is the availability of 110 V socket outlets. This power is required for general lighting, safety lighting and small tool operation. It is necessary to provide multiples of socket outlets for this purpose and the following types of equipment are available.

Socket outlet assembly 110 V single-phase (SOA/1)
The single-phase outlet unit is intended to be supplied from a 32 A single-phase source which in turn is derived from the local TA. The SOA/1/4 has four 16 A socket outlets, each pair being protected by a 20 A dp mcb; the SOA/1/6 has six socket outlets similarly protected. The SOA/1 units are in a portable and free-standing form, the cable being heavy duty TRS.

Socket outlet assembly 110 V three-phase (SOA/3)
The three-phase SOA/3 is not as common as the SOA/1/4−6, but the SOA/3/4−6 is normally provided with an additional 32 A three-phase and earth socket outlet for looping to a similar unit on a rising main system. This system of supply is more permanent and the flexibility is achieved by extensions to the system. Main control and protection is derived from a 32 A tp mcb and socket outlet on the local TA, and the attached four-core TRS cable is fitted with a matching plug.

Extension outlet assembly (EOA)
EOAs are available in both single-phase and three-phase designs, and are designated EOA/1 and EOA/3 respectively.

The versatile EOA/1/4 is intended, as the name implies, as an extension to the main single-phase 110 V distribution system and comprises a portable weatherproof enclosure fitted with four 16 A dp and earth weatherproof socket outlets to BS 4343. The flexible cable is connected via a plug top to an available socket outlet on the SOA/1/4−6, the SOA/3/4−6 or directly into the local TA. Some 2 kW is available at this point and a 20 A mcb protective device affords control.

Extension leads
Re-siting EOAs can be carried out by non-skilled instructed personnel and the use of extension leads speeds up this operation. Each extension lead comprises a length of heavy duty rubber insulated and sheathed flexible cable fitted with a 110 V connector (portable socket outlet), and matching plug.

DESIGN OF SYSTEM

This section provides a practical guide to the design of a distribution scheme for electrical services to plant, machinery, power tools, lighting and welfare facilities associated with construction sites.

Consultation with construction personnel

Before commencement of any design work the electrical design engineer needs to consult the main contractor's personnel to determine the methods they intend to use for the project construction. Various drawings should be made available, including an overall site plan detailing adjacent public roads, typical floor plan and an elevation drawing.

When the electrical designer becomes involved in a project of this nature much pre-planning by the main contractor will have been carried out. Underground service and drainage routes will have been planned and the overall site plan will be suitably marked with this information. Areas to be excavated should be ascertained in order to plan temporary cable routes.

An initial approach to the supply authority requesting available areas for supply intake is useful, and armed with this information the electrical engineer and main contractor can commence design.

The overall site plan should be marked up with major plant positions and equipment sizes, but if actual types of equipment are not decided upon at this stage, calculated assessments of loadings need to be made. Table 5.3 details electrical loadings associated with equipment employed on a construction project. The major electrical loads comprise tower cranes, batching plant, hoists, compressors and welfare facilities. At this stage of a project the position of these items and their expected arrival on site are more or less known and can be transferred to the site plan. An important topic for discussion is the proposed programme and whether working during hours of darkness is envisaged. Phasing of the contract affects the design of the electrical supply to the extent of load required and siting of equipment.

Table 5.3 Typical kVA requirements for plant and machinery. To be used when detailed information is not available.

Plant	Size	kVA
Hand tools	Average	1.2
Vibrators	Average	2
Concrete mixers	4/3	1
Concrete mixers	5/3½	3
Concrete mixers	10/7	7
Concrete mixers	18/12	10
Hoist	Mobile 350 kg	6
Hoist	Static 500 kg	9
Hoist	Static 1000 kg	13
Hoist	Static 1500 kg	18
Pumps	40 mm	1.5
Pumps	50 mm	2
Pumps	75 mm	4
Sawbench	300 mm	2
Sawbench	600 mm	5
Indirect fired space heater	337.62×10^6 joules	3
Indirect fired space heater	189.91×10^6 joules	1.5
Two-tool compressor		35
Saga jib crane		2
Belt loader		3.5
Tower crane (approx.)	2 t @ 15 m	30
	3 t @ 20 m	60
	4 t @ 25 m	85
	4 t @ 40 m	120
Dehumidifier	746 W	1.5
Dehumidifier	187 W	0.5
Kitchen equipment		
Cooker: industrial	Approx.	15
Four-way domestic		10
Hot cupboard		5
Fryer		3
Water boiler		3

Calculating peak demands

The peak demand occurs when the constructional work is at its highest level and/
or in the winter months when the more continuous power requirements of light,
heating and building drying must be provided. It is normally necessary to calculate
the maximum load in kW or kVA and the larger loads may require identifying,
e.g., lighting, heating, motors, etc.

This calculation is important as over-estimation results in unnecessarily high
connection charges, and under-estimation results in ultimate tripping of the main
protective device causing problems at times of peak production.

To arrive at the peak demand it is necessary to add the ratings of individual
appliances/machines together and apply a diversity factor. In the case of lighting
and space heating loads the full connected capacity should be allowed, although
office heating appliances with individual thermostatic control can be subject to a
10% allowance. Motor loads can be allowed a 50% diversity. The overall power
requirement for the contractors' equipment can be determined and the diversity
applied. The 110 V lighting and power requirements for the internal works can be
assessed by individually measuring lighting routes for each area and deciding on
the method of lighting, thus determining the loads, or allowing $3.5 \, \text{W/m}^2$ of floor
area. In each case 25% should be added for power tool usage.

Source of supply

The electrical supply can be obtained either from the local supply authority or
from petrol and/or diesel generating sets installed on site. The decision which to
use is not only one of comparative cost. Other factors involved include operational
requirements of power on site, cost and practical problems associated with distributing power safely and effectively over the site.

In remote areas where the mains supply may be unstable the most cost effective
method of supply is a combination of mains and generated power. In the event of
a mains failure the generating set can be automatically or manually started and
brought on line within a short period of time, thus assuring a constant supply. See
Chapter 1 about this combination.

Arranging for the supply

The supply authority requires as much time as possible to arrange for connection
and should be advised of the peak demand likely to be experienced during the
course of the contract, the contract period, and the position or positions required
for the supply intake. The types of electrical load, e.g., electrical heating, lighting,
motors, etc., will be helpful information to provide to the area board which may
in turn request details of motor types and methods of starting.

The power supply

Power supplies are required in the following forms, and distributed as detailed
earlier in this chapter.

415 V three-phase four-wire

This supply is used for major items of plant including tower cranes, compressors, goods and passenger hoists and other such equipment having electric motors above about 3.75 kW. All cables distributing this supply must be of armoured construction, unless in offices. When supplying semi-fixed plant, additional protection in the form of rcds is recommended.

240 V single-phase

A 240 V single-phase supply is recommended only for high fixed boundary or area floodlighting, dewatering pumps, small hoists, concrete mixers and site offices. All cables distributing this type of supply must be of armoured construction unless in offices. When supplying semi-fixed plant, additional protection in the form of rcds is recommended.

110 V single-phase

The safer voltage of 110 V is recommended for machines and hand tools driven by motors of up to about 2 kW and for all internal or portable lighting equipment. This supply is available from an isolating transformer which has its secondary centre tapped and earthed.

110 V three-phase

A three-phase 110 V supply should be used for machines up to about 3.75 kW. This supply is available from an isolating transformer, the star point of the secondary being earthed.

25/50 V single-phase

A 25/50 V single-phase supply should be used for dangerous situations, e.g. tunnelling work or inside boilers.

SUPPLY SYSTEMS

The two most commonly used supply systems currently in operation within the UK are the TN-S system, Fig. 5.5, and the TN-C-S system, Fig. 5.6.

TN-S system

The TN-S system was the most widespread in use until relatively recently and comprises separate neutral and protective conductors throughout the system. The protective conductor (PE) is the metallic covering of the cable (armouring, sheathing or conduit) supplying the installations, or a separate conductor. All exposed conductive parts of the installation are connected to this protective conductor via the main earthing terminal of the installation.

TN-C-S system

The TN-C-S system is more commonly referred to as protective multiple earthing

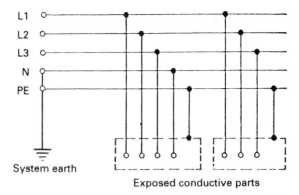

Fig. 5.5 TN-S system: separate neutral and protective conductors throughout the system.

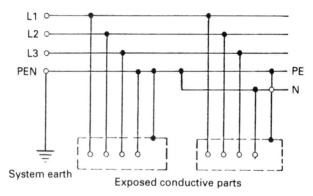

Fig. 5.6 TN-C-S system: neutral and protective functions are combined in a single
conductor in a part of the system.

(PME) with the PEN conductor as the combined neutral and earth (CNE)
conductor – and is the one generally used on construction sites. The supply
authority's regulations demand that all exposed metalwork is bonded to the
protective conductor, which in turn is connected to the neutral conductor at the
supply intake point. All phase-to-neutral faults are therefore effectively converted
into phase-to-earth faults. Owing to the impracticality of bonding all exposed
metalwork to form an equipotential cage an rcd is normally necessary at the mains
intake position.

The value of the rcd tripping current is dependent upon the earth loop impedance
of the sub-circuits (back to the supply authority's transformer), but values of less
than 250 mA should not be necessary, as an rcd installed at this point would be so
placed to satisfy the requirements of TN-C-S; providing that the maximum disconnection time of the main protection device does not exceed 5 seconds and the
product of the rated residual operating current and the earth fault loop impedance
does not exceed 25.

SELECTION OF EQUIPMENT

It is now possible to decide on the sizes and types of distribution units. Guidance is given below.

Once the load requirements are known and the supply details ascertained the appropriate type of distribution unit can be selected. The siting of this unit is invariably adjacent to the supply intake position and contains the main protection for this installation. This protective device must be capable of interrupting the relatively high levels of fault current that one would expect at the mains intake position.

Select for fault level

The supply authority will upon request provide details of the prospective fault level at the intake terminals and all the distribution equipment and associated cable installed in the site system must be sized accordingly. If the supply authority cannot provide an l.v. supply from the local network, it may install a power transformer, and it is fairly common for an 11 kV/433 V DY11 transformer to be used on the larger sites. Table 5.4 details typical transformers, fault level and winding data.

If an ISA is located at the main intake position, only one device has to be capable of handling the fault level at the incoming terminals. However, if subcircuit distribution is effected from an MDA at this position then each protective device in the MDA must be capable of interrupting the prospective fault current safely and quickly.

For example, if the supply transformer is rated at 500 kVA, the adjacent distribution board is subject to a prospective fault current of 13.5 kA (the impedance from transformer to incoming feeder terminals is negligible), and the outgoing feeder protection must be capable of interrupting this current. This can be achieved in a number of ways, the most economical being to provide mcbs/mccbs of sufficient interrupting capacity or a combination of mcbs/mccbs and current limiting HBC fuse links.

Upon establishment of the mains position, the sub-mains MDAs require siting. Load concentration areas affect the positions of sub-mains equipment and the tower crane positions are likely points. Accommodation and welfare buildings are

Table 5.4　Typical transformers for site supplies.

Size (kVA)	Fault level (kA)	Resistance/phase (Ω)	Reactance/phase (Ω)
1000	26	0.0023	0.0086
500	13.50	0.0049	0.0170
300	8.75	0.0098	0.0280
200	5.50	0.0158	0.0406
100	2.70	0.0380	0.0810

normally located towards the outer perimeter of the complex, and can be treated as an independent sub-mains. After calculation of the area load in the sub-mains vicinity a suitable MDA can be selected and the output requirements allowed for.

The supply cables to the sub-mains MDAs require careful selection and when the route has been determined the appropriate cable size can be chosen. Factors governing this selection are current rating of MDA, distance from supply incoming unit, method of installation, and the type of protection. The voltage drop should not be more than 2½% of the declared voltage.

Table 5.5 details typical conductor and armour resistances for PVCSWAPVC copper cables to BS 6346, and this information can be used to determine the prospective fault levels at the sub-mains position.

Using the information in Tables 5.4 and 5.5, and assuming a power transformer is situated adjacent to the main site protection, and a fault of negligible impedance, the fault current is calculated as follows. See Fig. 5.7.

$$I_f = 240/\sqrt{[(R_1 + R_2 + R_3)^2 + (X_1 + X_1)^2]}$$

where I_f = fault current
R_1 = transformer resistance
R_2 = phase conductor resistance
R_3 = armour resistance (negligible reactance assumed)
X_1 = transformer reactance
X_2 = phase conductor reactance.

Table 5.5   Details of copper conductor cables to BS 6346. All values are based on maximum conductor temperature (70°C).

Conductor area (mm²)	Conductor resistance (Ω/km)	Equivalent reactance (Ω/km)	Armour resistance (Ω/km)	
			2-core	4-core
1.5	14.478	0.1040	10.7	9.5
2.5	8.711	0.1010	9.1	7.9
4	5.516	0.0985	7.5	4.6
6	3.686	0.0935	6.8	4.1
10	2.190	0.0925	3.9	3.4
16	1.377	0.0880	4.5	2.6
25	0.870	0.0870	2.6	2.1
35	0.628	0.0815	2.4	1.9
50	0.464	0.0815	2.1	1.3
70	0.322	0.0785	1.9	1.2
95	0.232	0.0785	1.3	0.98
120	0.185	0.0755	1.2	0.71
150	0.150	0.0755	1.1	0.65
185	0.121	0.0755	0.78	0.59
240	0.0929	0.0750	0.69	0.52
300	0.0753	0.0750	0.63	0.47
400	0.0604	0.0745	0.56	0.34

If the supply authority advises on the fault level at the incoming terminals the prospective line/earth fault current can be determined at any point on the site.

The values of I_f and the voltage drop to the sub-mains positions should be noted as subsequent calculations and material selection involve these values.

The fault level provides the information necessary to select the protective device at the main intake position. The breaking capacity rating has already been ascertained relative to the prospective fault current and it is now necessary to decide the speed of operation of the main device.

The IEE Regulations stipulate that the maximum disconnection times under fault conditions are 5 seconds for fixed equipment and 0.4 seconds where the circuit feeds socket outlets. However, this latter requirement does not apply to reduced voltage circuits described in Regs 471−28 to 471−33. Similarly the Code of Practice 1017, *The distribution of electricity on construction and building sites*, advises that protection against earth fault may be obtained in two ways:

(1) by providing a low impedance path to enable the overcurrent device protecting the installation to operate in a short space of time;
(2) by inserting in the supply a circuit-breaker with an operating coil which trips it when the current due to earth leakage exceeds a predetermined safe value.

When the earth path is via cable armour as in (1) above and Fig. 5.7, extreme care is required in protective device selection. The armour impedance may result in insufficient current flowing to allow the device to trip within the required time, and changes to specifications do occur; such things as cable routes, equipment positions, etc., could be revised, and it may be necessary to install a cable with high conductivity armour, use an additional parallel protective conductor, or provide an earth leakage protection device, as in (2) above.

The types of load can vary enormously, so the area distribution unit needs to be versatile enough to incorporate a variety of sub-circuit protection devices. The largest single load is probably the local tower crane where the motors may total some 80 kVA. Various factors need to be considered such as motor starting or dampness, and because of these a sub-circuit device providing protection against earth leakage in addition to overcurrent is recommended, particularly if the main site protection includes earth leakage protection. Careful selection of rated tripping current is important wherever an rcd is chosen for sub-circuit protection as these devices are obtainable with various tripping characteristics, instantaneous, inverse

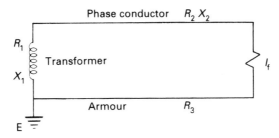

Fig. 5.7 Calculation of prospective fault current.

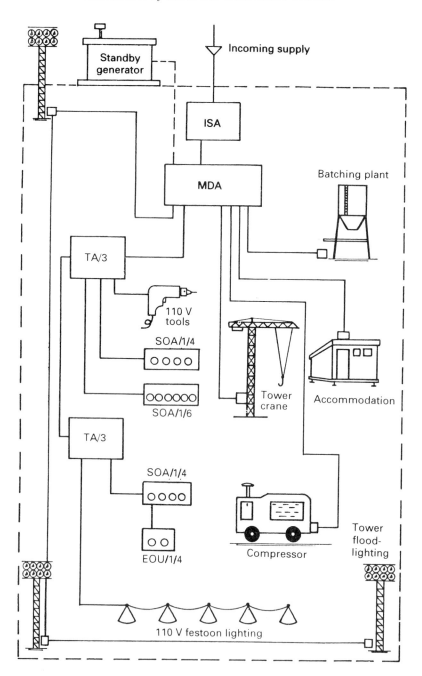

Fig. 5.8 Typical distribution network for a large site.

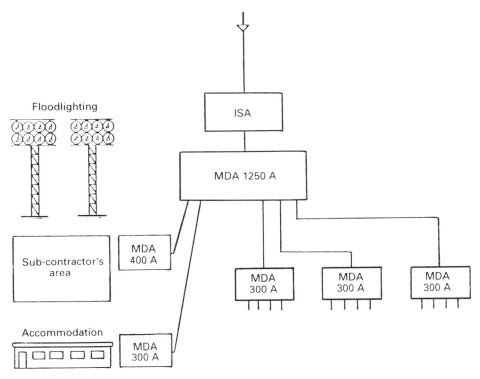

Fig. 5.9 Typical distribution system for a smaller site.

definite minimum time lag (idmtl), and time delayed. This variety will enable selection for discrimination to be achieved while minimizing spurious tripping.

Any socket outlets operating at voltages above 63 V to earth should also be fitted with rcds of the sensitive type to afford personnel protection, typically 30 mA to operate within 30 milliseconds. To reduce unnecessa.y cash outlay it is not essential to provide earth leakage protection for individual circuits, but care must be employed if a group of circuits is to be so protected, because a small leakage on one feeder can trip out adjacent healthy circuits.

All cables from the area MDA should be of armoured construction and flexible cables should have a separate earth conductor bonded to the earth screen. Additional protection using earth loop impedance monitoring devices requires consideration, particularly on flexible cables and cables subjected to frequent movement, so that should any increase in earth loop impedance occur the associated protection devices will trip. Correct protection of cables and loads is afforded by using mcbs and mccbs combined with back-up fuses to withstand the prospective fault current as necessary. It is important to ensure that in the event of a short-circuit or overload, the protective device on the load side of the fault operates. Correct discrimination is confirmed by comparing time/current characteristics of the associated protective devices under all fault conditions.

Figures 5.8 and 5.9 detail typical distribution networks in block form, and the relative cable sizes and fault levels can be recorded on the diagrams to assist in component selection for future additions to the network.

CABLE ROUTING

All cables operating at voltages exceeding 63 V to earth (three-phase system) or 55 V to earth (single-phase system) and routed across a site should be of armoured construction and buried wherever possible. Further protection should be afforded by sand and cable tiles. It is important to identify the cable route and also mark up the site drawings. Where cables are clipped to a wall or fence the distance between supports must not be so great as to cause unnecessary strain on the cable.

All cables within the construction area should be installed to prevent mechanical damage or frequent moving. Ceiling voids, ducts, risers and lift-shafts are the most common routing areas of these cables, reducing obstructions at floor level.

Cable terminations

To help maintain a satisfactory earth loop impedance it is essential to fit outdoor cable glands with earth tags for bonding to associated gland plate and adjacent cables.

Distribution at 110 V

Distribution at 110 V provides supplies for three different types of load, see Fig. 5.10: safety lighting/general movement lighting; task lighting; and power tool requirements.

The 110 V supplies are obtained from local single-phase or three-phase TAs and distributed by heavy duty flexible cables. Personnel protection by high sensitivity rcds is not necessary at this voltage but localized conditions such as confined and damp situations may require special attention, such as provision of a 25 V supply or high sensitivity rcds.

Safety lighting/general movement lighting

This is normally associated with access routes, for example stairways and corridors. Lift shafts are also included in this category because much work takes place in such locations. The ideal source of supply for this type of lighting is individual transformers with direct connection points, as opposed to socket outlet connections. Once installed, the lighting (festoon strings or bulkheads) operates continuously, the transformers being controlled from a switch on the area MDA. General movement lighting may also be treated in this manner, although as it is usually installed in open floor areas, it is more common to see festoon lighting strings fed by plug and socket outlet systems. If more than one festoon string is required per floor it is advisable to connect them to separate transformers.

Task lighting

This is usually in the form of portable floodstands, or handlamps plugged into a convenient transformer or socket outlet. The 500 W tungsten halogen type is among the most popular, the lumen output being 50% more than the equivalent gls lamp.

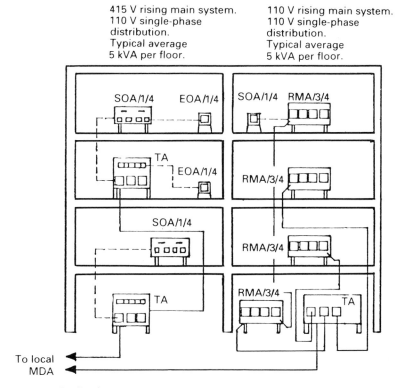

415 V rising main system.
110 V single-phase
distribution.
Typical average
5 kVA per floor.

110 V rising main system.
110 V single-phase
distribution.
Typical average
5 kVA per floor.

Fig. 5.10 110 V distribution system.

Power tool requirements

Whereas there is normally sufficient power available in an area to cater for the majority of portable tool demands, there is often a shortage of socket outlets. Work within a building under construction is periodically concentrated, but movement lighting requirement is widespread. It is therefore necessary to provide flexibility in available power points. This is achieved by ensuring that the area transformer has sufficient socket outlets for continuous area lighting and local power. For this reason the smallest transformer normally used for this task is rated at 5 kVA, with the 10 kVA unit probably the most popular. It can be seen from Table 5.2 that a 10 kVA transformer normally affords a minimum of five 16 A single-phase 110 V socket outlets, one 16 A three-phase 110 V socket outlet, and one 32 A single-phase 110 V socket outlet. If two of the 16 A socket outlets are utilized for general lighting (average 4 kW) the remaining socket outlets can be used for comprehensive power tool/task lighting applications, either directly from the transformer or via a socket outlet and an EOA/1. Sensible positioning of the portable EOA/1s provides comprehensive coverage and reduces the expensive tasks involved in searching for extra socket outlets.

CONSTRUCTION SITE LIGHTING

There are two main areas of consideration when planning a lighting scheme.

External floodlighting

The larger construction sites normally utilize floodlighting towers, positioned around the site perimeter to provide general movement, safety and security lighting. These towers take the form of mobile or fixed height static units. The factors affecting the decision for static or mobile arrangements are (1) duration of contract and (2) available siting positions.

If the contract is one of a relatively short duration, it may be more economic to use mobile towers. This type of tower is normally available up to 18 m in height and can be provided either with its own power source (generating set of up to 7 kW), or for direct connection to the mains. High efficiency luminaires are supported by the tower (four rated at up to 1500 W) and the illumination available effectively lights an area of 73 m by 50 m, the illuminance being dependent on the lighting source. Typically an 18 m tower carrying four 400 W high pressure sodium luminaires illuminates 3600 m^2 to an average of about 20 lux, the illuminance normally associated with clearing sites and handling materials.

The larger and, relatively speaking, more permanent site would probably warrant a more careful study of the lighting requirements (provided that space is available), and if there were fairly definite mounting positions, a proper engineering exercise could be implemented. This would employ lamp data, aiming angles and mounting heights in order to arrive at an overall illuminance with static towers supporting high intensity luminaires. The type of equipment now available can light areas to a level sufficient for evening working in the darker months, and reduce the contract programme time considerably.

High intensity discharge luminaires are more commonly used on this type of structure, the high pressure sodium lantern becoming increasingly popular owing to its high efficacy (about 125 lm/W for 1000 W), compared with the tungsten halogen (about 22 lm/W). The capital cost of high pressure sodium equipment is considerably higher than tungsten halogen, but many other factors are important, namely running costs, installation costs and lamp life. Careful dismantling of fixed height static towers at the end of a contract enables re-use and the only extra material required on another site are foundation stillages.

Internal lighting

For general movement within a building under construction an illumination level of 10 lux should be designed for. One festoon lighting string of 100 m fitted with twenty 100 W lamps at 5 m intervals lights an area of 750 m^2 to an approximate level of 10 lux.

TESTING AND INSPECTION

The IEE Regulations advise that a site installation should be periodically tested and inspected at intervals. As each section of the installation is covered a test certificate should be completed detailing circuit number, tests carried out and values of tests obtained. Reference can then be made to original test results when the periodic inspections take place. A typical testing form is shown in Fig. 5.11.

Contract No:	Location:	Site contact:				
Intake supply size:	A	Short-circuit ampere at intake				kA
Main protection type:		Fuses	mccb		mccb/rcd	/mA
Type of earthing arrangements:			TN-S	TN-C-S	TT	
Line/earth loop impedance at intake:				ohms		

Sub-circuit details			Fuses	mccb mcb	mcb mcb/rcd
MDA No:	Main protection		A	A	A/mA
Circuit No:	Sub-CCT protection		A	A	A/mA
Fixed or portable equipment:					
Line/earth loop impedance:		ohms Continuity:		ohms	
Polarity check		Cable type		Cable size	
Cable installation:	U/G	O/H	Surface	Cable condition	

Cable insulation resistance in megohms:

	L-E			
	L-L			
	L-N			

Recommended action:

Date of inspection: Inspector's signature:

Date of next inspection:

Fig. 5.11 Typical testing form.

INSTALLATION MAINTENANCE

The larger installations invariably have a resident electrical supervisor who should compile an ongoing maintenance programme to be implemented at frequent intervals. Liaison with constructional management enables this operation to be effected during off-peak periods. The preventive maintenance includes checking cable terminations and glands for security, cables above ground for possible mechanical damage and lack of support, rcd operation, and weatherproofing of distribution equipment and socket outlets.

Smaller installations which do not have resident electrical personnel require frequent visits to effect the foregoing checks.

Maintenance records should be kept and completed regularly giving details of tests carried out and readings achieved.

CHAPTER 6

Standby Power Supplies

D.E. Barber, CEng, FIEE

M.V.D. Taylor, MCIM

G.A. Lacey, AIEIE, AMIED
(Sales Manager, Gresham Power Electronics)

Reliability of electrical supply to industrial and commercial organizations is essential for continuity of production and safety of goods and personnel, and standby power systems have been in use for many years. The continuous increase in use of sophisticated electronic equipment, particularly computers and digital communications systems, has forced an even greater awareness of the need for standby power.

In the present context, standby power covers equipment installed specifically to supply electrical loads during failure of the prime power source, usually the public a.c. mains supply but occasionally some form of local generation.

This chapter discusses the two main types of standby power system – standby diesel generators, where the energy storage is in the form of fossil fuels, and battery-based uninterruptible power systems where the energy storage is in the form of an electrochemical reaction.

For certain critical installations which will not permit any break in supply, the standby equipment may be used in an on-line configuration, supplying the load normally as well as during failure of the prime power supply. This type of equipment is covered in this chapter, as well as composite installations using both standby diesel generators and battery-based equipment.

STANDBY DIESEL GENERATING SETS

Because oil is still a principal source of standby energy, diesel engine driven generating sets are used on a large scale as a back-up to the public electricity supply. Such sets comprise a diesel engine, coupled to a generator with appropriate control gear for operation, instrumentation and protection. Usually the sets are electrically started from 12 V or 24 V batteries and arranged to run up automatically on mains failure, take over the load and shut down again when the mains supply returns. In some cases the load is separated into non-essential and essential groups and only the latter is then supplied from the standby set during mains outage.

Range of sizes

Diesel generating sets are rated according to their electrical output expressed in kW at a load power factor usually assumed to be 0.8 lagging. The smallest units, up to say 10 kW, are often arranged for l.v. single-phase output but the greater majority, between say 10 and 1000 kW, have a three-phase output, some of the larger sizes generating at 3.3−11 kV. The most popular size for general industrial standby use lies in the 150 to 500 kW range and where greater standby capacity is required, two or more sets may be operated in parallel. Multiple set schemes frequently involve less capital outlay than a single larger unit and also provide greater flexibility, easier maintenance schedules and better system reliability.

To minimize initial cost, diesel sets are generally run at relatively high speeds and four-pole generators are very common running at 1500 rev/min for 50 Hz output or 1800 rev/min for 60 Hz. Very small sets may use two-pole generators (3000/3600 rev/min) while larger sets may sometimes use six-pole generators (1000/ 1200 rev/min).

Continuous and standby ratings

The electrical output is decided by the power available at the engine flywheel. Engines are usually rated on a continuous basis against Standard Reference Conditions, defined in BS 5514: Part 1 (ISO 3046) as 1000 kN/m^2 barometric pressure, 27°C air temperature and 60% relative humidity. Industrial diesel engines built to standard specifications will handle 10% overload for 1 hour in any 12 hours of operation at the continuous rated load. This overload rating is frequently used as a continuous standby rating on the assumption that normal industrial loads are of a fluctuating nature and standby duty is relatively infrequent. Some engines are assigned a special standby rating by the manufacturer which may be higher than the 10% overload rating but in such cases a restriction on total engine service life may be imposed.

Alternators to BS 4999: Part 101, do not have an overload rating as operation in excess of the continuous rating produces a winding temperature rise higher than the allowed figure. However, the reference temperature for generators is 40°C and BS 5000: Part 3, allows a 10% generator overload rating for 1 hour in 12, providing the cooling air temperature is not more than 27°C. As in the case of the engine, this overload rating can be used as a standby rating for fluctuating industrial loads.

The majority of engines used are of the four cycle type in which each cylinder contributes one power stroke for every two revolutions of the crankshaft; the two cycle engine giving one power stroke per revolution is not much used. In general terms the number of cylinders increases with the power output of the set and while twin-cylinder engines are frequently used for small sets, twelve and sixteen cylinder engines are popular for the larger outputs and six and eight cylinder engines predominate for the most popular middle part of the output range.

Development of diesel fuel pumps has resulted in the ability to inject increasing quantities of fuel into the cylinders to give greater power output. To burn this fuel a corresponding increase in the air input is required and natural aspiration, in which the air is injected at atmospheric pressure, severely limits the engine output. Consequently, much development has been done on pressure charging whereby the mass flow of air is increased by raising its pressure, usually achieved by an exhaust turbocharger. Turbocharging results in substantial power increase and because of this most diesel sets are turbocharged engines.

Figure 6.1 shows typical performance graphs for the diesel engine of a 200 kW generating set giving figures for the continuous and also the standby ratings, and it will be seen that fuel consumption per kWh is fairly constant over a wide range of speed and load, although generating sets operate at a fixed nominal speed which is maintained by a governor and fuel pump.

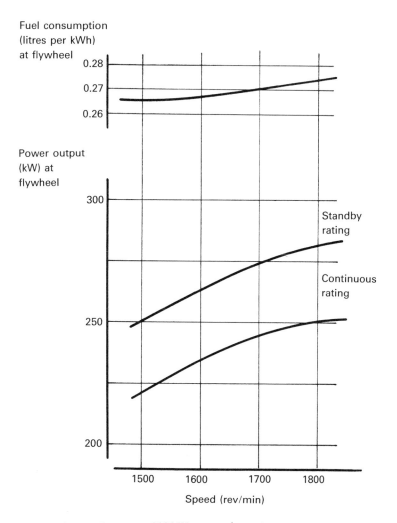

Fig. 6.1 Typical engine performance 200 kW generating set.

Governing and fuel supplies

Engines fitted with standard mechanical or hydraulic governors experience a considerable momentary reduction in speed when load is suddenly applied, but this is quickly corrected by the governor/fuel pump system. The momentary drop in speed is load dependent so that where prescribed frequency limits are important (e.g., standby for computers), it is necessary to ensure that only a partial electrical load is applied as a single step. Turbocharged engines tend to have lower load acceptance capability than naturally aspirated engines, owing to the response time of the exhaust turbocharger, and the maximum load step may therefore be restricted to 60% or 80% of rated load.

Standard governors usually give a speed drop of a few per cent between no load and rated load and this is advantageous in equalization of load between parallel running sets. Special governors are available which adjust the no load and full load speeds to be equal, and such isochronous governing may be specified for certain critical installations. Parallel running with isochronous governing requires load sensing input to the governors to ensure proper load sharing between individual sets.

A typical generating set will consume about 0.3 litres of fuel per kWh generated. Fire regulations limit the amount of fuel that can be stored in the generator room and the usual arrangement of fuel supply comprises a small service tank on or near the set which is topped up from a bulk tank situated outside.

Cooling and exhaust system

All internal combustion engines are relatively inefficient and Table 6.1 shows how the energy in the fuel is transformed into useful work and heat loss. Column A shows the figures relative to 100% fuel energy, while column B bases the figures on 100% rated set output, after allowing for generator inefficiency.

Small engines are frequently air cooled whereby an engine driven fan circulates atmospheric air around the hot surfaces of the engine. Water cooled engines dissipate heat through an air cooled radiator or in some cases remotely via a heat exchanger system. A considerable portion of heat loss is discharged through the exhaust, and installations must be designed to allow free flow of exhaust gases without restriction.

Table 6.1

	A	B
Total fuel energy	100	33.3
Heat loss to exhaust	30	100
Heat loss to cooling system	27	90
Mechanical and radiation losses	10	33
Generator losses	3	10
Energy to load	30	100

Derating of engines

It is unlikely that the UK environmental conditions of air temperature, pressure and humidity would be more adverse than ISO Standard Reference Conditions, and derating factors will not normally be applicable. Installations abroad will however frequently require reduction of standard ratings to allow for ambient air variations and BS 5514: Part 1 gives suitable adjustment factors.

Alternating current generators for standby sets

Alternating current generators fall into two broad groups, rotating armature and rotating field. Rotating armature machines are confined to the smaller ratings due to limitations of slip rings and brushes and most modern machines over about 20 kW rating are of the rotating field type.

To maintain a constant output voltage the field excitation of a generator must be adjusted automatically with changes in load. One method, used on many smaller generators, derives excitation current from the output terminals through an arrangement of current transformers (CTs) and chokes which compensate for changes in load current and power factor; this is called static excitation because the generator does not require a rotating exciter and Fig. 6.2 shows the schematic arrangement.

For most generating sets brushless excitation is used as shown in Fig. 6.3 in which the rotating field is supplied with excitation from a rotating armature exciter and a shaft mounted rotating rectifier. Adjustment of the exciter stationary field current by an automatic voltage regulator (AVR) produces the required change in the excitation of the main generator to compensate for load variations and maintain the terminal voltage at nominal value. The AVR is a solid state unit comprising a sensing input from the generator terminal voltage, a stable reference element, a comparator stage to detect the voltage error and an amplified output stage controlling the exciter field.

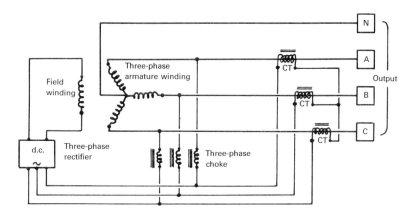

Fig. 6.2 A.c. generator with static excitation.

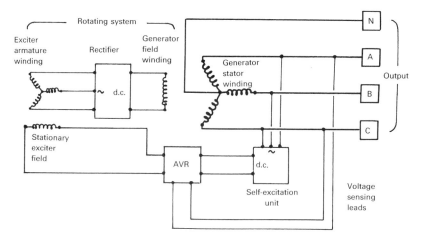

Fig. 6.3 A.c. generator with brushless excitation.

Performance

The performance of an engine driven generator includes factors such as temperature rise of windings, overload capacity, voltage waveform and voltage regulation under changing load conditions. BS 4999 gives permitted temperature rises for different classes of insulation with a cooling temperature of 40°C. If the cooling air temperature is less than 40°C a proportionate increase in winding temperature rise is permitted and this is the basis for the 10% overload allowance in BS 5000: Part 3. Most generators are capable of developing an output considerably in excess of their continuous rated value and this is frequently exploited for standby duty where the load is of a variable nature. However, operation for long periods under overload conditions may produce deterioration of the winding insulation leading to premature breakdown.

Most generators produce an output voltage waveform which approximates to an ideal sine wave. Certain types of load, i.e. thyristor loads, draw currents which have high harmonic content, resulting in distortion of the output voltage waveforms of standard generators which may affect the operation of other equipment fed from the same supply (see section on non-linear loads).

The voltage regulating system of modern generators will hold the terminal voltage within quite tight limits (typically ±2%) for any load from zero to rated value. However, at the instant of applying or removing load there will be a momentary voltage change which is outside these steady state limits and this is known as transient voltage regulation and is due to the transient reactance of the generator windings. Figure 6.4 shows a typical voltage change following sudden application of rated load. Initially, the AVR maintains the output voltage within the steady state limits but at the instant of applying load the current produces an internal reactance drop which subtracts from the terminal voltage causing the initial transient dip. The AVR immediately responds to this change and produces a large increase in field current which rapidly restores the voltage to steady state

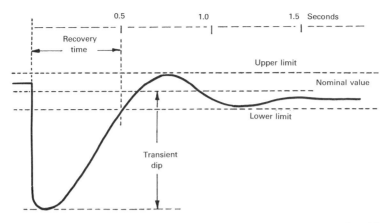

Fig. 6.4 Transient voltage recovery following application of full load on a brushless generator.

tolerances. The time required to reach the lower tolerance limit is known as the recovery time and is usually in the order of 0.3 to 0.7 seconds depending on the size of the machine.

The transient reactance of the generator is usually much greater that the source impedance of a mains supply. It therefore follows that the transient voltage dips which occur when supplying the load from the standby system will be greater than when mains powered, and the effect of this increased voltage depression on other running equipment may be significant even though recovery is rapid. For general industrial installations the transient voltage dip should be restricted to about 20% but, for more critical loads, 15% or even 10% is frequently specified. To meet these tighter requirements an oversize generator may be required, but the alternative approach is to reduce the magnitude of the load steps by suitable arrangement of the installation. For instance, direct-on-line started cage induction motors might be equipped with reduced voltage starters or multiple load elements might be arranged to start in sequence rather than all at the same time.

Mechanical construction

Most generators used in standby sets are horizontal, foot mounted, self-ventilated machines. In two-bearing machines the rotor is self-supported by ball and roller bearings and the shaft extension at the drive end then carries a flexible coupling driven from the engine flywheel, often with a spigoted coupling ring to ensure concentricity between the generator frame and the engine flywheel housing.

Single-bearing generators, common in the US for many years, are now finding favour in Europe. The drive end of the rotor shaft carries a number of flexible steel discs, the outer diameter of which fits closely into a recess in the engine flywheel and this arrangement provides positive radial location of the rotor, a smooth drive and a degree of axial flexibility to take care of the tolerances. The generator stator frame and flywheel housing are positively located by a spigoted coupling ring.

Control systems

The primary function of the control system for a standby generating set is to detect a mains failure situation, start up the set, transfer the load to the generator, supply the load for the duration of the outage, detect the restoration of the mains supply, transfer the load back and shut down the generating set. These functions may be carried out manually or automatically in accordance with pre-programmed instructions, usually contained in some form of solid state logic circuit.

The mains failure detection system comprises solid state devices which monitor the voltage on the individual phases on the mains supply to detect loss (or abnormal reduction) of voltage. It is customary to delay the activation of the run-up of a mains failure set for a second or two to prevent a momentary loss of mains voltage being interpreted as a genuine power failure.

The next stage in the sequence is the starting of the standby generator set. It has often been considered desirable to have up to three timed attempts to start, but many manufacturers are now changing the starting sequence. The first stage comprises part energizing of the starter motor system so that, at say one-second intervals, the starter pinion attempts to engage and as soon as this has been achieved, the second stage of starting, i.e. actual full power cranking, commences. An engine speed detection circuit indicates when the engine has fired and has started to run up to speed and at this point the starter motor is de-energized and locked out.

Acceleration of the engine to full speed is comparatively rapid, typically a few seconds, in which time the generator voltage and engine oil pressure build up, and as soon as speed, voltage and oil pressure are within acceptable limits the generating set is deemed to be ready for load transfer.

The load is fed from the mains supply system through one of a pair of change-over contactors or motorized circuit-breakers interlocked so that they cannot both be closed at the same time. When the generating set is ready the load is transferred from the mains to the generating set which supplies the load for the duration of the mains outage. The changeover system must ensure that in no circumstances can the generator be inadvertently connected in parallel with the mains, and four-pole contactors, isolating the neutral, are frequently specified.

Sometimes restoration of mains supply is momentary in duration and several attempts may be made to re-establish supply before becoming permanent. For this reason, many standby generating sets are fitted with some form of mains return delay circuit which ensures that mains voltage is within acceptable limits for a pre-arranged period before actual transfer of load back to mains. Small sets are then usually shut down immediately, but the larger engines may be run for a few minutes on a no-load cooling run which allows a degree of equalization of heat in the engine parts and prevents thermal stresses after shutdown. Most standby control circuits allow for the immediate repetition of the operation sequence should a second mains failure occur before the complete shutdown cycle has finished.

It will be apparent that from a functional operational point of view, the successful activation of the standby generating set and the takeover of load depends entirely on getting a good start and this depends on the reliability of the battery system. For this reason, all modern standby installations are fitted with automatically controlled battery chargers which keep the batteries in good condition and service schedules should make due allowance for proper maintenance of batteries to avoid deterioration (as discussed later in this chapter). Most service problems involving standby sets arise from a battery problem of one sort or another.

Instrumentation and protection

Some instrumentation is desirable even on unattended standby sets, as from time to time it is necessary to monitor conditions and check operating levels. Sets are therefore equipped with basic instrumentation such as engine oil pressure, engine temperature, engine speed and the usual electrical outputs. Measurement of engine speed is often based on digital techniques involving the counting of pulses obtained from a flywheel tooth magnetic pick-up device and these pulses can then be used in the mains failure control module for activating operational sequences and the overspeed protection circuit.

Serious faults such as loss of engine oil pressure, excessive engine temperature and overspeed require immediate shutdown of the set to avoid damage. Electrical overload requires disconnection of the load from the generator and other optional protective circuits may be added such as overvoltage, frequency out of limits, loss of engine battery charge current, etc. and LED signal lamps are used to indicate operational status.

Parallel operation

The use of parallel running sets as a standby service complicates the control equipment considerably because in addition to the normal mains failure sequences, the separate sets must be synchronized and adjusted for load sharing. Most automatic paralleling systems operate on the basis that on mains failure all sets start up, synchronize and take the load for the duration of the outage. As the loads are usually of a variable nature, the system must monitor the total load and the individual share taken from each set, so that automatic adjustment of the fuelling can take place to achieve satisfactory load sharing. The generators are equipped for good sharing of the reactive component of the total load as this is not affected by engine fuel adjustment.

In some cases it may be considered advantageous to shut down some of the paralleled sets if the load falls to a very low level. The automatic removal of generating sets from the parallel running group and their re-introduction should load levels increase further complicates the control system and is normally only justified in special cases.

It is customary to take the neutral connection from only one generator with the neutral of other parallel running machines left isolated. In such cases a neutral earthing selection panel may be required.

Peak lopping system

The capital investment in a standby generating set is sometimes offset by using the equipment for peak lopping duty whereby the highest electrical tariff charges can be avoided by running the set in parallel with the mains supply during peak periods. There are certain minimum requirements concerning standard of equipment and safety associated with this form of operation and local electricity boards vary in their attitude towards the idea, some allowing it readily while others forbid it completely. This is dealt with more fully in Chapter 1.

Non-linear loads

The general term non-linear applies to loads which do not have a constant impedance to the sinusoidally varying voltage. Variations in impedance may be due to the natural non-linear characteristics of the load, or may be due to the switching action of circuit components such as occurs in rectifiers or thyristors. The essential characteristic of non-linear loads is that the current waveform is not sinusoidal, even when the supply voltage is of ideal sine-wave shape. When the load current is not sinusoidal the internal reactance drop in the generator windings will follow the non-sinusoidal shape of the current, so that the actual terminal voltage of the generator will be the basic sine-wave voltage generated by the machine less the non-sinusoidal voltage drop, and the result of this will be a non-sinusoidal terminal voltage which may adversely affect the operation of other equipment fed from the same supply.

The amount of harmonic distortion of the output voltage will depend on how much of the applied load is non-linear and on the internal reactance of the generator windings. The acceptability of harmonic distortion will depend on how critical the total load is to supply voltage waveform. If the load on the generator is totally non-linear then, although maximum distortion of the voltage waveform will occur, this may be acceptable as a working condition. As a broad general rule, non-linear loads should represent no more than about 40% of the total load on conventional generators of normal reactance.

The adverse effects of any type of non-linear load can always be reduced by selecting an oversize generator and this is frequently specified by the generating set manufacturer. The effect of increasing the kVA rating of the machine is to reduce the per-unit reactance in the same proportion, and in this way voltage distortion can be minimized to any required extent. Some reduction in the reactance of standard generators can often be achieved by simple engineering changes and some generator manufacturers are prepared to undertake this work, which then leads to a lesser increase in frame size than otherwise required.

Many non-linear loads include their own closed loop control system with a high degree of amplification in the electronic circuits. It is well known that interconnection of two separate closed loop systems may produce interaction causing system instability or 'hunting' even though the two systems are individually stable. An automatically controlled thyristor system fed from a brushless alternator having the usual type of electronic AVR constitutes such an arrangement, and an unfortunate combination of time constants may produce this form of instability. However, the system can usually be stabilized quite simply by adjusting the time constant of one of the control loops. In the case of the generator the main time constants are in the machine itself and are therefore not easy to change, so it is customary to solve the problem of instability by adjusting the load time constants.

Battery charging loads

There is an increasing use of thyristor circuits in large automatic battery chargers. These use a thyristor circuit, frequently a three-phase bridge, with a closed loop controller which monitors the output current into the battery and adjusts it to a particular charging rate. Where such a battery charger constitutes the main load on the generator the harmonic distortion previously discussed will be present and there may be a risk of instability between the control system of the charger and the AVR. The effective impedance of the battery load is relatively high as it includes the back e.m.f. of the battery, but in dynamic terms, when the output of the charger is undergoing change, the effective impedance of the load is very low and a small change in the generator voltage will produce large changes in current. The handling of such instability problems is usually carried out by fairly simple adjustments to the battery charger control time constants.

Fluorescent lighting loads

For large areas of illumination it is normal practice on three-phase mains installations to power groups of fluorescent lights from different phases and this practice should be retained for installations supplied by three-phase generating sets.

Although most fluorescent lighting fittings have their own power factor correction capacitors incorporated in them, this point should be checked when the installation is to be fed from a generator because without the capacitor the power factor of the fitting may be as low as 0.4 lagging, which would be unacceptable for normal generators if the lighting installation is the main load.

Capacitive loads

Most industrial loads operate at a power factor which is less than unity and is most frequently lagging. It is generally agreed that a power factor less than 0.8 lagging is disadvantageous and steps are then taken to correct the power factor to a value in excess of this figure. The usual procedure is to connect capacitors across the load circuits which then draw a leading current which partially cancels out the lagging current of the main load, thus raising the overall power factor to the required value. In some installations arrangements are made to adjust the value of the capacitors automatically with changing loads, and this is the ideal situation as the overall power factor is then retained at its desired value.

However, in some commercial installations the capacitor is fixed at an average level of compensation based on the load expectations of the equipment, and if the actual reactive pattern of the load does not match this assumed value, over-compensation will occur and the total load on the supply will operate at a leading power factor. Two examples of such situations could occur with welding trans-formers and fluorescent lighting installations where the power factor correction capacitors are arranged for total correction and are not split into individual units to match the separate load elements.

The field excitation current requirement for a loaded generator depends very much on the power factor of the load current and in the usual case of lagging power factor the field current increases as the power factor reduces. On the other hand, with leading power factor loads the effect is opposite and as the power factor reduces the field current requirement for a given output will go down. If the leading power factor of the load circuit is reduced sufficiently, a condition may be reached in which the field current requirement becomes zero and this is known as self-excitation. At this critical point the AVR sensing the generator output voltage would lose control, and any further reduction of the leading power factor would cause over-excitation and the voltage would rise without any corrective action from the AVR. Such a condition is obviously most undesirable and could cause damage to the connected load.

Because self-excitation depends on armature reaction effects from the stator winding, modern a.c. generators which have high synchronous reactance will reach this point of voltage instability at a lower level of reactive load than older type machines which tended to have lower synchronous reactance values. Circuits which have any risk of self-excitation should be carefully examined before arranging to feed them from a generating set supply and in doubtful cases it may be necessary to adjust the values of the power factor correction capacitors.

Unbalanced loads

Ideally the phase currents of an a.c. generator should be equal, as in this way the line-to-line and line-to-neutral voltages are held within their normal tolerances. Large differences in the individual phase currents may cause voltage values outside normal limits and because of this it is recommended that efforts should be made to balance the loads to within about 20%.

The maximum load which is connected to any one phase of a three-phase generator should not exceed one third of the nominal kVA rating.

Lift and crane loads

As with all motor starting applications it is essential that the kVA rating of the generating set is sufficient to start and run the lift motor while still supplying the maximum other loads likely to be connected to the system.

In a few cases, however, the generating set is installed solely to operate a lift or crane and a problem might occur when the lift or crane is descending with a heavy load. This is due to the regenerative action of the lift or crane motor when energy is fed back to the generator which then acts as a motor and tends to drive the generating set. If the energy transferred exceeds the losses in the engine and generator, the speed of the generating set will rise since the engine governor can no longer exercise its normal control. If there is a risk of this situation occurring one solution is to fit equipment on the generating set which will detect the reverse power situation and connect a resistive load element across the generator to absorb it.

Components

The effectiveness of a standby system depends entirely on the reliability of the components used to monitor and operate the set. The modern tendency is to use solid-state circuitry in place of relays wherever power levels are low, but relays are still used to interface the electronics with power consuming components such as fuel solenoids, contactor operating coils, etc. The use of integrated circuit logic has made possible many options in mains failure control not previously available in moderately priced sets and for highly sophisticated mains failure systems with automatic paralleling, etc., microprocessor equipment is now being applied. Microprocessors can also be used for monitoring the state of set and degree of use and will, if required, carry out a regular routine test run.

Linear and torsional vibrations

By its nature a diesel engine produces a pronounced degree of linear vibration which is transmitted through its structure to the generator and other parts of the equipment and due allowance for this has to be taken at the design stage. The construction of the generator should conform to BS 5000: Part 3, which lays down vibration levels which it must be capable of withstanding. Resilient mountings may be used to isolate the vibration sources from vulnerable equipment such as control gear, and instruments. Pipe systems and cable runs should be sufficiently flexible to withstand the vibration without damage.

Torsional vibration of the engine crankshaft is produced by the firing of the engine cylinders and this is transmitted to the generator shaft through the coupling. The torsional system is complicated and requires detailed analysis by the engine manufacturer to ensure that no critical torsional frequencies are generated at or near normal operating speeds.

Enclosures

Although many standby generating sets operate indoors there are situations where, owing to space restrictions, the set is installed outside and manufacturers offer a range of enclosures designed to protect the set from weather conditions. The generator control gear and the mains changeover equipment are usually mounted indoors in free-standing control cubicles.

If an outside generating set is operated close to offices or residential areas, some form of soundproofing will certainly be required and in such cases the weatherproof enclosure can be lined with some form of sound attenuation material. To achieve low noise levels (85 dBA at 1 m is a common standard), the air inlet and outlet from the set require special attention to achieve the required sound attenuation without restricting air flow. The cost of soundproofing generating sets increases dramatically as the degree of attenuation is increased and it may be more economical to construct a special soundproof building rather than complicate the actual enclosure of the set.

Figure 6.5 shows a cut-away view of a 200 kW diesel generating set.

Assessing the load

There are three main factors which determine the choice of generating set for a standby installation. These are the value of the maximum continuous load, the maximum kW which can be applied in a single step and the maximum value of

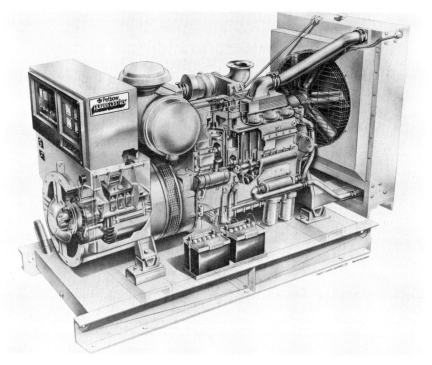

Fig. 6.5 A cut-away view of a 200 kW diesel generating set.

applied kVA in a single step relative to a given transient voltage dip. Clearly the two main types of load, i.e. passive loads such as heating and lighting and dynamic loads such as induction motors, need to be considered individually so that their effect on the generator when connection is made can be properly assessed.

Generating set manufacturers can quickly determine the most economical size of set providing they are given the basic data and one convenient way of laying out the load analysis is using the form illustrated in Table 6.2 which is a typical example of a mixed load comprising heating, lighting and motors. In this example the loads are assumed to be applied in sequence, but frequently the total load can be applied as a single step.

In the tabulation the final load, i.e. 365 kW, represents the minimum standby rating for the generating set but because the maximum peak kW exceeds this an adjustment of engine size is needed. Similarly the maximum starting kVA, i.e. 419 kVA, is used to determine whether the transient dip is acceptable and in critical cases this could lead to an oversize generator. It may be noted that the starting kW for the cage motor is taken as 0.5 times the starting kVA whereas for the slip ring motor the factor becomes 0.9 due to the high power factor of the starting arrangements normally used with slip ring machines.

Table 6.2 Load analysis.

Load element	Description
A	20 kW lighting and heating
B	55 kW cage motor – dol started (6 × FLC)
C	75 kW cage motor – star-delta started (2 × FLC)
D	150 kW cage motor – star-delta started (2 × FLC)
E	37 kW slip ring motor – rotor resistance started (1.25 × FLC)
F	

			A	B	C	D	E	F
Initial load		(1)	0	20	80	162	324	
	Start kVA	(2)		419	187	368	59	
Applied	Start kW	(3)	20	210	94	184	53	
load	Running kVA	(4)		70	94	184	47	
	Running kW	(5)	20	60	82	162	41	
Peak kW	= (1) + (3)	(6)	20	230	174	346	377	
Final kW	= (1) + (5)	(7)	20	80	162	324	365	

Maximum peak kW = 377	Maximum final kW = 365	Maximum starting kVA = 419

Installation

The plant room must be large enough to accommodate the sets with sufficient space around them to carry out proper servicing routines with some arrangements to permit lifting out sections of the plant for external repair. The ventilation of the plant room is vitally important in view of the large amount of radiated heat from generating sets and the need to circulate large volumes of cooling air through radiators. Ideally, radiators are mounted as close to an outside wall aperture as possible and if for any reason this cannot be done then a suitable ducting system should be provided. An additional extraction system may be provided to keep the plant room cool, particularly in the case of installations with heat exchanger systems or remote radiators.

If the plant room is at ground level or in a basement then the foundations are probably adequate to carry the set. It is recommended that concrete plinths are erected to lift the set above ground level to provide for drip trays and access beneath. If the set is installed above ground level the question of floor loading needs careful consideration and discussion with the architect and it may be necessary to fit resilient mountings so that vibration is not transmitted to the building. The height of the set also determines the fuelling supply arrangements. Bulk storage tanks are normally installed at ground level and circulation of oil to the generating set service tank may require fuel transfer pumps as an alternative to gravity feed.

Exhaust systems should be short with the minimum number of bends and the bend radius maximized to avoid back pressure. If a long exhaust run is necessary the cross-sectional area of the pipe must be increased to compensate. The connection between the exhaust run and the engine manifold must always include a flexible section otherwise vibration from the engine will damage the exhaust installation. For multiple set installations each engine must be provided with its own exhaust system and in no circumstances should exhausts be paralleled at any point. Exhaust pipe temperatures can reach 500°C close to the engine and exhaust system runs should be lagged to prevent radiated heat from raising the temperature of the plant room. Normal industrial silencers are usually satisfactory but for residential areas a special additional silencer may be required.

Safety codes must be strictly observed in the operation of standby sets and display notices should be erected giving the usual precautions against fire and electric shock. A special notice is also required advising personnel that a standby generating set can start automatically at any time without warning.

The electrical cable run from the generator terminal box should ideally use a flexible type cable to withstand vibration from the generator. Free-standing control panels should be mounted close to the set. Mains standby installations should include bypass switches to allow the load transfer contactors to be made dead for safe maintenance work.

Commissioning

Following familiarization with the plant by study of operating manuals, instruction books, diagrams, etc., preparation for the first run includes checking batteries, filling with oil and fuel, bleeding the injection system, checking air cleaners, fitting guards, etc.

The initial run should be short, for a minute or two, to check for leaks in the fuel and water systems followed by a 15 minute run ideally at about one-third load, under close observation. During this run the water temperature should reach normal value controlled by the thermostat. Following examination and oil check, the set can then be run up again and full load applied in 25% steps with a 5 minute run between each change. Full load can be continued for about 1 hour after which the set should be shut down and all joints tightened and a final inspection made for leaks or any other peculiarities. If a test load is not available the period of light load running should be limited to a few minutes and long enough only to check the essentials.

A check can then be made on the standby changeover functions and this, if possible, should be carried out when the service load is minimal. Disconnection of the electrical supply at the consumer's mains switch simulates a mains failure and should result in the automatic run up of the standby set and operation of the changeover contactor.

Service and maintenance of standby sets

Manufacturers' handbooks describe standard service and maintenance schedules and these should be followed strictly. The peculiarity of standby set operation is that the active service life in running hours is usually low and this makes it essential that a test run is carried out at regular intervals, say monthly. Simulation of mains failure during the test run is highly desirable and should be done whenever possible, but the engine starting system and general mechanical checks should be carried out on each occasion. Starting batteries must be well maintained and kept in a fully charged condition.

BATTERIES FOR STATIC SYSTEMS

The storage battery is the stored energy source of every static standby power system. Despite a continuing search for an alternative, the choice of battery type for standby systems is still between lead-acid and nickel-cadmium cells.

Table 6.3 summarizes the most important features of the main battery types. The lead-acid Plante cell was the preferred type for many years, offering a working life of 15 to 25 years when correctly maintained. Cell condition can be visually checked, and the storage state can be easily checked with a hydrometer.

Not all sites require the long working life of a lead-acid Plante type cell, and in more recent years there was first a swing towards flat-plate technology, offering a service life of up to 12 years, at approximately 80% of the cost, size and weight of Plante. However, with the introduction of gas recombination sealed lead-acid (GRSLA) standby batteries, even further benefits arose.

As GRSLA batteries are in essence sealed and designed not to emit gas or electrolyte vapour under normal operating conditions, they can be built into or placed in close proximity to the equipment they operate. This has obviated the need for separate battery rooms and expensive cabling. GRSLA batteries are classified as maintenance free, which in turn reduces operating costs because there is no need for electrolyte sampling or topping up with water.

Because of these major savings in cost, weight, size and resources, GRSLA technology is employed in a high proportion of new static power systems.

There is no real reason why lead-acid tubular cells should not be used with battery standby systems, although they are designed for repetitive charge/discharge duties such as providing the tractive power for electric vehicles. They do, however, have a relatively high internal resistance, which militates against high rate discharge applications.

The main benefit of nickel-cadmium cells is that they are not damaged if left discharged for long periods – although leaving the cells discharged for long periods is not recommended in standby power systems. They are not affected by extremes of heat or cold, or by vibration. There are some applications in the standby field where these considerations are important, although for many standby power applications, the high cost of nickel-cadmium cells effectively rules them out.

Automotive type batteries should not be used in standby power systems, except perhaps in a short-term emergency situation. These batteries are designed for high current discharge, for very short durations. Their working life would be reduced if used in longer-duration standby applications.

It may be tempting on the grounds of cost to install automotive type batteries for starting duties on a standby diesel generating set – this too is ill-advised. A standby diesel will typically be started up only a few times each year, and reliability is all-important.

Table 6.3 Comparison of the main features of lead-acid and nickel-cadmium cells.

	Lead-acid				Nickel-cadmium		
	Plante	flat plate	tubular plate	automotive	low performance	medium performance	high performance
Nominal volts per cell			2			1.2	
Float volts per cell*			2.25			1.45	
Boost volts per cell*			2.65			1.7	
Range (Ah)	15–2000	15–500	50–2200		5–285	11–316	11–900
Storage life			3–6 months			indefinite	
No. of cycles	600	600	1200		1500	1700	1100
% cap end life	100	80	80		80	80	80
Life (years)	25	12	12	2	25	25	25
Weight %	100	75	67		64	68	83
Volume %	100	82	51		90	78	85
Maintenance in months	9	6	6	1		12	

* Guide figures only.

D.C. STANDBY SYSTEMS

Figure 6.6 shows a d.c. system of the kind that is used for many industrial and commercial applications. Incoming a.c. power is converted to d.c. in a charger/rectifier. Under normal circumstances (that is, when the mains supply is present), this d.c. power performs two functions – it meets the power requirements of emergency lighting, switchgear, controls and other d.c. loads, and it float-charges a battery.

If the mains fails, these d.c. loads draw their power from the battery until the mains is restored, a standby generator is started, or the battery becomes fully discharged. The d.c. system is designed to recharge the discharged battery and continue supplying the standby loads.

Battery chargers

Unsuitable battery and charger combinations account for the majority of problems encountered by users, while inadequate maintenance accounts for many of the rest. To a large extent, these two aspects are linked, because the maintenance requirements of a standby battery installation are greatly influenced by the choice of the battery charger.

It is important that the charger is accurately set to the voltage levels appropriate to the type of cell in use, see Table 6.3. For sealed lead-acid batteries, typical float voltages are:

2.275 V/cell for 5-year design life types;
2.23 V/cell for 10-year design life types.

In flooded-cell types, a charger voltage of 10–15% above the recommended level can dissipate electrolyte within days, leaving the battery at best temporarily useless, and possibly with only a fraction of its nominal capacity recoverable. For sealed batteries this overcharging will cause gassing and shorten service life. Conversely, a charger voltage of 10% below the recommended level means that the battery will not be fully recharged.

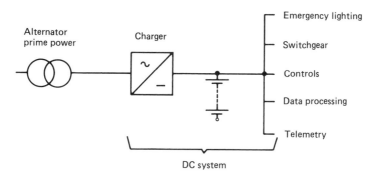

Fig. 6.6 Typical d.c. system.

If the charger output is poorly regulated, and unpredictable, and has no metering or alarm facilities, very frequent – perhaps daily – checks of charger voltage and current, and topping up of electrolyte are required.

Temperature management is more important for sealed lead-acid cells. Placement near or above heat sources should be avoided, and where necessary, temperature compensation of float voltage should be used, Figs 6.7 and 6.8.

The simplest form of charger is the taper charger, for which a typical circuit diagram is shown in Fig. 6.9. It should be noted that this type of charger is not suitable or recommended for use with sealed lead-acid or nickel-cadmium cells. In this type of circuit, the components are selected so that the battery is trickle-charged to balance any losses. An electromechanical switch (manual or automatic) connects a lower resistance to give the higher current needed for boosted charging.

The taper charger is a relatively simple and low-cost device, but it does have some serious disadvantages. If the mains varies, the unit has to be reset, otherwise under- or overcharge will damage the battery. Electrolyte consumption varies, so the level must be examined at regular intervals, and topped up when necessary.

If switching from trickle-charge to boost charge is automatic, the control unit must be 100% reliable, or there is a risk of not switching to boost charge after the battery has discharged. Worse still, there is also the risk of the charger not switching back to the trickle charge after a boost charge, so that the electrolyte will rapidly be boiled off.

A far better approach to battery charging is to use a constant-voltage charger. In this type of charger, the voltage applied to the battery is controlled automatically and kept constant, independent of any mains or load fluctuations. Constant-

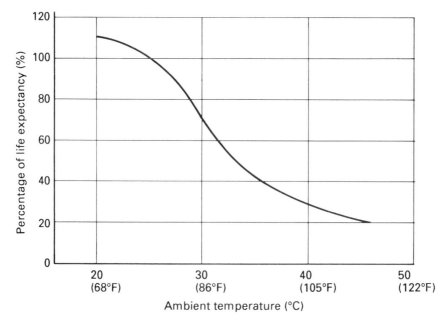

Fig. 6.7 Relationship between float service life and ambient temperature.

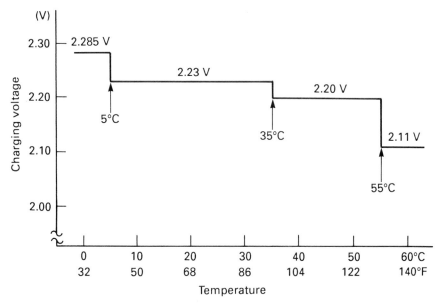

Fig. 6.8 Relationship between temperature and charging voltage.

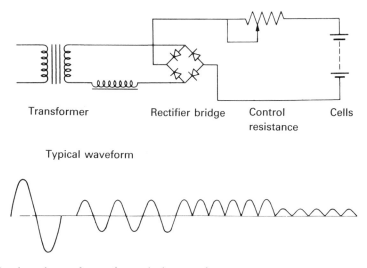

Fig. 6.9 Circuit and waveform of a typical taper charger.

voltage charging is recommended for both sealed lead-acid and nickel-cadmium cells.

Constant-voltage chargers use either transistor or thyristor control, with a current-limiting circuit to protect the components in the event of an excessive overload. The circuit of a transistor constant-voltage charger is shown in Fig. 6.10.

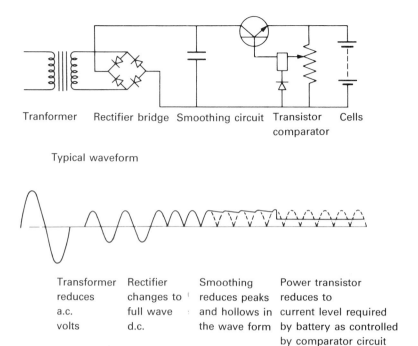

Fig. 6.10 Circuit and waveform of a typical transistor constant-voltage charger.

From a practical viewpoint, the main benefit of the constant-voltage charger is its ability to compensate automatically for variations in load and mains voltage. The battery is kept in a healthy state, and maintenance intervals are extended. There is no need to incorporate a battery load test circuit in order to check that the battery is fully charged.

Another benefit is that the size of the charger does not depend on the size of the battery installation. So extra battery capacity may be added to meet changing requirements, without necessarily incurring the cost of a replacement charger (as would be the case with a taper charger). The constant-voltage charger automatically compensates for the increased charge requirements of some batteries as they age.

Meters and alarms

Some meters and alarms are desirable on any battery charger, although it is not uncommon for users to request far more complex metering than is strictly justified. For most users, alarms are more useful than meters.

A charge-fail alarm is probably the most valuable alarm, because there are conditions where the charger can be working satisfactorily within current limits, but the battery is discharging. A high-voltage alarm, which indicates if the charger is left in the boost charge position, enables positive action to be taken to prevent possible permanent damage to the battery.

Boost charging is not usually required with sealed lead-acid cells. However, if external requirements call for this facility, care should be taken to follow the battery manufacturer's specifications.

A low-voltage alarm, set at a level towards the end-of-discharge voltage, reveals that the battery is approaching the end of its design performance. Other alarms that may be useful are an earth-fault alarm, and for flooded cell type batteries an electrolyte level-low alarm, which indicates that the electrolyte needs topping up.

As far as meters are concerned, a voltmeter indicates that the system is operating correctly; an ammeter, however, is an unnecessary expense in most cases. Current levels in constant-voltage chargers are factory preset, and in any case during the final float stage, the current would typically be only 0.005 the rated capacity of the battery.

Charger techniques for regulated d.c. loads

Where the load includes telecommunications equipment and instrumentation, and the d.c. standby supply must be regulated to within 5–10% of the nominal voltage, problems may arise unless precautions are taken to boost charge the battery in such a way that the system voltage does not rise to an unacceptable level.

There are several ways of achieving this. The cheapest and simplest approach is to use a technique known as end-cell tapping.

This involves arranging the system so that, during boost charging, the load is supplied from less than the full number of cells, by automatically switching out a certain number of cells using a voltage-sensing relay and contactor. Figure 6.11 shows this arrangement.

One potential problem is that during boost charge the tapped cells are charged at a higher rate than the maximum recommended by the manufacturer. In general, end-tapping is satisfactory where the load is small in comparison with the capacity of the battery (that is, where the system is designed for long standby operation), or where the duty cycle is such that the boost charge occurs only infrequently.

Where the standby system is designed to supply a relatively large load for a short duration, or where boost charge occurs frequently, or where system voltage regulation better than 5% is required, overcharging of cells can become a problem. A modified form of end-cell tapping, with a separate charger to boost charge the end cells, is one way of overcoming this. A typical circuit is shown in Fig. 6.12.

A completely different approach is to use diode series regulators, as shown in Fig. 6.13. The series diodes are switched into the load supply line during boost charge, so that the voltage drop across them compensates for the increased voltage caused by boost charging.

This is a more elegant engineering solution, as far as the battery is concerned,

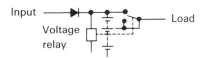

Fig. 6.11 End-cell tapping.

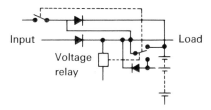

Fig. 6.12 End-cell charging.

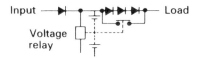

Fig. 6.13 Diode regulation.

often enabling a smaller battery to be used, and eliminating the overcharging associated with simple end-cell tapping. Against this must be set the fact that each diode can dissipate as much as 2 kW of heat, calling for large heat sinks, and perhaps necessitating a separate cubicle. The diode may be switched in and out in stages, to match the voltage characteristics of the system.

It may be that by deliberately over-sizing the battery, so that it achieves its desired performance with only an 80% charge, the need for boost charging can be eliminated. Whether or not it is acceptable to provide an extra 20% battery capacity that will never be used depends on the particular application. Apart from considerations of cost, the size and weight of the battery may be a problem – especially in restricted locations such as offshore platforms.

Another alternative, which avoids the voltage fluctuations caused by boost charging, is a split battery with a charger for each half. By ensuring that the boost charge only takes place on the battery that is off-load, a very closely regulated supply is assured. With this arrangement, shown in Fig. 6.14, it is necessary for each battery/charger to have sufficient capacity to supply the load on its own.

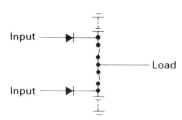

Fig. 6.14 Duplicate/split battery.

Sealed lead-acid batteries

With sealed lead-acid batteries, end-cell tapping and similar techniques are not required because:

(1) maintenance boost charges are redundant, and
(2) normal operating characteristics of sealed lead-acid batteries cover a much narrower voltage band.

Figure 6.15 illustrates the performance of these batteries.

This narrower band complies with the equipment regulation and as an additional benefit, removes the problems associated with possible switching spikes corrupting stored and transmitted digital data.

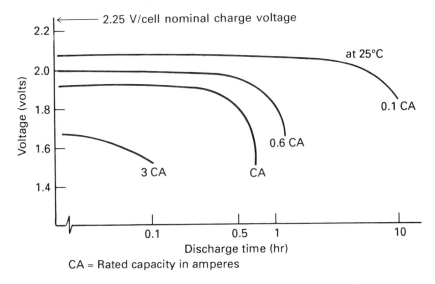

Fig. 6.15 Discharge performance (CA is the rated capacity in amperes).

ALTERNATING CURRENT SYSTEMS

A wide range of equipment in industrial and commercial premises requires a secure a.c. power supply, and with the increasing use of computers and other sophisticated electronic equipment, the demands placed on the quality and continuity of the a.c. supply have grown correspondingly.

Even in the industrialized nations, where the a.c. mains supply is very reliable, it is estimated that a computer will on average suffer from approximately 130 power surges, sags or minor interruptions per month. In addition between 80 and 90% of users will be subjected to voltage drops greater than 10% of the nominal. In developing countries, mains outages of three or four hours a day are not uncommon.

Where a battery-based a.c. standby supply is required, static inverters using thyristor or transistor techniques are used to convert the d.c. supply from a battery into a.c. at the voltage and frequency required by the load.

Simple inverter systems for non-critical loads

Some a.c. load, such as lighting systems and electric typewriters, do not require the same quality or continuity of supply as a computer. If the public a.c. supply fails, they can tolerate a few seconds' power loss without causing danger to human life or incurring financial penalties.

For this type of load, a standby inverter with a simple electromechanical changeover device is often adequate. Figure 6.16 shows a typical circuit arrangement. Under normal circumstances, the load is supplied direct from the a.c. mains. The mains also supplies a charger/rectifier which float charges the storage battery.

If the mains fails, the load is switched automatically to the a.c. output of an operating but unloaded static inverter, which gives the prerequisite voltage, frequency and output rating for the load. During mains failure, the static inverter powers the load, drawing its input from the rechargeable battery, until either the mains is restored, an alternative primary source is switched in, or the battery becomes fully discharged.

Switching the load to the static inverter in this way, only in the event of a power failure, is termed off-line or standby operation. Figure 6.17 shows a standby unit designed for non-critical desktop equipment such as personal computers and word processors. This system is more commonly known as a standby UPS.

Standby UPS

The main differences between the off-line inverter and the standby UPS are primarily to do with the type of system application and the resultant battery back-up time required.

The former is typically used for emergency lighting requirements in public

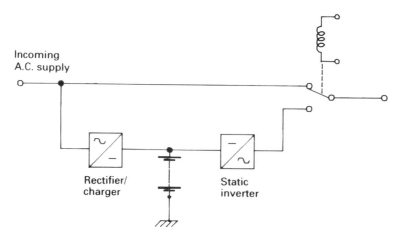

Fig. 6.16 Simple standby static inverter supply for non-critical a.c. loads.

Fig. 6.17 A standby unit designed to protect non-critical desktop equipment, such as personal computers and word processors, from mains-borne hazards like electrical noise, overvoltage spikes and blackouts.

buildings such as cinemas, where regulations dictate that the discharge time shall have a minimum period of 3 hours, whilst the latter tends to be designed for no more than 10 minutes duration.

In addition, the power supply module within the computer is more generally of the switch mode type. These can tolerate short periods of fairly crude input waveforms such as square-wave (quasi) or step-wave configurations with their resultant high harmonic contents. As the control circuitry for these types of inverters is simple, the overall cost of producing these units is kept to a minimum thus offering the various markets a cheap, if not crude, UPS system.

Some standby UPS systems have been designed with faster and more sophisticated static transfer switching arrangements to meet the needs of less critical computer and telecomms applications. However, the various manufacturers of standby UPS all carry one common flaw within their designs. This type of unit is not strictly uninterruptible as the output will have a break of at least 2 ms whilst the unit detects a mains failure and switches to inverter mode and battery back-up.

In order to counter this effect and still produce cost effective units, manufacturers have utilized the latest innovations to produce the now common line-interactive UPS which has the capability of supplying no-break power with off-line technology.

Line interactive UPS

As with all significant changes the concept of line-interactive UPS was born of a market requiring cheaper alternatives to the more traditional on-line designs whilst maintaining the integrity of no-break principles. Thus older more established and, in some cases, forgotten designs such as constant voltage transformers (or CVT) and ferro-resonant transformers were given a new lease of life and incorporated within the modern UPS packages.

In principle a CVT or ferro-resonant transformer provides the heart of the UPS as clearly defined in Fig. 6.18. This acts as a full time power conditioner filtering out any harmful spikes, surges or sags with the additional benefit of providing isolation between input and output sources. A rectifier ensures the battery is maintained during normal conditions with the inverter synchronized but not conducting.

When the mains fails, the off-line inverter is switched on and supplies the load via the CVT. This in turn discharges its residual energy during the few milliseconds it takes for the inverter to become active thus achieving an effective no-break transfer from supply failure to inverter mode. The inverter itself can be a simple square-wave system as the CVT will condition the waveform to provide a sinusoidal output. This concept can be known as single conversion as well as line interactive.

The static inverter on the input stage ensures there is no back feed on to the mains through the transformer.

Several problems exist in this type of design. The CVT will not correct for any significant frequency deviations in excess of ± 1 Hz without reverting to its inverter and free running on the output. Whilst this effectively deals with the immediate problem, this condition cannot be sustained as it is limited by the amount of available battery back-up and therefore actual running time. Some manufacturers overcome this difficulty by widening the input frequency tolerance to ± 3 Hz. However, as there is no effective control of the input stages, the output from the UPS and subsequently the load will be subjected to the same levels of variation. It is therefore necessary to check equipment tolerances before selecting this mode of operation.

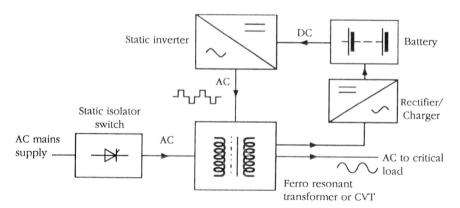

Fig. 6.18 Typical on-line interactive UPS system.

Equally, some designs have difficulty in discriminating between mains failure and some other faults on the line. For example, motors can induce a back e.m.f. into the circuit at the point of mains failure. The CVT seeing the back e.m.f. as a valid voltage may continue to try and function as normal, thus the detection circuit and transfer mechanism would be delayed. By the time the back e.m.f. has diminished and been recognized as a loss of supply the load would be dropped. Other types of line interactive designs use a multi-tap auto-transformer on the input stage and replace the CVT with a capacitive reservoir. This allows the input voltage to be boosted under low voltage conditions with typical tolerances of 164 V to 264 V and the output regulated between 187 V to 264 V. When the mains fails, the capacitive reservoir discharges in the same manner as the previous CVT system, thus providing a no-break transfer.

The alternative to standby inverter configurations is on-line operation, where the load is powered by the static inverter at all times. In the event of a mains power failure, the battery charging function ceases, but the inverter continues to drive the load without interruption. These systems are usually referred to as double conversion or continuous mode on-line.

Continuous mode on-line

This design topology is ideally suited for more sensitive operational environments. For example, ultra-critical computer systems require a much higher degree of input to output isolation in order to avoid transference of power disturbances such as common and normal mode line noise in addition to providing a totally secure output.

Figure 6.19 shows a schematic of a typical on-line UPS system, which provides continuous, conditioned a.c. power for the computer. Incoming a.c. from the public supply (or other prime power source) is rectified by a rectifier/charger. The d.c. output serves two purposes: it float charges a storage battery, and it supplies a high-performance static inverter. The output of the static inverter, at appropriate voltage and frequency, supplies the load directly.

This arrangement means that the quality of the a.c. supply to the load is determined solely by the characteristics of the static inverter. Any transients or mains-borne interference on the a.c. supply does not reach the computer load due to the principle feature of the continuous mode design, which is the full time operation of its static inverter. This negates the need for any switchover mechanism as required by standby UPSs or the use of residual energy sources for fly-wheel

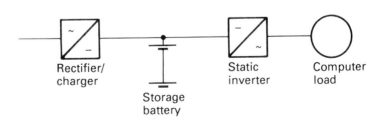

Fig. 6.19 Typical on-line UPS system.

effects as required by line interactive UPSs. As discussed previously, the line-interactive concept is unable to deal with excessive frequency fluctuations. However, the double conversion UPS can sustain tolerances of ± 5 Hz without switching to battery as the rectifier absorbs such variations.

In the event of an interruption to the a.c. supply, the load continues to draw its power from the static inverter, which in turn draws its d.c. input requirements from the storage battery. The capacity of the battery is chosen at the design stage to suit the power requirements of the load, and the standby duration it must provide. When the supply is restored, the rectifier/charger once again takes over as the source of d.c. power for the static inverter.

UPS systems above 1 kVA usually employ a static switch to transfer the critical load in the event of inverter failure, Fig. 6.20.

If the sensing circuitry in the static transfer switch detects that the output from the static inverter is about to exceed its limits, the switch transfers the load to an auxiliary a.c. supply. In many installations, this auxiliary supply is in fact the a.c. mains.

The possible risk of the static inverter experiencing a fault at the same time as a mains power failure is regarded as small enough to ignore. Where extra levels of security are required, the auxiliary a.c. supply could be another static inverter which would probably be maintained on active standby.

Matching the UPS to the load

Most of the problems experienced with static UPS systems arise because certain aspects of the load to be supplied by the system have not been fully understood at the time the system was specified. Apart from the voltage, number of phases, frequency and power rating of the load, the following characteristics are important.

Steady-state voltage tolerance
Most equipment designed for mains operation will tolerate a $6-10\%$ swing in voltage, although if computers form part of the load, only 1% or 2% may be specified, depending on the transient load response required.

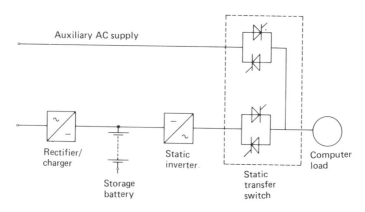

Fig. 6.20 An on-line static UPS system.

Transient load response

Time constants associated with the filter circuits in the inverter mean that the voltage regulating system cannot react immediately when a load is suddenly applied. The tolerance of the load in this respect varies considerably. Tungsten and fluorescent emergency lighting, for example, can tolerate complete loss of power for a few seconds, provided that a visible blink is acceptable. On the other hand, some major computer manufacturers specify a voltage deviation limit of +8% and −10% for a few milliseconds only, provided that the steady-state voltage limits are closely controlled, say to within 2%.

Load linearity

Resistive loads including tungsten lighting are linear loads. Switched-mode power supplies (often used in the power input circuits of computers) constitute non-linear loads, and as such can pose a problem for the UPS system. They draw current from the static inverter only in short pulses. The inverter must therefore be able to supply these spiky currents, high in harmonics, without producing unacceptable harmonic voltages.

Voltage distortion

The static inverter must provide a sinusoidal waveform of sufficient accuracy to meet the requirements of the load. The general industry standard for total harmonic distortion (THD) is 5% of the fundamental sine-wave voltage. Inverter specifications usually demand this for linear loads – non-linear loads can increase the harmonic content.

Frequency tolerance

It is relatively easy to achieve an inverter frequency tolerance of 1%, controlled by an electronic oscillator, independent of load or input voltage. Some computer loads require an accuracy of 0.05%, in which case crystal control is required.

Load inrush

Many loads take several times their rated current when first switched on to a power supply. The duration of this inrush may be a few cycles (as with tungsten lamps) or a few seconds (as with electric motors). A static inverter normally tolerates an output current only 25% to 30% above the rated figure, so this characteristic of the load needs careful consideration.

In most large installations, where the system is switched on rarely, and normally runs for prolonged periods, it is usually acceptable to allow the UPS to behave as if it were sensing an out-of-tolerance output from the static inverter. The static switch transfers the load to the auxiliary a.c. supply, which delivers the required current until the load settles down to its steady state, at which point the static switch transfers the load back to the inverter.

Dynamic loads

Some types of load, notably those containing constant-voltage transformers, exhibit a dynamic impedance characteristic, and from the viewpoint of the static inverter behave like a negative impedance. This can cause the voltage control loop of the inverter to become unstable, resulting in severe oscillations at a few hertz. In practice, this is not usually a problem if the dynamic portion of the load does not constitute more than 10% of the total load. Some types of output filter are more tolerant of dynamic loads than others.

Selecting a UPS

The revolution in personal computers and their increased use in networked form to do the job previously achieved with small mainframes or minicomputers has accelerated the demand for UPSs particularly below 5 kVA. Recent years have seen an upsurge in the number of suppliers and the variety of sytems available. This can make the selection of a UPS somewhat bewildering.

One of the most important decisions to make is whether to go for a single, large UPS to protect all critical loads in an office or building, or to install a number of smaller systems to protect individual loads or groups of loads.

Using a single, high-power UPS to supply all the critical loads in a building means that the 'clean' power has to be distributed to those loads. The UPS would provide power to the central processing system(s) and to remote units via dedicated, clean power feeds.

There are a number of potential problems with this arrangement. First, induced interference in the dedicated power feeds in the form of noise and spikes caused by the switching and operation of nearby electrical equipment – for example, lifts, power tools and even photocopiers – can cancel out the benefits of a conditioned power supply.

Another problem is ensuring that the clean power sockets do not get used for unintended purposes, for example during cleaning and maintenance of a building. The office cleaner or contractor may be looking for a socket for a floor cleaner or welding equipment at a time when computers are unmanned, but performing important search and sort or communications routines.

The centralized option also means that the failure or accidental disconnection of the UPS could mean the failure or malfunction of all the equipment connected to it.

A decentralized arrangement, involving several lower-power UPS units, usually offers increased security. The distance between the load and the UPS can be kept to a minimum so the risk of induced interference is greatly reduced. The failure of one UPS in a decentralized system affects only the equipment connected, and spare capacity in the other UPS units could be used to get the equipment up and running again while any faults are rectified.

The 'office cleaner' problem can be solved by installing UPS systems that have only computer-type IEC output sockets, so there is no risk of a vacuum cleaner or drill corrupting the power supply. The decentralized approach is more flexible than the centralized one. Because the UPS units are smaller, they can more easily be moved around if necessary, and the system is easily expandable and can be tailored to individual load requirements (power rating, support time, and so on).

Size and design

Generally, UPS units are getting smaller and smaller. Their size is dictated by the voltage and back-up time to be provided by the batteries, the full-load and continuous overload power rating, and the type of inverter employed. Off-line units are usually smaller than on-line ones, and square-wave systems are smaller than sine-wave ones.

The floor area, or footprint, of the unit is an important aspect. In the lower-power market, the newer systems employ a 'tower' design and are small enough to sit conveniently under a desk, Fig. 6.21. For larger systems, the weight and hence the floor pressure exerted by the units is an important consideration – some

Fig. 6.21 Modern UPS systems need to be designed for the office environment. They need to be compact, easy to use, quiet and stylishly designed. The unit shown is a 500 VA on-line UPS, which at full load provides 15 minutes back-up power – enough time to complete important tasks, save data and perform an orderly shutdown. Battery extension packs can be added to extend the duty period up to several hours running time.

floors may require strengthening with steel joists to prevent the UPS and batteries falling through them.

Manufacturers of low-power UPS units have benefited from improved, lower-cost power transistors and more reliable sealed, rechargeable batteries.

UPS systems are coming out of the plant room and are used more and more in the place of work. This means they ought to blend in with the 'high-tech' image of modern offices and not look like a smaller version of an industrial plant UPS. A range of on-line UPS systems ranging from 1 kVA to 5 kVA output rating is shown in Fig. 6.22.

Ease of use

The number of switches and controls actually needed on a UPS is fairly small — an output on/off switch should suffice. Any controls and indicators should be clearly labelled.

A UPS should include some form of audible and, in noisy environments, visual alarm and status indicators. The absolute minimum warning that should be given by the UPS is that the mains has failed and reserve power is being used.

Almost as essential is an indication that a certain amount of back-up battery time (say 10% of total full load support time) remains. Such an alarm allows the user to switch off any unnecessary loads when the mains supply fails, to extend

Fig. 6.22 An on-line modular UPS system with parallel connected power boards for system expansion or redundant configuration from 800 VA to 6.4 kVA.

support time, and gives the certainty that an alarm will signal when an orderly shutdown (as opposed to a sudden and catastrophic one) should be performed.

Computer interfaces

In many cases the element of manual control has become obsolete with the increasing use of computer software interfaces to ensure real-time monitoring of the various UPS parameters as well as providing a fully automatic and controlled shutdown of the user's equipment.

UPS designs have therefore taken on board the necessary elements in which to communicate on such a level with the incorporation of RS232 digitized outputs and LAN interface ports. These in turn are used with comprehensive software packages supplied in the various local area network environments such as UNIX, NOVELL, DOS and WINDOWS to name a few.

A further advantage to this type of communication is the increased use of microprocessor control circuitry in modern day designs. This can provide vital information concerning historical data on the UPS performance and its environment. With the addition of some minor equipment such as a modern, this information can be accessed and used either for service and maintenance reasons or for analytical purposes by a computer manager. With the use of a Simple Network Management Protocol (SNMP) module for reporting purposes the UPS will act as a normal proprietry item on the network and provide information to the management station such as mains failures, low battery warnings and general alarms on an automatic basis, thus allowing pro-active control of the UPS from any given location on the network.

Safety

Ingress protection to IP31 is sufficient to protect the unit from common office hazards such as spilt coffee, paper clips and other foreign matter. Virtually all low-power systems now have sealed, maintenance-free batteries, with no restrictions on proximity to naked lights, etc.

Acoustic noise

Larger UPS systems tend to be used in noisy environments – large control rooms, computer rooms, or plant rooms – so the acoustic noise generated by the unit is not important. However, with the trend towards 'office-compatible' UPS systems, a quiet unit is desirable.

As a rough guide, 45 dBA measured at one metre would be virtually imperceptible in a quiet office atmosphere with some other electronic equipment in use. Noise levels of 60 dBA and over would be uncomfortable.

Expandibility

With the advances of modern technology a limited number of manufacturers have made it possible to obtain low powered UPS from 800 VA through to 6.4 kVA which have the capacity of running their inverter outputs in parallel. This type of configuration allows the user to purchase a basic UPS for his immediate needs with the flexibility to increase the output power as and when load requirements are increased. This is simply achieved by the addition of complete UPS power boards fitted together in a modular format as seen in Fig. 6.23.

Another benefit of this type of system is the ability to configure the UPS for redundant operation. This means simply that the load can be engineered to be supplied from two boards but at a level not exceeding the rating of one powerboard, for example 800 VA/420 W. The two boards have their outputs coupled together in parallel equally sharing the load at 50%. In the event of one board failing, the remaining healthy board would immediately continue to supply the load without break at 100% capacity.

This type of configuration is also used by a number of manufacturers who produce three-phase high-powered UPS systems. Though the actual mechanics differ between designs the principles of operation remain unchanged. It is therefore possible to install UPS in the region of 2.4 kVA or higher, based upon a parallel redundant configuration

Battery extensions

For small UPS it is generally felt that for the type of application the unit will be used for, i.e. personal computers, a battery back-up time of between 5 and 10 minutes will be sufficient to meet most user requirements. This has naturally led

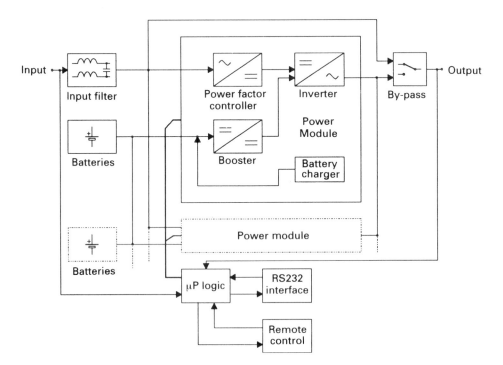

Fig. 6.23 A modular UPS schematic showing parallel power modules or boards.

to compact designs and low production costs. However, when larger UPS are employed, i.e. with critical network installations, longer battery runtimes may be needed to maintain system integrities. Therefore, we see UPS autonomies between 30 minutes and several hours as the norm. When such configurations are used it is important to ensure that the UPS rectifier or charger has sufficient current to effect a speedy recharge of the battery following its normal discharge period. In some cases it may be necessary to use an additional charger in parallel with the UPS rectifier to undertake this purpose. Thus, the amount of battery back up time one can use is not generally restricted other than through normal engineering constraints and manufacturers' technical specifications.

Versatility

If a UPS is to be used to drive several load types, care should be taken to ensure that the UPS is compatible with the loads. The output waveform from the UPS inverter can be distorted when non-linear, usually highly capacitive loads are being operated such as input power factor corrected power supplies. Under these circumstances a sine-wave output is preferable, as it can cope with a wider range of load types.

EMC Directive

As of January 1996 all power modules, including UPS must comply with a new EuroNorm directive on input harmonics (EN60 555) and radiated noise or EMC (EN55 022). The principal change is that power modules such as rectifiers and switch mode power supplies must have linear input current waveforms with a resultant power factor close to unity. This ensures that no harmful harmonics are fed back on to the mains supply to pollute other non-protected loads.

In the case of EMC, UPS inverter bridges with high frequency switching circuits produce radiated noise which can affect sensitive equipment or indeed interfere with certain radio bandwidths. The new directive limits the amount of permissible radiated noise which can be overcome with the addition of shielding or screening such as metal cabinets or covers.

All UPS systems from 1996 will carry a CE mark which indicates a manufacturer's self-certification of compliance with the new directive.

COMPOSITE STANDBY SYSTEMS

Although battery-based power systems are capable of meeting the normal and standby power requirements of the most demanding loads, the one problem with rechargeable batteries is the cost of installing sufficient battery capacity to provide a standby duration of more than a few minutes. The larger the electrical load, the more significant the financial implications.

For higher-power systems, it is common practice to use a standby diesel generator as a substitute supply for the battery until the public a.c. supply is restored. The role of the storage batteries then becomes that of supplying the

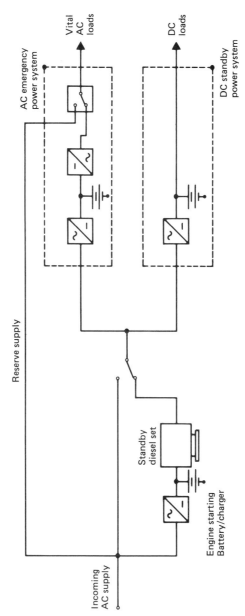

Fig. 6.24 A composite standby power system with standby diesel generator, on-line UPS and d.c. supplies.

various loads until the standby generator starts up. In practice, it is usual to specify a storage battery capable of meeting full load for 10 minutes. The cost savings associated with the smaller battery usually cover the cost of a standby diesel set.

Figure 6.24 shows a composite standby power system with a computer-grade static UPS, a battery d.c. system, and a standby diesel. The a.c. loads are divided into essential and non-essential, and it is important to segregate them so that the UPS does not supply non-essential loads that can tolerate mains interruptions.

Interfacing the static UPS to the standby generating set can pose some problems if not carefully considered. One is harmonic distortion caused by the use of thyristor-controlled circuitry in the rectifier/charger. Even with a sinusoidal voltage, the rectifier/charger takes a non-sinusoidal current waveform because the thyristor delays the start of conduction in each half cycle. The resulting harmonic currents, of the order of 20%, react with the diesel generator impedance to produce harmonic voltage distortion.

Another problem is the risk of instability because the rectifier/charger output is regulated. If the input voltage from the generator to the rectifier falls for any reason, the control circuitry in the rectifier causes it to draw an increased current to maintain a constant power output to its load. The generator sees this as a negative impedance – a drop in voltage results in an increase in current. Unless the voltage control loop of the generator is specifically designed to counteract this problem, instability will result.

Both these problems can be overcome by careful system design, and close collaboration between the manufacturers of the diesel generator set and the static UPS.

A facility sometimes requested when standby sets are used with static UPS systems is inhibition of battery charging when the diesel generator set is on-line. This avoids unnecessary load on the diesel generator in the event of the battery becoming fully discharged, for example after a standby duty.

Where both a.c. and d.c. battery-based power systems are installed on the same site, it is desirable to insist on separate storage batteries for each system.

In some applications, there is no alternative but to install a single battery for both systems. If a shared battery is implemented, the user must accept an increased risk of a fault on the d.c. line affecting a vital a.c. load. An additional limitation with a shared battery is the fact that in some applications a d.c. supply is needed at a different voltage to the d.c. input to the static inverter. If a shared battery is unavoidable, then the expert advice of a specialist in battery systems engineering should be sought.

It is increasingly common to see dual charger arrangements implemented, as a means of enhancing the integrity of the standby power system. As well as greatly reducing the chances of a charger fault putting the system at risk, a dual charger arrangement also means that one charger can be isolated for maintenance.

The battery/charger arrangements for starting duties on standby diesel are vital to the successful operation of the entire system, and the same comments made on battery chargers earlier in this chapter apply to chargers used for engine starting batteries. For maximum security, the battery should be charged by its own charger, rather than by an alternator on the machine.

CHAPTER 7

Ground Earthing

J.R. Wales, DFH, CEng, MIEE
(Consultant, Stemet Earthing Co. Ltd)

T.E. Charlton, BEng, CEng, MIEE, MBA
(Director, Strategy and Solutions Ltd)

INTRODUCTION

Earthing of electrical installations is primarily concerned with ensuring safety. In power networks the earthing system helps to maintain the voltage of any part of the network at a definite potential with respect to earth. For example, if designed correctly, it should allow enough current to flow under fault conditions to operate the protective devices installed. The rise of potential experienced during the fault combined with the speed of fault clearance, should be such as to minimize the risk of electrocution to individuals near the site of fault. The widespread use of electrical appliances, both in the factory and the home, introduces many situations where efficient earthing is of paramount importance to prevent electric shock under fault conditions.

The two main principles used in earthing are equipotential bonding and formal earth electrodes. Equipotential bonds seek to minimize the potential difference experienced across exposed metallic conductive parts. The formal earth electrodes normally consist of metallic components in direct contact with the soil. They are required to disperse any fault energy to ground in a safe, controlled and effective manner.

The installation of an earth electrode is an important factor in achieving a satisfactory earthing system. It involves burying conductive material which is in direct contact with soil or the general mass of the earth. Soil conditions can vary enormously from site to site and directly affect the resistance value of a given electrode. It is thus necessary to consider the soil and other factors which affect the actual resistance of the earth electrode at the design stage.

The earthing system consists of conductive material above ground, and metal electrodes within the soil and the surrounding soil itself. Each of these will contribute towards the overall resistance value. There are contact resistances, for example at joints and at material interfaces. The contact resistance of joints must be kept to a minimum by using appropriate materials and installation practice. In a new installation, the most significant contact resistance is likely to be at the interface between electrodes and soil. This arises mainly because the soil has not yet consolidated.

158

To understand the influence of soil on the resistance of an earth electrode, consider a vertical electrode of one metre in length inserted vertically into the ground. The soil immediately adjacent to the rod will influence its contact resistance, which will form an important part of the overall resistance. Let us assume that the soil is in perfect contact with the rod and is made up of an infinite number of cylindrical shells of equal thickness surrounding the rod. In uniform soil the shells nearest the rod will have most influence on the rods resistance value. Each subsequent shell will affect the resistance, but will have progressively less influence. In practice, the shells will not have the same resistivity value, so their effect will be a combination of their actual resistivity and distance away. An extreme case is if the natural soil is of low resistivity, but the rod is surrounded by soil which has dried out and has a high resistivity. The rod will have a high resistance dictated by the nearby high resistivity layer. A rod installed in a cavity in rock may be surrounded by low resistivity soil. It will still have a high resistance. This is because the shells some distance away have a much higher resistivity value than those close to the rod. This resistivity factor overwhelms the effect of distance. As the size of the cavity increases, so the effect of the surrounding rock will fall, but it will always affect the resistance value.

SOIL RESISTIVITY

It is important to know a little more about the electrical properties of soil because it is so critical to the eventual resistance value of the rod or earth system. The most significant electrical property is its resistivity value, which is expressed in ohm metres (Ω m). This reflects the resistance between the two opposite faces of a one metre cube of material. Some typical resistivity values of soils are shown in Table 7.1.

The two main factors which influence the soil resistivity value are the porosity of the material and the water content. Porosity is a term which describes the size and number of voids within the material, which is related to its particle size and the pore diameter. It varies between 80 and 90% in the silt of lakes, from 30 to

Table 7.1 Typical values of resistivity for different soils.

Type	Resistivity (Ω m)
Sea water	0.1 to 1
Garden soil	5 to 50
London clay	5 to 100
Clay, sand and gravel	40 to 250
Chalk	30 to 100
Rock	1000 to 10 000
Dry concrete	2000 to 10 000
Wet concrete	40 to 100
Ice	10 000 to 100 000

40% in sands and unconsolidated clay and by a few per cent in consolidated limestone.

An increase in water content causes a steep reduction in resistivity, until the 20% level is reached, when the effect begins to level out. Dissolved minerals and salts in the water may help further reduce the resistance, particularly where these are naturally occurring and do not become diluted over time. The effect of water content is clearly shown in the resistivity values of concrete in Table 7.1. This has a high resistivity when dry, which falls dramatically when wet to a value close to that of the surrounding soil.

Moisture has so great an influence that it is sensible to make use of this property if possible in the earthing system design, for example, by inserting earth electrodes into the water table to provide a consistently low resistivity. Site surveys seek to find such information.

RESISTIVITY SURVEYING

Some geophysical prospecting techniques based on electrical measurements of the subsoil to various depths, have been adapted to suit the requirements of electrical earthing systems. The method devised by Dr F. Wenner in 1915 has proved to be particularly suitable for the calculation of soil resistivity.

Equipment necessary

The necessary equipment comprises:

(1) an earth resistance/resistivity tester, having four terminals and a built-in source of power, Fig. 7.1;
(2) four metal test electrodes;
(3) four lengths of single insulated cable – used for connecting the test electrodes to the tester.

The leads should be as long as practically possible, commensurate with the size of the earth grid to be installed, the length of ground available, the accuracy of the instrument and the size of its energy source. For single vertical electrodes the test electrode separation would need to be several times the maximum depth anticipated. For earth grids, the separation should be at least the same as the diagonal across the site. Lengths of 62 m for voltage leads and 100 m for current leads should be sufficient for most rods and small grids. Usually operators prefer to develop their own cable reel equipment for this operation. For example, the cables can be of different colours, wound on reels and mounted on a stand. Often the cables are marked or have tee connections when a practice of taking measurements at a set number of spacings is adopted. This can save much time on site.

Method of use

The tester is positioned at a convenient point, often coincident with the location for a vertical earth rod or the centre of a proposed earth grid. The four test

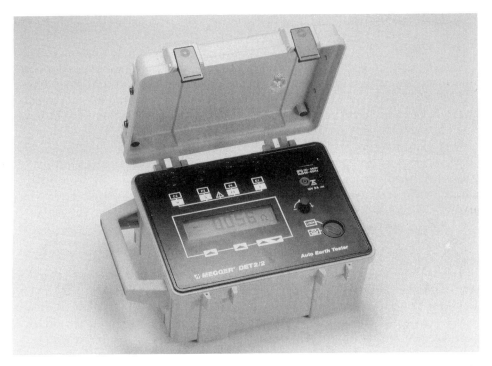

Fig. 7.1 Megger DET2/2 auto earth tester. For accurate measurements of earth electrode resistance or soil resistivity. Microprocessor control makes it virtually impossible to take a false reading. Menu driven in five languages via large alpha numeric display.

probes are positioned in as straight a line as practically possible and at equal distance from one another, Fig. 7.2. The probes do not have to be driven too deeply. For example, the current probes should have a resistance of less than 100 Ω, which can normally be achieved by less than 1 m depth. The voltage probes

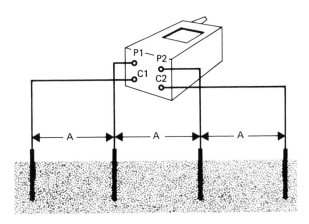

Fig. 7.2 Diagram of connections for an earth tester when used for resistivity surveying.

need only be 10–20 cm deep. The probe depth should not exceed 1/20 th of the spacing or a revised formula for the soil resistivity calculation must be used.

The probes are then connected to the tester as shown in Fig. 7.2. The connections between the probes and tester must have low resistance and some of the modern instruments can be used to check this prior to each measurement. To overcome polarization and interference effects, the polarity of the current source is normally reversed several times during the test. With older equipment this used to be a manual process, but in modern equipment the d.c. supply is converted into a variable a.c. waveform whose frequency can be varied to minimize electrical interference. A low frequency helps minimize induction effects between leads at longer measurement spacings.

The current produced by the instrument passes through the two end electrodes and the voltage between the two inner electrodes is measured. The relation voltage/current is read directly as ohms (R) from the tester. This reading represents the resistance presented by a hemispherical volume of soil beneath the tester. As a practical approximation on site of the depth of soil being measured, Stemet assume that the Wenner method is examining the properties of the soil to a depth of 75% of the probe separation. For example, if the spacing is 4 m, the depth examined is 3 m.

For a more accurate estimate in uniform soil, a graph is available. This shows the percentage of the total current flowing through a vertical plane beneath the tester. Approximately 50% flows through a plane of depth equal to half the electrode separation, 70% through a plane equal to the electrode separation and so on.

From the resistance figure obtained we are able to calculate the apparent resistance of the material contained within the hemisphere. This is via the following formula for resistivity (ρ):

$$\rho = 2\pi RA \; \Omega \; \text{m}$$

where R = earth tester readings (Ω)
       A = spacing used (m).

The above equation assumes that the soil is of consistent structure – i.e. consistent grain size, type and water content throughout the whole volume. This is very rarely the case and the reason why the calculation provides the apparent resistivity and not the true resistivity. In fact the soil is likely to have several horizontal, sloping or vertical layers. Each layer will have a different real resistivity and the volume of each layer within the measured hemisphere will contribute towards the apparent resistivity.

This measurement technique can be used to examine how the real soil resistivity is changing with depth. To achieve this a series of resistance measurements are taken, each at a different probe spacing. It follows that the depth of material investigated at each spacing will differ and will consist of different layers of soil. Typically the probe spacing would start at about 0.5 m and increase in stages to 30 or 50 m. The results of a typical survey are set out in Table 7.2.

Table 7.2 Results of a typical survey.

Spacing A (m)	Depth D (m)	Instrument reading, R (Ω)	Resistivity (Ω m)	Rod, r (Ω)
2	1.5	90	1140	676
4	3	21.5	538	182
8	6	10	502	94
12	9	5.5	415	55
24	18	3	450	33
40	30	2	502	23

INTERPRETING MEASUREMENTS

Normally the apparent resistivity readings from column 4 of Table 7.2 would be plotted against the Wenner probe spacing (column 1) on log/log graph paper. By examining the shape of the curve, much can be deduced about the type of soil structure. Normally each significant change in direction or gradient indicates a change in soil resistivity at a certain depth. Figure 7.3 is such a plot for a three-layer soil structure. For effective vertical earthing we are trying to identify layers where the soil resistivity is low – for example, in layers below the water table. Figure 7.3 illustrates that there is a low resistivity layer some distance beneath the surface. Software facilities are available which can derive a fairly accurate soil model from test data. For example, the soil model derived from the same data used in Fig. 7.3 is:

Surface layer: 105 Ω m, 1.1 m thick
Central layer: 265 Ω m, 2.5 m thick
Lower layer: 85 Ω m, at least 16 m thick.

This soil model would be used as the basis of accurate computer modelling of the proposed earthing system, if this is required. In multilayer soil the actual depth of the hemisphere of soil being tested can vary substantially, owing to the natural preference of the current to seek a route through low resistivity layers. It is only really possible to account for this by using computer packages or experience to derive the soil model.

Stemet Earthing have developed a graphical method as a first approximation of computer based techniques. The readings from Table 7.2 are plotted as diagonals across logarithmic scaled graph paper, as shown in Fig. 7.4. From this an approximation of the rod resistance can be read directly off the vertical scale. If the resistivity value is required, this can either be calculated (as shown in Table 7.2) or assessed graphically (Fig. 7.5). Diagonals have been superimposed on the grid to show the theoretical resistance of an earth rod driven into uniform soils of different resistivities.

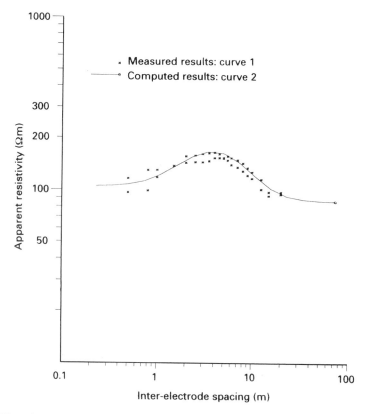

Fig. 7.3 Plot of apparent resistivity against Wenner probe spacing for a three layer soil.

Rod resistance formula

The next stage is to convert the resistivity figures obtained into a resistance value for the earth electrodes. For example, it is useful to be able to predict in advance the resistance value of a single earth electrode before it is driven to greater depth.
   The simplified formula is:

$$r = 0.366 \times \rho \, \log_{10} \frac{3D}{d} \; \Omega$$

where r = resistance of the rod (Ω)
        ρ = resistivity measurement taken at certain spacings (Ω m)
        D = depth of rod (m)
        d = diameter of rod (m).

Example from Table 7.2
Taking a reading from Table 7.2, the approximate resistance of a rod driven to a depth of 3 m is calculated as follows:

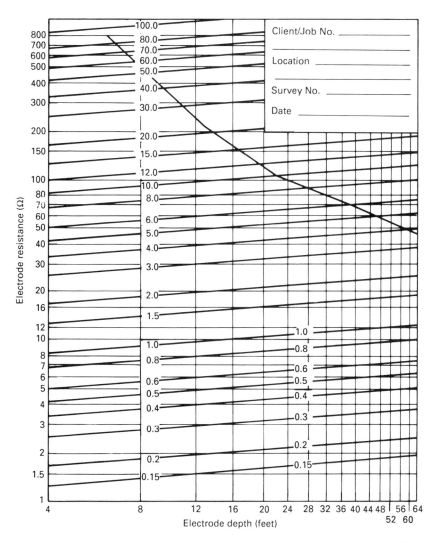

Fig. 7.4 Site chart: earth tester readings for different spacings enabling rod resistance values to be read directly off the vertical scale.

$$r = 0.366 \times \frac{538}{3} \times \log_{10} \frac{3 \times 3}{0.016} \; \Omega$$

$$= 180 \, \Omega$$

In general, if the curve showing the calculated resistance drops off at least as sharply as the diagonals in Fig. 7.5, it is more advantageous to carry on driving a single rod rather than use a number of shorter rods connected in parallel. However, it must not be assumed that deep driven rods are the answer in every case. Some sub-soils are quite unsuitable and it is not worth continuing to drive into these. In

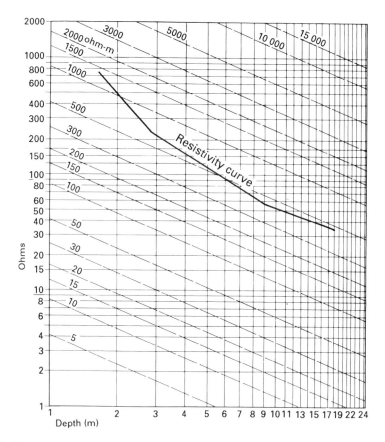

Fig. 7.5 Curves of specific resistivity and electrode resistance plotted on a logarithmic grid.

such cases, it is better to use a multiple number of rods connected in parallel, each of which is driven to about 4 m depth. Rods should always be at sufficient depth such that their resistance value is not significantly affected by seasonal changes which may change the soil resistivity. As discussed earlier, the variation in the water table is important in this respect and advice on this may he obtained via geologists or local knowledge. Ideally, the separation of these earth electrodes should not be less than their depth. For example, if an electrode is driven 10 m deep, then the earth electrode nearest to it should be positioned at least 10 m away, otherwise the interaction between them will reduce the overall effectiveness.

    Where multiple earth rods are used, they are normally connected together with bare copper tape or cable. A bare interconnection acts as an additional electrode and helps to reduce further the overall resistance value. The depth at which interconnecting tape or cable is installed should be sufficient to afford protection and mechanical strength, and maintain contact with damp soil. This is normally 0.6–1.0 m deep. In areas subject to extremely low winter temperatures, greater depth may be necessary. The top ends of the vertical earth electrodes are installed to the same depth (0.6–1.0 m) to reduce voltage gradients on the soil surface.

Large electrode networks should ideally be designed so that they can be split up into sections. Connections should be brought out from each section to a test position. This sub-division allows the installation to be tested whilst the associated equipment continues to be earthed.

The improvement in the earth resistance that can be expected from the installation of additional rods in terms of the percentage improvement each additional rod produces, is approximately as follows:

Second rod   60%
Third rod   45%
Fourth rod   35%.

Clearly, increasing the number of rods indefinitely has a diminishing benefit and is illustrated in Fig. 7.6.

The above method of calculating the resistance is limited to straightforward structures and conditions. For more complex earthing arrangements and soils which have several layers (particularly where one is rock), more detailed estimates or computer modelling is required. New standards are imposing the need to calculate touch and step potentials across a substation site. The earthing installation directly affects these potentials and computer modelling enables them to be calculated accurately throughout the site. To illustrate the power of the computational modelling tools available, Fig. 7.7 shows the potential on the soil surface above an earth grid, in three dimensions. The results obtained can also be presented in

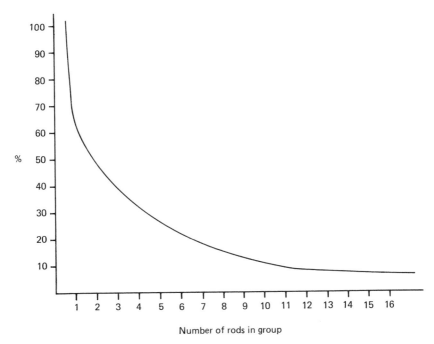

Number of rods in group

Fig. 7.6 Total resistance of multiple earth rods expressed as a percentage of a single rod.

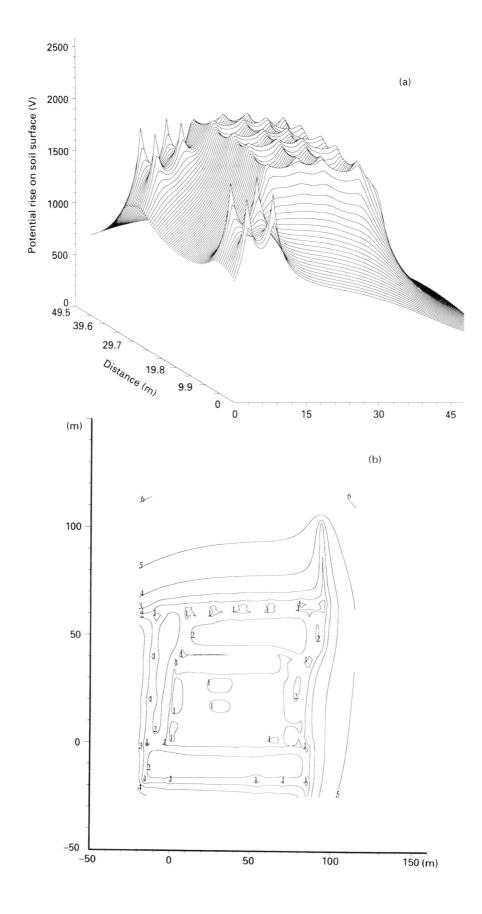

two dimensions, as in Fig. 7.7b. The two-dimensional facility is often used to produce the 430 V or 650 V hot zone contours required by telecommunication authorities.

Artificial method of reducing earth resistivity

Low resistivity materials

It is widely thought that adding low resistivity materials to the soil in the immediate vicinity of the earthing system or rod will have a dramatic effect on reducing its resistance. This is not normally true. There is always a contact resistance between the electrode and the soil. Where a new rod has been driven into the ground, the sideways movement will have increased the width of the hole in which the rod lies. The gap between the rod surface and the compressed soil to its side will introduce a large contact resistance which will be apparent when testing the resistance of the rod. However, experience shows that this resistance falls over time as the soil becomes consolidated around the rod due to rainfall, etc. One way to accelerate this is to add a low resistivity material, such as bentonite slurry, as the rod is driven in. Bentonite is a fine, naturally occurring clay powder. It is mixed with water to form a slurry. As the earth electrode is driven into the soil the bentonite is drawn down by the rod. By continuously pouring the mixture into the hole as driving proceeds a sufficient quantity is dragged down to fill most of the voids around the rod and lowers its resistance. Bentonite is thixotropic in character and therefore gels when in an inert state, so should not leach out. It has a low resistivity and does not dry out under normal conditions. However, its use in this manner does not necessarily produce a significantly lower earth resistance than that which would occur naturally over time. In some cases, installing the rods a little deeper can achieve the same or a better result than using low resistivity material.

Adding bentonite and similar materials, such as marconite, in a trench or larger drilled hole around the electrode has the effect of increasing the surface area of the earth conductor, assuming its resistance is lower than that of the surrounding soil. Figure 7.8 shows the effect on the resistance value of one rod when its radius (and hence surface area) is increased. Clearly there is an improvement, but this falls off rapidly with increasing radius and cost. Additional reasons for increasing the surface area may include a requirement to reduce the build up of electrochemical deposits on the electrodes which may reduce their efficiency.

Chemical treatment

Treating the soil surrounding the earth electrode with chemicals normally only provides a temporary improvement as the chemicals will usually be dispersed over time by rainwater. The environmental effects may also mitigate against this undesirable practice. In addition, one must ensure that the chemicals used do not have any corrosive effect on the electrode material.

Fig. 7.7 (a) Potential rise on the soil surface above an earth grid and terminal tower; (b) touch voltage contours above a buried substation earth grid.

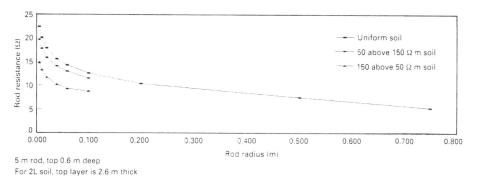

5 m rod, top 0.6 m deep
For 2L soil, top layer is 2.6 m thick

Fig. 7.8 Curves of rod resistance with increasing rod radius.

One interesting technique used in rocky sites in Hungary was to drill a hole several metres deep into the rock. Explosives were introduced and when detonated, formed a cavity and fissures. Low resistivity material was then introduced into the cavity and fissures and the earth electrode installed within it.

Marconite, concrete or similar materials are also now increasingly used around steel reinforcing structures where these are to be used as part of the earthing system.

RESISTANCE MEASUREMENT OF ELECTRODE SYSTEMS

Usually the same instrument that is used for soil resistivity surveys can be used for measuring the resistance of an earth electrode system. The most common technique is the 'fall of potential' test. It should be noted that safety precautions are necessary whilst carrying out this test if the earth system is connected to live equipment. For example, fused leads and an appropriate isolation routine are necessary. This is to ensure that an earth fault occurring during the test does not subject the individual handling the test probes to a transferred potential.

A potential and current terminal on the tester are connected together to the earth electrode under test. The remaining current terminal is connected to a current probe which is inserted into the ground some distance away. The remaining potential terminal is connected to a voltage probe which is situated 61.8% of the distance between the current probe and the rod (or centre of the earth grid). Current is circulated between the electrode under test and the current probe, via the ground. From the voltage detected at the 61.8% position, the apparent resistance is displayed on the tester. Under ideal conditions, this apparent resistance reading corresponds to the actual resistance value of the earth electrode. The reason for choosing this position is based on mathematical theory applied to uniform soil. However, the two probes must be sufficiently far away from the electrode under test so that they do not interact. For earth rods the current lead would normally be about 100 m long and the voltage lead 61.8 m or longer, but maintaining the same ratio.

The most common cause of error arises from placing the current electrode too close to the electrode under test. The effects of the two electrodes overlap and the measured resistance will give a lower value than that of the real one. The second most common mistake is placing the voltage probe too close to the test electrode, which normally produces a much lower reading than the true value.

The theory (and hence the 61.8% rule) does not hold if the soil is not uniform, the grid is large or (as stated above) the current electrode is too close. Computer simulation can again assist here, by predicting the distance from the grid at which the real resistance will be measured.

The fall of potential method can be used on larger grids, but it is suggested that the current electrode is positioned at a minimum distance of ten times that of the diagonal distance across the grid. This is not normally practical, so several derivatives of the fall of potential method have been developed. These include the slope method (where the gradients between adjacent measurement points are computed) and the intersecting curves method. In another derivative of the test, the voltage probe is moved out at right angles to the grid/current probe traverse. The distance of the probe from the grid is progressively increased until the resistance figure barely changes. This figure should then be just below the actual grid resistance.

There must be no metallic underground pipes or cables running in the same direction as the test route as this will again produce an incorrect earth reading.

Where the earth grid is large or has long radial connections, e.g. to cables or tower line earth wires, the resulting size of the earth grid is so large that the fall of potential is impractical. Measurement is then made via current injection techniques, which require specialist equipment.

TYPES OF EARTH ELECTRODES

Earth electrodes are made from a number of materials which include cast iron, steel copper and stainless steel. They may be in the form of plates, tubes, rods or strips. The most favoured material is copper. It has good conductivity, is corrosion resistant to many of the salts that exist in the soil and it is a material that is easily worked. Earth plates are normally of cast iron or copper, typically one metre or more square, with surface ridges to increase the surface area in contact with the soil. They have good current-carrying properties, but as they are often buried at a relatively shallow depth, their resistance tends to be high, thus increasing the number needed.

An alternative method, which is particularly suitable for hard ground plus a shallow layer of top soil, is to install lengths of copper tape in pre-dug trenches. This method is used both in power installations and in communication networks where it has a wider application.

The most versatile type of earth electrode is the driven rod. This may be of copper, copper clad steel, galvanized steel or austenitic iron. Standard sizes are 1.0−2.5 m in length and 16 mm diameter. The rods are extensible and are joined together by either an internal stud or an external sleeve.

While copper rods are an excellent material for earth electrodes, they do not have sufficient strength or hardness to enable them to be driven to great depths. If copper is used the rods are normally inserted in pre-drilled holes to overcome this problem. Copper clad steel rods are more suitable for driving and will penetrate most types of soil to a depth of 10–15 m. It is important when using copper clad steel rods, to ensure that the copper is actually bonded to the steel. If the copper covering is only extruded onto the steel core, then it has no adhesion, and damage may occur during installation.

Criticism has been made of the external sleeve referred to above. When this has a larger diameter than that of the rod it increases the size of the driven hole. This increases the difficulty of installation and makes the initial earth resistance higher. The latter may be overcome by time or by driving the rods through a puddle of bentonite thus assisting consolidation in the hole. This will become obvious as installation progresses.

Another method to alleviate this problem of contact with the ground is by using a stranded bare copper cable which is dragged into the ground giving very good

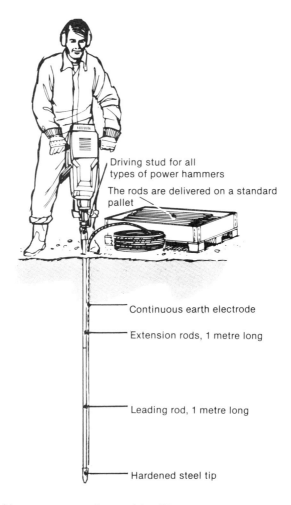

Driving stud for all types of power hammers

The rods are delivered on a standard pallet

Continuous earth electrode

Extension rods, 1 metre long

Leading rod, 1 metre long

Hardened steel tip

Fig. 7.9 Deep earthing system manufactured by *Elpress*.

contact as 'the slot' does not enlarge. The cable is fitted in a driving head and a leading steel tube holds it in position. This is driven into the ground and extension rods are added as required. The conductor is continuous and may be taken direct to a busbar without introducing any joints. The method of installation is shown in Fig. 7.9

There are examples of more 'exotic' earth rod designs used for special applications. For example a cylindrical copper/alloy casing which contains hydroscopic chemicals to attract water to the vicinity of the rod.

INSTALLATION

Some form of power hammer is needed to install impact driven rods. Electric, petrol, hydraulic oil, or air type driven tools have all been used successfully as shown in Figs 7.10 and 7.11. Air tools are the lightest, but require a large compressor to operate them. A petrol driven road breaker, with a special head to go over the end of the rod is extremely efficient for this work. It is entirely self-contained and can be used in cramped conditions.

While hammers can be hand-held if used for a short space of time, it is more usual for them to be used in conjunction with a rig which will take their weight whilst they are hammering the rod. On compacted ground installation can take some time, especially after the electrode has been driven some 10 m or so. Rigs vary from simple tripods to more elaborate vehicle-mounted models. A rig which can be easily assembled and dismantled is the most practical to use, as it is

Fig. 7.10 New method showing a copper-weld earth rod driven by hydraulic 'mole' Grundomat (*T T. UK Ltd*).

Fig. 7.11 Standard method showing copper-clad earth rod driven by a compressed air tool with compressor, as used by *Stemet Earthing Co. Ltd*.

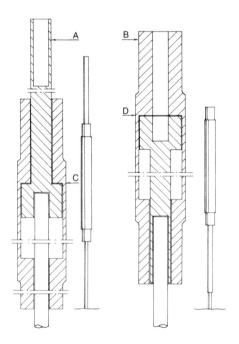

Fig. 7.12 Reversible pole hammer (*Elpress*).

often impossible to manoeuvre a vehicle to the position in which an earth electrode has to be installed.

Site conditions dictate the layout of the earth electrode system, but it has been found that many earthing problems can be solved by impact driven earth rods. A hand-held pole hammer may suffice on small installations where high resistance values are adequate, Fig. 7.12. A note of caution however: make a thorough investigation of structures or utility equipment in the area before installing vertical rods.

STANDARDS APPLICABLE TO EARTHING PRACTICE

Some of the main authoritative documents setting out the principles, rules and guidance are listed at the end of the handbook. In addition to these there are industry standards and codes of practice. In particular, those applicable to the quarrying, mining, petrochemical or chemical industries should be referred to. The Health and Safety Executive can normally assist in providing details of the most up to date, specific practices required.

CHAPTER 8
Cathodic Protection

J.D. Thirkettle, MICorr, MNACE

G. Clapp, MA, MSc

Revised by J.D. Thirkettle, MICorr, MNACE
(ACEL Group Ltd)

INTRODUCTION

Cathodic protection is an electrochemical process for the prevention of corrosion of steelwork and other metallic structures which are buried and immersed. Typical structures to which cathodic protection is applied include: pipelines, tank farms, plant pipework, steel work well casings, sheet and tubular pile systems, jetties, piers, offshore drilling rigs, production platforms, and semi-submersibles.

Standards

Cathodic protection in the UK is generally carried out in accordance with the appropriate British Standards. These are BS 7361, *Cathodic protection*: Part 1, *Code of practice for level and marine applications* and BS 5493, *Code of practice for the coating of iron and steel structures against corrosion*. In addition to the British Standards, practical theoretical and technical information is available from the Institute of Corrosion (UK) and the National Association of Corrosion Engineers (USA). Documents including state of the art reports and recommended practices are available for either general or particular reference with respect to cathodic protection technology and practices. Some large organizations and cathodic protection specialists have internal standards which take precedence over the standards for specific projects.

Theory of corrosion

Corrosion of a metal in contact with soil or water is fundamentally an electro-chemical process. The corrosion process involves anodic areas (anodes) at which metal is converted to positive ions and cathodic areas (cathodes) at which an oxidizing species, usually oxygen, is reduced. An electric current flows from the anode to the cathode through the electrolyte and the species formed at the anodic and cathodic areas combine to form the corrosion product. The circuit is completed by current flow within the metal between the cathodic areas and the anodic areas.

A particular area of metal is determined as anodic or cathodic by the reactions taking place on it. The following must be present for corrosion to occur: anode, cathode and electrolyte.

Electrochemical and galvanic series

When two different metals are in metallic contact and in the same electrolyte, in general, one metal will have a greater tendency to go into solution than the other. The electrochemical and galvanic series are attempts to make this predictable. The two series are the potentials adopted by different metals in an electrolyte. The electrochemical series comprises the potentials adopted by metals when in a solution of their own ions, thus it does not have much practical application. The galvanic series comprises the potentials adopted by metals when in sea water or the ground, thus this corresponds to a real situation.

It is only possible to measure potential differences; absolute values of potential cannot be obtained. In order to standardize potential measurements a particular potential value is taken to be zero. One such standard is the potential of the standard hydrogen electrode.

When two metals are in metallic contact in the same electrolyte the metal with the more negative potential in the galvanic series will be made more anodic and suffer accelerated corrosion; the other metal will become more cathodic and have reduced corrosion. Tables 8.1 and 8.2 show the electrochemical and galvanic series of some metals of interest in corrosion control and cathodic protection work.

The two series do not indicate the magnitude of the change in corrosion rate which will occur when two metals are connected together in the same electrolyte, but only the direction of any change.

Table 8.1 Electrochemical series of some metals.

Metal	Voltage
Magnesium	−2.37
Aluminium	−1.66
Zinc	−0.76
Iron	−0.44
Tin	−0.14
Lead	−0.13
Hydrogen	0.00
Copper	+0.34 to +0.52
Silver	+0.80
Platinum	+1.20
Gold	+1.50 to +1.68

Table 8.2 Galvanic series of some metals.

Metal	Voltage*
Commercially pure magnesium	−1.75
Magnesium alloy (6% Al, 3% Zn, 0.15% Mn)	−1.60
Zinc	−1.10
Aluminium alloy (5% Zn)	−1.05
Commercially pure aluminium	−0.80
Mild steel (clean and shiny)	−0.5 to −0.8
Mild steel (rusted)	−0.2 to −0.5
Cast iron (not graphitized)	−0.5
Mild steel in concrete	−0.5
Copper, brass, bronze	−0.2
High silicon cast iron	−0.2
Mill scale on steel	−0.2
Carbon, graphite, coke	+0.3

* Typical potential, normally observed in neutral soils and water, measured
   with respect to a copper/copper sulphate reference electrode.

Corrosion cell

There are two basic cell arrangements in which corrosion takes place, known as
galvanic and electrolytic cells.

Galvanic cell

A galvanic cell comprises anode, cathode and electrolyte, where the corrosion
current flows spontaneously. The anode is negative with respect to the cathode.
Galvanic cells exist not only when dissimilar metals are in contact but may be
caused by one or more of the following conditions: dissimilar surface conditions;
differences in heat treatment; differences in local stresses within metal; differences
in concentrations of ionic species; differences in oxygen concentrations; differences
in soil conditions. Examples where galvanic cells are created by dissimilar surface
conditions and differential aeration are shown in the Fig. 8.1.

Electrolytic cell

An electrolytic cell comprises anode, cathode and electrolyte, where the corrosion
current flows as a result of the application of an external electromotive force. The
external potential causes current to flow into the electrolyte from the anode and
back to the power source through the cathode. The anode is made positive with
respect to the cathode by the externally applied e.m.f.

This reversal of polarity of the external terminals of an electrolyte cell, as
compared with a galvanic cell is a source of confusion; this confusion can be
avoided if it is remembered that in all types of electrochemical cells, current
always enters the electrolyte from the anode and leaves the electrolyte at the
cathode.

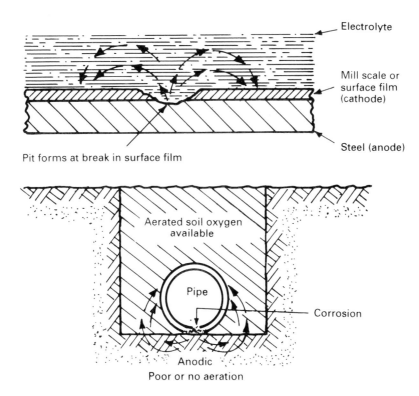

Fig. 8.1 Galvanic corrosion.

PRINCIPLES OF CATHODIC PROTECTION

In all corrosion cells, corrosion normally occurs only at anodes. There are, however, exceptions; these include copper and aluminium which are amphoteric. If an entire structure buried in the soil, or immersed in water, is made the cathode of an electrochemical cell so that the entire surface receives current from the electrolyte, the structure will not corrode. Under these conditions, the structure is said to be cathodically protected.

The cell, of which the protected structure is the cathode, may be either galvanic, with current entering the electrolyte from galvanic anodes more electronegative than the protected structure, or, electrolytic with the anode electrically connected to the positive terminal of an external power source. In any cathodic protection system, the structure to be protected is made more negative with respect to the surrounding medium so that it picks up current from the electrolyte, and this current is conducted through an external circuit to an anode system where it enters the electrolyte. The principles of protection by the galvanic (sacrificial) and electrolytic (power impressed) system of cathodic protection are illustrated in Figs 8.2 and 8.3.

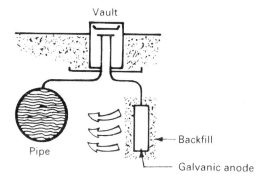

Fig. 8.2 Sacrificial cathodic protection system.

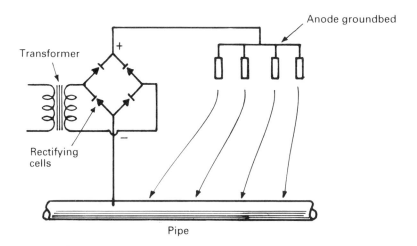

Fig. 8.3 Power impressed cathodic protection system.

Mechanism of cathodic protection

To understand cathodic protection one needs to consider both general and electrical concepts.

General concept

On the surface of any corroding structure there are cells consisting of anodic and cathodic areas. Current is discharged from the metal surface to the electrolyte at the anodic areas and returned to the metal structure in the cathodic areas. Qualitatively, it can be seen that when an external current is impressed on the structure, the current from an external anode tends to oppose the current leaving the structure at the anodic areas. If sufficient current is forced on to the structure to oppose completely the current leaving the surface at all anodic areas, the structure is cathodically protected and corrosion prevented.

In practice, a structure is deemed to be protected when a minimum structure/electrolyte potential of 850 mV (referred to copper/copper sulphate) or a minimum potential shift of 300 mV negative is achieved at all points on the structure. Where anaerobic conditions exist a minimum potential of 950 mV must be achieved. Potential levels vary however, dependent upon where the potential is measured.

Electrical concept

In order to understand more fully the mechanism of cathodic protection, consider an anode and a cathode immersed in a solution and connected in such a manner that the potentials and currents involved in the corrosion reaction can be measured. Such a condition exists in the circuit indicated in Fig. 8.4.

The potentials of the anode and cathode exist when no corrosion current flows. These values can be measured directly against a reference electrode. If switch S_1 is closed and the corrosion current I_a is varied, the anode and cathode potentials (E_a and E_c) vary accordingly with the flow of corrosion current, due to polarization of the electrodes. The polarization voltage always acts in the direction opposing the flow of current.

The steady-state corrosion current is that current which flows when the external resistance (R_1) is reduced to zero; this occurs when the sum of the polarization potential drops at the anode and cathode equals the open circuit potential difference between the anode and cathode. Under these conditions, the polarized anode and cathode potential are equal.

If switch S_2 is closed, protective current will be applied to the system through

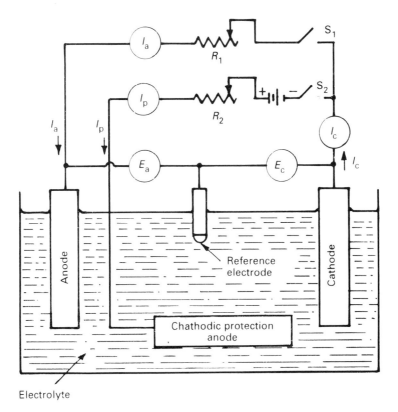

Fig. 8.4 Electrical concepts of cathodic protection.

the cathodic protection anode. As the protective current is increased, the corrosion current I_a will gradually decrease until it finally reaches zero. In the freely corroding condition, the cathodic current I_c is equal to the anode current I_a. When a small protective current is applied, the cathode current is increased, but the anode current is decreased; under these conditions, the cathode current equals the sum of the anode current I_a and the protective current I_p. When complete protection has just been achieved, I_a is equal to zero and the protective current I_p just equals the cathode current. This condition exists when the cathode potential is made equal to the anode open-circuit potential, i.e. cathodic protection is adequate when the cathode has been polarized to the open-circuit potential of the anode.

CATHODIC PROTECTION SYSTEMS

There are two types of cathodic protection systems which are used; these are sacrificial and power impressed systems.

Sacrificial cathodic protection

Sacrificial cathodic protection involves the use of a galvanic cell so that corrosion takes place away from the structure being protected. This is done by connecting the structure to be protected to a sacrificial anode. The sacrificial anode material is a metal which is more active in the galvanic series, i.e. it adopts a more negative potential than the metal of the structure to be protected.

Power impressed cathodic protection

Power impressed cathodic protection is an application of the electrolytic cell. An external source of e.m.f. is used to make the structure to be protected entirely cathodic. The current which must be passed through the electrolyte on to the structure comes from an electrode which is placed in the electrolyte. This electrode is the anode or groundbed.

From the viewpoint of the structure, the sacrificial and power impressed cathodic protection systems are very similar. In both cases current flows on to the surface of the structure from the electrolyte and this current has the effect of making the structure more cathodic and thus stopping the corrosion on the structure. There are, however, differences between the two types of system which mean that under different circumstances one or other method of application of cathodic protection is preferable.

Measurements

In practical terms the performance and effectiveness of a cathodic protection system are determined by electrical measurements. These generally comprise voltage and current measurement in the case of power impressed systems, current in the case of sacrificial anodes and in both cases structure/electrolyte potentials.

It is to be noted, however, that voltage/current measurements cannot be determined where sacrificial anodes are attached directly to immersed steelwork. In these cases the effectiveness of the cathodic protection system can only be quantified by structure/electrolyte potential measurement.

Structure/electrolyte potential measurements are required to determine the effectiveness of cathodic protection. They are necessary because in the field it is not possible to determine by current measurements when the current due to the anodic processes is totally suppressed by the cathodic protection current. Potential measurements are taken utilizing a reference electrode, the potential of which is constant, and which is robust enough for field use. The potentials measured may be converted to the Standard Hydrogen Electrode Scale. Reference electrodes in common use are as follows: land-based structures – copper/copper sulphate; marine/undersea structures – silver/silver chloride; other electrode types are calomel and zinc. Potential measurements are recorded by placing the reference electrode into the electrolyte close to the structure and connected to the structure via a high resistance voltmeter. When potentials are thus recorded the reference electrode that has been used should always be stated.

Isolation

The design of a cathodic protection system will always be based on the surface area that is required to be protected. It is essential, therefore, that only the surface area to be protected should be connected to the cathodic protection system.

The isolation of the structure may be obtained by the use of electrically insulated joints in the case of pipelines, and by other insulating materials. Certain installation practices may provide 'shorts' across otherwise insulated joints; examples include metal conduit and fittings, cable tray, non-insulating pipe supports, and earthing systems. These should be avoided. If the structure is not isolated electrically from all other buried or immersed metallic structures the cathodic protection system suffers from reduced effectiveness.

Earthing systems

When considering the provision of cathodic protection systems special attention should be given to the earthing system that will be installed. Such systems often comprise bare copper wire which is connected to the structure at numerous points. Apart from the risk that the earth system may electrically connect the protected structure to other buried structures it may itself constitute an increase in surface area and receive cathodic protection current.

INSTALLATION PRACTICE

Cathodic protection systems are always installed according to a particular specification. The installation work is of an electrical nature. The major items of equipment which will be installed are described in the following sections.

Sacrificial anode systems

Sacrificial anode systems represent the simplest form of cathodic protection system, and involve only the connection of sacrificial anodes to the structure and the placing of these anodes close to the structure.

Directly fitted sacrificial anodes

In certain cases the sacrificial anodes are directly connected to the structure. (Particular applications include offshore platforms, drilling rigs, semi-submersibles, etc.) The anode has a steel core which is welded or bolted to the structure. It is important to ensure satisfactory electrical contact with the structure. The anodes may be zinc, aluminium or magnesium alloy. The shape and size of the anodes depend on the structure to be protected. Thus a tubular structure (e.g., a submarine pipeline), being externally protected, may have bracelet type anodes. Anodes may be flush mounting or they may stand-off from the structure. Anodes will be flush mounting when streamlining is of importance, for example on ship hulls. Where streamlining is of less importance stand-off anodes are preferred since they provide a better distribution of cathodic protection. The anode material and quantities selected depend upon surface areas, coating quality and electrolyte resistivity.

Indirectly connected sacrificial anodes

Anodes for pipelines, pipework, tanks and other structures are cast generally from magnesium although in specific instances zinc or aluminium may be used. The anodes are cast in rod form with a cable attached to facilitate connection to the structure. Such anodes are usually buried alongside the structure and it is usual for the anodes to be prepacked in a cotton sack containing specially prepared backfill of bentonite, gypsum and sodium sulphate. The purpose of this backfill is to provide a homogeneous, rapid wetting, low-resistance electrolyte in which the anode can be activated.

The cable is routed from a brazed connection in the anode through the top of the cotton sack and is accessible for connection purposes. The anode(s) is laid in the soil close to the structure and connected to the structure by the cable. Anodes should be installed at the same depth as the structure and no closer than 1.5 m (ref. BS 7361). No other metallic structure should pass between the anode and the structure.

The cable from the anode is always taken to a test facility (see p. 186) and two cables taken from the test facility to the structure unless otherwise specified. One cable is for the anode current and the other for test purposes. Cables should be connected to the structure in an approved manner (see p. 187). It is normal for such anodes to be installed at numerous locations and connected to the structure to ensure full and effective protection.

Power impressed systems

The installation of power impressed systems is more complex but fewer are generally required. The configurations of power impressed systems are numerous and it is not practical to detail each possible variation. The following sections therefore describe the more common installation methods in use.

Marine applications

Anodes for protecting jetties, platforms, vessels, etc. are generally of platinized titanium, niobium or mixed metal oxides; other materials may also be used such as lead-silver alloys. They may take the form of rod or 'shield' fabrications and are usually supplied by manufacturers for installation on to or close to the structure. Shield anodes are either bolted or welded to the structure and rod anodes attached to the structure as a cantilever or suspended in the electrolyte. Cables are taken from the anodes via a distribution cable network and connected to the positive terminal of the transformer-rectifier.

A distribution system is necessary as a number of anodes are connected in this way to various locations throughout the structure. It is also usual for anodes at different locations to be fitted with cables of varying cross-sectional areas to compensate for the voltage drop on differing lengths.

In these networks distribution boxes and resistances are generally used. The function of the resistor is to provide adjustment in the amount of current supplied to each anode during operation. It is important when such structures are protected to ensure electrical continuity between all parts where protection is required. This is often inherent in the design, e.g., welded structures, and is accomplished during construction.

A cable capable of carrying the total return current of the system is connected between the structure and the negative terminal of the transformer-rectifier. In many cases these installations are required in potentially explosive areas and equipment must be approved by the relevant authorities such as British Approvals Service for Electrical Equipment in Flammable Atmospheres (BASEEFA).

The precise method and location of cable fixing and routing and distribution/junction box positions are invariably specified.

Onshore applications

Power impressed anodes for protection of onshore pipelines, pipework, tanks and other structures are generally manufactured from high silicon iron, graphite, magnetite or mixed metal oxides. In some cases short-life anodes may be constructed from scrap steel. A number of anodes are installed in a single position known as a groundbed. In order to increase the current capacity of a groundbed anodes are installed in a carbonaceous backfill generally of coke or calcined petroleum coke granules.

A structure may be protected by a single groundbed or a number of distributed groundbeds. The groundbeds should be located at positions electrically remote from the structure, generally some 50–100 m away.

A groundbed may be installed in a shallow horizontal trench, or vertically in a shallow borehole, or a single deepwell borehole arrangement. The precise location of the groundbed is specified together with the manner of its installation. In the case of a horizontal groundbed a trench is excavated for the anodes. The anodes are laid on a bed of carbonaceous backfill and covered with the same. They are provided with cable tails which are routed to a feeder cable which may be a single feeder or part of a ring main. Cable connections between the anode cable tails and the feeder cable are achieved using cable splicing kits and line tap connectors. The cable is buried, or laid along with other cables in ducting or cable tray and connected to the positive terminal of the transformer-rectifier.

It is usual for the transformer-rectifier to be located close to the structure to minimize d.c. cable lengths and facilitate adjustment and measurements.

All buried cathodic protection cables are installed at least 1 m deep and are protected by tiles, conduit or warning tape. Groundbeds are generally installed such that the anodes are between 1.3 and 1.5 m deep. It is important to ensure adequate depth of cable as the ground may be subject to further disturbance for agricultural or constructional reasons.

Test points

It is usual on structures which are buried for considerable lengths, e.g., pipelines, to provide test points. These are located at pre-determined intervals along the length of the structure. They comprise a cable connected to the structure and brought to an accessible point on the surface. The cable is then terminated and housed in a test box. Methods of termination are specified and include crimped lugs, sweated lugs or a glanded box with terminals.

EQUIPMENT

The equipment used for cathodic protection installation work is manufactured by specialists. Their major features are described.

Transformer-rectifiers

A transformer-rectifier comprises a mains transformer and a full-wave bridge rectifier. The rectifier elements are generally silicon diodes. The assembly may be oil or air cooled and housed in a steel or glass fibre case. It is fitted with a voltmeter and ammeter, step or continuous d.c. output controllers and equipped with fuses, oil-cooled unit breathers and other ancillary items. A transformer-rectifier is manufactured for wall, plinth or pole mounting and designed to operate continuously.

Generally transformer-rectifier units provide a fixed constant current output controlled by manually set switches or autotransformers. In certain circumstances a variable output automatic unit with thyristor control may be installed.

Anodes

Cathodic protection anodes, with the exception of platinized titanium and niobium are manufactured by a casting process and are generally rod shaped. The weight of an anode varies according to design but typically it would weigh approximately 50 kg. The cast anodes are brittle and great care should be taken when storing, transporting or installing them. When installed the anodes are in intimate contact with the electrolyte or backfill and cables are routed away from the anode for connection purposes. Cathodic protection cables should not be routed over the top of anodes, and the length of cable in carbonaceous backfill should be minimized.

Cable

Cable for cathodic protection purposes is usually single core and is insulated and sheathed (with armouring where necessary) with chemical resistant materials, e.g., cross-linked polyethylene. Cable is sized not only for electrical capacity but also for mechanical strength and therefore large diameter cable may be used in certain instances. Cables are generally laid in continuous lengths, but where cable joints are necessary they usually take the form of crimped or line tap splice joints which should be housed in a cable splicing kit filled with epoxy resin.

Carbonaceous backfill

Carbonaceous backfill is manufactured as coke breeze or calcined fluid petroleum coke. The material comprises chips of coke which must conform to sizing as specified. A maximum amount of 'fines' is permitted by specification and it may sometimes be mixed with slaked lime (usually 10% by weight). Carbonaceous backfill must be carefully installed around the anodes and tamped lightly so that satisfactory electrical contact is achieved. The calcined fluid petroleum coke is a round grained material.

Distribution boxes

Distribution boxes are manufactured of steel or glass fibre, and where located in flameproof zones, must meet BASEEFA requirements. Each box is equipped with terminals, resistance control and other facilities as required by specification.

Cable connections to structures

Several methods are used to attach the d.c. negative or test cables to the structure. A summary of the common types of cable attachment methods is indicated.

Thermit welding
Cables are attached directly to the structure using thermit welding techniques. This system is not favoured by all users as damage to high-strength materials may occur. Alternative brazing and soldering systems are now available for this purpose.

Cable attachment plates
A steel plate approximately 50 mm square is attached to the structure. The plate may be welded or alternatively affixed with an electrically conductive epoxy resin. A bolt is welded to the plate to accept a crimped or sweated lug cable connection.

Collar/bracelet connections
In the case of pipelines or pipework a collar or bracelet may be attached to the pipe to facilitate a cable connection.

Test stations

Test stations may be of concrete (in the form of a location designation post), glass fibre, steel, cast iron or plastics materials. They contain the necessary terminals, resistors and other items of equipment.

Test equipment

Test equipment needed includes voltmeters, which should have a high internal resistance, multimeters with voltage, current and resistance measuring facilities, low-resistance earth testers, reference electrodes, test cable and hand tools and associated equipment. Other more specialist equipment may be required in specific instances.

The quality of measuring equipment for cathodic protection purposes has shown significant advance in recent years. Digital instruments and microprocessor-based and computer-assisted measurement systems are now in general use. The advent of such instrumentation has allowed the cost effective collection of vast amounts of data which may be used to assess the operating performance of the system. It is likely that further advances will be made in this respect as measurement systems become more sophisticated, accurate and portable.

MONITORING, INSPECTION AND MAINTENANCE

Every installed cathodic protection system requires monitoring and maintenance. Monitoring will generally take the form of routine voltage, current and potential measurements to ensure the satisfactory performance of the system.

Specialized inspection procedures include close interval potential surveys, and current density and voltage gradient surveys to analyse applied cathodic protection performance. External coating systems are evaluated by Bearson and current attenuation surveys. The purpose of all monitoring systems is to provide the owner with detailed information with respect to the level of applied cathodic protection. Any deviation from required criteria or other defects that are determined are generally corrected under a maintenance programme. Maintenance will also include the physical and electrical inspection of installed equipment with remedial measures taken as appropriate to maintain the system in perfect working order. The importance of properly executed monitoring and maintenance schemes for cathodic protection systems is generally regarded as paramount by most owners and operators as an integral part of their corrosion control programme.

INTERACTION

Interaction is the presence of stray earth or other external d.c. or a.c. currents which may affect the performance of installed cathodic protection systems. The installation of a cathodic protection system itself, particularly of the power impressed type, may cause corrosion interaction on neighbouring or adjacent structures. The effect of such interaction may be to increase the corrosion rate thus leading to shorter service life. All newly installed cathodic protection systems should therefore be subject to interaction notification and testing. Where tests show unacceptable levels of interaction safeguard measures should be taken to correct the situation. Where cathodic protection systems are installed close to low-voltage d.c. circuits (e.g., railway track signalling systems) serious disruption of circuit balance and operation can occur.

Stray currents leading to accelerated corrosion effects may also be detected as a result of natural geomagnetic effects. In all cases a test programme is required so that appropriate mitigating action can be taken.

Reference should be made to British Standard Code of Practice 7361: Part 1 for interaction testing and other requirements regarding installation.

RECENT DEVELOPMENTS

The application of cathodic protection systems as a corrosion prevention measure is being extended to structures hitherto not considered appropriate. Significant areas of development have included the protection of steel in concrete and metallic structures which are not continuously in contact with an electrolyte but operate in corrosive conditions. Protection of steel in concrete is now widely used to ensure structural integrity whereby cathodic protection impressed currents are applied to reinforcing steel within concrete where chlorides are likely to be present. Anode systems generally comprise titanium or mixed metal oxide mesh or electrically conductive paint/coating systems laid on the external concrete surface so that current may flow uniformly to the reinforcing bars within the structure. Testing and monitoring are carried out via fixed reference electrodes installed at representative sites throughout the protected structure. These may be connected permanently to a centralized data logger or alternatively used with portable measurement equipment. The application of cathodic protection to steel within concrete is of a specialist nature and does not accord with the criteria set out in this chapter. Further information is available from The Society for the Cathodic Protection of Reinforced Concrete.

Cathodic protection is also applied to metallic structures which are intermittently in contact with an electrolyte of corrosive or potentially corrosive nature. This includes above-ground structures subject to wetting by either natural electrolytes (e.g., rain or salt water) or artificial electrolytes from process manufacturing. Impressed current anodes are attached to the structure via insulated pads and current is applied when the electrical resistance between the anode and cathode allows these currents to flow. This form of cathodic protection is not subject to the testing and monitoring procedures previously mentioned but is inspected on a frequent and regular basis to ensure continued integrity.

It is expected that cathodic protection will continue to find new applications with the development of new materials and methods to ensure technical and fiscal viability. In all cases, however, installation of cathodic protection equipment remains an electrical discipline.

CONCLUSIONS

Cathodic protection is a specialist technology used for corrosion control, but the installation of cathodic protection equipment conforms to the appropriate engineering standards. Specialist advice should be sought, however, where no clear or adequate specification exists.

REFERENCES

Further information with respect to cathodic protection is available from the following organizations:

Institute of Corrosion
PO Box 253,
Leighton Buzzard,
Bedfordshire,
LU7 7WB.

National Association of Corrosion Engineers,
PO Box 47,
Godalming,
Surrey,
GU7 1TD.

Society for the Cathodic Protection of Reinforced Concrete (SCPRC),
PO Box 72,
Leighton Buzzard,
Bedfordshire,
LU7 7EJ.

CHAPTER 9

Lightning Protection

J. Sherlock
(Technical Manager, W.J. Furse & Co. Ltd)

P. Woods
(Marketing Manager, W.J. Furse & Co. Ltd)

INTRODUCTION

In recent years, the subject of lightning protection has developed in two directions:

(1) protecting structures; and
(2) protecting electronic systems.

It will be appreciated that, although the topics are related, the sensitivity of the two elements to lightning strikes differs considerably.

It is essential that there should be no confusion as to the precautions that need to be taken for each condition. For this reason this chapter has been divided into two stand-alone parts, which in some instances duplicate important data.

PART 1. PROTECTION OF STRUCTURES

Characteristics of lightning

Lightning strokes are the visible discharge of static electricity accumulated in storm clouds created by meteorological conditions. Strokes may occur within the cloud, between clouds, or from the cloud base to earth or earthed structures.

Lightning is formed as a result of a natural build-up of electrical charge separation in storm clouds. The base of a storm cloud is commonly 5–10 km above the earth's surface, with the cloud 12 km high. The charge at the base of the cloud is usually negative and induces an equal and opposite charge on the earth's surface and earthed objects beneath the cloud. Buildings of masonry, concrete and timber are sufficiently conductive to reach the same potential as the earth's surface.

The earth's atmosphere contains drifting pockets of ionized air, which also take on a charge of opposite polarity to the cloud base, and the presence of such a pocket near the cloud can create a potential difference sufficient to cause electrical breakdown of the air and a downward leader stroke develops from the cloud, having a current of a few hundred amperes. A step progression can then develop from pocket to pocket of the leader stroke towards the earth in a few microseconds until the tip of the leader is within a distance of 50–100 m from some point on the earth. At this stage the potential difference is sufficient to initiate an upward discharge or streamer from the earth or some object on it (see Fig. 9.1). The distance at which this occurs is the striking distance of the stroke. The downward stepped leader and upward streamer rapidly advance until they meet, forming a continuous ionized path from cloud to earth, initiating the full visible discharge.

The fast rise time and large peak amplitude of the stroke can produce severe mechanical effects. Long-duration currents can cause fire, while short-duration high-current peaks tend to tear or bend metal parts when they make contact. A typical current amplitude against time waveform is illustrated in Fig. 9.2.

Current magnitudes can range from several kA up to 200 kA. Fortunately super bolts of over 100 kA are rare, the statistical average being 20 kA.

The core of the discharge may reach a temperature of 30 000 K causing explosive expansion of the air and the typical sound of thunder. The resulting shock wave close to this stroke readily dislodges tiles from roofs.

The 'rolling sphere' concept

Given the lightning process already described, it is logical to assume that a

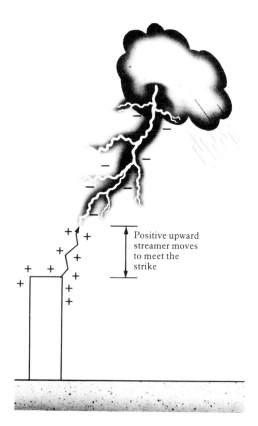

Positive upward
streamer moves
to meet the
strike

Fig. 9.1 Development of the downward stepped leader and upward streamer.

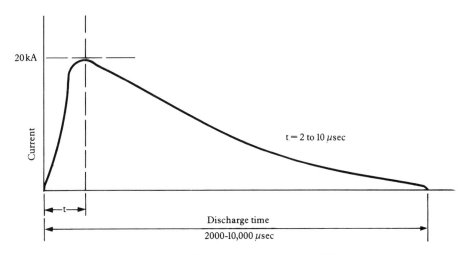

Fig. 9.2 Current amplitude against discharge time of a typical lightning strike.

lightning strike terminates on the ground (or on structures) at the point where the upward streamer was originally launched.

These streamers are launched at points of greatest electric field intensity and can move in any direction towards the approaching downward leader. It is for this reason that lightning can strike the side of tall structures rather than at their highest point.

The position of the greatest field intensity on the ground and on structures will be at those points nearest to the end of the downward leader prior to the last step. The distance of the last step is termed the striking distance and is determined by the amplitude of the lightning current. For example, points on a structure equidistant from the last step of the downward leader are equally likely to receive a lightning strike, whereas points further away are less likely to be struck (Fig. 9.3). This striking distance can be represented by a sphere with a radius equal to the striking distance.

Lightning frequency

Lightning is nothing more than a long spark. However, it is estimated that about 2000 storms exist at any one time in the world, hurling 30 to 100 flashes to the ground every second. If these estimates are correct, then each year over 3 billion lightning bolts bombard the earth.

STRIKE PROBABILITY

The route of a downward leader, the striking distance and the point where an upward streamer starts are all unpredictable, but hilly or mountainous areas naturally provide a shorter discharge path, as do tall, isolated structures. There is, therefore, a reasonable probability that taller structures are struck more often

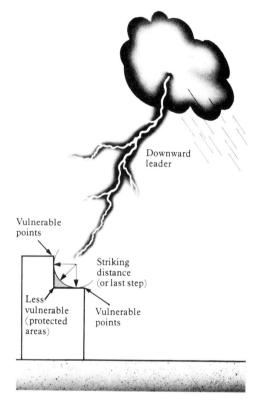

Fig. 9.3 Striking distance (last step).

than low structures. If a structure is sufficiently conductive in itself, or is provided with a low-resistance conducting path to earth, and is strong enough at the point of strike to withstand the impact of the expanding air, the lightning is discharged without damage to the structure.

Those structures that are non-conductive will require the installation of lightning conductors. A lightning conductor is incapable of discharging a thundercloud without a lightning flash. Its function is to divert to itself a lightning discharge which might otherwise strike a vulnerable part of the structure to be protected and to convey the current safely to earth. The British Standards Institution has published a code of practice, BS 6651:1985, *Protection of structures against lightning*. This code gives sound practical advice on how to protect a structure from lightning.

The need for protection

Before proceeding to design a detailed lightning protection system, first carefully consider if the structure needs protection.

In many cases, it is obvious that some form of protection is required. High-risk structures, i.e. explosives factories, oil refineries, etc., will require the highest possible class of lightning protection to be provided. In many other cases the need for protection is not so evident.

BS 6651 provides a simple mathematical overall risk factor analysis for assessing whether a structure needs protection. It suggests that an acceptable lightning strike risk factor is 10^{-5} per year, i.e. 1 in 100 000 per year. Therefore, having applied the mathematical analysis to a particular set of parameters, the scheme designer will achieve a numerical solution. If the risk factor is less than 10^{-5} (1 in 100 000), for example 10^{-6} (1 in 1 000 000), then in the absence of other overriding considerations, protection is deemed unnecessary. If however, the risk factor is greater than 10^{-5}, for example 10^{-4}, then protection would seem necessary.

A suitable analogy could be made with the odds in horse racing. The shorter the odds (e.g. 5:1), the more likely it is that the horse will win the race. The longer the odds (e.g. 100:1), the less likely it is that the horse will win.

The shorter the risk factor (e.g. 1 in 10 000) the greater the risk that a structure will be hit by lightning. The longer the risk factor (e.g. 1 in 1 000 000) the less likely it is that the structure will be hit by lightning.

It is acknowledged that certain factors cannot be assessed in this way and these may override all other considerations. For example, if there is a requirement that there should be no avoidable risk to life or a requirement for occupants of a building always to feel safe, then this will favour the installation of protection, even though it would normally be accepted that protection is unnecessary.

The factors which should be considered for determining an overall risk factor can be summarized as follows.

The geographical location of the structure

This pinpoints the average lightning flash density or the number of flashes to ground per km^2 per year. For structures sited within the UK this figure can be taken from the map in Fig. 9.4.

The effective collection area of the structure

This is the plan area, projected in all directions taking account of the structure's height. The significance is that the larger the structure, the more likely it is to be struck (see Fig. 9.5).

For a simple rectangular box shaped structure as illustrated in Fig. 9.5, the collection area (symbol A_c) is simply the product of:

$$A_c = LW + 2LH + 2WH + \pi H^2,$$

where: L = length.
W = width
H = height.

The probable number of strikes to the structure per year (Symbol P) is the product of the flash density and the collection area, thus

$$P = A_c \times N_g \times 10^{-6},$$

where N_g = the number of flashes to ground per cm^2 per year. (The number 10^{-6} is included because A_c is in m^2 whereas N_g is per km^2.)

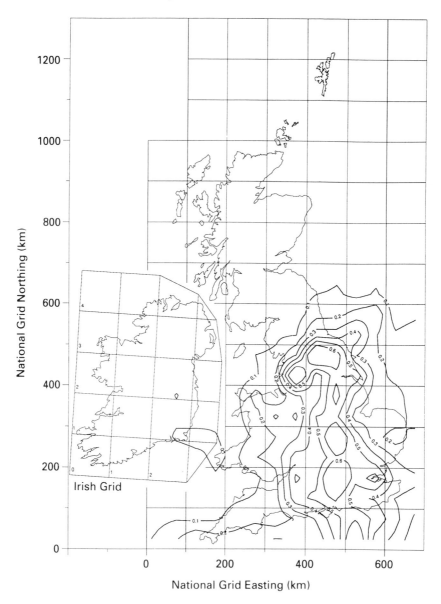

This lightning density map was compiled by Electrical Association Technology Ltd from data accumulated over four years from its Lightning Location System.

Fig. 9.4 Number of lightning flashes to the ground per km^2 per year for the UK.

General arrangement   Collection area and method of calculation

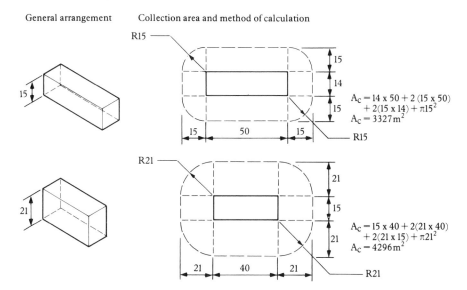

Fig. 9.5 Details of structures and collection areas.

The intended use of the structure
Is it a factory or an office block, a church or perhaps a hospital?

The type of construction
Is it built of brick or concrete? Does it have a steel frame or a reinforced concrete frame? Does it have a metal roof?

The contents of the structure
Does it contain valuable paintings, or a telephone exchange with important equipment, or is it maybe an old people's home?

The location of the structure
Is it located in a large town or forest or on an isolated hillside?

The topography of the country
Is the structure located in generally flat countryside or in a mountainous area?

The last five factors can be interpreted from Tables 9.1−9.5 and are termed 'the weighting factor values'. They denote a relative degree of importance in each case. These tables are taken from BS 6651.

A sample calculation of overall risk factor

A school building in the east of England with brick walls and a metal roof, and located close to similar types of buildings in a wooded area. The dimensions of the school are:

Table 9.1 Weighting factor A (use of structure).

Use to which structure is put	Value of factor A
Houses and other buildings of comparable size	0.3
Houses and other buildings of comparable size with outside aerial	0.7
Factories, workshops and laboratories	1.0
Office blocks, hotels, blocks of flats and other residential buildings other than those included below	1.2
Places of assembly, e.g. churches, halls, theatres, museums, exhibitions, department stores, post offices, stations, airports and stadium structures	1.3
Schools, hospitals, children's and other homes	1.7

Table 9.2 Weighting factor B (type of construction).

Type of construction	Value of factor B
Steel frame encased with any roof other than metal*	0.2
Reinforced concrete with any roof other than metal	0.4
Steel frame encased or reinforced concrete with metal roof	0.8
Brick, plain concrete or masonry with any roof other than metal or thatch	1.0
Timber framed or clad with any roof other than metal or thatch	1.4
Brick, plain concrete, masonry, timber framed but with metal roofing	1.7
Any building with a thatched roof	2.0

* A structure of exposed metal which is continuous down to ground level is excluded from the table as it requires no lightning protection beyond adequate earthing arrangements.

$L = 60\,\text{m}$
$W = 60\,\text{m}$
$H = 15\,\text{m}.$

The risk factor can be calculated as follows using the map in Fig. 9.4 and Tables 9.1–9.5.

Table 9.3 Weighting factor C (contents or consequential effects).

Contents or consequential effects	Value of factor C
Ordinary domestic or office buildings, factories and workshops not containing valuable or specially susceptible contents	0.3
Industrial and agricultural buildings with specially susceptible* contents	0.8
Power stations, gas installations, telephone exchanges, radio stations	1.0
Key industrial plants, ancient monuments and historic buildings, museums, art galleries or other buildings with specially valuable contents	1.3
Schools, hospitals, children's and other homes, places of assembly	1.7

* This means specially valuable plant or materials vulnerable to fire or the results of fire.

Table 9.4 Weighting factor D (degree of isolation).

Degree of isolation	Value of factor D
Structure located in a large area of structures or trees of the same or greater height, e.g. in a large town or forest	0.4
Structure located in an area with few other structures or trees of similar height	1.0
Structure completely isolated or exceeding at least twice the height of surrounding structures of trees	2.0

Table 9.5 Weighting factor E (type of country).

Type of country	Value of factor E
Flat country at any level	0.3
Hill country	1.0
Mountain country between 300 m and 900 m	1.3
Mountain country above 900 m	1.7

- Determine the number of flashes to ground per km^2 per year, N_g. From the map in Fig. 9.4 this is 0.2.
- Determine the collection area, A_c, of the school building.

$$A_c = LW + 2LH + 2WH + \pi H^2$$
$$= (60 \times 60) + 2(60 \times 15) + 2(60 \times 15) + \pi \times 15^2$$
$$= 3600 + 1800 + 1800 + 707$$
$$= 7907\,m^2.$$

- Determine the probability of being struck, P.

$$P = A_c \times N_g \times 10^{-6}$$
$$= 7907 \times 0.2 \times 10^{-6}$$
$$= 1.5814 \times 10^{-3}\ (or\ 0.0015814).$$

- Applying the relevant weighting factors from Tables 9.1–9.5:

$A = 1.7$
$B = 1.7$
$C = 1.7$
$D = 1.0$
$E = 0.3$

The overall weighting factor $= A \times B \times C \times D \times E = 1.4739$.
- Therefore, the overall risk factor

$$= \text{probability of being struck} \times \text{overall weighting factor}$$
$$= 1.58 \times 10^{-3} \times 1.4739$$
$$= 2.33 \times 10^{-3}\ (or\ 0.00233)$$

Or expressing this answer as a reciprocal we obtain 1 in 429.

As 10^{-5} (1 in 100 000) is the criteria for determining whether protection is necessary, we can see that 1 in 429 is 'shorter odds' and so protection is necessary.

Having determined that lightning protection is necessary, we must now consider the actual design of the installation. To do this we must understand the principles of the zone of protection.

Zones of protection

BS 6651 qualifies, in detail, the meaning of 'zones of protection' and the 'protective angle'. It is sufficient to say that the 'zone of protection' is simply that volume within which a lightning conductor provides protection against a direct lightning strike by attracting the strike itself (Figs 9.6–9.8).

As can be seen in Fig. 9.6, structures below 20 m are regarded as offering a 45° protection angle. For structures greater than 20 m in height (Fig. 9.7), the protection angle of any installed lightning protection conductors up to the height of 20 m would be similar to the structures in Fig. 9.6. For tall structures above 20 m in height (Fig. 9.8), BS 6651 recommends the use of the rolling sphere for determining the areas where lightning protection may be advisable.

Lightning protection design

There are available proprietary devices of air terminations where claims are made that they ionize the air, which in thunderstorm conditions could initiate the

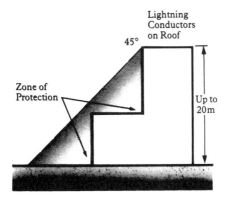

Fig. 9.6 Zones of protection.

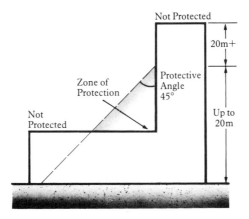

Fig. 9.7 Zones of protection.

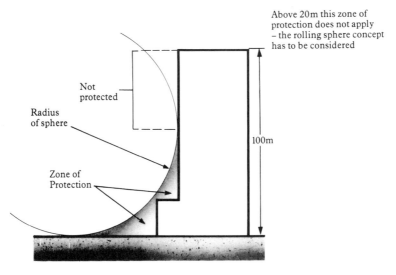

Fig. 9.8 Zones of protection.

upward streamer and, in effect, attract the downward leader, hence determining the point of strike.

The foreword of BS 6651 carefully words strict adherence to the geometry of conventional lightning protection systems throughout its pages – to the total exclusion of any other device or system which claims to provide enhanced protection. This is based on research that has been unable to substantiate the enhanced performance these devices or systems are claimed to give.

This view is supported by the draft IEC document on lightning protection currently being compiled, and by technical papers given at the 18th International Conference on Lightning Protection 1985 held in Munich.

If it has been established that a structure requires lightning protection, certain general design features need to be considered.

Could, for instance, any of the metallic components in or on the structure be incorporated into the lightning protection scheme? Could the metal in and on the roof be used? Should window cleaning rails, window frames and handrails surrounding the structure be incorporated in the protection network? The reinforcing bars or the steel frame of a structure may well provide a natural conductive path within the lightning protection system.

If metallic components in a building are not used, then the structure will require externally fitted conductors. A lightning protection system can incorporate all natural conductors, all externally fitted conductors, or a combination of both. BS 6651 does not, however, recommend the routeing of conductors inside the structure.

The principal components of a lightning protection system comprise the following:

(1) An air termination network to collect the lightning discharge.
(2) A down conductor, or series of down conductors, which provide a safe path from the air termination network to the earth.
(3) An earth termination, which dissipates the charge into the earth.
(4) Bonds and clamps, which connect all of the three segments together and prevent side flashing.

Air termination networks

The simplest form of air termination is the air rod, or terminal projecting above the structure, invented by Benjamin Franklin. It is intended to be a point from which an upward streamer can start, and should be sufficiently robust to intercept the full stroke current. It is most applicable to a tall structure of small cross section, such as a church spire, but a number of rods, correctly spaced and interconnected, protects a greater cross section.

The air rod is not essential to the design of air terminations, and horizontal conductors of copper or aluminium, forming a ring round the top of a structure or developed as a grid network on larger areas, are equally effective and reflect current practice.

As stated earlier it is now accepted that lightning can strike the upper part of tall structures. BS 6651 introduces the concept of air termination networks on all sides of tall buildings (i.e. vertical air termination networks). No part of the roof within the air termination network should be more than 5 m from a conductor.

For large flat roofs, this will be achieved typically by a network mesh of 10 m × 20 m. For high risk structures, e.g. explosive factories, the air termination mesh is reduced to 5 m × 10 m.

If a building's metal reinforcing bars are to be used as down conductors, these should be connected to the air termination network as shown in Fig. 9.9.

BS 6651 advises the use of a rolling sphere to determine zones of protection. To minimize the likelihood of a lightning strike damaging the side of a building, it is suggested that the rolling sphere method should be applied to identify those areas where an extension of the air termination network should be considered. Where the sphere touches the structure determines the extent of the air termination network. The recent amendments to BS 6651 advocate the use of a 20 m radius of sphere for all high risk structure categories and 60 m for all other types of structure.

Where structures vary in height and have more than one roof termination network, the lower roof network should not only be joined to its down conductors, but also joined to the down conductors of the taller portions of the structure. This will ensure that a lightning strike to a lower portion of the structure will not lead to side-flashing to other 'remote' down conductors and will provide a multi down conductor path for the lightning current to disperse.

Down conductors

The function of a down conductor is to provide a low impedance path from the air termination network to the earth termination network, to allow the lightning current to be safely conducted to earth.

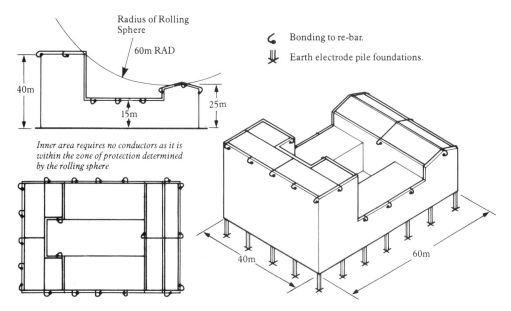

Fig. 9.9 Lightning protection scheme to BS 6651 using the reinforced concrete within the structure for down conductors.

BS 6651 advocates the use of various types of down conductors. A combination of strip and rod conductors, reinforcing bars, structural steel stanchions, etc., can be used as all or part of the down conductor system – providing they are appropriately connected to the air and earth termination networks and are known to offer good electrical conductivity.

External down conductors are usually copper or aluminium in tape or circular solid form.

They should, where possible, take the most direct route from the air termination network to the earth termination network. Ideally they should be symmetrically installed around the outside walls of the structure starting from the corners. Routeing to avoid side-flashing should always be given particular attention in designing any installation.

Down conductors should be positioned no more than 20 m apart around the perimeter at roof or ground level, whichever is the greater. If the structure is over 20 m in height, then the spacing is reduced to every 10 m or part thereof.

Sharp bends in down conductors at the edge of the roofs are unavoidable and are permitted in BS 6651; however, re-entrant loops in a conductor can produce high inductive voltage drops which could lead to the lightning discharge jumping across the side of the loop. To minimize this problem, BS 6651 recommends that the length of the conductor forming the loop should not exceed eight times the width of the open side of the loop (Fig. 9.10).

Down conductors must never be routed down lift shafts. If an external route is not available, an internal continuous, non-conducting, non-flammable duct may be used.

Each external down conductor must incorporate a test clamp installed approximately 3 m from the ground. This enables the continuity of the system to be checked as well as isolating each local earth network for testing purposes. The section of down conductor from the ground to the test clamp should have a non-metallic guard to protect from unauthorized interference. To minimize surface

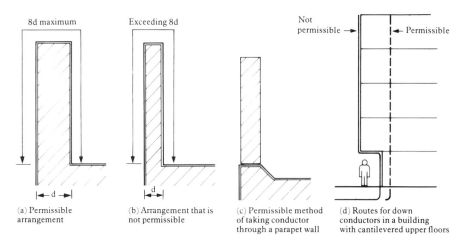

(a) Permissible arrangement

(b) Arrangement that is not permissible

(c) Permissible method of taking conductor through a parapet wall

(d) Routes for down conductors in a building with cantilevered upper floors

Fig. 9.10 Re-entrant loops.

voltage gradients the down conductor should ideally be insulated for the first metre of entry into the soil.

Earth termination networks

Each down conductor should have its own local earth electrode. Copper conductor should be used as the connection from the test clamp to the earth electrode. Copper-bonded steel-cored rods are by far the most popular form of earth electrode. These may be driven to considerable depths. Alternatively shallower driven rods interconnected by copper conductor can be used.

Aluminium and copper, the two metals most commonly used in lightning protection systems, are not compatible, so great care must be taken when both are used in a system – particularly where they come into contact with each other.

If aluminium is selected as the material for air termination networks and down conductors, it has to be converted to copper at or around the test clamp. This is because both BS 6651 and the Earthing Code BS CP 1013 do not permit aluminium to be buried underground due to the likelihood of corrosion.

A simple and effective means of joining the aluminium and copper conductors is with a friction-welded bi-metal clamp. This termination, if used in conjunction with an inhibitor grease, minimizes the effect of corrosion.

The contact surfaces of dissimilar metals should be kept completely dry and protected against the ingress of moisture, otherwise corrosion will occur. A particularly effective means of excluding moisture is to use inhibitor pastes, bitumastic paint, or approved protective wrappings.

As aluminium is prone to corrosion when in contact with Portland cement and mortar mixes, aluminium conductors need to be fixed away from the offending surface with an appropriate fixing.

There are two stages in testing an earth network for satisfactory resistance:

(1) With the test link removed from the down conductor and without any bonding to other services, etc., the earth resistance of each individual earth electrode should be measured. The resistance, in ohms, should not exceed 10 times the number of down conductors on the structure. For example, if there are 15 down conductors equally spaced around a building, then the resistance of each electrode with the test link removed should not exceed $10 \times 15 = 150\,\Omega$.

(2) With the test links replaced, the resistance to earth of the complete lightning protection system is measured at any point on the system. The reading from this test should not exceed $10\,\Omega$. This is still without any bonding to other services.

BS 6651 provides a guide to the minimum dimensional requirements of various electrode systems. For example, where earth rods are chosen, the minimum combined rod length to complete an earth electrode system should be 9 m – therefore a small structure with only two down conductors would have a minimum requirement of 4.5 m for each electrode. Each local earth rod electrode should be a minimum length of 1.5 m.

On rock foundations with shallow earth covering, it may be necessary to drill vertical holes, usually 75–100 mm in diameter, and using earth rod electrodes, fill the drilled hole with a suitable soil conditioning agent such as bentonite or conductive concrete.

For further recommendations on earthing see BS CP 1013: 1965 *Earthing*, and Chapter 7 in this book.

There is an increasing trend, encouraged by BS 6651, to use the foundation reinforcement of a structure as the earth electrode system whenever possible. The depths of the foundations usually ensure that there is more than sufficient electrodes available to satisfy the test criteria. Care must be taken to ensure that there is electrical continuity between the reinforcement foundations and the down conductors. Purpose-made mechanical clamps are available to suit this application.

Bonds

All metalwork on or around a structure must be bonded to the lightning protection system if side-flashing is to be avoided. When a lightning protection system is struck, its electrical potential with respect to earth is significantly raised, and unless suitable precautions are taken, the discharge may seek alternative paths to earth by side-flashing to other metalwork in or on the structure.

Typically, water pipes, gas pipes, metal sheaths and electrical installations which are in contact with earth, remain at earth potential during a lightning discharge. Even metal parts that are not in contact with earth will see a potential difference between them and the lightning protection system during a discharge, even if this potential is smaller in magnitude to the metal parts in direct contact with earth.

All exposed metalwork should be bonded into the lightning protection installation (Fig. 9.11).

There are two ways of preventing side-flashing. To isolate nearby metal from the lightning protection system, or if that is not possible, to connect the metalwork to the lightning protection system with an appropriate bond.

To determine whether the distance between the suspect metalwork and the lightning protection system is large enough for the metalwork to be considered 'isolated' or close enough to be 'bonded', BS 6651 provides a mathematical means of determining the minimum isolation distance for a given set of parameters.

Internal bonds need only be half the cross-sectional area of external bonds, as they are, at most, only likely to carry a proportion of the total lightning current.

The general view held by both the technical committee of BS 6651 and BS CP 1013 is the need to bond the lightning protection scheme to all metal services entering or leaving a structure. This connection, which includes the electrical power earth, should normally be made at the main earth terminal point. Permission should be sought from the appropriate authority before such a connection is made.

Material specification

The correct choice of material for a lightning protection system and the installation method adopted should ensure a satisfactory life span of at least 30 years.

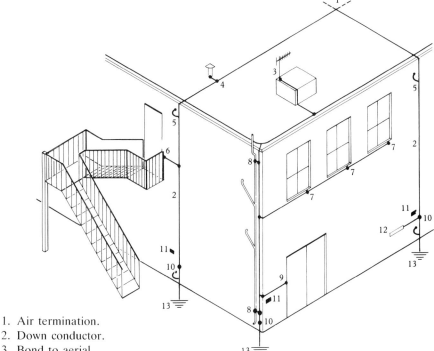

1. Air termination.
2. Down conductor.
3. Bond to aerial.
4. Bond to vent.
5. Bond to re-bar.
6. Bond to metal staircase.
7. Bond to metal window frame.
8. Bond to vent pipe.
9. Bond to steel door/frame.
10. Test clamp.
11. Indicating plate.
12. Main earthing terminal of electrical installation.
13. Earth termination point.

Fig. 9.11 Bonding to prevent side-flashing.

To ensure an effective system and a satisfactory long-term performance all fittings need to be mechanically robust and provide good corrosion resistance qualities in widely differing environments.

In addition, the system should provide a low electrical resistance path to earth and have the ability to carry high currents repeatedly over the lifetime of the installation.

Table 9.6 recommends the materials to be used for the manufacture of lightning protection components and Table 9.7 recommends the minimum dimensions of component parts. Both tables are taken from BS 6651: 1985.

Although the tables permit the use of galvanized conductors, their popularity, particularly in the UK, is very restricted. It is more common in Europe, particularly West Germany.

Table 9.6 Recommended materials for the manufacture of lightning protection components.

Materials and processes	BS No.	Grade or type
Ingots for cast components		
Leaded gunmetal	BS 1400	LG1, LG2
Aluminium silicon bronze	BS 1400	AB1, AB3
Aluminium alloy	BS 1490	LM6M
Cast iron	BS 1452	220, 260
Malleable iron	BS 6681	–
Aluminium alloy	BS 1490	LM25
Forgings and stampings (hot or cold formed)		
Copper	BS 2872	C101, C102
Naval brass	BS 2872	CZ112
Aluminium	BS 1474	6082TF
Steel	BS 970: Part 1	All grades
Pressings and fabrication (from strip, coil, foil and sheet)		
Annealed copper	BS 2879	C101, C102
Aluminium	BS 1474	6082TF
Aluminium	BS 1470	6082TF
Naval brass	BS 2870	CZ112
Stainless steel	BS 1449: Part 2	316S12, 325S21
Steel (for galvanizing)	BS 4360	43A
Steel (for galvanizing)	BS 1449: Part 1	–
Bars, rods and tubes (for machined components and fittings)		
Hard drawn copper	BS 2874	C101, C102
Annealed copper	BS 2874	C101, C102
Copper cadmium	BS 2874	C108
Copper silicon	BS 2874	CS101
Phosphor bronze	BS 2874	PB102-M
Aluminium bronze	BS 2871: Part 3	CA102
Aluminium	BS 1471	6082TF
Naval brass	BS 2874	CZ112
Aluminium	BS 1474	6082TF
Steel (general use)	BS 970	All grades
Steel (for galvanizing)	BS 4360	43A
Stainless steel (general use)	BS 970: Part 1	325S21
Stainless steel (austenitic)	BS 970: Part 1	316S12

Table 9.6 (*Continued*)

Materials and processes	BS No.	Grade or type
Nuts, bolts, washers, screws and rivet fixings for use on copper		
Phosphor bronze	BS 2874	PB102-M
Naval brass	BS 2874	CZ112, CZ132
Copper silicon	BS 2874	CS101
Copper rivets	BS 2873	C101
Nuts, bolts, washers, screws and rivet fixings for use on aluminium		
Aluminium alloy	BS 1473	6082
Stainless steel	BS 970: Part 1	325S21
Galvanized steel, galvanized to BS 729	BS 970: Part 1	220M07
Solid rounds, flats and stranded conductors		
Copper		
Annealed copper	BS 1432	C101, C102
Hard drawn copper	BS 1433	C101, C102
Copper (stranded)	BS 6360	Insulated
Copper (flexible)	BS 6360	–
Hard drawn copper strand and copper cadmium	BS 125 or BS 2755	–
Aluminium		
Aluminium strip/rod	BS 2897	1350
Aluminium strip/rod	BS 2898	1350, 6101A
Aluminium	BS 3988	–
Aluminium (steel reinforced)	BS 215: Part 2	–
Aluminium alloy	BS 3242	–
Aluminium	BS 215: Part 1	–
Steel		
Galvanized steel	BS 302	–
Galvanized strip (see note 1)	BS 1449: Part 1	–

Notes
(1) The recommended finish is galvanized in accordance with BS 729, which has to be done after manufacture or fabrication.
(2) Stainless steel in contact with aluminium or aluminium alloys is likely to cause additional corrosion to the latter materials (see PD 6484). In these cases, it is important to take protective measures, e.g. the use of inhibitors.
(3) This table applies to finials, adornments and projections such as crosses or weather-vanes on churches which are used as part of the air termination network.

Table 9.7   Minimum dimensions of component parts.

Component	Dimensions (mm)	Area (mm^2)
Air terminations:		
aluminium, copper and galvanized steel strip	20 × 2.5	50.0
aluminium, aluminium alloy, copper,		
phosphor bronze and galvanized steel rods	8.0 dia.	50.0
Suspended conductors:		
stranded aluminium	7/3.0	50.0
stranded copper	19/1.8	50.0
stranded aluminium (steel reinforced)	7/3.0	50.0
stranded galvanised steel	7/3.0	50.0
Down conductors:		
aluminium, copper and galvanized steel strip	20 × 2.5	50.0
aluminium, aluminium alloy, copper and		
galvanized steel rods	8.0 dia.	50.0
Earth terminations:		
austenitic iron	14.0 dia.	153.0
copper and galvanized steel strip	20 × 2.5	50.0
copper and galvanized steel rods	8.0 dia.	50.0
hard drawn copper rods for direct		
driving into soft ground	8.0 dia.	50.0
hard drawn or annealed copper		
rods or solid wires for indirect		
driving or laying in ground	8.0 dia.	50.0
rods for hard ground	12.0 dia.	113.0
copper-clad or galvanized steel rods (see		
notes to table) for harder ground	14.0 dia.	153.0
Fixed connections (bonds) in aluminium,		
aluminium alloy, copper and galvanized steel:		
external strip	20 × 2.5	50.0
external rods	8.0 dia.	50.0
internal strip	20 × 1.5	30.0
internal rods	6.5 dia.	33.0
Flexible or laminated connections (bonds):		
external, aluminium	20 × 2.5	50.0
external, annealed copper	20 × 2.5	50.0
internal, aluminium	20 × 1.5	30.0
internal, annealed copper	20 × 1.5	30.0

Notes
(1) For copper-clad steel rods, the core should be of low carbon steel with a tensile strength of approximately 600 N/mm^2 and of a quality not less than grade 43A of BS 4360. The cladding should be of 99.9% pure electrolytic copper molecularly bonded to the steel core. **The radial thickness of the copper should be not less than 0.25 mm**.
(2) Couplings for copper-clad steel rods should be made from silicon bronze alloy, grade CS101 of BS 2874, or aluminium bronze alloy, grade CA102 of BS 2871.

(3) The use of internal phosphor bronze dowels may give a lower resistance than the external couplings of diameter greater than the rod.

(4) For galvanized steel rods, steel of grade 43A specified in BS 4360 should be used, the threads being cut before hot-dip galvanizing to BS 729.

(5) Stranded conductors are not normally used for down conductors or earths.

(6) Greater dimensions are required for the following:
    (a) structures exceeding 20 m in height;
    (b) special classes of structure;
    (c) mechanical or corrosive reasons.

As mentioned previously, copper and aluminium in solid circular or tape form are the preferred types of conductor. Their particular choice is usually governed by factors such as overall comparative cost, the corrosive nature of the environment, and a requirement to blend in with the colour of the structure. Mainly to assist with the latter requirement, there are PVC-covered conductors available in varying colours. These can be used, and are equally effective as the bare conductor, as air termination networks and as down conductors.

Lightning protection design

In order to provide a lightning protection scheme for a structure certain parameters need to be available to the designer. These can be summarized as follows:

(1) Drawings of the structure requiring protection, showing the roof plan and at least two elevations. These drawings should be clear, precise and have the scale shown.

(2) The materials used in the construction of the structure should be stated, along with information on the type of fixings permissible (e.g. can the roof be drilled to take screw plugs).

(3) For what purpose the structure is being used (its use will determine the risk category of the structure).

(4) The proximity of other structures, trees and the general locality.

(5) Information regarding any unusual features, such as aerial masts on the roof of buildings, which may not be shown on the drawings.

(6) At what stage of construction is the structure (i.e. complete, partly built, etc.).

(7) Notification of code that the scheme is to be designed to e.g. BS 6651: 1985.

(8) Is there any soil resistivity data available?

INSTALLATION OF LIGHTNING PROTECTION

On new buildings and structures it should be possible for the installation of the lightning protection to progress with the building work. This can give cost reduction in the provision of scaffolding and in eliminating the need to break through masonry or concrete when all the civil work is complete.

Down conductors should follow the most direct path possible between the air termination and the earth termination. Right-angle bends when necessary are permissible, but deep re-entrant loops should be avoided.

The magnetic forces induced by a lightning strike on a conductor try to straighten a right-angled bend. It is impossible not to have such bends in a system, but where possible they should be avoided. The radius of such bends should not be less than 200 mm and should be firmly anchored at either side. A well-formed bend only stuck down by felt pads onto a bitumen roof would probably not hold in the event of a strike.

The earth rod electrode should be installed close to its down conductor wherever possible. This offers a direct path for the lightning discharge in the event of a strike.

Modern buildings have roofs and wall structures constructed of metal cladding. Provided they are of adequate thickness (guidance is given in BS 6651) and are electrically continuous, then they can be used as, or form part of the lightning protection system.

Supporting stanchions of steel-framed buildings can be utilized as the down conductors. They should be adequately earthed and correctly bonded to the roof network. This type of building normally contains many supporting stanchions in parallel, each of which has a very large cross-sectional area. The resulting impedance seen by the lightning strike is, therefore, low and the danger of side-flashing is greatly reduced.

When a structure is being erected, all large and prominent masses of steelwork should be effectively connected to earth. This applies to all metal items including steel framework, scaffolding and on-site cranes.

Due to its unique nature, the installation of lightning protection schemes should preferably by carried out by a company specializing in this branch of electrical engineering.

There is a federation within the industry – The National Federation of Master Steeplejack and Lightning Conductor Engineers (NFMSLCE) – that offers this service.

INSPECTION AND TESTING OF A SYSTEM

All parts of a lightning protection system should be visually inspected on a regular basis. This is particularly relevant after alterations or extensions to the structure, to ensure that the system still complies with the recommendations of BS 6651. Particular attention should be paid to the mechanical condition of all conductors and fittings.

Each local earth electrode should be tested in accordance with CP 1013: 1965 (see Chapter 7 for more details) to ensure that it still complies with the requirements of BS 6651. Tests should be repeated at fixed intervals, preferably not more than 12 months. Ideally, testing every 11 months (or 13 months) would vary the season in which tests are made, which would give a more accurate reflection of the earth electrode resistance.

PART 2. PROTECTING ELECTRONIC SYSTEMS FROM LIGHTNING

As well as posing a threat to the building itself, lightning also threatens damage to the building's electronic contents. Lightning activity, anything up to a kilometre away, can cause transient overvoltages capable of destroying electronic equipment. To counter this threat, additional and complementary protection measures are required. A number of measures can be taken at the project design stage to mitigate the threat, though these will need to be supported by suitable transient overvoltage (or 'surge') protectors.

Transient overvoltages caused by lightning

A transient overvoltage is a very short duration increase in voltage measured between two or more conductors. Although only lasting from microseconds to a few milliseconds in length, this increase in voltage may be of several thousand volts.

Lightning can cause transient overvoltages through direct strikes to incoming electrical services or, most commonly, through 'indirect' strikes which are coupled into electrical services through resistive, inductive and capacitive effects.

Direct strikes to h.v. power cables
Strikes to high voltage (h.v.) overhead power lines are quite common. It is often thought that the high voltage to low voltage transformer action eliminates transient overvoltages. This is not so. Although transformers protect against transient overvoltages between line and earth, line to line transients pass through unattenuated.

When h.v. lines are struck by lightning they flashover to earth. One line will flashover before the others, converting a line to earth transient into one between line and line – these will easily pass through the transformer.

Also, capacitance between the transformer's windings provides transients between any combination of conductors with a high frequency path through the transformer. This could have the effect of increasing the size of existing line to line transients, as well as providing a path through the transformer for line to earth transients.

Direct strikes to l.v. power or telephone lines
When lightning hits low voltage (l.v.) overhead power or telephone lines, most of the current travels to earth as a result of line flashover to ground. A relatively small (but devastating) portion of the lightning current is transmitted along the line to electronic equipment.

Resistive coupling
This is the most common cause of transient overvoltages and it will affect both underground and overhead lines. Resistively coupled transients occur when a lightning strike raises the electrical potential of one or more of a group of electrically interconnected buildings (or structures).

Common examples of electrical interconnections are:

- power feeds from substation to building;
- building to building power feeds;
- power supplies from the building to external lighting, CCTV or security equipment;
- telephone lines from the exchange to the building;
- between building telephone lines;
- between building LAN's or data communication lines;
- signal or power lines from a building to external or field based sensors.

Figure 9.12 shows two buildings. Each contains electronic equipment, which is connected to earth through its mains power supply. A data communication line connects the two pieces of equipment and hence the two separate earths.

A nearby lightning strike will inject a massive current into the ground. The current flows away from the strike point – preferentially through the path of least resistance. The earth electrode, electrical cables and the circuitry of the electronic equipment (once damaged), are all better conductors than soil. As the current attempts to flow, devastating transient overvoltages can be recorded across the sensitive components of the equipment.

Figure 9.13 provides an example of how resistively coupled transient overvoltages can occur on a mains power supply. Resistively coupled transients can occur when separately earthed structures are only metres apart. Resistive coupling will affect both underground and overhead cables.

Inductive coupling

This is a magnetic field transformer effect between lightning and cables.

A lightning discharge is an enormous current flow and whenever a current flows, an electromagnetic field is created around it. If power or data cabling

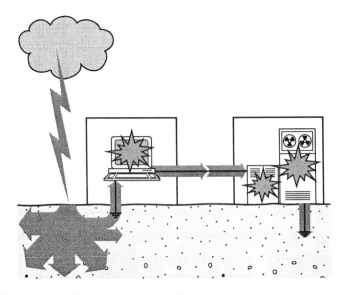

Fig. 9.12 Resistive coupling through a data line.

Fig. 9.13 Resistive coupling through a mains power supply.

passes through this magnetic field, then a voltage will be picked up by, or induced on to it. This frequently occurs when lightning discharges close to overhead power or telephone lines (Fig. 9.14). Much the same thing happens when a building's lightning protection scheme is struck. The lightning current flows to earth through the building's down conductors. The resulting magnetic field may well encroach upon cabling within the building, inducing transient overvoltages on to it.

Capacitive coupling
Where long lines are well isolated from earth (e.g. via transformers or opto-isolators) they can be pulled up to high voltages by capacitance between them and the charged thunder clouds. If the voltage on the line rises beyond the breakdown strength of the devices at each end (e.g. the opto-isolators), they will be damaged.

Size of transient overvoltages caused by lightning
The American Institute of Electrical and Electronic Engineers (IEEE) has collated extensive research into transient overvoltages caused by lightning. The American Standard IEEE C62.41 states:

- a typical worst case of 6000 V for transient overvoltages within a building's power distribution system – its sparkover clearance ratings will ensure that the transient does not generally exceed 6000 V.
- on mains power supplies within a building, secondary lightning currents are unlikely to be more than 3000 A and are certainly no greater than 10 000 A.

Worst case transient overvoltages for data communication, signal and telephone lines are less easy to quantify with certainty. However, we seem to be dealing

Fig. 9.14 Inductive coupling.

with a worst case of 5000 V, or so, and hundreds of amps (based on the test recommendations of the CCITT).

Characteristics of lightning
Lightning has a tendency preferentially to strike taller structures and objects. Strikes to ground are, however, quite common where there is a distance between structures of more than twice their individual height.

For lightning protection purposes an all conducting building, with metal cladding and roof, is an ideal structure. It effectively provides electronic equipment within it with a 'screened room' environment. Many steel framed or reinforced concrete buildings with metal cladding will approximate to this ideal. If lightning strikes the building, a 'sheet' of current will flow all over the surface and down to earth, provided that the cladding and roofing is correctly bonded together. Any small differences in resistance will have little effect on current flow – flow paths are dictated by inductance and not resistance, owing to the fast impulsive nature of the lightning return stroke and restrikes.

Current flows in steel framed or reinforced concrete structures show a similar preference towards external conductors. Figure 9.15 shows that even when lightning strikes the centre of a building's roof, the majority of the current will flow down external conductors rather than the nearer internal conductors. The current flow through the three internal stanchions is relatively small, creating small magnetic fields within the building.

Thus, buildings with large numbers of down conductors around the edge of the

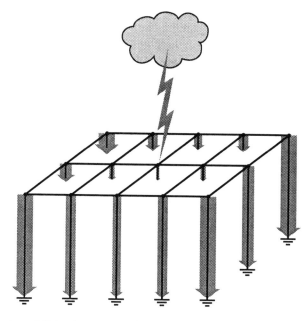

Fig. 9.15 Distribution of lightning current in a 15 stanchion structure.

building will have greatly reduced magnetic fields inside the building, minimizing the risk of transient interference to electronic equipment from the building's lightning protection system.

Problems that transient overvoltages cause

The effects that transient overvoltages have on unprotected equipment depend upon the size of the transient and the sensitivity of equipment. Transient overvoltages manifest themselves as disruption, degradation, damage and downtime. Generally, the larger the transient the worse are its consequences.

Disruption
Although no physical damage is caused, the logic or analogue levels of the systems are upset, causing data loss, data and software corruption, unexplained computer crashes, lock-ups, and the spurious tripping of residual current devices (RCDs). The system can be reset (often just by switching off and on) and will then function normally. Much of this nuisance may go unreported.

Degradation
This is somewhat more serious. Long-term exposure to lower level transient overvoltages will, unknown to the user, degrade electronic components and circuitry, so reducing the equipment's lifetime and increasing the likelihood of failures.

Damage

Large transient overvoltages can cause damage to components, circuit boards and I/O cards. Severe transient overvoltages can manifest themselves through burnt-out circuit boards, destroying equipment. Ordinarily, damage is less spectacular.

Downtime

Unnecessary disruption, component degradation and damage all result in equipment and systems downtime, which means:

- staff unable to work;
- staff overtime;
- senior managers and technical specialists unnecessarily tied-up problem solving;
- lost productivity;
- delays to customers;
- lost business.

Equipment sensitivity

There is no absolute level of vulnerability which can be applied to all pieces of electronic equipment. Much depends on the sensitivity of differing components and circuit designs. Susceptibility levels will therefore vary with different equipment and manufacturers. At present the transient susceptibility of particular pieces of equipment can often only be guessed at.

As a general rule 'few solid state devices can tolerate much more than twice their normal rating' (IEEE 1100−1992, *The Emerald Book*). A single phase 230 V power supply has a maximum normal rating of $230\,V \times \sqrt{2} \times 1.1$ (10% supply tolerance). This gives us a maximum normal rating of 358 V and hence susceptibility level of 700 V or so.

This is, of course, a rule of thumb only, and should be disregarded where equipment has a known susceptibility level. Ethernet systems, although having a signalling voltage of a few volts, have a susceptibility level of many times this. (ECMA 97 details a minimum susceptibility level of 400 V for Ethernet communication ports).

Is protection required?

This depends upon the importance of the electronic systems at the site in question. Consider:

- the cost of replacing damaged equipment;
- the cost of repair work, especially for remote or unmanned installations;
- the cost of lost or destroyed data;
- the financial implications of extended stoppages – sales lost to competitors, lost production, deterioration or spoilage of work in progress;
- potential health and safety hazards caused by plant instability, after loss of control;
- the need to safeguard the operation of fire alarms, security systems, building management systems and other essential services;
- the need to minimize fire risks and electric shock hazards.

Thus, the decision whether and what to protect will be heavily influenced by the associated costs of repair and the criticality of system operation. However, where uncertainty exists a risk assessment may prove helpful. BS 6651: 1992 outlines the basis upon which the risk of lightning disruption to electronic equipment can be calculated. It is important to note that this risk assessment differs markedly from the risk assessment for buildings and structures contained in BS 6651 and pp. 194–200 of this handbook.

The threat to electronic equipment from the secondary effects of lightning depends upon the probable number of lightning strikes to the area of influence, and the vulnerability of the system configuration.

Probable number of lightning strikes

The probable number of lightning strikes is derived by multiplying 'lightning flash density' (the probable number of lightning strikes per square kilometre per year) by the effective collection area. It is given by the formula:

$$P = A_e \times N_g \times 10^{-6}$$

where A_e = total effective collection area (m^2)
N_g = lightning flash density per km^2 per year
10^{-6} = a conversion factor to take account of the fact that N_g and A_e are in different units of area – multiplying square metre values (m^2) by 10^{-6} (1/1 000 000) gives us square kilometres (km^2).

Lightning flash density (N_g)

Details of lightning flash densities are contained in a number of national standards. Figure 9.16 shows a map of lightning flash density for the UK, derived from BS 6651: 1992.

Effective collection area (A_e)

The effective collection are A_e (in m^2) is given by:

$$A_e = A_S + C_{SG} + C_{AS} + C_{MP} + C_{DL}$$

where A_S = the plan area of the structure (m^2)
C_{SG} = the collection area of the surrounding ground (m^2)
C_{AS} = the collection area of adjacent associated structures and their surrounding ground (m^2)
C_{MP} = the collection area of incoming/outgoing mains power supplies (m^2)
C_{DL} = the collection area of data lines leaving the earth reference of the building (m^2).

Thus the overall collection area has a number of constituent elements, which we will consider separately.

Plan area of the structure (A_S). This is often a simple length by width calculation, requiring little further explanation. In practice this is often calculated together with the collection area of the surrounding ground (C_{SG}).

Collection area of the surrounding ground (C_{SG}). Lightning strikes to earth or structures cause large local increases in earth potential. The effect of a nearby

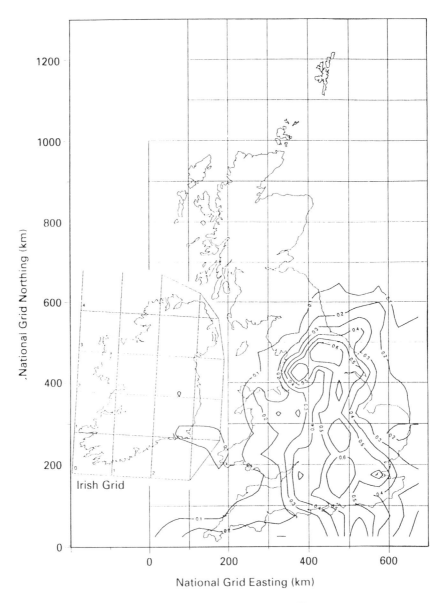

This lightning density map was compiled by Electrical Association
Technology Ltd from data accumulated over four years from its
Lightning Location System.

Fig. 9.16 Number of lightning flashes to the ground per km^2 per year for the UK.

ground strike on a building's earth potential will diminish the further this strike is
from the building. Thus, beyond a certain distance a ground strike will not
significantly increase the building's earth potential. This is the collection distance
D, measured in metres.

The distance D should be taken to be numerically equal to the soil resistivity value – up to a maximum value of $D = 500$ m for a soil resistivity of $500\,\Omega$ m or more. Where the soil resistivity is not known, a typical $100\,\Omega$ m resistivity soil should be assumed giving $D = 100$ m. The collection area of the surrounding ground is an area extending D m all around the building. Figure 9.17 shows this area for a rectangular building. This area is therefore calculated as follows:

$$C_{SG} = 2\,(a \times D) + 2\,(b \times D) + \pi D^2$$

Note, the four quarter circle areas have been combined to give a circular area which can be calculated using πr^2 where the radius r is equal to D. If the structures height h exceeds D, the collection distance is assumed to be h. Where a complex shaped building is being assessed, approximations and rough graphical methods will allow the area to be calculated with sufficient accuracy.

Collection area of adjacent associated structures and their surrounding ground (C_{AS}). Where there is a direct or indirect electrical connection to an adjacent structure, the collection area of this structure should be taken into account. Typical examples are lighting towers (supplied from the main building), radio transmission towers and other buildings with computer control and instrumentation equipment. Structures within a distance of $2D$ from the main building are considered to be adjacent. As before, the collection area is calculated as the area of the structure plus the area extending D m all around the building. Once again if the height of the structure h exceeds D, the collection distance is assumed to be h. Any part of this area which falls within the collection area of the main structure C_{SG} should be disregarded.

Collection area of incoming/outgoing mains power supplies (C_{MP}). The effective collection areas of all incoming and outgoing power supplies should be considered. This includes outgoing supplies to neighbouring buildings, CCTV equipment, lighting towers, remote equipment and the like. The effective collection areas of different types of mains power supplies are shown in Table 9.8.

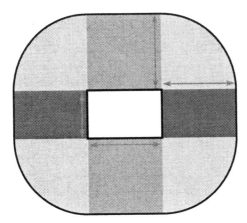

Fig. 9.17 The collection area of the surrounding ground for a rectangular building.

Collection area of data lines leaving the earth reference of the building (C_{DL}).
The effective collection areas of all incoming and outgoing data lines should be considered. This includes data lines to other buildings, security equipment, remote data logging equipment and the like. The effective collection areas of different types of data lines are shown in Table 9.9.

Vulnerability of system configuration

The overall risk to electronic equipment from the secondary effects of lightning will therefore depend upon P (the probability of a strike) and the type of structure F, the degree of isolation G, the type of terrain H. Weighting factors based upon the relative degrees of risk are assigned for F, G and H in Tables 9.10, 9.11 and 9.12, respectively.

Calculation of overall risk

We described earlier how the threat to electronic equipment from the secondary effects of lightning depends upon the probable number of lightning strikes to the area of influence, and the vulnerability of the system configuration. We have now seen how the former can be calculated from $P = A_c \times N_g \times 10^{-6}$ and how the latter is given by the weighting factors F, G, and H.

Thus, the risk R of a lightning strike coupling into electrical or electronic systems through incoming/outgoing mains power supplies or data lines is:

$$R = P \times F \times G \times H$$

The average number of years between transient overvoltages caused by lightning is given by $1/R$. It should be noted that average values such as these are based on periods of many years. This risk assessment R is intended only to guide the user's protection decision. Both the commercial impact of systems damage and downtime and the health and safety implications should be considered (see earlier). Often the decision to protect will be based upon a simple comparison of the cost of damage and downtime to computers and plant systems, with the cost of protection and prevention.

Protector exposure levels

This risk assessment can also be used to indicate the type of protector required at different types of installation. BS 6651 provides a four tier classification of installations based upon the consequential effects of damage to the installations' contents. This is shown in Table 9.13. Combining together the risk R with the consequential loss rating we can determine the exposure level for which transient overvoltage protectors should be designed (see Table 9.14).

These exposure level categories are based on a lightning risk assessment only. Where transients of another cause are likely to be present, you should consider upgrading the protectors. Mains power supplies are susceptible to switching transients, also. Where the presence of switching transients is anticipated, a protector from the next highest exposure level should be used (unless a high exposure level protector is already required). Protector exposure levels are explained in greater detail later.

Table 9.8 Effective collection area of incoming/outgoing power supplies.

Type of mains service	Effective collection area (m^2)
Low voltage overhead cable	$10 \times D \times L$
High voltage overhead cable (to on-site transformer)	$4 \times D \times L$
Low voltage underground cable	$2 \times D \times L$
High voltage underground cable (to on-site transformer)	$0.1 \times D \times L$

Reproduced from BS 6651: 1992.
Note 1: D is the collection distance in metres (see above). Under no circumstances should h be used in place of D.
Note 2: L is the length, in metres, of the power cable up to a maximum value of 1000 m. If the value of L is unknown 1000 m should be used.
Note 3: Where there is more than one power line/cable, they should be considered separately and the collection areas summated. Multicore cables are treated as a single cable and not as individual circuits.

Table 9.9 Effective collection area of incoming/outgoing data lines.

Type of data line	Effective collection area (m^2)
Overhead signal line	$10 \times D \times L$
Underground signal line	$2 \times D \times L$
Fibre-optic cable without a conductive metallic shield or core	0

Reproduced from BS 6651: 1992.
Note 1: D is the collection distance in metres. Under no circumstances should h be used in place of D.
Note 2: L is the length, in metres, of the data line up to a maximum value of 1000 m. If the value of L is unknown 1000 m should be used.
Note 3: Fibre-optic is a non-conductive means of data transmission. It therefore has a collection area of zero.
Note 4: Where there is more than one data line/cable, they should be considered separately and the collection areas summated. Multicore cables are treated as a single cable and not as individual circuits.

Table 9.10 Type of structure (Factor F).

Structure classification	Value of F
Buildings with lightning protection and equipotential bonding to BS 6651	1
Buildings with lightning protection and equipotential bonding to CP 326	1.2
Buildings where equipotential bonding for electrical or electronic equipment reference may be difficult (e.g., buildings over 100 m long)	2.0

Reproduced from BS 6651: 1992.

Table 9.11 Degree of isolation (Factor G).

Degree of isolation	Value of G
Structure located in a large area of structures or trees of the same or greater height, e.g., in a large town or forest	0.4
Structure located in an area with few other structures or trees of similar height	1.0
Structure completely isolated or exceeding at least twice the height of surrounding structures or trees	2.0

Reproduced from BS 6651: 1992.

Table 9.12 Type of terrain (Factor H).

Type of country (or terrain)	Value of H
Flat country at any level	0.3
Hill country	1.0
Mountain country between 300 m and 900 m	1.3
Mountain country above 900 m	1.7

Reproduced from BS 6651: 1992.

Table 9.13 Classification of structures and contents.

Structure usage and consequential effects of damage to contents	Consequential loss rating
Domestic dwellings and structures with electronic equipment of low value and small cost penalty due to loss of operation	1
Commercial or industrial buildings with essential computer data processing, where equipment damage and downtime could cause significant disruption	2
Commercial or industrial applications where loss of data or computer process control could have severe financial costs	3
Highly critical processes where loss of plant control or computer operation may lead to severe environmental or human cost (e.g., nuclear plant, chemical works, etc.)	4

Reproduced from BS 6651: 1992.
Note: The examples of structure usage are only intended to give greater meaning to the descriptions of consequential effect – they should not be seen as binding.

Table 9.14 Classification of exposure level.

Consequential loss rating	Exposure level, R			
	< 0.005	$0.005-0.0499$	$0.05-0.499$	> 0.5
1	Negligible	Negligible	Low	Medium
2	Negligible	Low	Medium	High
3	Low	Medium	High	High
4	Medium	High	High	High

Reproduced from BS 6651: 1992.
Note: Where the exposure level is negligible, protection is not normally necessary.

Protection techniques and basic considerations

There are a number of techniques which can be used to minimize the lightning threat to electronic systems. Like all security measures they should, wherever possible, be viewed as cumulative and not as a list of alternatives. The use of transient overvoltage protectors is covered in detail below.

BS 6651: 1992 describes a number of other measures to minimize the severity of transient overvoltages caused by lightning. These tend to be of greatest practical relevance for new installations. These measures are:

- extensions to the structural lightning protection system;
- earthing and bonding;
- equipment location;
- cable routing and screening;
- use of fibre-optic cables.

Extending structural lightning protection
We saw in Fig. 9.15 how lightning currents preferentially flow down external conductors. Thus, a building's lightning protection is enhanced by having many down conductors around the edge of the building. The greater the number of down conductors around the sides of the building, the smaller the magnetic fields inside the building, and the lower the likelihood of transient interference into electronic equipment. It follows from this that extra down conductors should be installed on buildings containing important installations of electronic equipment.

Many systems incorporate elements installed outside or on the building. Common examples of external system components include:

- aerials or antennae;
- measurement sensors;
- parts of the air conditioning system;
- CCTV equipment;
- roof mounted clocks.

Exposed equipment, such as this, is not only at risk from transient overvoltages caused by the secondary effects of lightning, but also from direct strikes! A direct lightning strike must be prevented, if at all possible. This can be done by ensuring that external equipment is within a zone of protection and where necessary bonded to the structural lightning protection. Figure 9.18 shows CCTV cameras safely positioned within the 45° zone of protection provided by an air termination point.

It may be necessary to include additional air termination points in the building's lightning protection scheme, in order to ensure that all exposed equipment is protected. For exposed parts of the air-conditioning system, it is possible to bond just its metal casing on to the roof top lightning conductor grid. Where air termination points cannot be used, for example with whip aerials, the object should be designed to withstand a direct lightning strike or be expendable.

Fig. 9.18 CCTV cameras located within the 45° zone of protection provided by an air termination point (left) or flat roof-top conductor (right).

Exposed wiring should be installed in bonded metallic conduit or routed such that suitable screening is provided by the structure itself. For steel lattice towers the internal corners of the L-shaped support girders should be used. Cables attached to masts should be routed within the mast (as opposed to on the outside) to prevent direct current injection.

Earthing and bonding

Detailed guidance on earthing is beyond the scope of this chapter. Additional and complementary guidance is given here to improve earthing, and to achieve an area of equal potential, ensuring that electronic equipment is not exposed to differing earth potentials and hence resistive transients.

All incoming services (water and gas pipes, power and data cables) should be bonded to a single earth reference point. This equipotential bonding bar may be the power earth, a metal plate, or an internal ring conductor/partial ring conductor inside the outer walls. Whatever form it takes, this equipotential bonding bar should be connected to the electrode(s) of the earthing system. All metal pipes, power and data cables should, where possible, enter or leave the building at the same point, so that extraneous metalwork and armouring can be bonded to the main earth terminal at this single point (see Fig. 9.19). This will minimize lightning currents within the building.

If power or data cables pass between adjacent buildings, the earthing systems should be interconnected, creating a single earth reference for all equipment. A large number of parallel connections, between the two building's earths, are desirable – reducing the currents in each individual connection cable. This can be achieved with the use of a meshed earthing system. Power and data cables between adjacent buildings should also be enclosed in metal conduits, trunking, ducts or similar. This should be bonded to both the meshed earthing system and also to the common cable entry and exit point, at both ends.

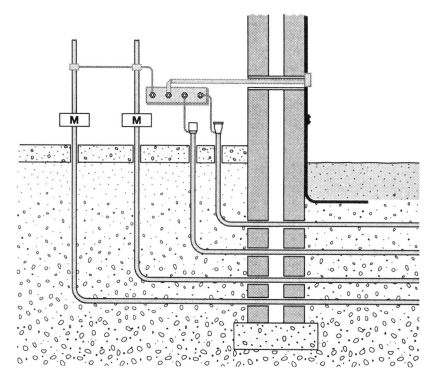

Fig. 9.19 Bonding of incoming services.

Equipment location
Electronic equipment should not be located where it will be close to large current flows and the threat of induced transient overvoltages.

(1) Equipment should not be located on the top floor of the building where it is adjacent to the roof-top air terminations and conductor mesh of the building's lightning protection system.
(2) Similarly, equipment should not be located near to outside walls and especially corners of the building, where lightning currents will preferentially flow.
(3) Equipment should not be located in buildings close to tall, lightning attractive, structures, such as masts, towers or chimneys. These tend to have a single route to ground, causing very large current flows and hence very large magnetic fields.

The issue of equipment location can only be ignored if the building corresponds to the screened room ideal (bonded metal clad roof and walls), described earlier.

Cable routing and screening
Power, data communication, signal and telephone cabling may also be at risk from induced transient overvoltages within the building. Equipment cabling should

avoid possible lightning carrying conductors such as those on the roof and, within or on, the building (see equipment location above). Large area loops between mains power and data communication cabling are, as a result of inductive coupling, effective at capturing lightning energy, and should therefore be avoided. To minimize loop areas, mains power supply cables and data communication, signal, or telephone wiring should be run side by side, though segregated. The cables can be installed either in adjacent ducts or separated from each other by a metal partition inside the same duct.

For a screened room type building the routing and location of cabling is not as critical; however adoption of the above is good practice. For buildings made from non-conducting materials the above practices are essential in order to minimize damage to equipment or data corruption. Cable screening or shielding is another useful technique which helps to minimize the pick-up and emission of electro-magnetic radiation. Power cables can be shielded by metallic conduit or cable trays, whilst data cables often incorporate a screening outer braid. The screen acts as a barrier to electric and magnetic fields. Its effectiveness is determined by its material and construction as well as by the frequency of the impinging electro-magnetic wave. For transient protection purposes the screen should be bonded to earth at both ends, although there are instances, particularly in data comms, where single-end earthing is preferred to help minimize earth loops.

Use of fibre-optic cable on building to building data links
Special care should be taken with the protection of data lines which:

- pass between separate buildings, or
- which travel between separate parts of the same building (i.e. not structurally integral) and which are not bonded across. Examples include parts of a building which are separated by settlement gaps or new wings which are linked by brick corridors.

Fibre-optic cable is the optimum method of protection for building to building data links. This will completely isolate the electronic circuits of one building from the other, preventing all sorts of EMC problems including transient overvoltages. The use of fibre-optic cable for data transmission does not diminish the need for the protection of the mains power supply to equipment.

Many fibre-optic cables incorporate metal draw-wires or moisture barriers and steel armouring. This can establish a conductive link between buildings, defeating the object of using a fibre-optic link! If this cannot be avoided the conductive draw-wire, moisture barrier or armouring, should be bonded to the main cable entry bonding bar as it enters each building, or stripped well back. No further bonding should be made to the fibre-optic cables 'metal'. Lightning protection for fibre cable is considered in greater detail by CCITT IX K25. The cost of fibre-optic cable may make it unattractive for low traffic data links and single data lines.

Where conductive data lines, such as unshielded twisted pairs or coaxial cables, are required, precautions should be taken to prevent transients from flowing along the line, threatening equipment at each end:

- install transient overvoltage protectors/lightning barriers (see below);
- interconnect the earthing systems of the buildings.

The earthing systems of the buildings can be interconnected via cable armouring, metal trunking, raceways or conduits. These should be bonded, at each end, to each building's earth system, thus providing an electrically continuous link. If many parallel data links exist between buildings a good interconnection of earths will be achieved, producing very low induced voltages. Also, it is beneficial to install earth cables linking each structure's earth.

Certain types of coaxial and screened cable should only be bonded to earth once (ECMA 97 details types of LAN for which this is the case). The use of transient overvoltage protectors will provide additional bonding, whilst retaining the integrity of the system's screening.

Deployment of transient overvoltage protectors

Transient overvoltage protectors should be used to supplement and support the protection techniques outlined above. When deploying transient overvoltage protectors, the objective is always to install the transient protection between the source of the threat and the equipment that is to be protected.

We saw earlier that transient overvoltages can be conducted into electronic equipment via:

- mains power supplies;
- data communication, signal and telephone lines.

We must therefore look at protecting both power and data communication/signal/telephone inputs.

Protecting mains power supplies against lightning
Both overhead and underground cables can be affected by lightning. Underground cables, which provide a link between two different earth references, are subject to resistively coupled transients. Overground or overhead cables are susceptible to resistive, inductive and capacitive transients.

Mains power supplies entering the building should be protected, to control large externally generated transients before they enter the building's power distribution system. A suitable protector should be installed on the main l.v. incomer or at the main l.v. distribution board. Some buildings have more than one main incomer or main distribution board – all should be protected. Mains power supplies leaving the building may also need to be protected.

(1) Protection as the supply leaves the building will prevent transients from outside being injected back into the building's power distribution system – often there will be a local distribution board or auxiliary services board feeding this external equipment, at which protection can be installed.
(2) Sensitive external or on-site equipment, such as CCTV cameras, should also be protected locally – although the equipment will share the same power

earth as the main building, it is likely to be inadvertently connected to a local earth (pole-mounted cameras may be earthed locally via its support mast) and hence vulnerable to resistive transients.

Mains power supplies between buildings should be protected at both ends. Supplies between buildings should be considered as an outgoing supply (from one building) and an incoming supply (to the other building).

We saw earlier how transient overvoltages can be induced on to power lines from the electromagnetic field caused by a large current, such as lightning. In a similar fashion, when the building's lightning protection system carries a current to earth, it creates a magnetic field. If electronic equipment or its power supply is located near to these current-carrying conductors, then an induced transient will result.

In new buildings it may be possible to ensure that this risk is eradicated. However, for most existing installations this particular source of transients is a very real threat. It is therefore advisable to install protection locally to important pieces of equipment in order to control lightning induced transients inside the building.

Protecting uninterruptible power supplies (UPS)

An uninterruptible power supply (UPS) is essentially a large battery providing several minutes of back-up power. Many UPSs incorporate a small filter, and on this basis many UPS manufacturers have loosely claimed that their devices provide transient overvoltage protection. Some manufacturers have gone as far as to suggest that their UPSs protect against transient overvoltages caused by lightning. However, the type of filters used provide protection only against quite low level transients and not against the larger transients which cause the disruption, degradation, damage and downtime we are trying to avoid!

Where equipment is connected to a UPS, good lightning and transient protection should still be used. First, to protect the equipment against transient disturbances on the power supply, whilst the UPS is on bypass. Secondly, the modern electronic UPS is itself a sophisticated piece of electronic equipment, and can be prone to the effects of transient overvoltages – though more traditional motor-generator (or rotary) designs are less prone to this problem. Thirdly, some UPSs have their neutral connected straight through from input to output, potentially allowing the unhindered passage of neutral to earth transients. Finally, where the UPS is in an external building separate from its load, its output, its input and the load will require protection.

The protector should be installed at the local power distribution panel, feeding the UPS. On large UPSs the protector can be installed within the UPS cabinet, on the power incomer. Some larger UPSs are supplied with protectors built in and therefore provide effective protection against transient overvoltages caused by lightning.

Protecting data communication, signal and telephone lines

Inside the building, screened cables can offer protection against voltages induced from current-carrying lightning conductors. Most data cables incorporate a metallic screen. Outside the building, data lines can be susceptible to inductively, capacitively or resistively coupled transients. Underground cabling, between separate earth references, will be susceptible to resistively coupled transients. Overground or overhead cables may be susceptible to resistive, capacitive and inductive transients. Thus, as a general rule, transient protectors should be installed on all incoming and outgoing data communication, signal and telephone lines. Spare cores and unutilized cables should be connected to an electrical earth, at both ends.

Most sizeable businesses and organizations have their own telephone exchange or private branch exchange (PBX). This is the link between the telephone company's incoming lines and the customers' own internal extension lines. Just like other communication lines, telephone lines which travel into the building or between buildings should be protected. PBXs can be protected against transient overvoltages with two tier (gas discharge tube plus semiconductor) protection. This typically consists of semiconductor protection installed at board level and gas discharge tubes installed on the main distribution frame. Certainly this is hardly ever done in the UK nor, we think, in most other parts of the world. If lines are correctly protected this tends to be true of the incoming (PTO) telephone lines only and not extension lines. These often travel from building to building and are therefore very much at risk from lightning.

The use of gas discharge tubes alone does not provide adequate protection. The PBX should be properly protected by devices with a suitably low let-through voltage. The PBX is an excellent place to install protection because all incoming and outgoing lines meet at this point.

Choosing and specifying transient overvoltage protectors

In selecting a transient overvoltage protector it is important that the device selected survives, has a suitable transient control level, and is compatible with the system it is protecting. It is vital that the protector chosen is capable of surviving the worst case transients expected at its intended installation point. Also, as lightning is a multiple pulse event, the protector should not fail after exposure to the first transient.

The protector should be able to control transients to a level below the susceptibility and vulnerability of the equipment to be protected. For example, if a computer's operation is unhindered by transients of up to 700 V, then the let-through voltage or transient control level of the protector should be less than 700 V. Allowing a suitable safety margin, the worst case let-through voltage of the protector should be 600 V, or less. (Note, the connecting leads of a correctly installed protector will cause an increase in the let-through voltage.) The protector should not impair or interfere with the protected system's normal operation. Communication systems and intrinsically safe circuits are particularly susceptible to this type of problem.

Location Categories
The protector's ability to survive and to achieve a suitable 'let-through' voltage clearly depends upon the size of transient to which it will be subjected. This in

turn may depend upon the protector's location. The American Standard IEEE C62.41 and subsequently BS 6651 outline three location categories.

As a mains-borne transient travels through a building, the amount of current it can source grows smaller (owing to the impedance of mains cables and current division). This is based upon the assumption that a typical mains transient caused by lightning has a $1.2/50\,\mu s$ waveshape.

Location Category C is defined as either:

- outside the building, or
- the supply side of the main, incoming, l.v. distribution board, i.e. the board bringing the power supply into the building from the electricity supply authority, l.v. transformer or from another building, or
- the load side of distribution boards providing an outgoing power supply to other buildings or to on-site equipment.

Location Category B is defined as either:

- on the power distribution system, downstream of Category C and upstream of Category A, or
- within apparatus which is not fed from a wall socket, or
- sub-distribution boards located within a 20 m cable run of Category C, or
- plug-in equipment located within a 20 m cable run of Category C.

Location Category A is defined as either:

- sub-distribution boards located more than a 20 m cable run away from Category C, or
- plug-in equipment located more than a 20 m cable run away from Category C.

Transient overvoltages on data lines are not significantly attenuated by the cable and so protectors should always be rated for Location Category C. Regardless of where they are installed in the building, the worst case will be similar. This is based upon a $10/700\,\mu s$ waveform transient.

Exposure levels

Transient overvoltage protectors are designed to protect against the probable worse case transient overvoltage. In high transient exposure level areas, very large transients (perhaps only occurring once in every few thousand events) can be anticipated over much shorter time-scales than in a low transient exposure level area. Thus, the probable worst case transient will be much smaller in a low transient exposure level area, than in a high exposure area.

The transient exposure level can be derived from the risk assessment – see earlier. However, if a risk assessment has not been done, it is probably wise to assume a risk, $R = 0.6$. This can be combined with the consequential loss rating (in Table 9.13) in order to derive the protector exposure level from Table 9.14. Probable worst case mains transients for high, medium and low exposure levels are tabulated in Tables 9.15–9.17, for all location categories. All data line protectors fall within Category C and probable worst case data line transients are tabulated in Table 9.18.

Protector performance

These location categories and their probable worst case transients, provide us with a yardstick with which to evaluate protectors. Subjecting the protector to an appropriate transient test enables us to determine whether it will survive. The

Table 9.15 Mains power supply – Category C.

System exposure level	Peak voltage (kV)	Peak current (kA)
High	20	10
Medium	10	5
Low	6	3

Derived from IEEE C62.41 and BS 6651: 1992.

Table 9.16 Mains power supply – Category B.

System exposure level	Peak voltage (kV)	Peak current (kA)
High	6	3
Medium	4	2
Low	2	1

Derived from IEEE C62.41, UL 1449 and BS 6651: 1992.

Table 9.17 Mains power supply – Category A.

System exposure level	Peak voltage (kV)	Peak current (A)
High	6	500
Medium	4	333
Low	2	167

Derived from UL 1449 and BS 6651: 1992.

Table 9.18 Data lines – Category C.

System exposure level	Peak voltage (kV)	Peak current (A)
High	5	125
Medium	3	75
Low	1.5	37.5

Derived from CCITT IX K17 and BS 6651: 1992.

protectors let-through voltage for this test tells us its transient control level. Let-through voltage is a measure of the amount of the transient overvoltage which gets past, or is let-through, the protector. Thus, if a protector is required for a mains power distribution board (Category B) in a high exposure level, it should be able repeatedly to survive a transient of 6 kV and 3 kA. Its let-through voltage when injected with this 6 kV, 3 kA transient, should be no greater than the desired transient control level.

The let-through voltage of a protector also takes into account its response time. Protectors with slow response times will have consequently higher let-through voltages. Response times are only of significant importance when dealing with very fast transients such as NEMP.

BS 6651: 1992 Appendix C, suggests that manufacturers of transient protectors should provide the following information about their products. The protectors let-through voltage should be quoted against a given test, e.g. 600 V (live−neutral, neutral−earth, live−earth), 6 kV 1.2/50 µs, 3 kA 8/20 µs, BS 6651: 1992 Category B-High.

The let-through value should be the result of tests conducted on the complete protective circuit − it should not be a theoretical value derived from the performance of one or more components.

Transients can occur between any combination of conductors. Manufacturers should therefore make quite clear which modes of transient propagation let-through performances relate to, e.g.:

- live/phase(s) to earth, live/phase(s) to neutral and neutral to earth for mains power supplies;
- line to line and line(s) to earth for data lines.

The maximum transient current the protector can withstand should be clearly stated, e.g., 20 000 A, 8/20 µs. Again, this is a test value for the whole protector and not a theoretical value. The maximum transient current of the protector also takes account of its energy handling capability. Energy ratings, as a sole indicator of protective performance, are misleading, as the energy deposited in a protector by a transient current source depends upon the suppression level (i.e. let-through voltages). Thus, lower energy ratings do not necessarily mean that the protector is less likely to survive.

The international telecommunication standard also calls for high impulse current testing of data line protectors for telephone lines. This test (within CCITT IX K12) is intended to test the capability of gas discharge tube(s) present in most data line protectors. Test levels are shown in Table 9.19.

Any factor that may interfere with the system during normal operation should be quoted. These may include the following:

- nominal operating voltage;
- maximum operating voltage;
- leakage current;
- nominal current rating;
- maximum continuous current rating;

Table 9.19 High impulse current tests.

System exposure level	High impulse current (kA)
High	10
Medium	5
Low	2.5

Derived from CCITT IX K17 and BS 6651: 1992.

- in-line impedance (or resistance);
- shunt capacitance;
- bandwidth;
- voltage standing wave ratio (VSWR) or reflection coefficient.

Table 9.20 summarizes potential causes of incompatibility or system impairment for different types of power and data line protectors. Protector manufacturers should therefore document values for their protector for all the characteristics indicated. The question of the protector being incompatible with the system it is trying to protect most commonly arises with data communication, signal and telephone systems. However, when a mains power protector is based on a gas-discharge tube connected across the supply, operation of the gas-discharge tube can short-circuit the supply. The large mains current flowing through the tube is likely to destroy the tube and disrupt the power supply.

Specification for lightning and transient overvoltage protection

A number of products are available which claim to protect electronic equipment against lightning and transient overvoltages. However, testing shows many of these to have an unexceptably high let-through voltage. This can prove disruptive and harmful to electronic equipment. Use of the specification which follows will reduce the likelihood of ending up with ineffective protection.

Key requirements
In order to provide effective protection, a transient overvoltage protector should:

- be compatible with the system it is protecting;
- survive;
- have a low let-through voltage, for all combinations of conductors;
- not leave the user unprotected, as a result of failure, and
- be properly installed.

Compatibility. It is important that the protector does not interfere with or restrict the system's normal operation.

(1) It is undesirable for mains power supply protectors to disrupt or corrupt the continuity of power supply and for them to introduce high earth leakage currents.

Table 9.20 General indication of potential system impairments which manufacturers of transient overvoltage protectors should provide.

	Protectors for mains supplies		Protectors for data lines		
	Parallel protectors	In-line protectors	Low frequency protectors	Network protectors	Radio frequency protectors
Nominal operating voltage	\	\	\	\	\
Maximum operating voltage	\	\	\	\	\
Leakage current	\	\	\	\	\
Nominal current rating	×	\	\	\	\
Max continuous current rating	×	\	\	\	\
In-line impedance	×	×	\	\	\
Shunt capacitance	×	×	×	\	\
Bandwidth	×	×	\	\	\
Voltage standing wave ratio	×	×	×	\	\

Derived from W.J. Furse, *Electronic Systems Protection Handbook*, 1994, W.J. Furse & Co. Ltd.

(2) Protectors for data communication, signal and telephone lines should not impair or restrict the systems data or signal transmission.

Survival. Although lightning discharges can have currents of 200 kA, transient overvoltages caused by the secondary effects of lightning are unlikely to have currents exceeding 10 kA. The protector should therefore be rated for a peak discharge current not less than 10 kA.

Let-through voltage. The larger the transient overvoltage reaching the electronic equipment, the greater the risk of interference, physical damage and hence system downtime. Thus, the transient overvoltage let through the protector should be lower than the level at which interference or component degradation may occur.

Modes of protection. A transient overvoltage can exist between any pair of conductors: phase and neutral, phase and earth, neutral and earth on mains power supplies; line to line, line to screen/earth on data communication, signal and telephone lines. Thus, the transient overvoltage protector should have a low let-through voltage for all combinations of conductors.

Protection failure.   When in-line protectors (such as those for data communication, signal and telephone lines) fail, they take the line out of commission, thereby preventing damage to the system. However, it is unacceptable for protectors on mains power distribution systems to fail short-circuit. If these protectors fail sudden death they will leave the system unprotected. It is therefore important that protectors for mains power distribution systems have a properly indicated pre-failure warning, whilst protection is still present.

Installation.   The performance of transient overvoltage protectors is heavily dependent upon their correct installation (e.g. the length and configuration of connecting cables). Thus, the transient overvoltage protector must be supplied with documentation detailing the installation practice. The installer is required to conform with these installation instructions.

Specifying protection for mains power distribution systems
(1) Transient overvoltage protectors should be installed on all power cables entering or leaving the building, in order to protect equipment connected to the power distribution system against transient overvoltages coming into the building from outside.
(2) Protectors should also be installed at local power distribution boards feeding vulnerable equipment, in order to protect these against transients generated downstream of the protectors in (1), above. (These transients may be the result of inductive coupling or electrical switching.)
(3) Protectors shall be tested in accordance with the requirements of:
   BS 6651: 1992, *Protection of structures against lightning* (Appendix C),
   BS 2914: 1972, *Specification for surge diverters for alternating current power circuits,*
   IEEE C62.41: 1991, *Guide for surge voltages in low voltage AC power circuits.*
(4) Protectors for a given Location Category shall be rated for a High Exposure Level (as defined by BS 6651: 1992), unless contrary information is available.
(5) The protector must not interfere with or restrict the system's normal operation. It should not:
   • corrupt the normal mains power supply;
   • break or shutdown the power supply during operation;
   • have an excessive earth leakage current.
(6) The protector shall be rated for a peak discharge current of no less than 10 kA (8/20 ms waveform).
(7) The protector shall limit the transient voltage to below equipment susceptibility levels. Unless otherwise stated, the peak transient let-through voltage shall not exceed 600 V, for protectors with a nominal working voltage of 230 or 240 V, when tested in accordance with BS 6651: 1992 Category B – High (6 kV 1.2/50 ms open circuit voltage, 3 kA 8/20 ms short-circuit current).

(8) This peak transient let-through voltage shall not be exceeded for all combinations of conductors:
   - phase to neutral;
   - phase to earth;
   - neutral to earth.

(9) The protector should have continuous indication of its protection status. Status indication should clearly show:
   - full protection present;
   - reduced protection – replacement required;
   - no protection – failure of protector.

(10) Remote indication of status should also be possible via a volt free contact.

(11) The status indication should warn of protection failure between all combinations of conductors, including neutral to earth. (Otherwise a potentially dangerous short circuit between neutral and earth could go undetected for some time.)

(12) The protector shall be supplied with detailed installation instructions. The installer must comply with the installation practice detailed by the protector manufacturer.

Specifying protection for data communication, signal and telephone lines

(1) Transient overvoltage protectors shall be installed on all data communication/signal/telephone lines entering or leaving the building, in order to protect equipment connected to the line against transient overvoltages. (Where data lines link equipment in separate buildings, transient overvoltage protectors should be installed at both ends of the line in order to protect both pieces of equipment.)

(2) Protectors shall be tested in accordance with the requirements of:
   BS 6651: 1992, *Protection of structures against lightning* (Appendix C).
   CCITT IX K17, *Tests on power fed repeaters using solid-state devices in order to check the arrangements for protection from external interference.*

(3) Protectors shall be rated for Location Category C – High exposure level, unless contrary information dictates a lower exposure level.

(4) The protector must not impair the system's normal operation. It should not:
   - suppress the system's normal signal voltage;
   - restrict the system's bandwidth or signal frequency;
   - introduce excessive in-line resistance;
   - cause signal reflections or impedance mismatches (high frequency systems only).

(5) The protector will have a low transient let-through voltage for tests conducted in accordance with BS 6651: 1992, Category C – High (5 kV 10/700 ms test).

(6) This let-through performance will be provided for all combinations of conductors:
   - signal line to signal line;
   - signal line to screen/earth.

(7) The protector shall be rated for a peak discharge current of 10 kA.

(8) The protector shall be supplied with detailed installation instructions. The

installer must comply with the installation practice detailed by the protector manufacturer.

(9) The protector manufacturer should allow for the facility to mount and earth large numbers of protectors through an accessory combined mounting and earthing kit.

ACKNOWLEDGEMENT

Extracts from BS 6651: 1992 are reproduced with the permission of BSI. Complete copies can be obtained from BSI Customer Services, 389 Chiswick High Road, London W4 4AL, telephone: 0181 996 7000.

CHAPTER 10

Special Installations or Locations

G.G. Willard, DipEE, CEng, MIEE
(Consulting Electrical Engineer)

The Wiring Regulations, (now issued as a British Standard BS 7671: 1992) makes particular reference to electrical installations in 'special locations'. Apart from the need to meet the general requirements of the British Standard, additional requirements or restrictions apply to these installations. This chapter expands and explains the main additional requirements of BS 7671: 1992. Particular Regulations are called up where necessary and referred to by their BS 7671 numbers.

In general, 'special locations' involve one or more of the following, which, in the absence of special arrangements, would give rise to increased risk of electric shock:

- wetness or condensation, i.e. reduced skin resistance;
- absence of clothing, i.e. greater opportunity to make direct or indirect contact through increased area of bare skin and bare feet;
- proximity of earthed metalwork, i.e. increased risk of indirect contact;
- arduous or onerous site conditions, i.e. conditions of use which may impair the effectiveness of the protective measures.

The special requirements for 'special locations' have been devised by assessing the relevant risks under each of the above categories and making adjustments to the protective measures accordingly. The intention is to remove or minimize the additional risks to users, presented by the electrical installation within the 'special location'. Designers, installers and operators of such installations should also be aware of current legislation in respect of electrical installations.

Installers should also be aware of the special legal relationship which may be created between themselves and the ultimate user of an installation. The work creates what is legally termed a 'duty of care' on the part of designers – to exercise skill and care, and on the part of installers and maintainers – to follow strictly the best practices of the profession and *actually* achieve a workable end product. These obligations are not unusual to special locations; however the inherently greater risks associated with such installations underscore the need for particular care and attention to detail.

Installations such as in explosive atmospheres or lightning protection are not referred to here because these are covered either by a British Standard or are the subject of HSE guidance. Furthermore, marinas have not been included because, although guidance exists through the International Electrotechnical Commission, this has yet to be accepted in the UK. In respect of 'restrictive conductive locations' and 'earthing requirements for the installation of equipment having high earth leakage currents', these installation types are covered in a prescriptive format within BS 7671.

As further work is done at international level, and applications develop, it is likely that more installation types will be categorized as 'special locations'. The list covered here should therefore be regarded as subject to amendment.

Reference has generally been made to the content of BS 7671, however IT distribution systems have not been considered here in particular, as it remains a specific UK requirement of the Electricity Supply Regulations 1988, that low voltage systems should be connected with earth.

LOCATIONS CONTAINING A BATH TUB OR A SHOWER BASIN

The risks

The risks are from:

- wet skin;
- absence of clothing;
- proximity of earthed metal.

Users of baths for medical treatment will potentially be subject to additional risks associated with infirmity and reduced tolerance to electric shock. Such installations will require further specialized treatment, and full consultation with the client's advisers and equipment suppliers is essential.

The protective measures

BS 7671 allows for no socket outlets in bathrooms unless they are part of a safety extra low voltage (SELV) circuit. The thinking behind this is undoubtedly the very real concern that persons using a bathroom cannot be made safe from the effects of direct or indirect contact from portable equipment supplied at low voltage. Some relaxation is permitted in rooms, other than a bathroom or shower room, where a shower cubicle is installed, provided that any socket outlet, other than a SELV device, is installed at least 2.5 m from the shower cubicle. It should be remembered here that in the case of a bath tub, adults tending the bathing of others are potentially placed at risk by hand-to-hand indirect contact with metalwork which may become live in the event of an earth fault. The 2.5 m referred to therefore provides for a restriction of access to items of substantial risk.

This 'inaccessibility' principle extends to electrical equipment generally and only switches using pull cords to operate BS 3676 devices, or specially designed controls to BS 3456 instantaneous water heaters are permitted within this zone. It is accepted however that a shaver supply unit with a BS 3535 transformer may be installed, as may switches supplied by SELV where the nominal voltage does not exceed 12 V r.m.s.

Fixed equipment within the 2.5 m zone should be selected and erected according to its likely duty. The likelihood is for splashing water to be present and this requires an Index of Protection of IPX4. Power showers can generate penetrating jets of water and may require IPX5 equipment.

BS 7671 does not actually prohibit the use of Class I fixed equipment but states that stationary appliances with heating elements which can be touched should be mounted outside the 2.5 m arm's reach zone. Similarly, luminaires to be fixed within this zone should be either totally enclosed, or if of Type B22, should be fitted with a BS 5042 (Home Office) shield.

In general, those responsible for selection and erection of equipment will find that Class II devices more readily satisfy the requirements for such rooms. However, consideration should be given to the possible deterioration of the Class II insulation due to the prolonged effects of moisture. Furthermore, eventual use cannot be controlled and inadvertent replacement by a Class I device in service must be considered a possibility.

With the increased use of pumped water in bathrooms for spa-baths, power showers, etc., there is a need, on occasions, to provide for motive power within the area of the bath. In order to protect against direct and indirect contact, supplies for such equipment must be by SELV with the nominal voltage not exceeding 12 V r.m.s. Where it is necessary to mount the SELV source within the bath enclosure, then it may only be accessible by means of a tool. The reason for this is concerned with the now familiar definition of skilled person, i.e. it is assumed that persons using a tool to access the SELV source under a bath will be sufficiently informed and skilled to avoid danger and that the circuit will be dead and isolated before work is commenced.

SELV circuits apart, there is a fundamental requirement to clear earth faults on electrical equipment within 0.4 seconds. Strictly speaking, the test is whether such equipment is simultaneously accessible with exposed or extraneous conductive parts. It is unlikely that any bathroom in a modern dwelling could pass this test.

A fundamental point which gives rise to considerable confusion is supplementary bonding. It is a requirement to supplementary bond all simultaneously accessible exposed and extraneous conductive parts. This should include all incoming metallic services and any structural metalwork – including a metallic floor grid, if provided. The definition of 'extraneous conductive part' has changed over the years and is now recognized as 'a conductive part liable to introduce a potential, and not forming part of the electrical installation'.

Installers should note that it is acceptable to employ the protective conductor within a short length of flexible cord as a supplementary bonding conductor for an appliance, provided that connection is by a flex outlet. There is currently no requirement to provide a separate electrical connection from the bathroom back to the main bonding conductor's marshalling block.

BS 7671 does not make specific recommendations regarding wiring systems for bathrooms, excepting that systems employing exposed surface metalwork such as metal conduit may not be used. Apart from introducing unnecessarily large areas of exposed conductive parts, surface metal would also attract considerable condensation in use. This restriction, however is a harmonization requirement from CENELEC. It should be remembered that much of Europe does not permit metallic wiring systems and their prohibition in areas of additional risk is to be expected.

Installers who are aware of current European practice will know that socket outlets in bathrooms are permitted on the continent. This issue is one where there remains disagreement at international level with no alignment in IEC or CENELEC at this time. Because of this, this 'special location' should be regarded as subject to continuing review.

SWIMMING POOLS

Swimming pools are similar in some respects to domestic baths, however their scale and inevitable use of electrical equipment requires a more rigorous approach.

The risks

The risks are from:

- wet skin;
- absence of clothing;
- proximity of earthed metal;
- arduous conditions (possible corrosive effects of chemicals).

The protective measures

Three volumetric zones A, B and C are defined by BS 7671 in order to provide an analytical basis for the design and erection of the electrical installation. It also recognizes the need in a swimming pool environment for portable cleaning equipment and the desirability of fittings such as special pool luminaires. The zones are shown in Figs 10.1 and 10.2. As with rooms containing a bath or shower tray, the 2.5 m inaccessibility limitation applies.

General requirements in all zones extend to the local supplementary bonding of all extraneous and exposed conductive parts. Requirements for an equipotential grid have changed with the first Amendments to BS 7671. Where a grid is installed within the solid floor, then in order to ensure that the equipotential principle is maintained over what could be a substantial area of the building it is to be supplementary bonded as any other extraneous conductive part. The connections (at remote ends of the grid) should be accessible for inspection and testing, by means of a tool.

Requirements within the Zones

Zone A

- No socket outlets are permitted.
- SELV system may be employed, but subject to 12 V maximum nominal voltage and the provision of additional protection against direct contact, comprising:
  - barriers or enclosures to IP2X, or

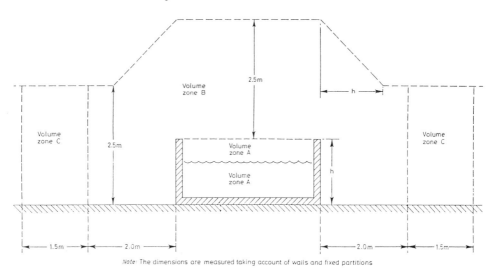

Fig. 10.1 Zone dimensions for basin above ground level. The dimensions are measured taking account of walls and fixed partitions (IEE Wiring Regulation, 16h edn).

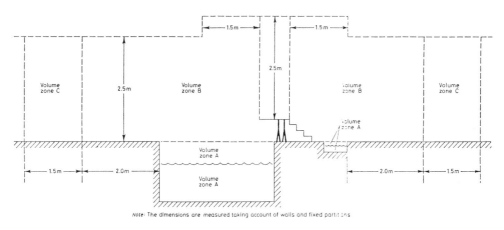

Fig. 10.2 Zone of dimensions of swimming pools and paddling pools. The dimensions are measured taking account of walls and fixed partitions.

- ● a wiring system capable of withstanding a test voltage of 500 V a.c. r.m.s. for 1 minute.
  The safety source must not be installed within Zones A, B, or C.
- ● Floodlights must be supplied from individual transformers, mounted outside Zones A and B, with a maximum open circuit voltage of 18 V.
- ● Equipment must have an index of protection of at least IPX8.
- ● Equipment must be *specifically* intended for use in swimming pools.

Note: Although requirements for swimming pools are aligned through CENELEC, this does not necessarily mean that all imported equipment can be incorporated into UK swimming pool installations and it is necessary to check that the country of origin is a signatory to CENELEC in addition to checking the standard to which the equipment is manufactured.

- Only wiring necessary to supply equipment in Zone A may be run in Zone A.
- No accessible metallic junction boxes are permitted.
- As with bathrooms, surface wiring systems must not employ metallic sheaths or enclosures.

Zone B
The above items apply *except*:

- IPX8 is relaxed to IPX5 and permits waterjets to be used for cleaning purposes – IPX4 applies where waterjets are unlikely to be used.
- It is recognized that due to constructional limitations, it might not be possible to mount socket outlets outside Zone B. As a relaxation, BS EN60309-2 (was BS 4343) socket outlets may be installed, but at least 1.25 m from the Zone A border (single arm's reach restriction of access) and 0.3 m above the floor. However, they must be protected by an appropriate residual current device (RCD).
- An electric heating element may be laid into the floor. This must incorporate a metallic sheath, which is to be bonded to the metallic grid (if supplied). Alternatively, if a grid is not supplied, the metallic sheath should be connected to the equipotential bonding.
- Water heaters are permitted.

Zone C

- Electrical equipment must have minimum degrees of protection of IPX4 for outdoor pools and IPX2 for indoor pools – but only on the basis that water jets are not to be used for cleaning purposes.
- Alternatives for providing protection against direct and indirect contact for electrical equipment are the use of RCDs, SELV or individual electrical separation.
- A heating element with a metallic sheath may be laid into the floor. Bonding arrangements are to be as in Zone B.
- Installers should be particularly careful over the selection and erection of equipment in Zone C. It is likely to be used by persons who are soaking wet – having just emerged from the pool. Although the environs of Zone C will seem mainly dry and apparently give rise to little risk, the bather's wet skin would otherwise create a risk when using electrical equipment. Instantaneous water heaters to BS 3456 are permitted without special earth fault protection; however, hand dryers, sun-beds, etc. will require additional protection as above.

Hot air saunas

The risks

The risks are from:

- wet skin/high humidity;
- absence of clothing;
- proximity of earthed metal;
- arduous conditions of service.

The protective measures

The various areas of hot air saunas are classified by volumetric zones, as with swimming pools but with limits also defined by maximum temperatures. These zones are illustrated in Fig. 10.3.

Zone A
Only the sauna and equipment directly associated with it may be installed here, i.e. heater, flexible cord and junction box to fixed wiring.

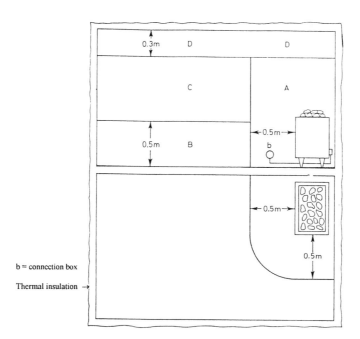

Fig. 10.3 Hot air saunas – zones of ambient temperature.

Zone B

There are no special heat resistance requirements, due to thermal gradient within the sauna enclosure.

Zone C

Equipment should be suitable for ambient temperatures of 125°C.

Zone D

Equipment should be suitable for ambient temperatures of 125°C and must be restricted to sauna control devices (e.g. thermostat, cut-out) and their associated wiring.

Wiring systems within the sauna must employ 150°C flexible cords complying with BS 6141.

Care should be taken in the selection and erection of luminaires to avoid overheating due to the accumulation of hot air combined with the heating effect of the luminaire itself.

CONSTRUCTION SITE INSTALLATIONS

The risks

The risks are from:

- wetness/wet skin;
- proximity of earthed metalwork;
- arduous conditions;
- temporary nature of connections.

It is important to remember that temporary installations on construction sites, demolition sites, major building renovations and works of a similar nature can greatly accelerate deterioration on wiring systems, through the arduous nature of the works. The protective measures indicated below should therefore be supported by regular and frequent reports on the installation. Of particular importance in such reporting is the physical and electrical condition of cable sheaths and terminations. A detailed examination should be carried out to establish whether these have been subject to excess physical strain in service. The scope of these notes does not include portable tools or their associated flexible cords.

The protective measures

The Regulations allow for the use of the following methods of electrical protection.

Supplies for general use

(1) Automatic disconnection and reduced low voltage, i.e. earthed centre-tapped 110 V secondary or earthed star point 110 V secondary transformers.

Note: BS 7671 prescribes a maximum disconnection time of 5 seconds, which should be ascertained by reference to the earth loop impedance for the system and the characteristics of the particular protective device.

(2) The use of a 30 mA residual current device – provided the inequality $Z_s I_{\Delta n} \leq 25$ V can be satisfied, i.e. the touch voltage which could be experienced in the event of an earth fault would be restricted to safe limits.

(3) SELV.

(4) Electrical separation – but on the basis of individual socket outlets protected by individual separated sources.

(5) Automatic disconnection – applies only to 400 V three-phase and 230 V single-phase supplies to the fixed installation and movable equipment such as floodlights.

It is understood that it is common continental practice to use RCDs on construction sites. However the UK safety record with reduced voltage systems is good and it should be borne in mind that the very nature of construction sites is potentially damaging to delicate devices such as RCDs which should be regularly checked for physical damage and to confirm their electrical characteristics.

Note: BS 7671 requires a maximum disconnection time of 5 seconds and 0.2 seconds disconnection time for fixed and moveable equipment respectively at 230 V. Moveable equipment supplied at 400 V must be disconnected in 0.05 seconds. Maximum values of earth loop impedance to achieve these times, for 230 V are tabulated in BS 7671.

If the required disconnection times are not achievable by this method, then supplementary bonding can be applied and the inequality $R \leq 25/I_a$ may be used to ensure contact with exposed conductive parts does not give rise to danger. R is the resistance of the supplementary bonding conductor between simultaneously accessible exposed conductive parts and extraneous conductive parts and I_a is the operating current of the protective device. In the author's view, this is not a satisfactory means of achieving compliance with BS 7671, due to the changing circumstances which prevail on construction sites.

Supplies for portable hand lamps in confined or damp conditions

(1) SELV is suitable but the extra low voltage must be no greater than 25 V.

(2) Reduced low voltage using 50 V single-phase and earthed centre tapped transformers.

Standards for equipment

- Distribution assemblies should be to BS 4363 and BS EN60439-4 (was BS 5486).
- All other equipment should have an Index of Protection of at least IP44.
- All plugs, sockets and cable couplers must be to BS EN60301-2 (was BS 4343).

Electricity supply arrangements

Installers should note that public electricity suppliers (PES) may be unwilling to make available a PME earthing terminal to construction sites. This means that in the event of a conventional TN-S system, or a dedicated PES local transformer not being available, it will be necessary to establish a reliable connection with the general mass of earth through the use of earth rods, or other independent means of earthing.

AGRICULTURAL AND HORTICULTURAL PREMISES

The risks

The risks are from:

- wetness/wet skin;
- proximity of earthed metal;
- arduous conditions of service;

plus

- fire;
- rodent attack on wiring systems.

The risks associated with agricultural premises are accentuated where livestock are concerned, owing to their apparently greater sensitivity to electric shock. This has long been held to be due to the distance apart of the limbs of a quadruped, which can give rise to a greater substantial touch voltage in the event of an earth fault. A voltage gradient across the soil in the vicinity of an earth rod can easily approach levels which are fatal to livestock. An earth stake or other extraneous conductive part will set up concentric lines of constant potential in the soil under earth fault conditions, with maximum voltage gradient (and therefore touch voltage) at the earth rod, reducing to zero, many metres away.

The plastic used in modern cable systems is subject to attack by rodents. Apart from causing functional failure, this can contribute to risk of shock and fire.

The protective measures

Installations for general use

SELV may be used to provide protection against direct and indirect contact. However, supplementary protection against direct contact must be provided by barriers or enclosures to IP2X – or insulation capable of withstanding a test voltage of 500 V for 60 seconds.

Measures for protection against direct contact such as the use of enclosures, obstacles and placing out of reach, are not excluded, although it is a requirement that all socket outlets should be further protected by an appropriate residual current device.

In respect of protection against indirect contact, maximum disconnection times are reduced and in these installations, disconnection must occur within 0.2 seconds for a nominal system voltage of 230 V. This should be achieved by designing and constructing within stated earth loop impedance maxima for a given protective device. For distribution circuits and circuits supplying fixed equipment a disconnection time of 5 seconds is permitted. A further relaxation is available for circuits connected to distribution boards, otherwise requiring lower disconnection times, provided that either specific limitations are placed on protective conductor impedances, or extensive equipotential bonding is applied.

Where a residual current device is installed, BS 7671 gives limiting values for circuit parameters such that the product of earth loop impedance and the rated residual operating current for the device, should not exceed 25 V.

The above comments for indirect contact apply to TN systems. Where TT systems are concerned, there is a further requirement for each circuit that:

$$R_a I_a \le 25 \text{ V}$$

R_a is the sum of the resistances of the earth electrode and the protective conductors connecting it to the exposed conductive parts, I_a is the current causing the automatic operation of the protective device within 5 seconds. In practical terms, this will frequently be a residual current device and the current should then be taken as the rated residual operating current.

Installations for the management of livestock
SELV may be used to provide protection against direct and indirect contact. However, where livestock is to be protected, the nominal extra low voltage of 50 V may be too high and will require reduction according to the type of livestock.

In order to protect livestock against the effects of indirect contact where low voltage is distributed, earthed equipotential bonding and automatic disconnection of supply should be employed, as described in (1) above, together with supplementary bonding of all exposed and extraneous conductive parts accessible to livestock.

Protection against fire
A residual current device with a rated residual operating current not exceeding 0.5 A should be installed to protect all equipment other than installations essential to the welfare of livestock. This should be regarded as an absolute minimum standard of protection against fire caused by earth fault currents and may require supplementing with smoke and heat detection systems, according to the type of installation. Installers should note here the use of two residual current devices in series, Regulation 531-02-09 requires the characteristics of the devices to be co-ordinated in order to achieve descrimination.

Wiring systems
Vermin may attack and seriously damage PVC wiring systems. In addition, urine from livestock and some agricultural chemicals can cause rapid corrosion and failure of steel enclosures. Wiring systems must therefore be carefully selected in relation to the risk of damage from agricultural type external influences.

Switchgear and accessories
Class II equipment should be used wherever practicable in this environment, but care should be taken over enclosure ratings because the prolonged effects of animal urine, damp and dusty conditions may reduce the effectiveness of the

Class II insulation. Individual circumstances will dictate standards, but as a guide IP44 will give adequate protection against ingress of solid bodies and liquids, unless dust or jets of water are present when IP55 or IP56 should be used.

Arrangements for isolation and switching

It should be remembered that agricultural installations are, in many respects, industrial installations with the additional hazard of livestock. Great care should therefore be given to the provision and siting of emergency switches for machinery.

Electricity supply arrangements

As with construction site electricity supplies, public electricity suppliers may be reluctant to provide a PME earthing terminal to agricultural and horticultural installations. This will be mainly due to the difficulty in maintaining the equipotential zone throughout the installation and the rare but potentially serious situation which could arise from a loss of neutral conductor.

CARAVANS AND MOTOR CARAVANS

These notes are intended to be applied to caravans, and motor caravans. They are not applicable to mobile homes and other moveable structures supplied through the use of a permanent electrical connection.

The risks

The risks are from:

* wetness;
* minimal clothing;
* arduous conditions;
* temporary nature of connections.

The protective measures

General

BS 7671 disallows certain protective measures for caravans and, in practical terms, installations will usually employ insulation and enclosures as protective measures against direct contact and earthed equipotential bonding and automatic disconnection of supply, (with supplementary protection by residual current device) as the most likely protective measure against indirect contact.

Because a caravan may require to be fully functional when low voltage is not available, extra low voltage circuits are normally employed. The separation of low voltage and extra low voltage, and the temporary nature of the electricity supply connection arrangements within a potentially metallic structure, require special consideration – in relation to the protective measures to be employed.

BS 7671 no longer requires a double pole residual current device to be installed within the caravan to provide supplementary protection by automatic disconnection of supply. This is in consideration of the fact that there is now a requirement for caravan parks to protect caravans, by means of a suitable residual current device. The view of the author is that the provision of a residual current device in the caravan gives the user protection against indirect and direct contact, irrespective of the caravan park supply arrangements. Without this independence, the user relies upon the caravan park operator to provide and maintain an essential protective measure on his behalf and this, in some circumstances, may not be acceptable.

It is a requirement that extraneous conductive parts should be bonded to the circuit protective conductor. However, on the basis of the revised definition of an extraneous conductive part being that which is '...liable to *introduce* a potential ..., etc', metal sheets forming part of the structure of the caravan, are not required to be bonded.

Selection and erection of equipment

Wiring systems are required to be of flexible or stranded conductor construction, with a minimum cross section area of $1.5\,mm^2$. Furthermore, cables are to be supported at intervals of not more than 0.4 m vertically and 0.25 m horizontally. Special attention is drawn to the need to bush or similarly protect cables where they pass through metalwork. These requirements are all due to the movement and vibration likely to be incurred by a caravan in motion. Generally, there may not be any wiring accessories with accessible conductive parts – excepting fixing screws.

Extra low voltage circuit arrangements are required to conform to the BS 7671 requirements for SELV. This places restrictions on the type and nominal voltage of the ELV source, and the selection and erection of ELV conductors. SELV nominal voltages are restricted to the preferred values of 12 V d.c., 24 V d.c. or 48 V d.c. with a dispensation to allow a.c. sources of the same (r.m.s) voltages in 'exceptional cases'. In order to secure the satisfactory electrical separation of low voltage and extra low voltage systems, cables of these two systems should be run physically segregated. Extra low voltage and low voltage sockets must not be compatible. Furthermore, extra low voltage sockets are required, as added security, to have their nominal voltage clearly and indelibly marked.

In respect of some accessories, such as dual voltage luminaires, special requirements are prescribed in BS 7671 to ensure the separation is maintained. In addition, specifiers should consider carefully the functional needs of the caravan user when planning the installation. Many caravan users are practical people who would encounter little difficulty in altering an existing installation by the addition of non-standard luminaires, without realizing the potential hazard which may be introduced.

In recognition of the fact that continual movement could damage pendant flexible cords, BS 7671 actually recommends luminaires to be fixed direct to the body of the caravan. Where it is necessary to fit pendant luminaires, the practitioner is required to ensure that the means of suspension of the luminaire are sufficiently robust for the mass to be supported and to provide adequate anchoring of the luminaire, to prevent damage whilst in motion.

As is to be expected, the means of connection of caravans is standardized, and plugs, sockets and connectors are all to be to BS 4343 standard (now re-numbered as BS EN60309-2), with the inlet socket not more than 1.8 m above ground level, but in a readily accessible position and mounted in an enclosure with a suitable cover to the outside of the caravan. The flexible cord or cable must not be longer than 25 m and BS 7671 requires it to be in accordance with BS 6007 or BS 6500. Recent convention has expected a cable rated capacity of 16 A, however modern caravans and motor caravans are fitted with substantial electrical appliances and in recognition of this BS 7671 provides for cables of up to 100 A capacity.

Vital to the safety of caravan users is the condition of the electrical installation – bearing in mind the arduous conditions and the potential for 'do it yourself' additions. BS 7671 recognizes this and prescribes a standard format notice to users advising that the installation should be inspected and tested at least once every three years.

CARAVAN SITE SUPPLY ARRANGEMENTS

The following notes will apply equally to caravans, motor caravans and tents where temporary connections are made to moveable leisure accommodation.

It is expected that supply will be via underground cable, where practicable, and suitably protected or distant from areas of potential damage from ground anchors, etc. The individual supply terminal equipment, or distribution unit, is to be adjacent to each site pitch and BS 7671 makes specific requirements in terms of the terminal equipment in so far as the enclosure positioning, socket outlet specification and related overcurrent protection are concerned. It also recognizes that supplies in excess of 16 A may be required.

It is a requirement to provide protection against indirect and direct contact by means of a suitable residual current device. It is convention to group pitches together electrically; if this is done then groups of up to three socket outlets may be protected by the same device. Groups of caravans must be fed from the same phase.

A network which feeds a large caravan site will normally be an underground system which will require considerable care in design. The use of small distribution pillars can allow for a main distributor system on either a ring or radial system which can be combined with smaller service type cables to individual groups of pitches.

It should be noted that in the event of a PME supply being provided for a caravan site, the caravan sockets must not be connected to the PME terminal. Instead, earthing arrangements should be as with a TT system with the protective conductors of caravan supply sockets connected back to an earth rod. This is underscored by a special reference to caravans in The Electricity Supply Regulations 1988, which instructs that where PME systems are employed, the supply neutral conductor must not be connected to any metalwork in any caravan.

HIGHWAY POWER SUPPLIES AND STREET FURNITURE

The risks

The risks are from:

- wetness;
- proximity of earthed metal;
- proximity of live parts;
- arduous conditions;
- potential for working at high level;
- potential for working amidst traffic.

The protective measures

In view of the traditional means employed of mounting some items of street furniture on public electricity suppliers' (PES) poles, BS 7671 makes some concessions as to the use of protection against direct contact by placing out of reach. It is, however, a requirement that operatives shall be skilled, specially trained persons where items of street furniture to be maintained are within 1.5 m of a low voltage line. Furthermore, placing out of reach may only be employed where the low voltage line is constructed to the standard required by The Electricity Supply Regulations 1988. The designer must therefore consider safe means of maintenance and a suitable methodology. The Electricity at Work Regulations 1989 place considerable obligation on designers and operators of systems requiring work to be carried out on or near live conductors. The PES will in any case have local requirements regarding design, operation and maintenance of systems to be erected in the vicinity of live low voltage conductors.

Measures may not include the use of non-conducting locations, earth free equipotential bonding or electrical separation as protection against indirect contact. Street furniture must survive in an arduous environment and is often subjected to harsh treatment, through road traffic accidents or vandalism. Protective measures should therefore be provided on the basis of a low maintenance robust system. It is most likely that these will be a combination of insulation, barriers and enclosures as protection against direct contact, with earthed equipotential bonding and automatic disconnection of supply as protection against indirect contact. It should be noted that doors in street furniture are not to be regarded as a barrier or enclosure. They are generally intended to provide access for skilled persons and do incorporate the use of a tool. However, such doors are sometimes seen in a dilapidated state of repair in service and do not in any case generally provide sufficient protection against ingress of moisture. Intermediate barriers are required by BS 7671, to prevent contact with live parts. These are to provide protection to IP2X and may only be removable with the aid of a tool.

A maximum disconnection time of 5 seconds is stipulated for fixed equipment and this should be determined through calculation of earth fault current from maximum earth fault loop impedances. Disconnection times may then be interpolated from time/current curves for the particular protective device.

Extraneous conductive parts are to be bonded back to the main earthing terminal, but this should not include metallic structures not connected to or a part of the highway electrical equipment.

If the PES gives permission to allow the cut-out in an item of equipment to be operated by others during maintenance, then such operation will require skilled or trained persons. It is likely that the PES will make requirements as to the level of skill to be employed and may wish to exercise some screening of personnel.

Where highway street furniture feeds out to other equipment, or the maximum load exceeds 16 A, then a double pole isolator (four-pole in the case of three-phase equipment) shall be installed as a switching device.

Underground highway cables should be marked or colour coded either by means of marker tape or other identification on ducts or tiles. This is to ensure they are identified separately from other services.

Accurate drawings and records should accompany the Completion and Inspection Certificate. This is to provide a sound basis for maintenance and help ensure that accidental excavation of live cables is prevented.

CHAPTER 11

Electrical Safety

K. Oldham Smith, CEng, MIEE, MCIBSE

(Electrical safety consultant and formerly an HM Senior Electrical Inspector of Factories)

The electrical hazards of shock, burn, fire and explosion, exact an annual toll of accidents ranging from electrocutions to minor burn injuries. Fires and explosions account for most of the dangerous occurrences which do not involve personal injury but destroy plant and property. Yet safety, which is of paramount importance to everyone, is often treated with indifference and is hardly a popular subject, except perhaps, with safety officers, the enforcement authorities and others with a professional interest in it.

Since 1833 when the Factory Inspectorate began, various Acts and Regulations for safety, health and welfare, have been promulgated to enforce safe working practices and conditions and to impose on all a duty of care for the safety of themselves and other people. Non-compliance with Regulations is a criminal offence, the penalties are severe and ignorance of the law is not an excuse.

The comments in this chapter relate specifically to UK installations and those following British practice.

LEGISLATION

Electrical work is subject to the following legal requirements:

(1) Factories Act 1961. This Act is gradually being superseded by other legislation and there are now no significant electrical requirements left in it.
(2) The Health and Safety at Work etc. Act 1974. This Act is applicable to everyone at work except those in domestic employment. It also protects others in the vicinity of the work. There are no specific electrical requirements but Section 6 is considered to be applicable to electrical installation work and imposes onerous duties on designers and constructors. In Schedule 3 of the 1987 Consumer Protection Act, Section 6 was made applicable to fairgrounds.
(3) The Petroleum Consolidation Act 1928. This deals with the storage and dispensing of petroleum spirit and requires the electrical installation to be designed to avoid ignition. The requirements are detailed in the Health and Safety Executive's guidance notes which are:

HS(G)41   *Petrol filling stations: construction and operation.*
HS(G)50   *The storage of flammable liquids in fixed tanks (up to $10\,000\,m^3$ total capacity).*
HS(G)51   *The storage of flammable liquids in containers.*
HS(G)52   *The storage of flammable liquids in fixed tanks (exceeding $10\,000\,m^3$ in total capacity).*

(4) The Low Voltage Electrical Equipment (Safety) Regulations 1989. These Regulations came into force on 1 June 1989 and superseded the Electrical Equipment (Safety) Regulations of 1975 as amended in 1976. They cover almost all electrical equipment operating on low voltage.

(5) The Electricity Supply Regulations 1988 and the Electricity Supply (Amendment) Regulations 1990. The amendment was needed following the introduction of the 1989 Electricity Act. They impose safety obligations on the electricity supply authorities and also affect contractors and consumers as they empower the suppliers to refuse to connect or terminate the supply to an unsafe installation. They came into force on 1 October 1988 and supersede the Electricity Supply Regulations 1937 – For Securing the Safety of the Public and for Insuring a Proper and Sufficient Supply of Electrical Energy and also the Electricity (Overhead Lines) Regulations 1970. The requirements have been up-dated to include protective multiple earthing of the neutral and interconnected supplies and the suppliers have to advise, on demand, the maximum fault level and earth loop impedance and the type and rating of their cut-out or circuit-breaker, at the intake. Otherwise, there are no significant differences to the former Regulations.

(6) The Building Standards (Scotland) Regulations 1990. These replace the Building Standards (Scotland) Regulations with Amendment Regulations 1971–80. Electrical installations have to meet 'Technical Standards', e.g. British Standards and/or CENELEC harmonization documents.

(7) The Cinematograph (Safety) Regulations 1955. The special requirements include separate circuits for the general and safety lighting and the projection room installation. A separate power source has to be provided for the safety lighting.

(8) Electricity at Work Regulations 1989. These took the place of the Electricity (Factories Act) Special Regulations 1908 and 1944 on 1 April 1980. Whereas the latter applied essentially to work in factories, the former are made under the provisions of the HSW Act and protect everyone at work (except in domestic employment) and others in the vicinity of the work place. They extend to mines and quarries and supersede the Coal and Other Mines (Electricity) Regulations 1956, the Quarries (Electricity) Regulations 1956 and Regulation 44 – Electricity of the Construction (General Provisions) Regulations 1961.

Other legislation

In the last few years, there have been a number of EC Directives on worker safety and there are more to come. The member states have to implement them by introducing their own legislation. Although they are not electrical, contractors need to be familiar with these new Regulations. Some of the more relevant ones are:

Management of Health and Safety at Work Regulations 1992.
Provision and use of Work Equipment Regulations 1992.
Personal Protective Equipment (PPE) Regulations 1992.
Health and Safety (Display Screen Equipment) Regulations 1992.
The Construction (Design and Management) Regulations 1994.

Non-statutory regulations

The electrical contractors' 'bible' is the Institution of Electrical Engineers Wiring Regulations – BS 7671, 16th edition, 1991. These Regulations are based on IEC Publication 364 and some parts are CENELEC harmonized. They superseded the 15th edition on 1 January 1993. Compliance is required by most specifications for installation work, to meet the requirements of the Electricity at Work Regulations 1989 and for approval by the National Inspection Council for Electrical Installation Contracting.

This list is not comprehensive but it covers the principal legislation affecting electrical work in the UK. Electrical safety for transport, aviation and marine applications is outside the scope of this book.

Enforcement authorities

Most of the safety legislation is enforced by the Health and Safety Executive (HSE) but certain work is delegated by HSE to the local authorities which are, for example, responsible for enforcing the 'Offices, Shops and Railway Premises Act' and for small workshops attached to retailers. The local authorities are also responsible for enforcing Part 1 of the Petroleum Consolidation Act which applies to fuel dispensing on garage forecourts.

The work of the HSE is divided on an industry basis and as electrical installation work is part of the construction industry, queries about electrical safety are referred to the Principal Construction Inspector at the local area office. Maintenance queries, however, are referred to the Principal Inspector for the relevant industry.

The delegated inspection work is usually handled by local authorities' Environmental Health Departments whence advice may be obtained. Enforcement of the Low Voltage Electrical Equipment (Safety) Regulations is carried out by the local authorities' Trading Standards Officers and by the HSE Inspectorate. The former deal with consumer equipment, i.e. apparatus used in domestic premises, and the latter with equipment for use at work. The Trading Standards Officers are not likely to be electrically qualified and so refer suspected apparatus for test and report to an approved testing laboratory.

The Electricity Supply Regulations are enforced by the DTI's Engineering Inspectorate from whom advice may be obtained.

Although the enforcement authorities provide some advice, they are limited by the statutes which brought them into existence and which define their powers and also by their financial and manpower resources. Independent safety consultants are available to provide more detailed advice and their help should always be sought in the event of a prosecution by the enforcement authorities.

Application of the regulations

Some authorities have issued guidance notes on the regulations to enable the layman to better understand them. One of the most helpful of these publications is HSE's *Memorandum of Guidance on the Electricity at Work Regulations 1989.* See also DTI's *Guide to the Low Voltage Electrical Equipment (Safety) Regulations 1989* and HSE's *Essentials of Health and Safety at Work* which deals, *inter alia*, with the 1992 Regulations listed above under 'Other legislation'.

HSE have also issued a free booklet about the new Construction (Design and Management) Regulations 1994 entitled *CDM Regulations, how the Regulations affect you, PML 54* and will issue an Approved Code of Practice. The Regulations came into force on 31 March 1995. These Regulations are of major importance to all construction industry contractors including electrical contractors who need to understand them and implement their provisions. Every contractor should acquire the IEE's Guidance Notes on the Wiring Regulations 16th edition which are published as separate documents and entitled:

Inspection and testing
Isolation and switching
Selection and erection of equipment
Protection against fire
Protection against electric shock
Protection against over currents
Special installations and locations.

together with the *On-site guide* which deals with small installations. Regulations are obtainable from HMSO. HSE and DTI publications respectively from HSE Books, PO Box 1999, Sudbury, Suffolk C010 6FS or from some Dillons Bookstores and DTI, Consumer Safety Unit, 10–18, Victoria Street, London SW1H 0NN. IEE publications are obtainable from the Institution of Electrical Engineers, PO Box 96, Stevenage, Herts SG1 2SD.

The Construction (Design & Management) Regulations 1994
These Regulations detail and extend the requirements of the Health and Safety at Work etc. Act 1974 for the construction industry. The onus for safety begins with the designer who has to provide a design which can be safely constructed. He has to co-operate with the planning supervisor, who can be the principal contractor, to ensure safety during construction.

The planning supervisor has to prepare a health and safety plan, incorporating safety rules, prior to the start of the work on site and ensure that it is understood by the principal contractor who has to implement it. The planning supervisor has to maintain a health and safety file.

The principal contractor has overall responsibility for safety and to ensure this gives directions to sub-contractors and sees that they co-operate with each other.

On a development, the electrical contractor is usually a sub-contractor but would be the principal contractor for a rewiring project, for example, where no builder is employed.

Safety policy statement

Under Section 2(3) of the HSW Act, an employer has to prepare a statement of his health and safety at work policy and the organization and arrangements for carrying it out. This entails the same risk assessment which is required in the 1992 Management of Health and Safety at Work Regulations. In both cases, reassessments are required to cater for hazard changes consequent, for example, on new business practices or the introduction of new equipment. If the safety policy statement is written on the form, provided for the purpose, by HSE entitled 'Our health and safety policy statement' it should be sufficiently comprehensive to satisfy the Act and the Regulations. HSE have a number of guidance booklets which are helpful to the compiler of the statement.

Safety rules

A copy of the firm's safety rules should be attached to and be part of the statement. During the required safety training, employees should be made conversant with these rules and provided with a personal copy of them. Electrical contractors are confronted with two types of hazard, those endemic to the trade and those due to site conditions. The safety rules should cater for the former. To provide for the latter, which are mainly the responsibility of the occupier or main contractor on construction sites, the contractor's safety officer should ascertain what the hazards are and agree the appropriate safety precautions with the occupier/main contractor and then instruct his work force.

Although senior management bears the ultimate responsibility for safety, the senior executive concerned should not concoct the rules himself and then impose them on the employees. He would be better advised to chair a small committee, comprising the safety officer, electrical engineer, an electrician (preferably the EETPU shop steward) and in the case of large concerns with their own electrical installation and maintenance staff, technologists familiar with the processes and associated hazards.

The rules should start by stating the objective, which is to have an electrical installation which is safely constructed, maintained and used and which complies with the law and that those concerned are suitably trained to avoid danger. They should also state who is responsible for safety and safety training and the duties of employees and others working in the premises, self-employed or employed by someone else, to observe them. It may be convenient to sectionalize the content by grouping the rules which apply to users, on the one hand, and to those responsible for installation and maintenance, on the other. The rules should cover *inter alia*:

(1) The duty to know and observe any relevant statutory regulations such as those already listed and the non-statutory IEE Regulations.
(2) Prohibition of the use of defective apparatus.

(3) Reporting accidents.
(4) Training in artificial resuscitation.
(5) Appointment of persons authorized to issue permits to work.
(6) Limitation of 'live' working to competent persons.
(7) The 110 V, CTE system for portable tools.
(8) Planned maintenance.
(9) Use of gloves, boots, mats, screens, insulated tools, etc.
(10) Test procedures.
(11) Test instruments.
(12) Permits to work.
(13) Emergency drill.

This list covers items common to contracting work and factory installations but does not purport to be comprehensive. Additional items will be needed to cater for non-electrical hazards and hazards peculiar to the processes and type of work involved in a particular organization.

SAFE DESIGN AND INSTALLATION

Section 6 of the Health and Safety at Work etc. Act refers to articles for use at work but in the definitions 'article' means any plant or component of such plant. 'Plant' is defined as any machinery, equipment or appliance. So, it would seem that electrical installations are covered by the term 'article for use at work' and are, therefore, subject to Section 6 which imposes onerous requirements on designers and installers. They are required to design and construct the installation so that it is safe at all times, test to prove its safety and provide the user with adequate information so that he can use it without risks to safety and health. There is also an obligation to ensure that components supplied for use in the installation are safe and to warn the user should any hazardous defect be discovered subsequently in any of them.

Conditions of use

In order to comply with the design safety requirements, the designer must ascertain the conditions of use and should consider the following.

Environmental conditions
Is the installation or any part of it, subject to damp, dusty, corrosive or flammable atmospheres, extremes of temperature, high altitude, vibration, impact, infestation by mould growth or insects, any type of radiation, wind, lightning, floods, earthquake or situated in a subsidence or high fire risk location? If so, the installation must be designed to be safe under these conditions.

Other legal requirements
Adverse environmental conditions may be the subject of other legislation so the designer must have regard to them. For example, Regulation 6 of the Electricity at Work Regulations has a similar list of adverse environmental conditions for

which the equipment has to be suitable to avoid danger. The designer must also be aware of non-electrical regulations which may have electrical requirements such as the Highly Flammable Liquids and Liquified Petroleum Gases Regulations 1972. In Regulation 10(8), for example, axial flow fan motors in exhaust ducts are prohibited. The designer should, therefore, specify a centrifugal fan with an external motor.

Human factors

Human factors also have to be considered. Is the installation likely to be subject to rough use or vandalism or does it have to be suitable for use by unskilled or handicapped persons or made child-proof? It must be as safe as is practicable.

Suitability of equipment

When specifying the equipment for the installation, the designer has to ensure its suitability and for this purpose should nominate the relevant standard (BS, IEC, CENELEC, etc.) and/or detail the appropriate safety requirements, e.g. equipment for use in a flammable atmosphere must be of the explosion protected type and comply with the appropriate standard or be intrinsically safe, purged or pressurized. It must also be suitable for the prospective fault level. See Chapters 1, 4 and 5. The contractor, or whoever is responsible for the construction of the installation, is also responsible for the safety of the equipment which he supplies and for ensuring that it has been tested to prove its safety. If there is any doubt about this, the installer should obtain copies of the test certificates from the maker or an assurance that the equipment has been tested and meets the requirements of Section 6 of the HSW Act or he will have to test it himself.

Safe construction

Adequate supervision of the installation work is essential to comply with the requirement that the installation shall be constructed so as to be safe. The completed installation must be tested to prove its safety. UK installations, designed and constructed and tested in accordance with the IEE Regulations should meet the requirements of Section 6 and should also satisfy the employer's obligation under Regulation 4 of the Electricity at Work Regulations, to provide a safe electrical installation.

User's instructions

There is a requirement in Section 6 of the HSW Act to provide adequate information about the installation, so that it can be used safely. The installer needs to consider this requirement carefully and will probably find that, except for the simplest installations, it will be necessary to provide a schematic and/or wiring diagram and to label all controls so that the user can identify them on the drawing. For most items of equipment, there will be a maker's instruction leaflet or manual and in most cases, it will only be necessary to hand this to the user. Supplementary instructions may be necessary sometimes, e.g. where sequence switching is used. The user should be notified of the need for periodic inspection, testing, and repair, in accordance with the IEE Regulations, to fulfil the requirement to maintain the installation in a safe condition.

Alterations

If there are alterations in the conditions of use, e.g. a change in a manufacturing process whereby a flammable solvent is introduced, then corresponding alterations are required to the electrical installation to preserve its safety. Any part of the installation, located within the flammable zone, would either have to be removed, explosion protected, pressurized, purged or made intrinsically safe.

MAINTENANCE

There is a provision in Section 2 of the HSW Act and in Regulation 4 of the Electricity at Work Regulations which imposes a duty on employers to provide and maintain plant in a safe condition and without risks to health, so maintenance of the installation must be adequate to ensure it remains safe. The 'fire party' system of maintenance whereby the repair party rush from one breakdown to the next and effect hasty repairs, does not satisfy the requirement. Planned maintenance is the best way of not only meeting the legal requirements but also of limiting breakdowns. It is suggested that all electrical equipment should be numbered and listed in a register and the frequency of maintenance decided for each item. There should be a schedule of inspection for the maintenance staff so that each item is regularly inspected and tested and any necessary maintenance and/or repairs effected. A record of the work done and the test results provide the data which enables the maintenance engineer to decide on any changes needed in the frequency of inspection, maintenance procedure or replacement time for a component or item.

TRAINING AND SYSTEMS OF WORK

Under Section 2 of the HSW Act, there is an employer's duty to provide a safe system of work and to train and supervise his employees and so ensure their safety, health and welfare. There is also a requirement in Regulation 16 of the Electricity at Work Regulations which reserves dangerous work to competent persons.

Fortunately, there is an established system of training in the electrical industry for electricians, whereby the apprentice is trained in both practical and theoretical work before becoming a skilled electrician, so problems of competence are not too difficult to resolve.

'Danger' in the electrical context refers, mainly, to electric shock or burn from contact with 'live' conductors, so work which can be carried out on a 'dead' installation is electrically safe. Most new installation work is in this category. Where, however, extensions or alterations to a 'live' installation are required then danger arises at the interface between the 'live' and 'dead' parts of the installation, for example, a new circuit fed from an existing distribution board with bare busbars. In this case, Electricity at Work Regulation 14 requires adequate precautions to be taken to prevent danger and a competent person would have to isolate and lock off the supply to the board so that the new circuit could be connected safely.

Regulation 14 permits 'live' working up to 1000 V but only in exceptional circumstances and when suitable precautions are taken to prevent injury. The precautions are not defined but would include the use of insulating stands, screens, boots, gloves, etc. and where the danger merits it, the presence of a second person to render assistance in the event of an accident. The precautions must prevent electric shock from direct contact with 'live' parts and also prevent an accidental short-circuit which could cause arcing and burn injuries to the operator.

Testing

Some testing, such as fault finding, often requires working 'live' and under pressure to find the fault, repair it and get the system back into operation as soon as possible. The urgency, however, does not justify relaxing safety precautions and the employer should have already assessed the risks of such work and have devised a safe system of work for it. As it is not likely that he can anticipate exactly the degree of danger which might arise, he would probably be well advised to stipulate that accompanied working is required and that the protective safety equipment has to be readily available for immediate use. How and when it is used is a matter for decision by the person in charge of the work who could decide, for example, that to check the voltage at an isolating switch, it is sufficient to use shrouded probes on the instrument leads with HBC fuse protection but it is not safe to do the same thing on the back panel inside a deep cubicle with 'live' components on the side panels. In this case, temporary insulating screens would be needed for each side panel.

During 'live' testing at potentially dangerous voltages, non-authorized persons should be excluded from the test site for their own safety and to avoid distracting the test personnel. For this purpose, barriers, with warning notices, should be provided.

Permits to work

Because of the enhanced danger when working on h.v. systems, it is necessary to take additional precautions to minimize the hazard. In the electricity supply industry, where there is a great deal of such work, the safety precautions are set out in their safety rules which include prescribed forms for limitations of access, sanctions for test and permits to work. In those plants where there is h.v. equipment, the permit to work system should be employed and the ESI safety rules observed whenever the h.v. apparatus is inspected, tested and overhauled. Appropriate guidance is available in BS 6626: 1985 and BS 6867: 1987 which are the codes of practice for the maintenance of h.v. switchgear and controlgear. The use of permits to work is also advocated for the more complex l.v. switchboards and control panels where, for example, they have more than one supply source and/or have double busbars and/or section switches.

The permits to work should be in triplicate and numbered. One copy is for the person in charge of the work on site, one copy is for the authorized engineer and the third is the office record.

For maintenance work the procedure is for the authorized engineer to:

(1) perform the necessary switching operations, apply the temporary earths and where applicable, tape off the safe zone and erect warning signs;
(2) record the details on the permit and illustrate with a diagram if needed;
(3) instruct the workers on the site and obtain the supervisor's signature in the receipt section of the permit which acknowledges that he understands what has to be done and which parts of the apparatus are 'dead' and safe to work on. The supervisor retains one copy of the permit.

On completion of the work, the supervisor withdraws his labour, tools and materials, signs the permit in the clearance section and returns it to the authorized engineer who removes the tape, warning signs and safety earths, restores the supply and cancels the permit.

It is advisable to appoint more than one authorized engineer so that there is always someone available to exercise the necessary authority where 'live' working is required and/or permits to work have to be issued.

Other safe systems of work

Apart from permits to work, there are a number of other safety precautions which can be employed. For example, an electrician may want to work on a circuit fed from a remote source of supply. If the circuit is fed from an isolator, switch or switch-fuse, he should padlock the switch in the 'off' position and pocket the key so that no-one else can operate the switch. He should also put a cautionary notice on the switchgear so that others will know the circuit is out of operation.

For circuits fed from distribution fuseboards he should, at least, remove and pocket the fuses to isolate the relevant circuit and again, display a warning notice. It is also advisable to padlock the door but if this is done, the warning notice should indicate the whereabouts of the electrician in case the fuses of another circuit operate.

If there is no provision for locking off, some switches which are interlocked with the hinged cover can be switched off and the cover tied open to prevent the switch being closed. Again, a warning notice is necessary and in this case, barriers are required to keep people out of reach of the exposed 'live' parts.

Less elaborate precautions may be taken where the control switchgear is at hand and within sight of the electrician.

Electricians should be wary of relying on single-pole isolation in the case of lighting circuits, as Fig. 11.1 shows. If an electrician opens the switch in room B to work on the lighting point and disconnects the looped-in neutral, operating the switch in room C would make the broken return conductor 'live' at 240 V to the neutral and to the protective conductor. If the line conductor is looped at the ceiling rose instead of at the switch, as shown in room C, there is an additional risk from the 'live' loop-in terminal.

All these precautions are necessary to comply with Electricity Regulation 22 which calls for adequate precautions to be taken to prevent any conductor or apparatus from being accidentally electrically charged when persons are working on it.

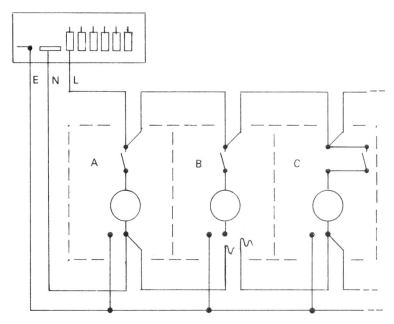

Fig. 11.1 Remove fuse for safe isolation of lighting points.

TOOLS

Electricians use a range of portable tools which can be obtained for operation on mains voltage or 110 V. Portable mains voltage apparatus is more likely to cause electric shock accidents than fixed apparatus. Although not mandatory, it is advisable to counter this risk by using 110 V portable tools, supplied from a circuit which eliminates the possibility of a dangerous electric shock. This is done by transforming the mains voltage supply to 110 V single- or three-phase and earthing the mid-point or star point of the secondary winding. See Chapter 5. As this system has been successful in virtually eliminating portable apparatus shock accidents on building sites, it is being adopted to an increasing extent in factories where the transformers are located at appropriate load centres and feed a 110 V distribution system of socket outlets.

Confined, conductive and/or wet locations enhance the risk of electric shock and in these places additional precautions are needed. The risk can be avoided altogether by using pneumatic or hydraulic tools but if these are not available and 110 V tools have to be used, further protection can be afforded by using Class II tools (double or all insulated) and/or by protecting the circuit with a residual current device (red), designed to operate on fault currents of about 30 mA in not more than 30 ms.

As tungsten filament lamps are readily available for very low voltages, hand-lamps should be supplied, at not more than 50 V, from a single-phase transformer with the secondary winding centre tapped to earth or from an unearthed safety extra low voltage circuit.

Electric arc welding sets should be fitted with an l.v. device so that the electrode holder is never energized at a voltage exceeding the maximum arc voltage of about 40 V, Fig. 11.2. On open circuit, the control circuit transformer (1) provides a signal voltage of about 20 V on the electrode (2). To start welding, the operator touches the workpiece (3) with the electrode thus completing the control circuit. The contactor (4) closes and completes the welding circuit enabling the arc to be struck. Breaking the arc de-energizes the contactor control circuit and the contactor opens and the signal voltage is restored by the auxiliary contacts (5) on the contactor.

Operators working in these potentially dangerous locations, i.e. confined, conductive and/or wet places, should be accompanied.

BURIED CABLES

Over half the accidents in the construction industry are burn injuries, from the arcing consequent on damage to underground cables. Damage occurs during excavation work, usually from pneumatic drills and hammers but also from picks, forks, the steel rods used in bar-holing and even from the sharp edges of shovels.

When excavating trenches for cable laying, the contractor is obliged to take adequate precautions to avoid danger from cables already there for compliance with Section 2 of the HSW Act and Regulation 14 of the Electricity at Work Regulations. The precautions are detailed in HSE publication HS(G)47 *Avoidance of danger from underground services* and are given below.

Before excavating:

(1) Obtain from the local authority the maps of buried services in the highway including cables. If however this information is not available or incomplete, the contractor will have to get it from the service owners.
   Note: Under the New Roads and Street Works Act 1991 local authorities have to keep a register of information of underground services in the highway and are supposed to co-ordinate the activities of excavators.

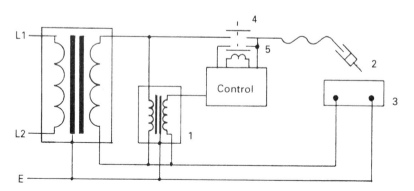

Fig. 11.2 Low voltage device.

(2) On private property, contact the site owner for the information.

(3) Mark the ground surface to indicate the locations of the buried services.

(4) Use a cable and pipe locator as a further and more precise check having regard to the limitations of the instrument. Unloaded cables are more difficult to detect so lack of a reading does not necessarily mean that there is no cable there.

Excavating:

(5) Do not use power tools within 0.5 m of any indicated cable except with short bits to break up hard surfaces.

(6) Dig trial holes to expose the cable(s) using careful manual methods.

(7) Use the cable locator as each layer of material is removed.

(8) Where existing cables clash with the excavation, expose them, using careful hand methods, working from the trial holes along the route.

(9) Support the exposed cables where necessary and protect them against damage while they are exposed.

(10) Before backfilling, contact the cable owner and agree on the procedure for protecting his cables.

(11) The contractor has a duty to safeguard others, apart from his own employees and should, therefore, take suitable precautions when laying the new cables, to avoid creating a trap for unwary future excavators.

(12) If the location is likely to be subject to frequent re-excavation, as on developing estates, all possible precautions should be taken, i.e. armoured cables should be used, protected by cable tiles and further protected by marker tape.

(13) At other locations, less elaborate precautions are needed but unarmoured cables should always be protected by cable tiles.

(14) The location of the cable should be shown on the installation 'as fitted' drawing and handed to the customer on completion of the work.

Apart from underground locations, cables are concealed elsewhere, such as in wall chases which are subsequently plastered over. Such cables could be a shock and/or burn hazard if damaged, so before drilling or cutting into walls, floors or ceilings, enquiries should be made as to the location of any wiring and a cable locator used as a check. Again, when concealing cables in walls, floors and ceilings, their position should be shown on the record drawings.

OVERHEAD LINES

Every year, overhead lines are the cause of a number of electrocutions in the construction industry. The victims are usually handling loads suspended from cranes and are electrocuted when the crane jibs touch an overhead line conductor and the victim becomes a conductor between the line and earth. A similar type of accident occurs when the victim is standing on the ground and touching the metalwork of a piece of civil engineering plant, such as an excavator, which becomes live when the jib is accidentally swung into contact with an overhead line conductor. Most of these accidents occur on construction sites.

When designing the temporary distribution system for a construction site, the temptation to economize by using an overhead line system should be resisted if possible and an armoured cable system used instead in accordance with the recommendations of BS 7375: 1991, *Code of practice for distribution of electricity on construction and building sites*. See also Chapter 5.

There may be, however, special circumstances which justify an overhead line distribution system. An example might be a chemical works on a large site with considerable distances between the load centres and where it is economical to install the permanent distribution system for use both during construction and subsequently after completion. In these circumstances, the design and construction should meet the requirements of the relevant parts of the Electricity Supply Regulations 1988 to ensure compliance with the Electricity at Work Regulations and the HSW Act. The contractor is also under the obligation of Section 6(1)(c) of the HSW Act to make available adequate information about any conditions necessary to ensure safety and should, therefore, advise the occupier, usually the main contractor, of the need for barriers to fence-off the overhead lines as shown in Fig. 11.3 and keep the civil engineering mobile plant at a safe distance from

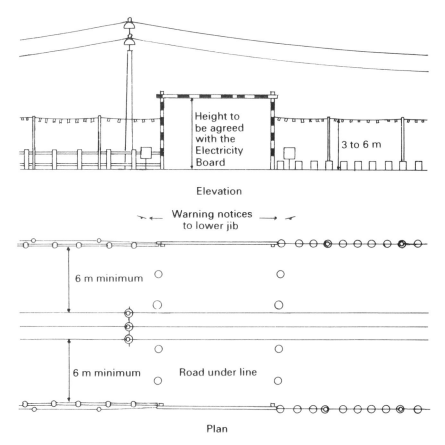

Fig. 11.3 Fencing-off overhead lines on construction sites.

them. See also HSE guidance note GS 6, *Avoidance of danger from overhead electrical lines*.

ELECTRIC SHOCK

The number of electrocutions per annum of people at work is, fortunately, small and has never yet attained three figures but electric shock accidents are common and are frequently accompanied by burn injuries at the point(s) of contact.

Shock parameters

Electric shock data have been compiled by a number of researchers who have experimented mainly on animals. Their results are not entirely satisfactory because of the difficulty of relating animal responses to humans. Some work, up to 'let go' current severities, has been carried out on human volunteers and has been useful but there are still unexplored areas and the data is incomplete. What follows, therefore, is based on present knowledge and logical probability.

The body is a non-linear resistor with an inverse characteristic with respect to the applied voltage. The resistance varies also with sex, state of health and between individuals. Most of the resistance is in the skin. Under dry conditions, point contacts and safety extra low voltages, the resistance may be thousands of ohms and there may be no shock sensation but a wet skin considerably reduces the resistance and a slight shock may be felt if sufficient current flows between the contact points. The shock sensation is frequency dependent and is at a maximum at about the mains frequency of 50 Hz and diminishes at higher or lower frequencies. The shock sensation disappears at frequencies exceeding about 1.0 MHz and also on a ripple-free d.c. However, contact with a d.c. conductor produces an initial shock as the current rises from zero to a steady state value and another when contact is broken and the current again drops to zero. Any sudden change in current produces a shock sensation, so shocks are experienced from pulsating or oscillating currents but not from a pure d.c. and not anomalously from very high frequency a.c.

There is a considerable difference in response to body currents between individuals but in general and for mains frequency, i.e. 50 Hz in the UK, the responses to sustained currents are as follows: 0.5 to 2.0 mA is the threshold of perception with no deleterious effect or unpleasant sensation. Over 2.0 mA the shock sensation increases with increases in current from a mild to a painful shock and at about 9.0 mA for a man and 6.0 mA for a woman, it begins to become difficult to break contact, due to partial muscular paralysis. At 20 mA and over, muscular paralysis inhibits the ability to 'let go' and the victim is liable to become unconscious through paralysis of the motor nerves. He may stop breathing and/or his heart may stop beating. At currents in excess of about 50 mA ventricular fibrillation, i.e. unco-ordinated operation of the heart valves, may occur and the victim's chances of survival are poor as this condition needs immediate specialized treatment.

Time is the other factor in electric shock danger. It is known that considerable shock currents can be tolerated for a very short time and it is possible to produce safety curves of shock currents plotted against time. These will be found in BS PD

6519: 1988, *Guide to the effects of current passing through the human body*, and are used by the designers of rcds to determine the performance characteristics of their products.

Harmless momentary shocks, however, can cause accidents because of the involuntary reaction of the victim. For example, a momentary shock can cause the victim to fall off a ladder and injure himself.

Rescue

To render assistance to a person undergoing an electric shock, a rescuer should first interrupt the circuit by switching off the supply or pulling out the plug. If he can do neither, he should try to break the victim's contact with the 'live' apparatus by either dragging him clear, using insulating material, e.g. his clothing, gloves, a rope or plank, or if the 'live' apparatus is portable, by pulling it away from him by the insulating sheath of the cable. On no account should the rescuer touch the victim's skin or he may become a victim himself.

Next, if the victim is unconscious, the rescuer should send someone else for medical help and immediately start artificial resuscitation and keep it up until the victim revives or medical help arrives and takes over the task.

As electricians are most vulnerable to and suffer from more electrical accidents than other people, they should be trained in simple resuscitation techniques so that in an emergency, they can render skilled assistance.

The techniques are described and illustrated on display placards which are available from a number of sources listed in a free leaflet, Form F731, obtainable from HSE. Such a placard, however, is a poor substitute for the services of a trained person. Success is very dependent on time and it is vital to commence the correct treatment as soon as possible and preferably without the delay engendered to peruse and absorb the placard instructions. There are various resuscitation methods. The most favoured is the mouth-to-mouth method with heart massage.

PROTECTION FROM ELECTRIC SHOCK

Supply systems

In the UK, the l.v. distribution system, i.e. 415 V three-phase 50 Hz, has a neutral which is earthed at the source, usually the star point of the secondary winding of the supply transformer.

The declared voltage was changed in January 1995 to a new EU standard of 400 V three-phase, 230 V single-phase, within the limits of +10% to −6% which will be increased to ±10% in January 2003 but the actual statutory variation permitted in the UK was ±6% and the supply authorities can maintain the actual voltage within the EU limits without changing present practice and will do so for the present.

In urban areas, most of the systems are TN-S, i.e. separate neutral and protective conductors but since the inception of PME (protective multiple earthing of the neutral) TN-C-S systems are coming into use and superseding the TN-S systems. In TN-C-S systems the supply authority uses CNE (combined neutral and earth) cables. The CNE conductor is earthed at several points including the star point of the transformer. In some rural areas, overhead TT systems are used, having no protective conductor between the substation and consumer, but the neutral is earthed at the transformer star point and each consumer has to provide an independent earth electrode for his own installation. These systems are being replaced by TN-C-S systems.

However, irrespective of the system of supply, the consumer's installation relies on the same safety principle for protection against electric shock which is that 'live' conductors, i.e. phases and neutral, are insulated from earth and from each other and that touchable metal parts are earthed. This ensures that in the event of an earth fault, there will be a minimum potential difference between the metalwork and anything else in the vicinity at earth potential, thus minimizing the possibility of an electric shock. A further safety precaution is provided by the operation of the fuse or circuit-breaker, rendering the faulty circuit dead.

Earth faults

In the event of an earth fault, the two main safety requirements are (1) to protect persons from electric shock and burn and (2) to minimize damage to the installation and avoid a fire.

When a fault occurs the metalwork of the installation attains a potential with respect to the star point of the transformer and the main body of the earth to which it is connected. These potentials are not the same because of the value of the volt drop, due to the resistance of the earthing electrodes of the star point, at the substation and the equipotential bonding at the fault location. The potential attained by the metalwork is dependent on the volt drops in the circuit between the supply transformer and the fault and these are proportional to the circuit impedance. Figures 11.4 and 11.6 show typical circuits for TN-S and TN-C-S systems and Figs 11.5 and 11.7 the corresponding impedance diagrams.

If, for example, the line impedance, i.e. $Z_1 + Z_2$, is the same as the return impedance, i.e. the resultant of the series and parallel impedances between the fault and the star point, then the potential on the case of the appliance will be half the supply voltage, say 120 V. If, however, the line impedance is greater than the return impedance, the case potential will be less and vice versa. The shock voltage should be only a fraction of the case potential and is represented by the volt drop between the appliance case and the building steelwork. This volt drop is proportional to the impedance of Z_5 and Z_M in parallel. The determining factor is Z_5 which should be a fraction of an ohm whereas Z_M will be in excess of the asymptotic value of 650 Ω. Provided that the equipotential bonding, the earthing contacts of the plug and socket and the several connections are making effective contact, the value of Z_5 should be low and the corresponding shock voltage harmless.

If the appliance were to lose its protective conductor, however, the case would attain a potential approaching that of the supply and the victim would receive a dangerous shock.

The shock duration and damage to the installation, consequent on the fault, are dependent on the operation of the overcurrent protective device, fuse or circuit-breaker which in turn, is dependent on the value of the earth loop impedance.

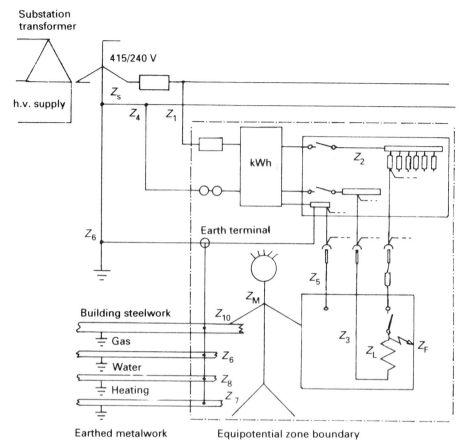

Fig. 11.4 TN-S system (using the notation outlined in Fig. 11.5).

For example, the earth loop impedance at the socket-outlet in Fig. 11.4 consists of the sum of the impedances of the phase conductor from the socket to the transformer, the source impedance and the impedance of the earth return, made up of the cable armour and/or metal sheath in parallel with the impedance of the ground and bonded metalwork to the earth terminal and thence to the socket outlet, i.e. $Z_2 + Z_1 + Z_S +$ the impedance of Z_6 in parallel with Z_7, Z_8 Z_9, and $Z_{10} + Z_5$. The actual earth loop impedance is slightly less because $Z_L + Z_3 + Z_4$ is in parallel with $Z_2 + Z_1 + Z_S$ and Z_M is in parallel with Z_5.

If an earth fault occurs at the socket outlet, the loop impedance must be low enough to ensure the distribution board fuse will interrupt in not more than 0.40 seconds. Where however it is not possible to attain a sufficiently low earth fault loop impedance, the interruption time may be increased to a maximum of 5 seconds provided that additional bonding is provided to ensure that, in the event of an earth fault, no dangerous potential differences can occur between items of touchable metalwork within the equipotential zone. See IEE Regulations 413–02–13 to 16.

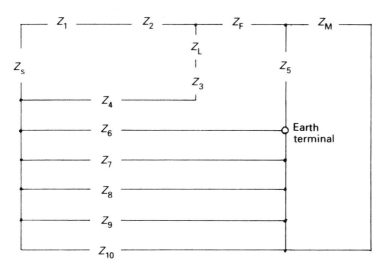

Fig. 11.5 TN-S system impedance diagram:
Z_s = source; Z_1 = source phase to main switch; Z_2 = main switch to appliance; Z_L = load; Z_3 = appliance neutral to main switch; Z_4 = main switch to neutral to source; Z_5 = appliance case to earth terminal; Z_6 = earth terminal to source; Z_F = fault; Z_M = victim; Z_7 = victim to central heating to source*; Z_8 = victim to water pipe to source*; Z_9 = victim to gas pipe to source*; Z_{10} victim to building steelwork to source*. (*Through the ground.)

To determine the earth loop impedance, the usual practice is to ignore the fault impedance (which is unknown anyway) and the parallel circuits referred to in the preceding paragraph, and use the reading of the earth loop impedance tester for existing installations and for new installations obtain the earth loop impedance figure, at the intake, from the supply authority and add the calculated value for the internal circuit.

It is evident that as the rating of the protective fuse or circuit-breaker increases, the permissible value of the earth loop impedance, to ensure operation in the event of a fault, decreases. The other factor to be taken into account is the time/current operating characteristics of the fuse or circuit-breaker. So the earth loop impedance must be low enough to ensure that in the event of an earth fault the protective device operates quickly.

The maximum disconnection time should, ideally, ensure circuit interruption before the victim suffers a harmful shock. Taking the worst case, a 240 V shock and a body resistance of say 1000 Ω, the shock current is 240 mA and the curve in BS PD 6519 would require a disconnection time of about 30 ms. However, 240 V shocks due to earth faults are unlikely, so for most applications, a much longer disconnection time is compatible with safety and the time/current operating characteristics of conventional fuses and circuit-breakers, see Table 11.1.

The required earth loop impedance value may be unattainable for heavy current circuits exceeding about 300 A and in these cases it is necessary to use earth leakage protection. A residual tripping current value should be selected to

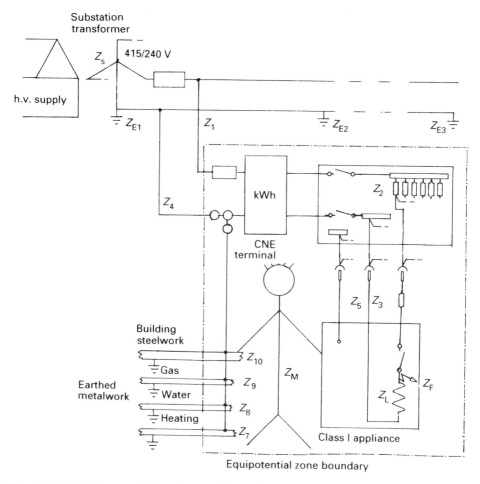

Fig. 11.6 TN-C-S system (using the notation outlined in Fig. 11.7).

Table 11.1 Maximum disconnection time for electric shock protection at mains voltage.

Location	Conditions	Maximum (s) fixed apparatus	portable apparatus
In equipotential zone	Normal	5.0	0.40
In equipotential zone	Abnormal	0.4	0.04
Outside equipotential zone	Normal	0.4	0.04
Outside equipotential zone	Abnormal	0.4	0.04

Note: Abnormal conditions exist where there is an increased risk of electric shock as in confined, conductive or wet locations.

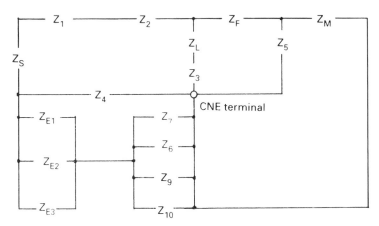

Fig. 11.7 TN-C-S system impedance diagram:

Z_S = source; Z_1 = source phase to main switch; Z_2 = main switch to appliance; Z_L = load; Z_3 = appliance neutral to main switch; Z_4 = main switch to neutral to source; Z_F = fault; Z_5 = appliance case to CNE terminal; Z_M = victim; Z_7 = victim to central heating to Z_{E1}, Z_{E2}, Z_{E3} to source*; Z_8 = victim to water pipe to Z_{E1}, Z_{E2}, Z_{E3} to source*; Z_9 = victim to gas pipe to Z_{E1}, Z_{E2}, Z_{E3} to source*; Z_{10} victim to building steelwork to Z_{E1}, Z_{E2}, Z_{E3} to source*. (*Through the ground.)

avoid nuisance tripping, not exceeding 500 mA and complying with the IEE Regulations, whereby the product of the earth loop impedance and the residual tripping current does not exceed 50. A time delay may be employed to avoid tripping on transients but the tripping time should not exceed the values in Table 11.1.

The required earth loop impedance may also not be attainable where TT systems are used, particularly in rocky areas and other high impedance ground conditions. Again, earth leakage protection is necessary, as the high earth loop impedances usually found in these conditions preclude protection by overcurrent devices. Only residual current-operated devices should be used as the IEE Regulations have been amended to preclude the use of the fault voltage operated devices with effect from 1 January 1986.

The maximum permissible disconnection times in Table 11.1 include those recommended in the IEE Regulations and are determined by the degree of risk. For example, a fixed installation is unlikely to lose its earth connection completely and the fault potential on the metalwork is usually substantially less than 240 V. Moreover, the metalwork is not being constantly handled so the permissible disconnecting times can be longer. Portable apparatus, however, connected to the supply via a flexible cable, is more likely to lose its earth connection and is subject to more handling, so a shorter disconnecting time is required.

As a safeguard against the simultaneous occurrence of an earth fault and loss of earth continuity which sometimes occurs, for example, when the protective conductor is detached from its terminal in the plug and touches the line terminal, sensitive earth leakage protection is necessary. An rcd, tripping on earth fault currents through the body down to about 30 mA, should be used, see Fig. 11.8.

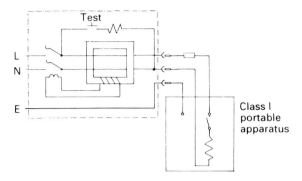

Fig. 11.8 Sensitive residual current circuit-breaker protection.

To safeguard Class I portable apparatus against losing its earth connection, circulating current earth monitoring protection may be employed. The circuit is shown in Fig. 11.9. An open-circuit in the pilot or protective conductor of the flexible cable, connecting the apparatus to the monitoring unit, trips the circuit-breaker or contactor and cuts off the supply to the apparatus. The diode and slugged relay are used on some designs to guard against a short-circuit between the protective and pilot conductors which might occur if the flexible cable is damaged. The slugged relay will not remain closed on a.c.

The protection is limited to the circuit shown, i.e. between the monitoring unit and the portable apparatus and does not cover a loss of earth continuity between the supply transformer star point and the monitor.

This type of protection is not as popular as protection by rcds, mainly because the plug and socket need as extra way to cater for the pilot conductor. If one monitoring unit is used to control a number of socket outlets, it is necessary to provide for the pilot circuit to be maintained at each unused socket outlet.

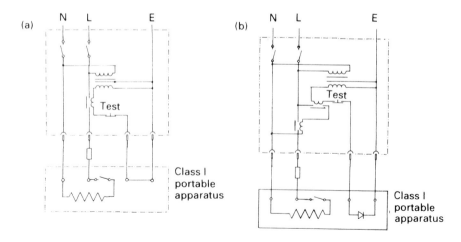

Fig. 11.9 (a) Circulating current earth monitoring protection; (b) with slugged relay and diode in the apparatus protected.

It is possible to obtain a combined earth leakage and circulating current monitoring device which can be employed where additional safety precautions are needed to cater for dangerous conditions. Two examples are the control of mains voltage supplies to the wet grinding machines used on terrazzo floors and to the steam/water pressure cleaners used to prepare vehicles for plating inspection and/ or corrosion resistance treatment and for a variety of other uses in industry and agriculture where wet conditions enhance the shock danger.

An alternative safeguard is to use Class II apparatus which is available in two categories, double insulated and all insulated. As it is unearthed, enclosed ventilated Class II apparatus should never be used in wet environments or where there is conductive dust. IP54 or IP6X enclosures are necessary. Class II apparatus has a very good record and there have been practically no shock accidents from the apparatus itself but at mains voltage there remains the hazard from a damaged flexible cable.

The use of apparatus operating at a lower voltage is a better way of reducing the shock hazard and for portable tools and other low power apparatus, 110 V, 50 V or 25 V is recommended as described under 'Tools'.

There is also another l.v. system which is used in conjunction with Class III (unearthed functionally insulated) apparatus. In the UK, it is used for special applications where an earthed system would be technically impracticable such as supplies to wire feeding devices on MIG/MAG welding sets. The system comprises a safety transformer which feeds the Class III apparatus at safety extra low voltage (SELV).

A somewhat similar system is known as 'safety by separation'. In this case, the safety isolating transformer or equivalent provides an unearthed supply of up to 500 V to the equipment. There has to be good insulation between live parts and earth, flexible cables have to be visible wherever they are vulnerable to damage and there must be no contact between the exposed metalwork of the separated circuit and the protective conductor and exposed conductive parts of other circuits. If more than one piece of apparatus is supplied, the exposed metalwork has to be bonded together by means of an equipotential bonding conductor in the flexible cables. Some of these requirements make the system impracticable for use in most circumstances but it could have applications for certain temporary installations such as location lighting and power supplies, fed from mobile generators, used by television and film companies and the rescue services and where it may be difficult to arrange for satisfactory earthing. Other safety precautions are outlined below.

Safety by position

Bare conductors may be made safe by locating them out of reach. Overhead lines and trolley wires are examples. In the latter case, however, it should be recognized that maintenance personnel may need access to various high-level services or may need to carry out building repairs or redecoration in the vicinity of the trolley wires. To safeguard them, temporary portable screens would have to be provided or the trolley wires made 'dead'. This could be a nuisance and/or affect production if it occurs frequently, so for these situations the exposed trolley wires should be replaced by the enclosed type where access to the conductors is through a slot designed to prevent anyone touching them.

Interlocking

Selection of the most appropriate system is dependent on the degree of risk, type of personnel to be protected, efficiency and facilities for maintenance and testing. The following examples indicate the various systems available.

Entry into an h.v. switchgear cubicle is a high-risk situation and in this case a trapped key system is appropriate to operate the interlock between the external remote isolator and the cubicle door.

Trapped key systems are also useful when non-electrical production personnel need access to apparatus where there is an electrical shock hazard. For example, poultry stunning cabinets, particularly the wet variety, have to be accessible for cleaning. If the cabinet is within an enclosure, the access door can be interlocked with the isolator, in the same way as the switchgear cubicle, thus making it impossible for operatives to gain access when the stunner is energized.

If it is necessary to 'lock off' a number of isolators to make a system safe to work on, this should be done by an authorized competent person. He then puts the keys in the key safe, locks it and retains its key or lodges it in a secure place, accessible only to him or other responsible persons. None of the isolators can then be unlocked and operated until a responsible person has satisfied himself that it is safe to do so and has unlocked the key safe and made the isolators' padlock keys available.

Multiple padlocking is appropriate when there is only one isolator controlling a scattered system which is made 'dead' so that a number of electricians can work on it simultaneously. The isolator is secured in the 'off' position by a caliper-shaped device with a number of holes in it, for the insertion of a padlock by each operative. Each operative has to unlock and remove his own padlock before the device can be withdrawn and the isolator reclosed.

Production line testing of domestic, electrical product, part-assemblies is a common application of interlocking using interlock switches. In these cases, the test personnel are likely to be relatively unskilled and are not competent persons and so they must not be able to touch exposed parts when 'live' or have access to 'live' terminals, yet the testing must be simple and quick to avoid disrupting production efficiency. If the test piece has exposed 'live' conductors, the testing should take place inside an interlocked box or enclosure. The operator inserts the test piece which is automatically connected or is connected by the operator who then closes the door or lid and, in so doing, automatically operates the test supply switch which is interlocked with it. The switch is 'on' only when the door or lid is closed. The interlock switch must be of the 'fail safe' type as shown in Fig. 11.10 where the circuit is made by the operation of the spring but is positively broken by the actuator. Care is required to ensure there will be no mechanical failure of the cam plunger mechanism and that failure is limited to a spring breakage or loss of temper. It must not be possible to defeat the interlock without the use of tools. Positive make switches are not suitable because the interlock can be defeated by operating the switch plunger and, moreover, such a switch can fail to danger if the spring breaks. Two interlock switches, connected in series, are sometimes used as an additional precaution.

Test pieces, having no exposed conductors, apart from the ends of the leads to

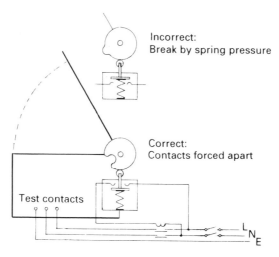

Fig. 11.10 Interlocked enclosure for production line testing.

be connected, can be tested on the bench. In this case, the test supply terminals are enclosed in a box with a hinged lid. When the lid is open, an integral switch opens the supply circuit, enabling the leads to be connected to the supply terminals which are usually spring loaded clamps. Closing the lid, automatically operates the integral switch and connects the apparatus to the supply. Again, a positive break interlock switch must be used with a suitable actuating mechanism, to prevent defeat of the interlock feature and to ensure mechanical reliability.

Interlock switches are often used on doors and the removable panels of control gear where access is required for maintenance and test purposes by electrically skilled persons. The necessity for this is dubious as such panels should not be openable without the use of a key or tools and so access is restricted to authorized and competent persons who are well aware of the hazards and who will have to defeat the interlock if they need to carry out 'live' testing. However, if such switches are used, they should preferably be of the 'fail safe', positive break type and be provided with a 'lock off' facility which can be used if the switch is in a position where it could be accidentally operated during maintenance or testing.

Other systems

There are two earth free systems called respectively 'Protection by non-conducting location' and 'Protection by earth free local equipotential bonding' which depend on their locations being earth free. As it is virtually impossible to guarantee that an earth free location is always likely to remain so, there is not much scope for these systems in the UK except for special applications such as the testing of radio and television equipment.

For electronic equipment, encased in earthed metal, there is a special high-resistance earthing system for testing, designed to eliminate the shock hazard when a 'live' conductor and the earthed metal case are both touched. The system comprises a safety isolating transformer, feeding a socket outlet via a circuit-breaker. The test piece is connected to the socket outlet. A centre tapped high resistance is connected across the circuit-breaker output terminals. The centre point is earthed via a rectifier bridge and relay coil. If either pole of the test circuit is touched when in contact with the metal case, the fault current is limited by the high resistance to say 0.50 mA which, although insufficient to produce a shock, is enough to operate the sensitive relay and trip the circuit-breaker, Fig. 11.11.

The principal application of energy limitation is in h.v. circuits used in: (1) TIG welding to ionize the air gap between the electrode tip and the workpiece, for arc initiation; (2) static elimination, also to ionize the air gap between the electrode and workpiece and to provide a conducting path for static dispersal; (3) electrostatic paint and powder spraying where the h.v. is used to charge the particles of material and induce them to adhere to the workpiece. By means of a series impedance, the potential shock current can be limited to a value below the threshold of perception.

BURNS

There are more reportable accidents from burns than from electric shock. There are, however, more shock incidents but unless the shock is accompanied by a burn injury, the victim usually recovers spontaneously and is unlikely to be off work or need medical attention and so the incident does not become a reportable accident. The burns are caused by contact or near contact with an electric arc and the victim may or may not receive a shock at the same time. Many victims claim that they received an electric shock which, when investigated, proves to have

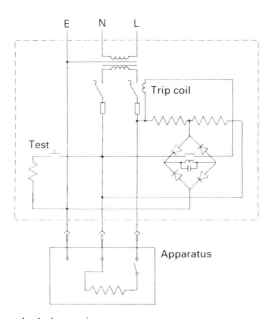

Fig. 11.11 Testing earthed electronic apparatus.

been a combination of fright and pain from the burn, i.e. a physical rather than an electric shock.

Radiation burns arise from the non-ionizing radiation produced by a variety of electrical apparatus and in the electromagnetic spectrum, range from radio frequencies to ultraviolet. Internal tissue damage can arise from the heating effect of microwave radiation exceeding $10\,mW/cm^2$. Infrared radiation is not considered dangerous as it produces surface heating of the body and is perceptible as sensible heat and so is readily avoided in excess. Intensive coherent radiation, however, whether infrared or visible light, in high-energy beams, can inflict damage, particularly to the eyes. Ultraviolet radiation produces surface burning of the skin and conjunctivitis of the eyes.

Causation and avoidance of burn injury

The worst injuries, including one or two fatalities per annum, occur when there is a low-impedance short-circuit on systems having considerable prospective fault energy levels, such as the public supply transmission and distribution networks.

Underground cables are often damaged during excavation work by a pneumatic drill, pick fork or shovel. The tool penetrates until it contacts a phase conductor, causing a short-circuit with the earthed armouring or outer CNE conductor. This low-impedance fault causes arcing which may burn the operator. High voltage cables are sometimes less hazardous than l.v. because the protection system often clears the fault before arcing develops whereas on l.v. systems, protected by excess current fuses, the arcing may continue until the fault burns itself clear as the loop impedance may well be too high for the fuses to operate. The appropriate precautions are detailed under 'Buried cables' earlier in this chapter.

The use of unsuitably insulated probes and crocodile clips, when carrying out 'live' testing, causes short-circuit accidents when the probe or clip bridges between 'live' conductors of different polarity or phase or between a 'live' conductor and earthed metal. The insulating covering should shroud the probe or clip so that only the probe tip or clip teeth are exposed.

Incorrectly switched multimeters can cause accidents. For example, if the instrument is switched to measure current when voltage is required, arcing will occur when the probes are applied to the 'live' conductors. The obvious remedy is to restrict such work to competent persons, have a checking system to ensure the instrument is correctly switched, insist on the use of properly shrouded probes and connectors and use fused leads for voltage measuring.

Maintenance and other work on 'live' apparatus can cause arcing and burn injuries when short-circuits are caused by tools bridging between 'live' conductors of different polarity or phase or between such conductors and earthed metalwork. The tools should be coated with a tough insulating material with the minimum of bare metal exposed.

Short-circuits also occur when components are being fitted or removed. For example, when adding or removing a fuse assembly on a distribution fuseboard, the bolts, nuts and washers used to secure the busbar connections can be accidently dropped inside and will frequently cause short-circuits and arcing. Magnetized and purpose-made tools to handle these components are useful safety aids. Flexible plastics sheets are useful, not only for screening off 'live' parts that are not being worked on, but as barriers to prevent components falling into dangerous areas. The sheets can be cut to shape on site and taped in position.

Plugs, sockets, connectors and appliance inlets cause hand burn injuries when a portable appliance is being connected and the component is damaged or a conductor has become detached from its terminal and is in contact with another terminal or the appliance or flexible lead is short-circuited. Under these circumstances, arcing may occur when the plug or connector is inserted into the socket or appliance inlet. Similar accidents occur when disconnecting if the cable is pulled and a conductor parts from its terminal and makes contact with another. The remedy is improved inspection and maintenance procedures to detect damage and incipient faults before danger can arise.

The sheaths of flexible cables often get damaged so that a 'live' conductor is exposed which can cause hand burn injuries to anyone handling the cable when in contact with earth. Again, the remedy is an adequate inspection and maintenance programme.

Burn injuries, on safety l.v. circuits, are usually caused by carelessness and thoughtlessness. Such circuits are often handled 'live' unnecessarily because operatives know there is no shock hazard and forget that short-circuits on such systems can be the cause of unpleasant and sometimes severe burn injuries. The remedy is to ensure operatives are aware of the hazard, make the circuit 'dead' when 'live' working is not necessary and take similar precautions as would be employed when working on higher voltage circuits.

Radiation burns arise from the leakage radiation from microwave heating and cooking equipment. The remedy is to prevent such leakage by the use of seals and wave traps at doors and other openings, careful maintenance to preserve their integrity and frequent checking with a leakage detector.

Intense coherent beam radiation is produced by lasers. To avoid burns personnel must be kept out of the beam path by suitable guarding. To permit safe access for maintenance or other purposes, the guards should be interlocked with the supply.

Ultraviolet radiation, produced by certain types of drawing office printing machinery and arc welding, for example, should be screened or filtered to prevent irradiation of exposed skin or eyes.

FIRES

The Home Office's annual fire statistics attribute about one third of all fires to electricity, but this includes electrical cooking equipment where the food is ignited but the cooker is not faulty. If cookers are excluded, about 12% of fires are caused by defective apparatus and 6% are due to a faulty wiring installation.

Design defects in a new installation, likely to cause a fire, are fortunately few because most installations are designed to comply with the IEE Regulations which ensure safety. Subsequently, however, DIY modifications, misuse and failure to maintain, may render the installation unsafe and result in a fire.

Causes and avoidance measures

There are still pre-war installations in use, using natural rubber insulated cables. The most dangerous utilize rubber-insulated conductors, tough rubber sheathed (TRS), usually two-core and earth and of rectangular section. A lot of these cables were used, without further protection, for wiring dwellings and small commercial properties. In most cases, the rubber has hardened, perished and cracked and when disturbed falls away exposing the conductors. Short-circuits are then likely to occur between 'live' conductors or to earth and the arcing will sometimes ignite the remains of the insulation and/or anything else flammable nearby.

Other installations have vulcanized rubber insulated (VRI) conductors, taped, braided and compounded and run in steel conduit. Again the rubber will have perished but is sometimes retained by the outer covering of tape and braid. The hardened material tends, however, to crack when bent, exposing the conductor. The conduit protects most of the wiring but the exposed ends which are sometimes in wooden switch boxes or in wooden pattresses supporting switches, socket outlets or ceiling roses, are vulnerable and again a short-circuit may occur and the arcing cause a fire. This old wiring should be replaced by the PVC insulated or insulated and sheathed equivalents.

Overloaded circuits are a fire hazard as the hot conductors damage the insulation which will eventually fail and permit a short-circuit to occur. Examples are sometimes found in old radial socket-outlet circuits where the DIY operator has sought to increase the supply capacity by replacing old 2 and 5 A by double 13 A socket outlets or by adding additional socket outlets to a circuit by looping without regard to the size of the cable. He then finds that the protective rewireable fuse on the distribution board is inadequate, and so increases the size of the fuse wire so that the inadequately sized conductors are no longer protected.

Again, where there are insufficient socket outlets, adaptors are used and loads, in excess of the rating of the socket outlet, connected. The multiple plugs and adaptor(s) assembly may not only overload the socket but very often the socket receptacles are damaged by the adaptor pins when the assembly is subjected to lateral pressure from mishandling or a blow. The result is a bad connection, minor arcing at the contacts, corrosion of the surfaces and overheating which ignites the insulation and anything else flammable nearby. The remedy is to eschew the use of adaptors and install an adequate number of socket outlets on ring mains.

Conductor overheating, resulting in fires, can also occur from loose connections. Pinch screw terminals in plugs and switches are prone to slacken in time, probably due to vibration, resulting in bad connections and an eventual fire. Checking terminal screw tightness should be part of the maintenance schedule.

EXPLOSIONS

Electrical explosions occur when there is a short-circuit between conductors on different phases or between a phase conductor and the neutral or earthed metal and sufficient energy is expended to cause an audible bang. In domestic installations where the line/neutral and line/earth loop impedances are comparatively high usually around $0.3\,\Omega$ to $0.5\,\Omega$, the short-circuit current will be 240/0.3 or $240/0.5 = 800$ or $480\,A$, which is enough to cause an audible bang at the fault location and at the fuse or circuit-breaker protecting the circuit which clears the fault. Such explosions do not usually cause any significant damage but if someone is close enough to the fault, a minor burn from the momentary arcing is a possibility.

At the location of the short-circuit, the arcing vaporizes the metal conductors and their insulation and these hot gases cause an immediate, local pressure rise. In a confined space, the pressure rise is more than in the open because the gases cannot disperse so readily and a more audible explosion results. In an HBC fuse, the fault is interrupted rapidly and the energy release curtailed so that the bursting pressure of the hot gases does not break the container. Where, however, the fault persists more energy is released and the resultant explosion may burst the container. Some of the worst explosions occur from street excavations when a tool accidentally penetrates a low-voltage distribution main causing a short-circuit between phases and to the earthed metal cable armouring or sheath. Very often, a fault of several megawatts causes an arc to be emitted which burns the tool operator.

A short-circuit between turns in an oil-insulated power transformer or between a 'live' conductor and the metal core, will cause arcing and vaporization of the metal and oil at the fault. The pressure rise is likely to burst the tank with a loud explosion if the pressure operated relay is ineffective. Oil circuit-breakers and oil-immersed control gear, are susceptible to explosions from carbonized oil. Arcing, at the contacts when on load and during operation, causes oil contamination and when this is excessive the carbon is deposited on the surfaces of internal insulators and tank and can form a tracking path between phases and/or between a phase and the earthed metal tank. A flashover can then occur across the contaminated surfaces or sometimes through the oil when the insulating value is sufficiently impaired by this contamination. In either event, the arcing causes a rapid pressure rise and may burst the tank with a loud explosion. The spilt oil is usually ignited and can then involve flammable materials nearby and spread the fire or set the operator on fire.

Remedies

There is much to be said for avoiding oil-filled equipment altogether, but if it is used, apparatus with pressure relief vents is preferable to minimize explosion damage together with regular inspection and testing to ensure that the apparatus is maintained in a serviceable condition. Connections should be clean and tight and contacts kept clean, adjusted and replaced when worn or damaged to ensure that arcing does not occur when on load. Some arcing, at the contacts of transformer tap changers, oil switches, circuit-breakers and controllers, takes place in normal operation and contaminates the oil – hence the need for regular maintenance. For heavy-duty operation, it may be worthwhile considering the use of a filter, through which the oil is passed continuously, to remove the carbon. Alternatively, samples of the oil should be inspected and tested periodically to ascertain when the oil needs changing or reconditioning. The tank, housing the contacts, should be opened whenever the oil is changed, the contacts maintained and contaminated surfaces cleaned. The maintenance procedures for switchgear are detailed in Codes of Practice BS 6423, 6626 and 6867 for the maintenance of switchgear for voltages up to and including 145 kV and are applicable also to on-load tap changers and controllers. Apparatus, containing large quantities of oil, should be situated in a fire-resistant location over a soakaway and surrounded by dwarf walls so that any spillage is contained and dispersed.

The procedure, for avoiding the hazard from buried cables, is described earlier in this chapter.

To minimize explosion hazards, from the wiring installation and other apparatus, ensure the protective devices are of adequate rating for the fault level, that their operation is as rapid as possible, and that they and the installation and apparatus, are regularly inspected, tested and repaired.

FLAMMABLE ATMOSPHERES

These are part of the electrical fire and explosion hazards. They are dealt with separately because of the complexities and problems peculiar to the subject. They occur when flammable gas, vapour or dust is present in a concentration in air between the upper (UEL) and lower (LEL) explosive limits. Such mixtures ignite if there is a source of ignition also present such as incendive arcs or sparks from electrical apparatus. The risk areas are divided into zones.

For flammable gases/vapours:

Zone 0 Where the flammable atmosphere is always present.
Zone 1 Where the flammable material is present during normal operations and a flammable atmosphere is likely to occur.
Zone 2 Where the flammable atmosphere is unlikely during normal operation and if it occurs it will exist only for a short time.

There is some international difference of opinion about the dust zones but the likely outcome is that the corresponding zones for dust will be labelled zones 21 and 22, Zone 21 being equivalent to Zone 1 and Zone 22 to Zone 2. There may be eventually a Zone 20 equivalent to Zone 0.

Zonal boundaries

These have to be determined by the occupier's specialist staff and marked on the site drawing in plan and elevation. It is a difficult task and requires a knowledge of the characteristics of the material and how it disperses from the source of emission. Some guidance is available in BS 5345: Part 2; 1983 which shows how to tabulate the relevant data but it does not indicate how to utilize it in zonal boundary definition. ERA Research, however, have evolved a method of doing this based on computational fluid dynamics.

Pits, trenches and ceilings, within flammable zones, should not be overlooked.

If unventilated, pits and trenches are Zone 1, because heavier than air gases can enter and remain, as are also coffered ceilings for lighter than air gases such as hydrogen. Garage pits, where petrol may be spilt, are an example of the former and battery rooms of the latter.

Zone 2 boundaries are particularly difficult to decide because they have to cater for accidental discharges from leakage, spillage or a burst container.

For dusts the problem is easier as the present Zone Z (21) usually occurs within 1 m of the point of release, e.g. emptying bags of material into hoppers and the present Zone Y (22) within about 15 m from the release point. Of course if the housekeeping is bad and a layer of dust is allowed to settle, Zone Z would have to include such locations as those where the flammable hazard would arise from contact of the layer with hot apparatus or from the dust cloud if the layer is disturbed during normal operations.

Installation design

Location of equipment
Armed with the zonal boundary plans, the next step is to determine the position of the electrical apparatus. Wherever possible, the risk should be avoided or minimized by locating the electrical equipment outside the risk areas or in Zone 2 areas rather than in Zones 0 and 1.

Zone 0 is usually inside reaction vessels or storage tanks. There is not much electrical apparatus available for use in Zone 0 and as the risk is high it should be avoided by locating the equipment outside, e.g. a container can have glazed portholes so that the interior can be viewed through one and lit from a luminaire shining through another. Zone 1 areas often occur only in the immediate vicinity of where the flammable material is processed. They are invariably surrounded by Zone 2 areas where the electrical apparatus should preferably be situated. It is sometimes possible to light outside site zones by luminaires located at high level in safe areas above the vertical zone boundaries. For flammable interiors, lighting can often be provided by positioning the luminaires outside in a safe location and shining the light through glazed panels in the roof or walls. Again, motors can be outside and drive an internal machine through a long shaft extension passing through a gas-tight seal in the wall. In some plants, it is convenient to locate the motor starters together on a starter panel in a motor control room. This room could either be in a safe area or if within the flammable zone, fed with clean air, from outside the zone, maintained at a positive pressure to ensure that gas cannot enter.

Equipment selection
Within the flammable zones, the equipment has to be explosion protected and matched to the zone as shown in Table 11.2.

Higher risk zone equipment may be used in a lower risk zone. For each category there is a British Standard and apparatus which may be tested and certified accordingly by an approved testing house. To be safe, only certified apparatus should be employed.

Table 11.2 Zones for gases and vapours.

Zone	Type	Category
0	EEx ia	Intrinsically safe
	Ex s	Special protection
	EEx m	Encapsulated
1	EEx d	Flameproof
	EEx ib	Intrinsically safe
	EEx p	Pressurized
	EEx e	Increased safety
2	EEx n	Non-sparking
	EEx o	Oil immersed
	EEx q	Powder filled

For dusts, the present position is that when the standards are revised dust protected apparatus will be marked as follows:

DIP	Dust ignition protected
A or B	European or American practice
21 or 22	Zone
Ta or Tb	Maximum permissible apparatus surface temperatures

The existing British Standard BS 6467 uses Zone Z for the more dangerous locations where dust clouds occur in normal operation and Zone Y where they occur occasionally and are of short duration but these designations will be changed to Zone 21 and Zone 22 respectively. The protection method is to have a dust-tight or dust protected enclosure to IP6X or IP5X and limit the exterior temperature of the apparatus casing when under a dust layer to prevent ignition.

Wiring

The installation work should comply with the IEE Regulations and BS 5345: Part 1 for gases and vapours and BS 7537 for dusts. Conduits should be solid drawn or seam welded with screwed joints. Where flammable gases have to be excluded from apparatus, barrier seals may be necessary. Flexible metallic conduit should be lined to prevent abrasion of the cable insulation. Aluminium and aluminium alloy conduits and accessories should not be used where frictional contact with oxygen-rich items such as rusty steel might occur and cause sparking.

Gas detectors

These devices sample the atmosphere in their vicinity and in the event of the presence of the relevant flammable gas can be used to provide a warning signal and/or operate a relay to trip the supply to electrical apparatus and/or start a ventilation fan. They are of particular use in unattended areas where a leak may not otherwise be detected, allowing the flammable material to spread perhaps into an otherwise safe area and be ignited by a source of ignition therein.

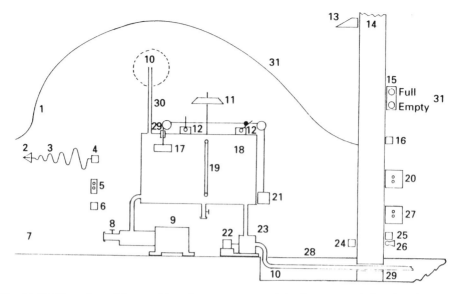

Fig. 11.12 Elevation of flammable solvent storage tank in factory yard showing the zonal boundaries, Key:

(1) Zone 2 boundary,
(2) earthing clip;
(3) earthing flexible lead,
(4) earthing switch,
(5) remote push button control for intake pump motor starter,
(6) local light switch,
(7) road tanker hard standing,
(8) intake pump,
(9) intake pump motor,
(10) Zone 1 boundary,
(11) local light,
(12) limit switches,
(13) area floodlight luminaire,
(14) factory wall,
(15) tank indicator lights,
(16) floodlight switch,
(17) float,
(18) flammable solvent tank (Zone 0),
(19) sight glass,
(20) intake pump motor starter,
(21) float counterweight,
(22) off-take pump motor,
(23) off-take pump,
(24) gas detector sensor,
(25) gas detector control,
(26) gas detector siren.
(27) off-take pump motor starter,
(28) trench (Zone 1),
(29) gland seal,
(30) vent pipe,
(31) safe area.

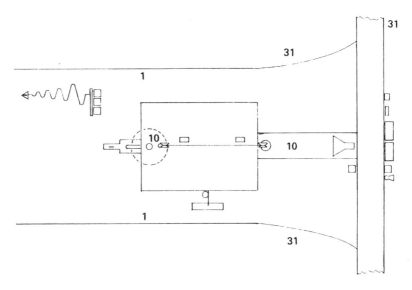

Fig. 11.13 Plan of flammable solvent storage tank in factory yard to show the zonal boundaries. Key: (1) Zone 2 boundary, (10) Zone 1 boundary, (31) safe area.

Lightning
Certain vulnerable installations may need lightning protection which should be installed to BS 6651: 1992 together with surge suppressors to prevent possible flashovers from any associated high voltage transients.

RF induction
The risk of sparking from radio frequency induction is relatively small and will be minimized by any suppressors fitted for EMC and lightning protection but if there are any RF transmitters in the vicinity, the installation should be checked to BS 6656.

An application
Figures 11.12 and 11.13 show a factory yard where there is an unattended flammable solvent storage tank. The solvent vapour is heavier than air. The tank is replenished by a road tanker through a hose connected to the intake pump. The tank contents are dispensed via the off-take pump. When the road tanker visits, any static charge is safely discharged to earth via the earthing clip, which is secured to the tanker metalwork, the flexible cable and earthing switch which is closed after the clip is applied. This has to be done before any flammable liquid is transferred.

The storage tank interior is Zone 0. There are two Zone 1s which are in the immediate vicinity of the vent and in the pipe trench. The Zone 2 boundary extends from the factory wall to the other side of the road tanker hardstanding. To avoid gas leaks, the float wire rope passes through a gland seal in the tank lid and the off-take pipe is in a sealed duct where it passes through the factory wall.

To minimize costs and maximize safety, the security floodlight is mounted in the safe area above the Zone 2 vertical boundary; its switch together with the pump motor starters, the liquid level indicator and the gas detector control panel and alarm are inside the factory which is also a safe area. The rest of the equipment is all in a Zone 2 area and has to be non-sparking so the local light and the squirrel cage pump motors can be EEx n. As the earthing and limit switches, the remote control push buttons for the intake pump motor starter, the local light switch and gas detector sensor are not non-sparking devices, they should be type EEx d excepting the sensor which should be a type Ex s.

CONCLUSION

Electrical engineers, technicians and electrical operatives have a legal responsibility to ensure that their installations are properly designed, constructed, maintained and used. If they conscientiously fulfill this responsibility, the present trend for a decline in the annual electrical accident figures should continue and be a beneficial contribution by the electrical industry to the well-being of society.

CHAPTER 12

Standards, Specifications and Codes of Practice

J.M. Faller, BA, IEng, MIEIE
(British Standards Institution)

Throughout the other chapters of this handbook on installation practice, there are numerous references to British Standard specifications, British Standard codes of practice and (in the legal sense) Regulations. It is the purpose of this chapter to explain in rather general terms the significance of standards (at national, regional and international levels), their relationship with legislation (both British and European) and the philosophy of the preparation of standard-type documents. The general principles will be illustrated by examples from the subject area covered by the handbook. In preparation it is necessary to understand clearly the difference between a 'standard' and a 'regulation'. The formal definitions of the two words are as follows:

- *Standard* A technical specification or other document available to the public, drawn up with the co-operation and consensus or general approval of all interests affected by it, based on the consolidated results of science, technology and experience, aimed at the promotion of optimum community benefits and approved by a body recognized on the national, regional or international level. It should be noted that a distinction is drawn between measurement standards of mass, length, time, etc. ('etalons' in French) and documentary standards as defined here ('normes' in French).
- *Regulation* A binding document which contains legislative, regulatory or administrative rules and which is adopted and published by an authority legally vested with the necessary power.

These two definitions taken together point to the fundamental difference between standards and regulations – the former are established essentially by voluntary co-operation and agreement between interested parties and their use is, in most cases, also voluntary. On the other hand, regulations have legal force such that failure to comply with them leaves the transgressor open to whatever penalties the law provides.

Having said that the use of standards is voluntary, it is nevertheless important to realize that, once a standard has (albeit voluntarily) been written into a contract, its provisions become binding on the parties to that contract. In such a case, failure to comply with the standard may make the offending party liable to damages in a civil action. The question of the status of those standards to which reference is made in the law, and the use of which therefore becomes more than purely voluntary, is dealt with later in this chapter.

As can be seen from its definition, the term 'standard' covers a wide (even open-ended) range of documents designed for use in a whole variety of circumstances. Almost all standards bodies (whether they are recognized at national, regional or international level) have found it necessary to distinguish between various types of standard according to the purpose for which they are intended. The two major classes of standard, and those which are of greatest interest to those concerned with electrical installations, are standard specifications (which describe a particular product or range of products in terms of dimensions, performance or safety) and standard codes of practice (which describe recommended techniques of design, installation, maintenance, etc.). Both of these types of standard inevitably include or rely on more basic types, such as glossaries, classifications, methods of testing, methods of measurement, methods of sampling, etc.

BRITISH STANDARDS INSTITUTION (BSI)

Within the UK, national standards are prepared and published by the British Standards Institution (BSI), an independent non-profit-making association incorporated by a Royal Charter granted originally in 1929 and now operating under the provisions of a consolidated Charter dated 1981, as amended in 1989 and 1992. In fact, BSI began its life as long ago as 1901 as the Engineering Standards Committee, set up as a co-operative joint venture by the leading professional engineering bodies, including the Institution of Electrical Engineers (IEE). The objectives of BSI, as laid down in the Royal Charter, are as follows:

(1) to co-ordinate the efforts of producers and users for the improvement, standardization and simplification of materials products and processes, so as to simplify production and distribution, and to eliminate the national waste of time and material involved in the production of an unnecessary variety of patterns and sizes of articles for one and the same purpose;
(2) to set up standards of quality for goods and services, and prepare and promote the general adoption of British Standards and schedules in connection therewith and from time to time to revise, alter and amend such standards and schedules as experience and circumstances may require;
(3) to register, in the name of the Institution, marks of all descriptions, and to prove and affix or license the affixing of such marks or other proof, letter, name, description or device;
(4) to take such action as may appear desirable or necessary to protect the objects or interests of the Institution.

Stripped of the solemnity of this legal prose, it will be seen that BSI's main functions are to draw up voluntary standards by agreement among all the interests concerned and to promote their adoption.

One indication of the independence of BSI is given by the breakdown of the major sources of income for standards work. Approximately 60% is derived from sales of BSI standards and other publications, approximately 25% from subscriptions paid by some 29 000 firms, trade associations, local authorities, professional institutions, etc. and the remaining 15% from a Government grant-in-aid based on subscription income. The total income (approximately £28 m for the financial year 1993/94) is a measure of the magnitude of the operation.

In accordance with the Royal Charter and the Bye-laws created under it, the Board of BSI is responsible for the policies of the Institution, subject always to the ultimate authority of the subscribing members of BSI exercised at a general meeting. The members of the Board together represent a broad range of national interests, all of which are in some way affected by the results of standards work. As permitted in the Bye-laws, the Board has delegated its responsibility for the preparation of British Standards to a Standards Board which has divided its activities into five functions:

(1) business development which assesses the needs of industry and members;
(2) an international department to lobby and to coordinate all international standards related policy issues;
(3) production which covers editing, electronic production, print production, stock management and distribution;
(4) sales which combine sales and delivery;
(5) standards development which contains all the technical departments.

Five technical departments exist within the Standards development function:

(1) Construction and Materials;
(2) Engineering;
(3) Services and Consumer Products;
(4) Electrical;
(5) Disc, the information technology specialist department.

Of these departments, the one of major interest to electrical engineers is the electrical department which, in fact, has a dual function. It is on the one hand, as the Electrotechnical Sector Board, responsible to the Standards Board for the preparation of British Standards within its subject area. On the other hand, it acts as the executive of the British Electrotechnical Committee (BEC). The BEC is the UK member in its own right of two important international standards bodies, IEC and CENELEC. Like the Standards Board, the Electrotechnical Sector Board is fully representative of all interests likely to be affected by standardization in the electrical and electronic engineering fields.

The Electrotechnical Sector Board is at the apex of a pyramid-like structure of committees responsible for the wide range of work falling within its terms of reference. The next inferior level consists of the four section committees, each responsible for planning the standards work within a major section of the electro-technical industry as follows:

- Electrical Power Engineering;
- Electronic Components and Equipment;
- Electrical Consumer Products;
- General Electrical Engineering.

Responsibility for the technical content of standards, and for participation in corresponding international work, is in turn delegated to the 250 or so technical and sub-committees, each responsible for a specific product area. For obvious reasons, great care is taken to ensure that no two technical committees have overlapping terms of reference. Depending on the size of the area of work involved, technical committees may in turn establish subcommittees and/or drafting panels.

It would be inappropriate (and possibly confusing) to include here a detailed description of all the procedures used in BSI committees to establish the content of those British Standards for which they are responsible. Nevertheless, it is important to emphasize that all of these procedures are aimed at under-pinning the two fundamental safeguards of BSI standards work, which between them ensure that the resulting standards satisfy the rigorous criteria implied by the definition of 'standard' given at the beginning of this chapter. These two safeguards are *consultation* and *consensus*.

When a new technical committee is established, every effort is made to involve all those interests likely to be affected by its work. Thus the constitution of each technical committee is intended to bring together all those with a substantial interest in the subject being discussed, wherever possible through organizations representing manufacturers, users, professional institutions, government departments, research organizations, certification bodies, trade unions and so on. Once the committee is established, the next most senior committee appoints an independent chairman, and a member of the BSI technical staff becomes its secretary and a full member of the committee in his or her own right.

In fact, the careful constitution of a representative technical committee is not of itself regarded as a sufficient guarantee that the requirement for consultation has been met. At a suitable stage, its work is exposed to public consultation by the issue, for each British Standard, of a draft for public comment. The availability of all drafts for public comment is announced in BSI's monthly journal *BSI News*, and any organization or individual may obtain a copy of any draft for a nominal fee on application. Any comment, from whatever source, received before the closing date (which is usually eight weeks after the date of announcement) must be properly considered by the responsible technical committee. Thus, by means of this double consultation (of a representative committee and of the public), every effort is made to ensure that the views of all interested parties are canvassed.

The other formal safeguard on the content of British Standards is provided by the system of decision-making by consensus. This implies that differences of opinion within a technical committee on the content of the standard are never resolved by means of a formal voting procedure; rather, discussion continues until all significant objections have been met or withdrawn. In the rare event that this process of compromise fails to produce an agreed text, recourse can be had to a formal appeal procedure, through all intermediate committees right up to the

Standards Board itself. If no other solution can be found, the Board will establish a small panel of its members to recommend a decision, which will be binding on all parties to the dispute.

The purpose of these twin safeguards (consultation and consensus) is to ensure that, once it has been approved by the responsible technical committee, a British Standard will have been drawn up with the co-operation and a general approval of all interests affected by it, thereby satisfying the definition of 'standard' given at the beginning of this chapter.

While the foregoing may explain the preparation and resulting status of British Standards, it gives no impression of the breadth or scale of the operation. Suffice it to say that there are some 3000 technical committees (of which about 1000 are likely to be active at any one time) and that there are more than 10 500 British Standards, covering almost the whole range of industrial products, from steam turbines to microprocessors, and of industrial processes, from deep-sea diving to tea-tasting. The continuing applicability of each British Standard is ensured by subjecting it to review by the responsible technical committee at least once every five years, at which time it must be confirmed, revised, amended or withdrawn.

If fast-moving areas of technology, this review is carried out much more frequently, as and when the need arises. Further, new projects concerning things which have not previously been the subject of standardization are continually being added to the programmes of work of the various technical committees to take the place of those completed. Proposals for new work are considered by the appropriate Section Committee, in response to requests from any individual or organization having an interest in the subject in question. In fact, planning the work is the major task of the Section Committee, and the division of available resources between them is a primary responsibility of the Councils of BSI.

THE INTERNATIONAL ELECTROTECHNICAL COMMISSION (IEC)

There are two world-wide standardization bodies. The older (dating from 1906) is the International Electrotechnical Commission (IEC), responsible for standardization in the fields of electrical and electronic engineering, while the younger (post-war) International Organization for Standardization (ISO) is responsible for standardization in all other fields of science and technology. In view of the subject matter of this handbook, only the former will be discussed here.

The IEC is an international non-governmental organization, whose members are the national electrotechnical committees (like the British Electrotechnical Committee in the UK) of 49 countries. Between them, these countries represent more than 85% of the world's population and account for more than 95% of the world's total electricity generating capacity. The permanent seat of the IEC, consisting of the office of the General Secretary and his staff of some 100 persons, is located in Geneva, Switzerland. The activities of the IEC are governed by a Council, which meets annually at an IEC General Meeting, consisting of representatives of all of the member national committees. Acting as an advisory and executive group on behalf of Council in the management of the technical work of the IEC, is a smaller Committee of Action, whose members are representatives of

12 national committees elected by all IEC members for six-year terms. Continuity of membership of the Committee of Action is ensured by a system whereby these terms of office overlap.

The essential constitutional duties of the IEC Council are the election of its officers (President, Vice President(s), Treasurer and Auditor) and the approval of the annual expenditure budget. This in turn determines the subscriptions payable by each national committee, distributed according to a formula based on population and gross national product which ensures that the greater portion of the financial burden is borne by those countries most able to do so.

Because they are responsible for overall IEC policy, Council and the Committee of Action may establish committees and working groups to study particular questions. Among the standing committees set up in this way are the Advisory Committee on Safety (ACOS), the Advisory Committee on Electronics and Telecommunications (ACET) and the Advisory Committee on Electromagnetic Compatibility (ACEC).

However, the main work of the IEC is carried out in more than 90 technical committees, each charged by the Committee of Action with the preparation of international standards in a specific field of electrical or electronic engineering. Each national committee has the right of participation in the work of every technical committee, so that the principle of consultation enunciated above applies equally to the international standardization work of the IEC. It is important to note that the technical work of the IEC is, in fact, decentralized as each technical committee has one of the national committees as its secretariat. Once again, the greater portion of this contribution of expertise is borne by those national committees with the greatest resources.

In order to carry out the work in the most efficient manner, technical committees may establish sub-committees to deal with specific parts of its work programme. At present, over 120 sub-committees are active and as with the technical committees, each national committee has the right of participation. Most projects are started in preparatory working groups consisting of individually appointed experts charged with the task of preparing a first draft international standard for discussion. When such a draft has been completed, it is circulated (as a 'Secretariat' document) to all national committees for comment. All comments received are discussed at the next available meeting of the parent technical committee or sub-committee and, if a sufficient measure of support is indicated a revised draft is circulated to members as a 'committee draft for voting'. The document is approved as suitable for circulation as a 'draft international standard' if a two-thirds majority of the participating members voting are in favour and not more than one-quarter of the total number of votes cast are negative. Editorial and minor technical comments are allowed at this stage. If sufficient support is indicated, the document, amended as necessary, is formally circulated for voting as a draft international standard under the six month rule. The criteria for acceptance is as before, however comments at this stage are not allowed. Approved final drafts are edited and published as international standards by the IEC. From this it will be seen that the principle of consensus is not applied in IEC work; indeed it would be almost impossible to do so. Nevertheless, the two-thirds majority required in the ballot of national committees for the approval of each IEC standard is such that major objections are overruled on only rare occasions.

Moreover, it is important to recognize the status of IEC standards as a record of the best technical solution that can be achieved by majority agreement between the member national committees spread all over the world and with their different economic and technological stages of development. In fact, the foreword of each IEC standard explains that it is recommended to national committees as a basis for the preparation of national standards on the same subject. This question leads naturally to that of the use (or, in formal language, 'implementation') of IEC standards. Although their direct use as such is not unknown, particularly by reference in contracts between manufacturers and customers in different countries, the more usual route to implementation is the indirect one, whereby use is made of a national standard whose content mirrors that of the IEC standard. It is for this reason, as well as for the more obvious practical ones, that participation in the work of an IEC technical committee on a particular subject is delegated by the British Electrotechnical Committee to that BSI technical committee which is responsible for national work on the same subject.

As far as the implementation of international standards is concerned, it has been customary to adopt, wherever practicable without change, the results of IEC work as British Standards. When the responsible BSI technical committee decides that this can be done, the resulting British Standard takes the form of an exact copy of the IEC text and carries the identification numbers of both the British and IEC Standards, e.g. BS 7638/IEC 781 *Application guide for calculation of short-circuit currents in low-voltage radial systems*.

To give some idea of the extent of IEC work in the area covered by this handbook, it is necessary only to list the titles of a handful of IEC technical committees whose standards are of particular relevance to electrical installation practice, as follows:

TC 8 Standard voltages, current ratings and frequencies
TC 16 Terminal markings and other identifications
TC 23 Electrical accessories
TC 64 Electrical installations of buildings
TC 71 Electrical installations for outdoor sites under heavy conditions (including open-cast mines and quarries).

EUROPEAN COMMITTEE FOR ELECTROTECHNICAL STANDARDIZATION (CENELEC)

Between the national (BSI) and world-wide (IEC) levels of standardization, there is an intermediate, regional level represented by the European Committee for Electrotechnical Standardization, commonly known as CENELEC from the acronym of its French title, Comité Européen de Normalisation Electrotechnique. This organization, which has its headquarters in Brussels is formally a non-profit-making association (*Association sans but lucratif*) constituted in accordance with Belgian law, formed of the 18 electrotechnical committees representing the following countries within Western Europe: Austria, Belgium, Denmark, Finland, France, Germany, Greece, Iceland, Ireland, Italy, Luxembourg, Netherlands, Norway, Portugal, Spain, Sweden, Switzerland and the UK. CENELEC also has seven affiliates from Eastern and Central Europe, the National Committees of the Czech Republic, Hungary, Poland, Romania, Slovakia, Slovenia and Turkey.

CENELEC is governed by a twice-annual General Assembly of all its member national committees and by a Technical Board consisting of one permanent delegate from each national committee (in most cases the secretary of the national committee). For reasons which will emerge later, there are comparatively few CENELEC technical committees, although constitutional provision is made for them to be established.

The main purpose of CENELEC is to remove, by mutual agreement between the CENELEC member committees, any technical differences between their national standards so that these national standards can become 'harmonized'. The process of harmonization of national standards is deemed to have been achieved when equipment manufactured in one country in conformity with the national standard of that country also conforms with the requirements of the national standards of the other member committees and vice versa. It will be seen that this rather formal definition of 'harmonization' permits editorial differences (particularly of language) between national standards but requires that they be technically equivalent. In this respect, the UK is perhaps fortunate in having as its mother tongue one of the three official languages of CENELEC (English, French and German), so that no difficulties of translation are encountered.

From this statement of the main purpose of CENELEC, it can be seen that the underlying aim of the work is the removal of technical obstacles to trade in electrotechnical products within the whole of Western Europe. There is another European body (CEN, the European Committee for Standardization), analogous to the ISO, which is responsible for standardization in fields other than those of electrical and electronic engineering. The process of the removal of technical barriers to trade by both CEN and CENELEC is obviously supportive of the objective of Article 100 of the Treaty of Rome, which governs the European Economic Community; nevertheless, the process extends further than the 12 members of the EEC, as it includes also the countries of the European Free Trade Association (EFTA), each of which has a reciprocal trade agreement with the EEC.

For practical purposes, CENELEC at its inception decided as a matter of policy to 're-invert the wheel' as rarely as possible. Therefore, most CENELEC harmonization work rests on the foundation of the results of worldwide (IEC) standardization work. Unless there are overriding reasons of impracticability, CENELEC seeks to reach agreement on the uniform application of IEC standards by its member committees, all of which will have had at least an opportunity to contribute to the IEC work.

In order to speed up the standards making process a cooperation agreement was reached between IEC and CENELEC on the planning of new work and parallel voting. The objectives of the agreement are:

(1) to expedite the publication and common adoption of International Standards, i.e. timely results prevail over an excessive degree of perfection;
(2) to ensure rational use of available resources. Full technical consideration of the standard should therefore preferably take place at international level;

(3) to accelerate drastically the standards preparation process in response to market demands.

New work items arising from the CENELEC General Assembly or from proposals received by the Technical Board are offered initially to the IEC in order that the work is done at the international level. However, if the time scale in the IEC exceeds CENELEC's planning requirements then the work is given to a CENELEC committee or working group. At the other end of the standards making process, mature texts produced by IEC committees are submitted concurrently for vote in both IEC and CENELEC. If the IEC and CENELEC results are both positive, the draft is published by the IEC and formally ratified as an EN by the CENELEC Technical Board. If the IEC result is positive and the CENELEC result is negative, the Technical Board will decide on what action to take, the adoption of common modifications for example. Only in exceptional circumstances does CENELEC embark on the preparation of a standard in anticipation of, or in contradiction to, an IEC standard. If this process becomes necessary, a CENELEC technical committee is established; as with IEC work, each CENELEC member is entitled to membership, and one national committee undertakes secretariat duties.

The situation is complicated slightly by the fact that CENELEC harmonization agreements are published in two different forms, either as Harmonization Documents (HDs) or as European Standards (ENs). In either case, the existence of a harmonization agreement imposes an obligation on each national committee to withdraw any conflicting national standard, so that any existing technical barrier to trade is removed. The appearance of new technical barriers to trade is prevented by the remaining obligation that, if a national standard on the same subject is considered necessary by a national committee, then that national standard must be technically equivalent to the Harmonization Document (if the agreement is published in that form) or identical in all respects with the text of the European Standard, as the case may be.

It is extremely important to recognize the fact that, within CENELEC, the principle of consensus decisions does not apply, in that significant objections can be over-ruled. This situation arises because approval of harmonization agreements is subject to a ballot of the national committees under a weighted voting procedure, analogous to that required for a 'qualified majority' within the EU. The number of weighted votes for each of the national committees is as follows:

France	10	Portugal	5
Germany	10	Sweden	5
Italy	10	Switzerland	5
United Kingdom	10	Denmark	3
Spain	8	Finland	3
Austria	5	Ireland	3
Belgium	5	Norway	3
Greece	5	Luxembourg	2
Netherlands	5	Iceland	1

Agreement is deemed to have been reached on technical questions provided that a number of conditions are simultaneously satisfied:

- that there is a majority of numbers in favour;
- that there are at least 25 weighted votes in favour;
- that there are at most 22 weighted votes against, and
- that no more than three members vote against.

The motivation behind this apparently complicated system is to provide safeguards, needed because the resulting decision is binding even on those national committees which voted against the draft in question. Thus the obligations arising from the approval of a harmonization agreement bind all CENELEC member committees, whether or not they voted in favour.

The further implications for standards work of Directives of the European Economic Community, in so far as they relate to those CENELEC member committees representing member states of the EU, are discussed later.

THE INSTITUTION OF ELECTRICAL ENGINEERS (IEE)

Of particular importance to the subject matter of this handbook are the *Wiring Regulations* published by the Institution of Electrical Engineers (IEE). Since the publication of the first edition of the Regulations in 1882, this authoritative set of requirements for the design and installation of electrical systems in buildings has been amended and revised from time to time to keep it up to date. In 1981 the 15th edition of the Regulations was published, under the new title *Regulations for Electrical Installations*, and marked a significant step towards the alignment of installation practices in the UK with those of the rest of the world. Its structure and detailed requirements are based closely on IEC Publication 364 *Electrical installations of buildings*. The various parts and sections of this large IEC standard have been the subject of harmonization agreements between the CENELEC member committees, although in many cases it has been found necessary to include common CENELEC modifications designed to make the requirements suitable for application within Western Europe. In 1991, the 16th edition was published which included further harmonized texts and amendments.

It is important at this stage to clear up a potential misunderstanding. The use of the word 'Regulations' in the description of the IEE document, although it is time-hallowed, does *not* indicate that it contains legal regulations in the formal sense implied by the definition given in the introductory section of this chapter. The use of the word antedates the international agreement on its meaning by almost a century. Indeed, the publication of its 1st edition preceded the formulation of the precursor of the British Standards Institution by almost a quarter-century.

On 2 October 1994, following significant pressure from industry, a formal agreement was signed between the British Standards Institution and the IEE establishing the IEE Wiring Regulations as a British Standard. BS 7671: 1992 *Requirements for electrical installations* came into being. The British Standard also bears the original title *IEE wiring regulations, 16th edition*. The technical authority for the new standard is vested in the BSI Standards Board and the Council of the IEE. The technical work is assigned to a joint committee whose constitution and procedures are consistent with the practices of both BSI and the IEE. This joint committee has an IEE staff member as secretary and is fully representative of all appropriate interests, including contractors, manufacturers, users, government departments, inspection authorities, research associations and so on, including of course the BSI. Decisions of the Joint Committee are taken by continuing discussions until all significant objections have been met or are withdrawn. Thus the acceptability of the text is safeguarded by the twin pillars of the democratic standardization process, consultation and consensus.

The British Electrotechnical Committee has also delegated the task of participation in international work on the subject to the Joint Committee. Thus the latter performs the usual functions of a BSI Technical Committee by advising the British Electrotechnical Committee on the UK stance in discussions within IEC TC 64 and CENELEC TC 64. To summarize, the Joint Committee is directly analogous in its functions and responsibilities, both national and international, to a BSI Technical Committee.

The close relationship between BS 7671, which originally was identical to the IEE 16th edition, and the corpus of British Standards is demonstrated by the fact that it makes reference to 75 British Standard specifications and codes of practice. The suitability of the content of these British Standards for the purpose of reference in BS 7671 and the co-ordination of their various requirements is ensured by the fact that all interested organizations, including the IEE itself, have a voice in the appropriate BSI Technical Committees.

It is not the purpose of this chapter to discuss the implications or implementation of BS 7671, subjects which are covered extensively in other chapters of this handbook.

STANDARDS AND THE LAW

Although it was stated in the introductory section of this chapter that the application of most British Standards is purely voluntary, resting usually on a decision by contracting parties, exceptions do occur when standards are referred to explicitly in the law. This technique is usually adopted by the legislature as a device to avoid the inclusion of detailed technical requirements in the body of the law. Thus an Act of Parliament (or a Statutory Instrument or Regulation made under an empowering Act) may make a particular standard part of the law, in the sense that it *must* be followed in order to obey the statutory requirement. In such a case, the use of the standard becomes mandatory rather than voluntary.

This technique has not often been used within the UK, although the legal systems of certain other European countries have forms that enable them to exploit it extensively. Mandatory reference to standards in British law is restricted almost entirely to cases in which the personal safety of a large section of the population needs to be protected. Examples include the legal references to British Standards for automobile seat belts, motor-cyclists' crash helmets and child-proof containers for analgesic drugs.

An alternative, and much commoner, technique is that used when the Act, Regulation or Statutory Instrument embodies a generalized technical provision supplemented by an indication that compliance with a particular British Standard or Standards is one (but not the only) way of fulfilling the legal obligation. In such a case, alternative designs which comply with the general legal provision are not precluded, but a solution which does comply with the standard in question is deemed to satisfy the legal requirement. It is important to realize that, although compliance with standards to which 'deemed-to-satisfy' reference is made in the law is voluntary, anyone choosing another route may be required to offer proof in court that his product, system or installation complies with the Regulation despite its non-compliance with the appropriate standard.

Although there is at present (1994) no legislation in the field of electrical installations which makes any particular standard mandatory, there are many 'deemed-to-satisfy' references to standards in relevant legislation. The most important such reference is made in the Electricity Supply Regulations 1988, for securing the safety of the public and for ensuring a proper and sufficient supply of electrical energy. The effect of Section (2) of the Regulations is to confer upon BS 7671, by reference to the IEE Wiring Regulations, 'deemed-to-satisfy' status, in that under closely constrained circumstances the supply of energy to a customer whose installation complies with the provisions of BS 7671 cannot be refused by the electricity supply authority on grounds of safety alone.

This reference to BS 7671 provides further confirmation of their status as a national standard. In order to avoid the delegation of law-making powers to a body which is not subject to direct parliamentary authority (in this case, the Institution of Electrical Engineers), the Regulations provide for the approval of any amendment or addition to BS 7671 by the Secretaries of State for Energy and for Scotland. As a matter of record, this approval has been granted by both Secretaries of State for all editions of the Wiring Regulations since 1937 up to and including the 16th edition.

Another example of the use of British Standards in the law relevant to electrical installations is provided by the Low Voltage Electrical Equipment (Safety) Regulations of 1989, made under the Consumer Protection Act of 1987. In part, the purpose of these Regulations is to implement the EEC Directive on low-voltage electrical equipment. Although the regulations do not in themselves include references to equipment standards, Regulation 6 states that electrical equipment which satisfies the safety provisions of harmonized standards shall be taken to satisfy the requirement that the equipment is safe and constructed in accordance with the principles generally accepted within the member states of the European Economic Community as constituting good engineering practice in relation to safety matters. To clarify the Regulation, the Department of Trade and Industry – Consumer Safety Unit has issued *Guide to the Low Voltage Electrical Equipment (Safety) Regulations 1989*. The guide is intended to assist suppliers of electrical equipment and enforcement authorities to understand the effects of the regulations. Although not an authoritative interpretation of the Regulation, which is a matter for the Courts, the guide seeks to explain the Regulations in general terms and does not attempt to address detailed issues. The guide gives the following order of precedence for the conformance of equipment to the Directive:

(1) conformance with the safety requirements of a harmonized European standard;
(2) where a harmonized European standard has not been drawn up or published,

conformance with the safety provisions of an IEC publication which has been published in the *Official Journal of the European Communities*;
(3) where no European or International standard exists, conformance to the safety provisions of a British Standard (i.e. a purely national standard).

Whilst it is not mandatory for electrical equipment to conform to any published standard, perhaps because it is an innovative product, this equipment must nevertheless comply with the basic requirement to be safe paying due regard to the principal elements of the safety objectives. There will be no presumption of conformity, hence these products will be open to challenge by the enforcement authorities, at any stage in the supply chain.

A third legal requirement of interest to those concerned with electrical installations stems from the Health and Safety at Work etc. Act 1974 as amended, which is applicable to all places of employment. This Act lays a general duty on every employer to ensure the health, safety and welfare at work of his employees and not to expose members of the public to health or safety risks. An employer can go a long way towards satisfying the law by ensuring that all equipment on his premises is manufactured, installed and maintained in accordance with appropriate British Standard specifications and codes of practice. Proof of this fact would be a powerful defence against any prosecution under the Act. Indeed, it is commonly considered to be most unlikely that the Health and Safety Executive would bring such a case before the courts. Indeed, to do so could be regarded as illogical as the constitutions of the BSI committees responsible for all of the relevant standards include representation from the Health and Safety Executive.

EUROPEAN UNION (EU)

The final link in the chain which binds standards and regulations together arises from the fact that British law is, to a specific and carefully controlled extent, subject to the law of the EU, formerly the European Economic Community (EEC), to which the UK was admitted on 1 January 1973. One of the fundamental purposes of the EU is that expressed by Article 100A of the Treaty of Rome, aimed at the development of a true Single European Market.

This is not the place for a detailed exposition of the functions and procedures of the various organs of the EU; all that needs to be said is that, after preparatory work by civil servants, the governments of member states of the community may agree by qualified majority to issue Directives intended to abolish or prevent technical barriers to trade. The obligations of such Directives rest with the governments of the member states which have to take whatever legal action is necessary to comply with the requirements of the Directive in question. The principle of reference to standards in European legislation is well established, and a number of Directives make reference to European standards published by CEN and CENELEC, either on a mandatory or 'deemed-to-satisfy' basis.

In the area of electrotechnical products, harmonization of national legal requirements was achieved in 1973 by Directive 73/23/EEC on the harmonization of the laws of member states relating to electrical equipment designed for use within certain voltage limits – commonly known as the Low Voltage Directive (LVD). Although the LVD has no direct bearing on legislative requirements for electrical installations, it does have an indirect effect, as it is concerned with the removal of barriers to intra-Community trade in any electrical equipment designed for use with a voltage rating of between 50 and 1000 V (for alternating current) and between 75 and 1500 V (for direct current) with only the very limited exceptions of electrical equipment for use in explosive atmospheres, electromedical equipment, electrical parts for goods and passenger lifts, electricity meters, domestic plugs and socket outlets, electric fence controllers and specialized electrical equipment for use on ships, aircraft and railways. Also excluded is radio-frequency interference, which is the subject of separate Directives.

The central provision of the LVD (contained in Article 2), is that

'the Member States shall take all appropriate measures to ensure that electrical equipment may be placed on the market only if, having been constructed in accordance with good engineering practice in safety matters in force in the Community, it does not endanger the safety of persons, domestic animals or property when properly installed and maintained and used in applications for which it was made.'

Although the Directive includes (in Annex 1) eleven very basic safety objectives, it effectively established formal criteria for the definition of the concept of safety by reference to harmonized standards drawn up by common agreement between the national electrotechnical committees meeting together in CENELEC. While the LVD contains many other administrative provisions, particularly those relating to the legal burden of proof that equipment is safe, its essential effect is that no member state may by administrative action raise a technical barrier to the import of electrical goods manufactured within the EU in compliance with an appropriate CENELEC harmonization agreement.

One consequence of the rules of procedure of CENELEC is that this protection extends to manufacturers whose goods comply with harmonized national standards. Thus, in the field of electrotechnical products, the LVD has created a true 'common market', resting on the foundation of agreements between the national electrotechnical committees of the Community countries meeting together in CENELEC. In fact, the practical consequences of the LVD run wider than the ambit of the EU members, as the CENELEC member committees also include those from EFTA countries.

It is a demonstration of the success of the LVD in removing barriers to trade that it provided the model for the 'new approach' to European standards and legislation launched by the ECC Commission in May 1985 and enshrined in the Single European Act of 1988.

CONCLUSION

The purpose of this chapter has been to explain the significance of standards and their relationship with legislative requirements. It has been demonstrated that national standards, European harmonization agreements and international standards are closely intertwined, and that those affected by these standards have a major part to play in their preparation. Further, it has been explained that the provisions of standards are subject to the overriding requirements of both British and European laws. If this exposition has any message beyond the purely didactic, it is that industry has a major rôle to play in the consultative processes both of law-making and of standards-making. Neither process can respond to the needs and best interests of the UK unless the requirements of that industry are promoted at appropriate times.

CHAPTER 13

Distribution Transformers

G.P. Harvey, CEng, MIEE

Revised by K. Frewin, CEng, MIEE
(Power Engineering Services)

Power distribution throughout most of the world, before the final consumer voltage, is at voltages usually between 10 and 13.8 kV and although some old systems of 5.0, 5.5 and 6.6 kV still remain these are rapidly being replaced with a higher voltage on economic grounds. The declared no-load voltage of the low voltage distribution network in Great Britain has been subject to harmonization with the European Union, and was re-declared as 230 V phase to neutral, +10% and −6% on 1 January 1995, and will be further revised in the year 2003 to 230 V ±10%.

Oil immersed distribution transformers, for use by British electricity supply authorities, are usually designed to comply with Electricity Supply Industry Standard 35−1. This specification covers ratings from 16 kVA to 1000 kVA, but ratings of up to 2500 kVA are not uncommon. Distribution transformers may be suitable for ground mounting, or pole mounting in rural electrification schemes. The rating of pole-mounted designs is limited by weight considerations, but units of up to 315 kVA have been supplied for use by British Area Boards. In order to retain simplicity of design, the majority of liquid-immersed distribution transformers are naturally cooled. Forced air cooling, when required, is applied to liquid-immersed transformers having ratings in excess of 5000 kVA and occasionally to dry-type transformers, particularly resin encapsulated types. It is possible to increase the ratings of certain dry-type transformers by up to 50% by the application of forced air cooling. For transformers supplying loads which have occasional peaks, this type of design can offer advantages in both cost and physical size.

TYPES OF TRANSFORMER

All transformers covered by this chapter should comply with the requirements of BS 171, which is equivalent to IEC 76: 1976, for liquid-immersed types and IEC 726: 1982 for dry types. The British Standard defines a transformer as a static piece of apparatus which by electromagnetic induction transforms alternating voltage and current between two or more windings at the same frequency and usually at different values of voltage and current.

There are two basic types of distribution transformer in common use today. One in which the core and windings are enclosed in a liquid-filled tank which provides cooling and insulation and the other where the core and windings are cooled directly by air. Liquid-filled transformers can again be sub-divided into units using flammable substance (mineral oil) and those using various types of fire-resistant liquid (silicone liquid or synthetic hydrocarbon).

Dry-type transformers are also available in two distinct types: those with the insulated turns of the winding directly in contact with the cooling air and those with the complete windings encapsulated within a moisture-resistant cladding of epoxy resin. Preferred ratings are listed in Table 13.1, but dry-type distribution transformers are rarely used in ratings much below 500 kVA. The ratings of individual transformers within installations are determined by load requirements, the physical space available, the distribution system voltage regulation, and the high voltage switchgear fault rating.

The heart of the conventional three-phase distribution transformer is a laminated electrical sheet steel core of three limbs. Each limb carries two concentrically wound coils; the secondary (low voltage) inside nearest the core and the primary (high voltage) on the outside, the whole being contained within an enclosure of steel. In the case of the liquid-filled transformer the enclosure consists of a leak-proof tank while the dry-type transformer merely requires a ventilated casing to enclose the live parts. Some designs of dry-type transformer are totally enclosed in non-ventilated sheet steel housings, and have an IP rating of 55 to BS 5490. These transformers are used in installations where environmental contamination would preclude the use of conventional dry-type designs, but their superior resistance to fire is required.

Tappings are usually included on the h.v. windings of distribution transformers to alter the turns ratio between the high- and low-voltage windings and thereby to compensate for variations in primary supply voltage in order to maintain the consumer voltage within the statutory limits. On distribution transformers taps are selected by means of an off-circuit device; the transformer must be disconnected from the supply before the taps can be changed. The normal variation in supply

Table 13.1 Preferred values of rated power for three-phase transformers (BS 171).

kVA	kVA	kVA
5.0	31.5	200
6.3	40	250
8	50	315
10	63	400
12.5	80	500
16	100	630
20	125	800
25	160	1000 etc.

voltage provided for by high voltage tappings is ±2.5% and 5%. Dry-type distribution transformers are normally provided with bolted link arrangements for
primary tapping selection. It is not recommended that tapping switches are used
for the adjustment of high voltage tapping on dry-type transformers, as they are
subject to contamination, which has led to failures in service.

Winding conductors are usually copper strip, or round wire. Some designs of
distribution transformer use copper sheet, sometimes called foil windings, for the
l.v. coils. Aluminium sheet is used for both the l.v. and h.v. windings of some
designs of resin encapsulated transformer. Conductor insulation materials are
chosen for the particular application, ranging from cellulose papers for liquid-
immersed types, to synthetic materials, capable of withstanding continuous high
temperatures, for dry-type designs.

Mineral-oil-filled transformers

This is the most common type of distribution transformer. It can be found on
electrical supply systems in every country of the world. Although the insulating oil
is a flammable liquid, the reliability of the oil-immersed transformer has been
proven over many years on supply systems where security of supply is of the
utmost importance. However, mineral oil is flammable and although most fault
conditions that occur within the windings of a transformer result in no more than
a discharge of oil it is possible, particularly when an electrical arc occurs just
below the surface of the oil, for it to be ignited. For this reason oil-immersed
distribution transformers are usually positioned outside buildings within a suitable
fence enclosure or in separate brick-built buildings away from personnel.

Where transformers filled with mineral oil are sited within an occupied area, it
is usual to find some type of safety provision. This may take the form of bund
walls, soak-away pits, or automatic fire-extinguishing equipment. However, the
additional costs of the civil engineering work required can make the installation of
a mineral-oil filled transformer uneconomic. In these instances, a dry-type transformer, or a transformer filled with a high fire-point liquid is often selected for the
installation.

The integrity of the insulation system of an oil-immersed transformer relies
partly on the condition of the oil. It has been common practice on most established
supply networks in the world to let transformers breath naturally as the insulating
liquid expands and contracts with load. However, it has also been recognized that
some form of protection system that prevents the contamination of the liquid by
air-borne pollutants has the advantage of a longer insulation life, particularly
when load factors are high.

The simplest form of oil protection system, which is perfectly adequate for most
installations of distribution transformers below 500 kVA installed in the temperate
zones of the world, is a silica-gel dehydrating breather. Here, the air drawn into
the transformer tank during reduced load conditions is passed through an oil bath,
to reduce solid contaminants, and then through the dehydrating crystals of silica-
gel to remove the moisture. It is essential, however, that the silica-gel crystals are
maintained dry and replaced as soon as the colour changes from blue to pink.

Probably the most common form of oil protection system is found with the conservator or expansion vessel; this has a sump which traps most air-borne pollutants, Fig. 13.1.

The most obvious method of eliminating oil contamination is to seal the tank from the outside air and design it to tolerate the pressures developed by the expanding liquid coolant, Fig. 13.2.

Because of gas solubility in oil these pressures are never very high and rarely exceed 0.43 kg/cm^2 under steady load conditions.

The development of special machinery which automatically folds and welds the steel plate into deep corrugations to form the transformer tank side, has allowed the corrugated tank to become more cost effective. The steel plate is usually between 1.2 and 1.5 mm thick, resulting in a tank which is both light and compact, the mechanical strength being obtained by the deep closely spaced corrugations. Plate widths of up to 2000 mm and depths of 400 mm allow transformers of up to

Fig. 13.1 A 500 kVA 11 kV/380 V mineral-oil-filled transformer with conservator (*Bonar Long Ltd*).

Fig. 13.2 A sealed oil-filled transformer rated at 1600 kVA for use in a cement works (*Parsons Peebles Distribution Transformers*).

5000 kVA to be cooled by this method, but their general application is in distribution ratings of up to 1600 kVA. The ability of the corrugated panels to flex has enabled the development of the fully sealed corrugated tank design. In this instance the tank is completely filled, expansion of the liquid being accommodated by flexing of the tank walls. The liquid within the tank has no contact with the atmosphere, which assists with the preservation of the transformer insulation system, as well as reducing the maintenance requirements. Corrugated transformer tank designs have been in service for in excess of 30 years and are now established as a reliable method of tank construction.

Askarel-filled transformers

Askarel-filled transformers were introduced about 50 years ago when the demand arose for a fire-resistant liquid to replace the mineral oil in units that were to be installed in occupied buildings. The increasing concern for the safety of people should a fire occur in a mineral-oil-filled transformer placed pressure on manufacturers to find an alternative coolant.

Askarels appeared on the market under a variety of names such as Pyroclor, Inerteen, Aroclor and Pyranol. They comprise a mixture of polychlorinated biphenyl (pcb) and trichlorobenzene (tcb), the latter being used to reduce the viscosity of the pcb. Basically, apart from a few synthetic resin products, the design of the transformer remained unaltered from the mineral-oil-filled unit but the use of this liquid has introduced hazards not appreciated at the time. The implementation of EEC Council Directive 85/467/EEC banned the sale of electrical equipment containing greater than 0.01% by weight of pcb. This directive became operative in 1986, and effectively stopped the sale of askarel-filled distribution transformers within the EU.

During the last few years it has become widely realized that pcbs are toxic and are also resistant to biological and chemical degradation. Research has shown that if ingested they can persist in the fatty tissues and cause damage to organs of the body. In the event of a fire involving an askarel-filled transformer, any of the coolant subject to partial combustion at low temperatures will form toxic by-products, which will contaminate the surrounding area.

Many of the transformers that were filled with pcbs has been in service for more than 15 years. Some of the older ones have developed slight leaks, usually at gasketted joints, and are easily repaired. There have been reports of welds leaking after 15 to 20 years of service and although there is no substantive evidence to show that pcbs attack metal it is of concern that so many tanks have started to leak. To guard against the danger of spillage due to leaks and internal faults some transformer users are having the pcbs removed and replaced by an alternative fire-resistant liquid. Special techniques have been adopted to carry out

Fig. 13.3 Retrofilling of an askarel-filled unit in situ by *R.F. Winder Ltd*. The liquid is being replaced by silicone under carefully controlled conditions. Protective clothing is being worn by the operators.

this replacement, the operation being termed retrofilling, Fig. 13.3. Because this is a specialist function requiring a high degree of expertise it must be performed only by organizations trained and equipped to do it.

Once removed the pcb has to be taken to an incinerator for destruction and again this is a specialist operation requiring skills and plant available from only a few organizations. The toxicity of pcbs and their handling are described in the UK Department of Environment publication *Waste Management Paper No. 6.*

Another school of thought insists that the pcbs cannot be completely removed by any retrofilling process and that the only sure way of removing pcbs from the environment is for the transformer to be completely replaced by one containing a harmless liquid.

A number of countries have banned the use of pcb in any form. Others continue with its use in transformers and capacitors. All recognize the potential danger that exists if a large local spillage does occur.

Other fire-resistant types

There are a number of liquids on the market for which suitability as a transformer coolant and resistance to fire is claimed. It is not possible to examine these liquids in detail but one is described as a synthesized ester and is marketed under the trade name of Midel 7131.

Another is described as a highly saturated paraffin oil and is marketed under the name of Rtemp.

Probably the most well known, because of its long time use in other industries is silicone fluid. This is marketed by a number of companies under their own trade names. Other fluids that are available for use in transformers, but which are non-flammable, include Formel, Wecasol and Ugilec T. These materials contain solvents, such as trichlorobenzene and perchloroethylene, and have not found such ecological acceptability as silicone or synthetic ester (Midel) materials. The general criteria for a low fire-point fluid, as defined by Factory Mutual Research Corporation, is that the fire point be at least 300°C. All the above-mentioned fluids meet this requirement, but other aspects of fire performance, such as heat of combustion, etc., must also be considered when selecting a high fire-point fluid for use in transformers.

The most favoured high fire-point fluids, at present, are the silicone based materials and the synthetic ester types (Midel). The silicone fluids used in transformers have a viscosity of approximately twice that of mineral transformer oils at 20°C; other characteristics make their performance not dissimilar to transformer oil. Accordingly, there is very little difference in the external appearance of a silicone fluid cooled transformer, or synthetic ester-cooled designs with their equivalent mineral-oil-filled units, Fig. 13.4.

The cellulosic insulation materials commonly used in the manufacture of oil-immersed transformers are in general compatible with silicone and synthetic ester fluids. When impregnated with these materials they can be operated at higher temperatures than is associated with mineral-oil filled transformers before thermal degradation takes place, but this feature has yet to be recognized by the relevant specification producing bodies. Care should be taken when specifying gaskets and

Fig. 13.4 Silicone-cooled distribution transformer; 2200 kVA complete with output circuit-breakers (*Goodyear Transformers Ltd*).

fittings for synthetic fluid cooled transformers, as materials and components used with mineral based oils may not be compatible.

Transformer manufacturers are looking at the future designs of synthetic fluid filled transformers in order to take advantage of the high temperature capabilities of these transformer coolants. Such transformers may well result in savings in weight, size and cost, as it will be possible to make better economic use of these expensive materials.

Dry-type transformers

Dry-type transformers are defined by IEC 726 as designs where the core and windings are not immersed in an insulating liquid. Cooling is usually by natural air circulation through the windings. Permissible winding temperature rise depends on the type of insulation used on the winding conductors and between the windings.

Transformer windings with Class A insulation are limited to a temperature rise of 60°C. When Class C insulation is used 150°C is the limit. All dry-type transformers with Class B insulation and above are considered to be fire resistant to some degree because the volume of combustible material is small and it does not maintain combustion. The modern Class C transformer, Fig. 13.5, is almost completely fire proof because the Nomex insulation used is self-extinguishing. Temperature rise limits are discussed in detail later.

Two types of dry transformer are available. The conventional arrangement was developed from the old Class B transformer. It uses high-temperature aromatic Polyaramid paper both for conductor covering and solid insulation. Large air cooling ducts between the windings and between the layers of the windings provide the cooling. This arrangement has the advantage of low cost and many years of satisfactory service close to load centres in the normally dry environment of an indoor installation.

It is not advisable, however, to leave this type of dry transformer unexcited for long periods of time in a damp atmosphere without some form of warm air circulation through the windings. Impulse voltage levels of 60 and 75 kV are available for the modern Class C transformer.

Because of the large cooling ducts between windings and the subsequently high surge impedance the current chopping characteristics of some vacuum circuit-

Fig. 13.5 A 1000 kVA 11 kV/433 V Class C dry-type transformer (*Parsons Peebles Distribution Transformers*).

breakers can produce very high transient voltages in dry-type transformers. It is advisable that transformer manufacturers are made aware of the intention to use vacuum circuit-breakers.

The newer development of dry-type transformer employs windings that have been completely encapsulated in epoxy resin, Figs 13.6 and 13.7. Designs are impervious to the ingress of moisture to the windings and damp environments have very little effect. Cast-resin transformers can be manufactured by several different methods. Coils may be cast in a mineral-filled resin system, the wall thickness of the casting being in the order of 5−8 mm, which gives the winding considerable mechanical strength. Modern practice with cast resin transformers of this design is to incorporate coarse weave glass fibre matt reinforcing material into the inner and outer walls of the coils, thus allowing the thickness of resin to be reduced, hence reducing the resin content, and improving both the heat dissipation and fire resistance of the transformer. Another commonly used manufacturing

Fig. 13.6 A 1000 kVA resin-cast transformer (*Goodyear Transformers Ltd*).

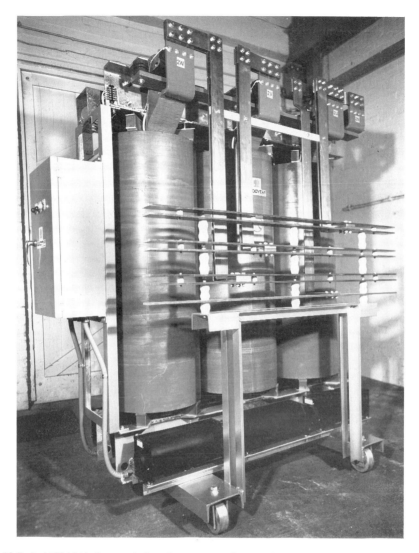

Fig. 13.7 A 1600 kVA fan-cooled resin-cast transformer (*Goodyear Transformers Ltd*).

method is to wind glass fibre sheet and rovings into the coils and to impregnate the complete assembly with a low-viscosity non-filled resin system. Both methods of manufacture produce robust windings, well able to withstand the effects of short-circuit forces. It is important in the winding design of resin encapsulated transformers that the conductor material and the resin system have very similar rates of thermal expansion. This is achieved by the mineral filler added to the resin, or by the effect of the glass fibre materials wound into the coils prior to impregnation. The epoxy resin encapsulated transformer is generally more costly to manufacture than the conventional dry-type transformer, and therefore in many indoor installations the additional cost may be difficult to justify.

Both types of dry transformer need to be housed within a waterproof building and sited within an enclosure to prevent physical contact with any live parts in order to comply with the IEE Wiring Regulations for safety.

It has become practice in America and the UK to build dry-type transformers as self-contained units within a metal casing with h.v. and l.v. cable boxes mounted on to the sides as with the liquid-immersed type.

The development of the resin-encapsulated transformer in Europe led to these transformers being positioned inside wire mesh enclosures without any casing. When costs of these two units are considered, the form of protection for the general public must be taken into account.

The windings of the Class C dry-type transformer are usually quite conventional, although with the introduction of newer and more sophisticated winding machinery, l.v. windings wound with wide strip aluminium or copper are becoming more popular with manufacturers.

The windings of cast resin transformers may be of copper or aluminium; where aluminium is used it is generally in the form of thin sheet, generally termed foil-windings. Many designs of l.v. winding, either copper or aluminium, are of the foil-wound type, but the designs using glass fibre as the filling material for the resin are usually wound with copper strip conductors for both l.v. and h.v. windings.

PERFORMANCE

When a transformer is selected for a particular application prime cost should never be the only consideration. In many cases it plays a very small part in the overall cost. Factors which also govern choice of a particular transformer should include load factor, cost of losses and efficiency, maintenance costs, fire-resistant qualities and associated building costs, space limitations and ambient temperature as well as prime cost. These matters are discussed below but not necessarily in order of importance.

Prime cost is always a consideration. The amount of capital available for an electrical distribution network often governs the type of equipment that is purchased, irrespective of the many advantages and long-term financial benefits accruing from buying more expensive plant. The first cost of the various transformers discussed is shown by the nomogram, Fig. 13.8, in which it will be seen that the mineral oil unit is the cheapest and the cast-resin design the most expensive at some 80% more. Locating the mineral-oil-filled unit close to load centres inevitably involves the installation of pits and drains and automatic fire protection. Special floor, roof and door construction are necessary. Even so, the cost of mineral-oil-filled transformers is so low relative to the fire-resistant alternatives that the decision to have the risk and limit the effect is taken over eradicating the risk. In any event there is some risk, even with fire-resistant transformers.

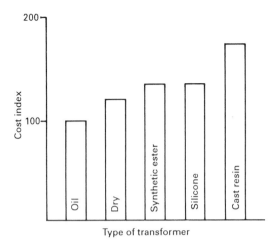

Fig. 13.8 A nomogram showing comparative capital costs of various types of distribution transformers.

Losses

No-load and load losses in a transformer result in loss of efficiency. They are the reason for the major running cost of a transformer. They have to be paid for, yet result in heat which is normally dissipated to the atmosphere.

The comparison between various manufacturers' liquid-immersed transformers can be made quite simply. It will be obvious that a transformer with a no-load loss of 700 W consumes 2628 units of electricity in a year less than a similar transformer with a no-load loss of 1000 W, if they are both excited continuously.

The cost of load losses of course depends on the load factor (LF) and does not vary greatly between manufacturers for the same transformer rating.

Quite large differences in losses do occur, however, between the various types of fire-resistant transformer and running costs could well be a deciding factor in the choice of a particular fire-resistant transformer.

Table 13.2 gives a comparison of losses between two types of fire-resistant transformer. The cast-resin values were taken from European manufacturers' published literature. The liquid-immersed losses are those which are available from a number of companies in the UK. It is quite evident that load factor and tariff structure play a most important part in any economic comparisons.

When examining this table the importance of the iron loss is more obvious at low load factors. For example at 50% load factor lower iron loss of the silicone-liquid-filled designs compared with the cast-resin makes them much more attractive than at higher load factors. At 80% there is not much to choose between the designs while at 100% load factor the cast-resin transformers are more cost effective. In the UK the average industrial load factor on a transformer is probably between 50 and 60% but where security of supply is of supreme importance the use of two transformers reduces this value to below 50%. Even lower load factors can apply where both load growth and supply security have to be taken into account.

Table 13.2 Comparison of losses of cast-resin and silicone-liquid-filled transformers.

Rating (kVA)	Cast-resin				Silicone-filled			
	no load loss (W)	load loss (W)	total loss 80% LF (W)	total loss 50% LF (W)	no load loss (W)	load loss (W)	total loss 80% LF (W)	total loss 50% LF (W)
315	900	4000	3460	1900	470	4600	3414	1620
500	1250	6000	5090	2750	700	6800	5052	2400
800	1800	8700	7368	3975	1130	9700	7338	3555
1000	2200	10800	9012	4800	1380	11800	8932	4330

The life of a transformer is an essential part of the overall cost equation. Mineral-oil-filled transformers have been in use on electrical supply networks for a great many years and the life expectancy of the average distribution transformer is known to exceed 30 years.

The silicone-liquid-filled transformer with normal temperature rise characteristics has an insulation system virtually identical to mineral-oil-filled transformers and the viscosity of the fluid is not significantly different. The transformer life of the silicone-filled transformer can be assumed to be very similar.

The insulation system of the modern dry-type Class C transformer dates back to about 1960 in the UK. More than 30 years experience have shown this to be a very reliable unit.

The resin-encapsulated transformer was first introduced more than 25 years ago and its life has been favourably assessed.

A Canadian manufacturer (Polygon Industries) has produced figures of electrical losses for all types of distribution transformers considered in this chapter and they are given as Table 13.3. While it must be remembered that these values apply to the American 60 Hz system, it will be seen that no difference is made between the losses of any of the liquid-filled types and that these are considerably lower than the dry-types at all loads between 25% and 100%.

It is interesting to note the difference between the values given in Tables 13.2 and 13.3 for the cast-resin and silicone-liquid-filled 1000 kVA transformer with regard to no-load and load losses. The 60 Hz system has a significant effect on the no-load loss.

The importance of assessing the value of the losses of the transformers for any contract cannot be over-emphasized. Most British manufacturers are in a position to offer low-loss designs as standard because of their familiarity with those parts of the world where losses are highly capitalized.

These low-loss units do have a higher prime cost but it may well be an economic proposition to pay a higher cost and take greater advantages in reduced running costs.

There are two sides to every story and although it may be proved that paying 20% more for low loss transformers can be justified by lower running costs, 20% fewer customers may be reached or the system may have a 20% lower capacity for

Table 13.3 Electrical losses comparison of 15 kV 1000 kVA transformers with a BIL* of 95 kV.

	No load	Losses in kilowatts at operating temperature								
		1/4 load		1/2 load		3/4 load		Full load	
Oil ⎫		No load	2.8	No load	2.8	No load	2.8	No load	2.8
Askarel ⎬	2.8	Load	0.6	Load	2.3	Load	5.2	Load	9.1
Silicone ⎭		Total	3.4	Total	5.1	Total	8.0	Total	11.9
Dry-type, 150°	3.2	No load	3.2	No load	3.2	No load	3.2	No load	3.2
		Load	0.8	Load	3.3	Load	7.4	Load	13.2
		Total	4.0	Total	6.5	Total	10.6	Total	16.4
Epoxy dry-type	3.2	No load	3.2	No load	3.2	No load	3.2	No load	3.2
		Load	0.7	Load	3.0	Load	6.7	Load	11.8
		Total	3.9	Total	6.2	Total	9.9	Total	15.0

* BIL = Basic insulation impulse level.

revenue earning. In areas with a high rate of growth this second consideration may be of greater importance.

Temperature rise

In temperate climates the differences in allowable temperature rise of liquid-filled and dry-type transformers are not important for most installations. There are instances, however, where the high ambient in which the transformer is sited may well limit the rating of a standard transformer to something below the nameplate value. There are also instances where the heat of the losses of a transformer could affect the functioning of sensitive electronic equipment. An example of both applications can be found in the first case where an oil-immersed transformer is used to supply power to an induction furnace or an annealing furnace. An arc furnace transformer must, because of the very low voltage, be close to the furnace where ambient temperatures are very high.

An example of the heat of a transformer affecting the operation of sensitive equipment can be found when power supplies to computers are considered. A large dry-type transformer in an open installation might well produce enough heat to create difficulties. Even power diodes and thyristors have a very limited temperature range and care is needed when transformers are sited for these applications.

Tables 13.4 and 13.5 are reproduced from BS 171 and indicate the permissible limits of temperature rise for the two types of transformer. Where ambient temperatures are very high and transformers are to be mounted indoors one might look favourably on the dry-type Class C transformer with its capacity for operation at high temperatures. But in making such judgement other factors need to be taken into account.

Table 13.4 Temperature rise limits for dry-type transformers from BS 171: Part 2.

1 Part	2 Cooling method	3 Temperature class of insulation*	4 Maximum temperature rise (°C)
Windings (temperature rise measured by the resistance method)	Air, natural or forced	A E B F H	60 75 80 100 125 150†
Cores and other parts: (a) adjacent to windings (b) not adjacent to windings	All		(a) Same values as for windings (b) The temperature shall, in no case, reach a value that will damage the core itself, other parts or adjacent materials

Note: Insulating materials may be used separately or in combination provided that in any application each material will not be subjected to a temperature in excess of that for which it is suitable, if operated under rated conditions.

* In accordance with IEC Publication 85. Recommendations for the Classification of Materials for the Insulation of Electrical Machinery and Apparatus in Relation to their Thermal Stability in Service.

† For certain insulating materials, temperature rises in excess of 150°C may be adopted by agreement between the manufacturer and the purchaser.

Table 13.5 Temperature rise limits for oil-immersed type transformers from BS 171: Part 2.

1 Part	2 Maximum temperature rise (°C)
Windings: temperature class of insulation A (temperature rise measured by the resistance method)	65, when the oil circulation is natural or forced non-directed 70, when the oil circulation is forced and directed
Top oil (temperature rise measured by thermometer)	60, when the transformer is equipped with a conservator or sealed 55, when the transformer is neither equipped with a conservator nor sealed
Cores, metallic parts and adjacent materials	The temperature shall, in no case, reach a value that will damage the core itself, other parts or adjacent materials

Note: The temperature rise limits of the windings (measured by the resistance method) are chosen to give the same hot-spot temperature rise with different types of oil circulation. The hot-spot temperature rise cannot normally be measured directly. Transformers with forced-directed oil flow have a difference between the hot-spot and the average temperature rise in the windings which is smaller than that in transformers with natural or forced but not directed oil flow. For this reason, the windings of transformers with forced-directed oil flow can have temperature rise limits (measured by the resistance method) which are 5°C higher than in other transformers.

The susceptibility of the dry-types to dust-laden atmospheres and the exclusion of vermin therefore requiring frequent maintenance checks might well exclude dry-types from overseas installations.

It should also be borne in mind that the temperature rise limitations specified in BS 171 assume the use of cellulosic materials with a Class A limitation in mineral oil to BS 148. Tests on standard solid insulating materials normally used with mineral-oil-immersed transformers have shown that when these materials are impregnated with silicone fluid they can operate for longer periods of time at elevated temperatures without loss of transformer life and short time overloads without thermal expansion problems.

Where transformers are situated in areas of very high ambient temperature, or where compact dimensions are essential, it is possible to design suitable transformers to suit these circumstances. By using special high-temperature insulation, and taking advantage of the high-temperature withstand properties of silicone or synthetic ester fluids, transformers having high-temperature withstand properties, as well as very compact dimensions, can be produced.

Loading guide

The British Code of Practice CP 1010, *Loading guide for oil-immersed transformers* indicates how oil-immersed transformers may be operated in different conditions of ambient temperature and service, without exceeding the acceptable limit of deterioration of insulation through thermal effects. Although specifically related to oil-filled units similar reasoning can be applied to transformers cooled by other liquids.

The object of the guide is to give the permissible loadings, under certain defined conditions, in terms of the IEC Publication 76 rated power of the transformer, so that planners can choose the required rated power for new installations. Basically, the cooling-medium temperature is 20°C, but deviations from this are provided for, in such a way that the increased use of life when operating with a cooling-medium temperature above 20°C is balanced by the reduced use of life when it is below 20°C.

In practice, uninterrupted continuous operation at full rated power is unusual, and the guide gives recommendations for cyclic daily loads, taking into account seasonal variations of ambient temperature. The daily use of life due to thermal effects is indicated by comparison with the 'normal' use of life corresponding to operation at rated power in an ambient temperature of 20°C.

Two examples of the use of the curves are given in the Code and are reproduced here, see Table 13.6. It is necessary to define the four symbols used in the tables in order to understand the examples. These symbols are:

K_1 = initial load power as a fraction of rated power
K_2 = permissible load power as a fraction of rated power (usually greater than unity)
t = duration of K_2 (h)
θ_a = temperature of cooling medium (air or water) (°C).

Table 13.6 Loading guide for oil-immersed transformers (CP 1010).

	$K_1 = 0.25$	$K_1 = 0.50$	$K_1 = 0.70$	$K_1 = 0.80$	$K_1 = 0.90$	$K_1 = 1.00$
$t = 0.5$	+	+	1.93	1.83	1.69	1.00
$t = 1$	1.89	1.80	1.70	1.62	1.50	1.00
$t = 2$	1.59	1.53	1.46	1.41	1.32	1.00
$t = 4$	1.34	1.31	1.27	1.24	1.18	1.00
$t = 6$	1.23	1.21	1.18	1.16	1.12	1.00
$t = 8$	1.16	1.15	1.13	1.12	1.09	1.00
$t = 12$	1.10	1.09	1.08	1.07	1.05	1.00
$t = 24$	1.00	1.00	1.00	1.00	1.00	1.00

ONAN and ONAF transformers: $\theta_a = 20°C$. Values of K_2 for given values of K_1 and t.
Note: In normal cyclic duty the value of K_2 should not be greater than 1.5. The values of K_2 greater than 1.5, underlined, apply to emergency duties. (See Clause 3.)
The + sign indicates that K_2 is higher than 2.0.

Note that $K_1 = S_1/S_r$ and $K_2 = S_2/S_r$ where S_1 is the initial load power, S_2 is the permissible load power and S_r is the rated power.

Example 1

A 1000 kVA ONAN transformer has initial load power 500 kVA. It is required to find permissible load power for 2 h at an ambient temperature of 20°C.

Cooling: ONAN $\theta_a = 20°C$, $K_1 = 0.5$, $t = 2$ h.

From Table 13.6, $K_2 = 1.53$ but the guide limit is 1.5. Therefore permissible load power for 2 h is 1500 kVA (then returning to 500 kVA).

Example 2

With $\theta_a = 20°C$ an ONAN transformer is required for 1400 kVA for 6 h and 800 kVA for the remaining 18 h each day.

1400/800 $(S_2/S_1) = 1.75$

From the curve drawn, using the data of Table 13.6, on the $t = 6$ line, the values of K_2 and K_1 giving $K_2/K_1 = 1.75$ are $K_2 = 1.18$ and $K_1 = 0.68$, see Fig. 13.9. Therefore the rated power is:

$S_r = 1400/1.18 = 800/0.68 = 1180$ kVA

It will be appreciated that the Code of Practice contains many more tables covering different thermal conditions and designs of transformers and the reader is advised to obtain a copy for use as required.

It must be pointed out that the preceding examples relate entirely to the transformer and the effect of cyclic loading on transformers. If cyclic loading is anticipated, however, at the time of installation consideration must be given to the extra demand put on the switchgear. Mention must also be made that if settings of protective relays are to be adjusted to account for cyclic loading then

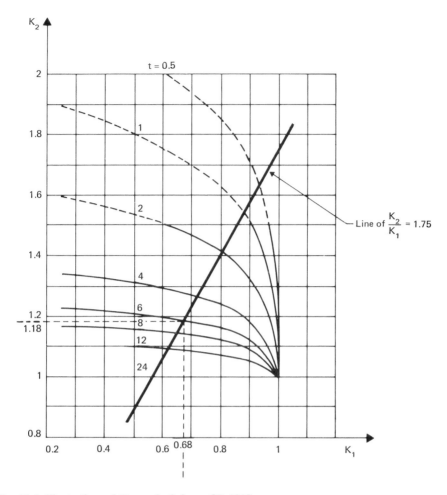

Fig. 13.9 Illustration of Example 2 from CP 1010.

the degree of protection against continuous overloads or faults is very much reduced.

Fire resistance

Dry-type and liquid-filled units already mentioned (except mineral oil), are designated fire-resistant designs but that does not mean that the materials used will not burn. Rather it refers to their high fire point (the temperature at which the material continues to burn when subject to a flame on the surface). This is the significant figure when considering fire resistance; it must be well above the maximum temperature reached by a transformer operating at its maximum overload condition in a high ambient. For reasons stated earlier askarel-filled units are not acceptable and so one is left with designs incorporating other fire-resistant liquids and the two dry-types.

Table 13.7 indicates the fire point temperature of some fire-resistant designs and shows that there is not much to choose between the different materials. But one must also consider the nature and toxicity of the smoke given off by materials when they are burning, together with other characteristics, to form a judgement as to the best to select.

The rate of heat release of a burning substance is also important for it governs the size and nature of the enclosure surrounding the transformer.

Most fire officers agree that smoke is the most serious hazard to life in any fire for it blinds, confuses and suffocates and can cause deaths well away from the centre of the fire. Epoxy resins, like most solid polymers, produce thick black smoke when burning, while silicone liquid generates a white vapour less dangerous in content and much easier to see through. Midel 7131 contains petroleum oils and certain synthetic oxyhydrocarbons but these are classed as of negligible toxicity when burning.

The rate of heat release of burning materials consists of two components, convective and radiative, the former being the higher value. The convective value provides a measure of the damage the burning material can do to a roof or any structure above the fire; the radiative indicates how much damage a fire could do to the walls and surrounding equipment. Table 13.8 shows these values for some fire-resistant materials.

Table 13.7 Fire-point temperature of some fire-resistant transformers.

Material*	Fire point (°C)
Silicone liquid	360
Midel 7131	310
Cast resin	450
Class H	†

* For comparison purposes mineral oil is 170°C. Askarel is non-flammable.
† These designs are virtually fireproof.

Table 13.8 Values of rate of heat release (RHR) for some fire-resistant materials.

Material	RHR convective (kW/m)	radiative (kW/m)
Silicone 561	53	25
High fire point hydrocarbon	546	361
Epoxy resin	—	—

Also important is how a material reacts when burning. Silicone liquid forms a silica crust and this assists in extinguishing flames. Some tests on the burning characteristics of epoxy resin showed that when heated with a bunsen burner, the material continued to burn when the flame was removed but did not when a welding torch was the medium of the heat. Manufacturers of cast-resin transformers claim that their designs do not make any appreciable contribution to the fire hazard because of the very high non-flammable filler content. Many have been installed within the past few years and are giving satisfactory service.

Class H dry-type transformers generally have superior fire resistance to most other types, the use of suitable flame-retardant grades of insulation giving further improvements of safety. The utilization of an open winding construction, in conjunction with flame-retardent insulation materials, has resulted in the development of the Securamid design of dry-type transformer, Fig. 13.10. Combustion tests, conducted on sample windings, have shown this type of transformer to have the best fire-resistant properties of all the available dry-types.

TAPPINGS AND CONNECTIONS

Part 4 of BS 171 deals with tappings and connections of power transformers. This

Fig. 13.10 A 500 kVA dry-type Securamid distribution transformer (*Goodyear Transformers Ltd*).

chapter, dealing as it does entirely with distribution transformers, will consider only tappings normally selected with the transformer disconnected from the supply.

A few years ago the standard distribution transformer had no tappings. It is argued that, with a fully regulated 10 or 11 kV supply system as exists in most western countries, tappings on distribution transformers are entirely unnecessary. However, tappings on the h.v. windings have become part of the distribution scene and in areas where long feeders remain adjustment for h.v. variation can be useful.

The standard off-circuit tapping range is 10% with two taps of 2½% above normal and two taps of 2½% below normal. In especially difficult areas most transformer manufacturers would be quite happy to supply this range as all negative taps.

It is generally agreed that there should be some movement toward voltage standardization within the EU. Inevitably the recommendations for the standard voltage are almost as numerous as the countries involved and any rationalization may well involve an extended tapping range.

Off-circuit taps on distribution transformers are normally selected by a switch on liquid-immersed transformers and links on dry-type transformers. Three- and five-position tapping switches are the most common in Europe. In certain difficult areas of the world it is not unusual to see off-circuit tapping ranges of 15% or even 20% specified. The costs of the special tapping switch and tapping arrangement make these non-standard arrangements very expensive and they should be avoided if at all possible.

Connections

Most 10, 11 or 13.8 kV distribution systems are supplied through system or network transformers where the l.v. winding is star connected. The system is therefore earthed via the neutral of the network or system transformer. For the satisfactory elimination of triple frequency harmonics it follows that the 10, 11 or 13.8 kV winding of distribution transformers should be delta connected. It is necessary on l.v. domestic distribution systems that phase-to-neutral voltages are available and that the neutral operates at or near to earth potential.

The most common winding connections are therefore either Dy 11, Dy 1, Dy 5 or Dy 7 with Dy 11 or its equivalent being widely used throughout the world. These connections are defined in BS 171 and are shown in Fig. 13.11. This diagram also includes other winding connections including the delta zigzag which is discussed below. The vector relating to the h.v. winding is taken as the vector of origin and the secondary phase relationship is related to the clock face. For example, Dy 1 where the secondary star lags the primary by 30° corresponds to one o'clock.

The choice of phase relationship between the primary and secondary windings is unimportant if only one transformer is used for a given site network. However, if more than one transformer is involved then they must all have the same phase relationship or else it is impossible to parallel them or to switch over the supply to the network from one transformer to another. More is said about parallel operation below.

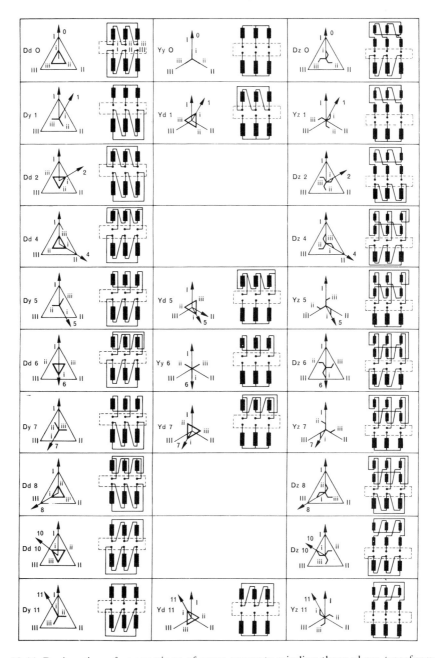

Fig. 13.11 Designation of connections of some separate winding three-phase transformers by connection symbols (BS 171: Part 4).

Delta/zigzag connections

There are a number of special applications where other winding connections are important.

Local power supplies are often obtained from diesel engine or gas turbine generating sets. Voltages generated are often too low for economic power distribution and transformers are used to raise the generated voltage to a satisfactory value for distribution then reduce it back to domestic levels. Contractors often order two identical delta-star transformers for this application which results in an unearthed and floating distribution system and a phase displacement between generated voltage and user voltage of 60 degrees.

It is necessary in this case to arrange for the generator-transformer to have a delta-connected winding connected to the generator and a star connection for the transmission voltage. With the connection Yd 1 there would be no phase displacement between the generated voltage and the user voltage and the higher voltage distribution system would have a star point for earthing.

There are still occasionally cases where unusual phase relationships exist from old privately owned systems. Linking these into modern networks can be made easier with the delta-interconnected star transformer or the star/interconnected-star transformer. The interconnected star winding, by varying the values of the winding sections, can be made to produce any phase relationship between primary and secondary windings.

Large supplies of d.c. are essential for certain industrial application. Furthermore 12- and 24-phase systems of rectifier supply are necessary if harmonics in the supply are to be kept to a minimum. The interconnected star transformer winding provides the solution.

Terminations

Electrical connections to transformers are made with cable which is either carried overhead on poles or pylons or buried in trenches in the ground. The former uses fairly simple steel reinforced copper or aluminium conductors and relies on air spacing between conductors for insulation.

Electrical cable for underground power distribution has solid insulating material between conductors and usually has a metal protection on the outer surface which is again protected against corrosion by an outer covering. At system voltage above 3300 V the underground cable becomes very much more complex and is beyond the scope of this chapter.

Generally speaking, distribution systems above 33 kV are by the overhead system and require bushing terminations on transformers. Distribution voltages of below 600 V in most of the cities and towns of Western Europe are by the underground method.

Underground cables have to be terminated by a method which does not expose the inner conductor insulation of the cable to the elements and this inevitably means some form of cable terminating box.

In the early days of power distribution the insulation most common to l.v. underground cables usually consisted of paper, cotton, jute, wax and hemp, etc. Although thoroughly impregnated with wax and oil these materials were quite prone to moisture absorption. Cable terminating boxes were compound filled.

Cable jointing and cable box filling was a fairly complicated business which required considerable skill. In spite of every effort failures in cable boxes were fairly common.

The development of modern plastics such as polyvinylchloride and polyethylene, etc. and their use in the cable industry has done much to simplify cable manufacture, cabling systems and termination. Almost all l.v. cable boxes are now air-filled with a simple compression gland securing the metal sheath of the cable. Low-voltage transformer terminations are usually an 'all-in-one' epoxy resin bushing plate assembly with bolt-on connections.

The compound filling of 6.6, 10 and 11 kV cable boxes on distribution transformers has remained in many areas of the UK because the most common 6.6 and 11 kV cable is still paper insulated. Developments in the field of plastics have produced cross-linked polyethylene cables where air-filled boxes with heat shrinkable sleevings eliminate the need for compound filling and greatly simplify the cable termination to transformers.

Dry-type transformers supplied in their own ventilated enclosures are generally treated as liquid-immersed transformers and have cable boxes mounted on the outside of the ventilated enclosure. Where the dry-type transformer forms part of a complete installation on h.v. and l.v. switchgear it is often required for the transformer to be supplied without an enclosure; h.v. winding connections are terminated in bolted connections just above the core and windings and l.v. connections are generally terminated on busbars.

Cable-end terminations for installations of this type can still follow the same form as the tankside cable box but with the three-phase bushings arranged and spaced for copper connections in air.

The resin-encapsulated transformer was developed originally in Germany where open installations are more commonplace. Special plug-in epoxy or moulded polymer connectors and bushings, incorporating electrical stress relief screens, have been developed by some companies and the incoming h.v. cable is simply laid into a moulded cable holder and the bared conductor bolted to metallic inserts.

COOLING

Transformers are identified according to the cooling method employed and the letter symbols used are indicated in Table 13.9. The simplest form is where the windings are cooled by natural air flow over the heated surface of the windings and core. These are heated by the load and no-load losses respectively and the heat is transferred to the surrounding air by convection and radiation. This type of cooling is described as air natural or AN by reference to Table 13.9.

The natural movement of air over heated coil surfaces is not particularly efficient. Winding conductors, disc- and coil-separator winding shapes, and the roughness of conductor insulation all help to create eddies in the air flow over the windings and reduce the heat transfer from the winding to the air. Even a small amount of force-directed air flow over the windings improves the heat flow, and ratings are increased significantly.

A dry-type transformer cooled by direct forced air is designated as AF. A dry-

Table 13.9 Letter symbols (BS 171: Part 2).

	Symbol
Kind of cooling medium	
Mineral oil or equivalent flammable synthetic insulating liquid	O
Non-flammable synthetic insulating liquid	L
Gas	G
Water	W
Air	A
Kind of circulation	
Natural	N
Forced (oil not directed)	F
Forced-directed oil	D

type transformer which has natural cooling and the facility for automatic fan operation should the temperature of the windings increase beyond normal limits has two ratings specified followed by AN/AF. For example 1000/1250 kVA AN/AF dry-type Class C transformer.

The oil- or liquid-immersed transformer must, by this definition, have two sets of letters; one which describes the cooling of the winding and one which describes the cooling of the surface of the liquid. Hence the most common distribution transformer in which the windings and core are naturally cooled by oil and the oil is subsequently naturally cooled by air has the designation ONAN.

Forced-cooling equipment and its control are costly additions to any transformer and this cost can rarely be justified on distribution transformers of the ratings covered in this chapter. However, the same ruling that applies to forced cooling of dry-type transformers also applies to the oil in an oil-filled transformer or the silicone fluid in a fire-resistant transformer. Ratings can be increased significantly if air is forced over the cooling surface of the tank. An oil-immersed transformer arranged to have fans automatically switched on when the oil exceeds a certain temperature level would have the dual designation of ONAN/ONAF. Yet a third condition of forced cooling is obtained when an oil pump is built into the oil flow system of the transformer. The designation becomes ONAN/ONAF/OFAF.

The naturally cooled ONAN or LNAN distribution transformer above the rating of approximately 50 kVA requires cooling surfaces in addition to the tank surface that would normally contain the core and windings. At one time this additional cooling surface was provided by tubes welded into the tank wall and, theoretically carried the hot oil from the top to the bottom of the tank. In more recent years it has become fashionable to manufacture plate radiators from pressings similar in pattern to the domestic hot water radiator and arrange these in banks on the tank side. The plate radiator has the advantage of reduced oil content and possibly results in lower manufacturing cost.

As early as 1925 manufacturers produced oil-immersed transformer tanks with the tank walls made from deep corrugations of thin steel plate. It was not until very many years later that machines (see p. 311) were developed to form deep corrugations into a continuous strip of plate automatically, weld the edges of the corrugations automatically, insert and weld corrugation strengthening strips and shear to a predetermined length.

The transformer tank with all four sides formed from thin (1.2 mm) corrugated steel plate with terminations and tapping switch all mounted on the cover has become the most standard arrangement throughout Europe and the USA, although standard supply industry terminations in the UK make this cooling arrangement less advantageous than the tank with radiators.

Ventilation of transformer enclosures

Transformers operating within an enclosed area inevitably reach a higher temperature for the same loading conditions than they would when operating in free air. It is important to the life of the transformer that this fact is appreciated and that substations or enclosures are designed so that this excess temperature is limited.

The problem of substation ventilation when fan extractors are used is very much more simplified than for naturally ventilated enclosures. Natural ventilation does not rely on the functioning and maintenance (or lack of it) of fans and is therefore preferred.

The excess temperature of the substation or enclosure is a function of:

(1) the total losses of the transformer;
(2) the net area of the inlet and outlet ventilation areas;
(3) the effective vertical distance between the inlet and outlet ventilation areas.

The inlet ventilation area is positioned ideally low down, below the centre line of the transformer radiators with the transformer fairly close. The outlet ventilation area is required to be high, not immediately above the transformer, but on a wall remote from the inlet so that cooling air passes over the transformer.

The minimum height of the outlet above the inlet area should ideally be equal to 1.5 times the height of the transformer.

The net area of the inlet or outlet has been shown by empirical means to be approximately:

$$A = 0.06P$$

where P is the total loss dissipated from the transformer in kilowatts and A is in square metres.

With these conditions met the temperature of the substation air should not be more than 7−8°C above the outside ambient.

IMPULSE WITHSTAND

The main three-phase 11 kV distribution system in the UK is by underground cables, and ground-mounted distribution transformers connected to this system are not subjected to high-level transient voltages of atmospheric origin because of the attenuating effect of the cable to steep-fronted waves. However, it is recognized that for a variety of reasons transient voltages do occur in cabled systems. Transformers can also be connected very close to an overhead system and an agreed system of impulse voltage testing and impulse voltage levels does give indication that a level of insulation strength has been achieved in the basic design.

Table II of IEC 76, Part 3 gives the impulse voltage withstand levels agreed by most countries of the world. Distribution voltages of 3.6, 7.2 and 12 kV are covered by two values of impulse withstand level. List I is for the transformer that is considered to be electrically not exposed to high voltage transients and list II with a higher level for the electrically exposed. The 11 kV system of the UK is considered by the supply authorities to require the 75 kV level, while pole-mounted transformers connected to overhead lines are usually tested to a higher level of 95 kV. Europe and Middle Eastern countries standardize generally on the 75 kV level for distribution systems between 10 and 11 kV.

At one time it was considered that dry-type Class C transformers could be classified as electrically unexposed to transient voltages because they could never be operated out of doors close to an overhead distribution system.

The increasing use of vacuum circuit-breakers and their current-chopping characteristics produce exceptionally high voltages in the low-capacitance, high-surge impedance windings of the dry-type transformer. Impulse levels therefore became necessary for the dry-type transformer and the list I level has been accepted as satisfactory for the unexposed transformer.

OPERATION IN TROPICAL CLIMATES

A number of special problems exist when transformers are operated in tropical climates. The effects of increased ambient temperature are well documented and require little elaboration in this chapter. If the average ambient temperature at site is 10°C higher than that specified in normal operating conditions given in BS 171 then the temperature rise of the transformer would either need to be designed for 10°C lower or the transformer would need to be derated to a level which produced a temperature rise 10°C lower. Other hazards exist which are, perhaps, not so well documented. High isoceraunic levels exist in some tropical countries and electrical storms can persist for long periods of time. Standard impulse voltage levels may well be insufficient for these areas and the added protection of surge arresters considered to be advisable. Standard bushings may again give insufficient creep and a higher voltage class considered necessary in these areas.

Temperature, intensity of solar radiation, rainfall, high wind, dust storms and humidity, all affect the life of paint applied as a protection against corrosion to the tank surface of liquid-immersed transformers. A standard paint finish generally consists of three coats. The first, the priming coat, is applied to the prepared metal. The second, the undercoat provides the key for the main protective top coat. This final coat is usually long lasting high gloss, the gloss providing the main protection. Generally this three-coat system provides adequate protection for a reasonable time, and is easy to maintain and replace when it has been damaged or worn. Thicker coatings can give extended life to any paint system. Special top coatings are available which again give extended life. It must be pointed out, however, that the more complicated the protective system, usually the more difficult it is to maintain and the integrity of any surface coating relies on adequate pretreatment of the metal.

Dry-type transformers can be vulnerable to inadequate maintenance in tropical countries. Dust and sand are a natural hazard in many areas and in countries where the temperature drops to quite low values at night the warmth of a transformer installation probably seems to be quite a haven to a host of small creatures. Extra maintenance and vermin proofing of enclosures might, therefore, militate against their use.

PARALLEL OPERATION

Satisfactory operation of transformers in parallel means that each transformer will carry its share of the load according to its rating; and for this condition to be met the voltage ratio, phase displacement and impedance must all be the same. Transformers in parallel must have the same secondary voltage for a common primary input.

The importance of phase displacement is obvious from Fig. 13.11. A pair of three-phase transformers of similar characteristics and having the same connection symbols can be operated both physically and alphabetically. For example Dy 1 and Yd 1 can be safely connected together. The impedance (which governs the regulation) decides the proportion of the total load which is taken by each transformer. The resistances of each unit must be similar.

When connecting units in parallel or paralleling supplies from two separate transformers the phase rotation must be the same.

Other points to remember when parallel operation of transformers is being considered are:

(1) The tested impedance of transformers can vary by ± 10% of the guaranteed value. Two transformers to the same guaranteed value of impedance can have test values which vary by as much as 20%.
(2) The length and type of cable connections must be considered if additional transformer capacity is added to an existing system and sited away from the original unit(s).
(3) Transformers that have tapping ranges larger than 10% need to have the impedance variation throughout the tapping range considered. Very large variations between manufacturers can occur owing to different winding arrangements.

PACKAGED SUBSTATIONS

A distribution substation consists essentially of h.v. switchgear (in many cases a ring main unit), a transformer and a fused l.v. distribution panel. Until recent years these were always supplied by different manufacturers as individual components. They were often sited in a small compound or building, and each component had its own termination system, usually cable boxes. The three equipments were electrically connected by short lengths of cable. Lack of available space, particularly in city areas, rationalization in the electrical supply industry and manufacturing industries and growth in demand for electrical power all gave effect to the development of the British unit substation. The three equipments are still separate components, which can be purchased separately from different suppliers, each with its own particular characteristics and be brought together to form a substation because of the standardization of terminal heights and flanges. Although a unit substation is often supplied with a ventilated steel or glass-fibre housing each equipment is of weatherproof design and requires no more than a wire fence surround. Ventilated housings are preferred by many authorities at home and abroad and can be of very pleasing appearance. Lockable doors afford access to l.v. distribution panels and h.v. switchgear. As many as 12 three-phase fused outlets are available on the l.v. distribution panel. Housing doors can be interlocked with the h.v. switchgear.

The packaged substation is described in greater detail in Chapter 2.

PROTECTION

Transformer protection is described more fully in Chapter 22 but for completeness the various systems available are enumerated. Two systems of protection, peculiar to transformers, are described in more detail; they are gas and oil relays and winding temperature indication.

Differential protection

Differential protection is based on the principle of comparing the primary and secondary currents of the transformer and if these balance any fault is external to the unit. The winding connection of the transformer primary and secondary are usually different (delta-star for the most part in the power range we are considering) and have to be compensated for by connecting the appropriate CTs in star-delta. Both balanced current and balanced voltage systems are used.

Restricted earth fault

The three CT secondaries on each side of the transformer are paralleled together with a relay connected across them. A fourth CT is connected in the neutral of a star-connected winding. The relays only operate for an internal earth fault as it is only under these conditions that the CT outputs do not sum to zero, causing an unbalanced current to flow in the relay circuit.

Unrestricted earth fault

A single CT in the neutral of a star-connected transformer provides a measure of protection against earth fault but the relay also operates for earth faults outside the transformer.

Overcurrent

A standard idmtl relay can be used to provide overcurrent protection but it will of course cover the whole of the network beyond the transformer. Overload settings can be adjusted to discriminate with protection on the load side of the transformer. This type of relay is often installed to act as back-up protection.

Gas and oil relay

The double-float gas and oil relay is fitted in the pipe between the main tank of the transformer and the conservator and is more commonly found on oil-immersed transformers above 2.0 MVA. The two pivoted floats carry switches which can be normally open or normally closed. One float is actuated when the oil level in the conservator and hence the relay falls to an unacceptably low level. The switch on the low-level float is usually connected to an alarm circuit that gives warning of low oil level in the transformer. The other float operates when there is a sudden production of gas within the transformer. The float then operates on surge. Switch contacts on this float are usually connected to the trip circuit on the associated switchgear which then disconnects the transformer from the supply. A sudden production of gas is usually indicative of serious fault conditions, hence the need for shutdown.

Flashover between connections, flashover to earth, breakdown between parts of the same windings, etc. produce different mixtures of gases. The most significant gases generated by the electrical breakdown of the oil are hydrogen, methane, ethylene and acetylene. Cellulose insulating materials, when broken down by an electric arc produce mainly carbon dioxide and carbon monoxide. An analysis of the gas collected from the gas and oil relay can give a very good indication of the materials involved in a gas-producing fault which initiated the operation of the relay.

It has been known for small quantities of gas to be released within a transformer immediately following installation, and over a period of time the gas collects in the gas relay and eventually an alarm is given of low oil level. In this case an analysis of the gas will often show it to be no more than trapped air.

Pressure-relief device

One of the most useful developments in recent years is the 'snap-action' pressure-relief device. Manufactured by Qualitrol, and since copied by various other companies, this device, which is mounted on the tank, wall or cover, operates when a predetermined pressure is exceeded within the tank. The seal snaps open and the large orifice allows gas to be discharged at the rate of $283 \, \text{m}^3/\text{min}$. The Qualitrol pressure-relief device was developed for the oil-immersed sealed transformer but it has become widely accepted as a reliable explosion vent and has almost replaced the old-fashioned diaphragm type. The Qualitrol device can be supplied with two single-pole double-throw switches. A brightly coloured plastics pin, located in the centre of the device, gives mechanical indication that the device has operated.

Winding temperature indicators

Unless the temperature of a winding can be measured by direct contact with the winding conductors, winding temperature indication can be no more than a close approximation and accurate over only a fairly narrow band of transformer loading.

The two main methods used by manufacturers in the UK to give indication of winding temperature are:

(1) a direct method whereby the temperature sensor of the instrument is held in close proximity to the l.v. winding;
(2) an indirect method whereby a 'thermal image' device simulates the winding-to-top-oil temperature differential.

Method (1) is used almost exclusively on dry-type transformers where large cooling ducts allow for the positioning of the instrument's temperature sensor and where the integrity of the winding insulation system is not impaired. The most recent practice is to fit standard platinum resistance sensing devices into the low voltage windings, and feed the output to electronic control devices. The instrument measures the resistance of the sensing device, which changes directly with the winding temperature. These devices are able to display the actual winding hot-spot temperature, some are able to store highest temperature values for future retrieval and display.

The indirect method of the thermal image device uses the standard mechanism of a dial-type temperature indicator. A current transformer mounted in the live connection to one winding supplies a proportion of the live current to a heater coil wound on to the operating bellows of the instrument. A calibrating resistor adjusts the current in the heater coil to a value that produces the correct winding-to-oil differential.

All winding temperature indicators can be fitted with contacts to operate alarms and trip mechanisms. The instrument in common use can be fitted with three or four switches to operate fans and pumps for forced-air and forced-oil circulation; sometimes they are connected into a building or installation management system, so that any untoward operating conditions may be noted, and corrective action taken.

SHIPMENT OF TRANSFORMERS

The preparation of a transformer for transportation depends on its size, type, destination, the distance and method of transport and the length of time the transformer is to remain in store before installation.

Liquid-filled transformers

Liquid-filled outdoor distribution transformers fitted with cable boxes and for delivery within the UK are invariably dispatched by road transport, securely anchored to open-backed vehicles with no more packing than perhaps some protection for the cooling radiators or corrugations. These, being of thin steel plate are easily damaged in transit if vehicles are not carefully loaded. Smaller, pole-mounted transformers, with exposed porcelain bushings require some protection if damage to terminals is to be avoided during transport. A careful arrangement of identical units with packing pieces spacing one unit from another is normally sufficient for UK road conditions.

Larger units, even for delivery within the UK, require more consideration. The removal of conservators, bushings and radiators is advised if the total load exceeds the width of a low loader. The route to be taken by the vehicle, sharp bends (particularly at the bottom of hills), road camber and low bridges are all hazards which have to be considered when units larger than 5.0 MVA have to be transported. It is advisable to cover and protect dial-type instruments, gas and oil relays, etc. from damage.

All transformers manufactured to BS 171 have lifting lugs as a standard fitting. The weight of an 11 kV industrial ground-mounted oil-filled transformer, with standard fittings is very approximately $0.022 \, (kVA)^{0.75}$ tonnes and requires lifting capacity of 6 tonnes or less for units of 1600 kVA or lower. Removal from the vehicle to site or store in the UK or anywhere within the industrialized west should not present problems.

When oil-immersed transformers are shipped to overseas locations, particularly where handling arrangements are restricted and long distances have to be travelled overland on poor roads, then the type of packing employed and the method of transportation have to be given much more consideration.

The single isolated unit is the most vulnerable because most distribution transformers and unit substations, supplied in large numbers to the high growth areas of the world, are transported from manufacturers to users by containerized shipping. Many units are packed and sealed into 10 m long 40 m³ volume containers by packing experts and the complete container is transported direct to site.

Larger transformers require the vulnerability of components, bushings, radiators, conservators, etc. all to be considered and it is often advisable to remove them and pack them separately in wooden crates.

Weight and freight restrictions often demand that oil-immersed transformers are dispatched dry with the oil packed in drums and shipped separately. No great problems exist here if the transformers are sealed with dry air or nitrogen and moisture is prevented from entering the tank and affecting the windings.

Sealed transformers are of course filled with oil and sealed at the manufacturer's works and are always dispatched filled with oil.

Dry-type transformers

Due consideration must be given to the shipment of dry-type transformers. Class C, and similar types, are susceptible to the ingress of moisture, and must be suitably protected during shipment. Cast-resin transformers are moisture resistant to a far greater degree than conventional dry-type transformers, but will still require appropriate protection during shipment. For dry-type transformers housed in sheet metal enclosures, it is possible for transportation to be by road, with the units covered by waterproof sheeting. For export, or where the transformers will not be immediately placed under cover, it is usual for close-boarded crates, lined with polythene, to be used. A quantity of moisture absorbing silica-gel desiccant is often placed within the enclosure as an extra precaution.

INSTALLATION

It is important to note that before installation transformers should be checked for any damage that may have been caused during transit. Liquid-filled units dispatched fully filled should be checked for correct oil level and for any leaks that may have occurred. The paint finish should be examined carefully for signs of damage. The cores and windings of Class H and C transformers should be examined for signs of mechanical damage to leads, connections and risers. Coils should be examined for signs of insulation damage. The cast-resin of encapsulated transformers can be easily chipped or cracked by being knocked and thus needs to be examined very carefully.

Liquid-filled transformers

It is usual to have some form of containment and soakaway for all liquid-filled transformers, be they mineral oil, silicone liquid or any other design, irrespective of whether they are installed indoors or in the open. In some countries this is required by law. The soakaway is usually a pebble base contained within a low brick wall enclosure. When installed in a building or substation without additional fire protection it may take the form of a catch pit, designed as a pitched or lowered floor, pit or trench as shown in Fig. 13.12. It should be large enough to contain all the liquid to cater for the possibility of a severe rupture of the tank. The American Factory Mutual Research organization has detailed this information in its Factory Mutual Loss Prevention Data Sheet 5−4A/14−85.

The shape, size and materials used for the building should be based on the rate of heat release of the transformer liquid when on fire and discussion of these is outside the scope of this chapter. The above-mentioned Data Sheet also gives guidance on this aspect of the subject.

Dry-type transformers

All dry-type transformers must be installed indoors with an earthed metal enclosure around them. In the UK this usually takes the form of a substantial metal framework, but there is no reason why the lightweight mesh-screened panels, as used by the European continental countries, should not be employed. This reduces the overall cost of the installation.

Class C or H designs are generally not as tall as their cast-resin counterparts, although other dimensions are similar. This slight difference in headroom might be important if the unit is to be installed in an existing substation. The larger dry-type transformers, 500 kVA and above, are usually built on a framework of metal channels and are usually provided with wheels to allow movement of the unit for cleaning and maintenance.

Curbing options

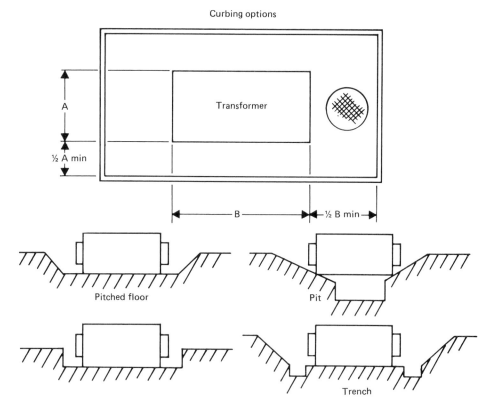

Fig. 13.12 Some different arrangements of soakaways in a building or substation housing a liquid-filled transformer.

CABLING

For the distribution transformers covered in this book cabling is usually PVC insulated and sheathed, although in special cases EPR or other forms of insulation may be employed. Copper or aluminium conductors are available. Armouring is usual for outdoor installations or where onerous soil conditions exist. A three-core design of h.v. cable is normal for both dry-type and liquid-filled design. Termination is by cable box, either compound-filled or air-insulated, with suitable heat-shrink termination kits; the choice is dependent upon the cable type and the preference of the transformer purchaser. Moulded plug-in type connectors are finding increasing application, particularly on cast-resin and other dry-type transformers.

Depending on the rating of the transformer, l.v. cables may be single-core or three-core. Termination methods are similar to those appertaining to the h.v. side, possibly using heat shrink connections for dry-type units.

COMMISSIONING

Before proceeding with commissioning tests a careful examination of the transformer and its surrounds must be carried out, particularly if the unit has been installed for some time. It is not unusual to find tools and other extraneous matter lying on the unit in a position that could be highly dangerous if a high voltage were applied to the transformer. For dry-type transformers, particularly those of Class C or H designs, the windings should be checked with an IR tester and then dried out as necessary. The drying out can be done electrically by passing a small current at l.v. through the windings or by the application of a 'gentle' external source of heat.

Where it is known that the dry-type Class C transformer is to remain in store for a long period of time or is to remain unexcited for long periods of time then it is advisable for the unit to be fitted with small heaters on the core and windings to circulate warm air during the time that the transformer remains inactive.

For liquid-filled units sent overseas it is important to ensure that the unit has in fact been filled with the correct fluid if the transformer was dispatched unfilled. Depending on the circumstances it may also be necessary to institute some form of winding drying-out procedure. In any case the fluid should be checked for electrical strength before commissioning.

Where relays or instruments are associated with transformers they should be examined to make sure there is no packing or safety lock incorporated as a protection against damage during transit. This should be removed. The position of the off-load tap change device must be set to correspond to the network conditions. If there is a conservator valve, ensure that it is open. If there are plastics caps covering breather pipes, ensure that they are removed.

When all the connections and other matters have been checked the transformer can be energized and allowed to run on no-load for a period long enough to ensure safe operation. Gradual loading can then be introduced until full load is reached. During the commissioning procedure instruments and protective relays should be observed for correct functioning and it should be checked that readings are of the expected order.

MAINTENANCE

It is not the intention to give full and detailed maintenance instructions here but to highlight the more important features related to the various types of transformer included in the chapter.

Mineral-oil-filled units

Oil is subject to deterioration or contamination in storage, in the course of handling or in service. Accordingly it requires periodic examination and possibly treatment to maintain it in a good condition. After many years in service it may even require replacement. BS 5730: 1979, *Code of practice for maintenance of insulating oil*, gives specific advice relating to the testing of oil in order to determine its fitness for further service.

Provision is usually made on all mineral-oil-filled transformers for samples of the oil to be drawn off, and this operation should be done while the oil is warm. A limited but useful amount of information can be obtained from the colour and odour of the oil. Cloudiness may be due to suspended moisture or suspended solid matter such as iron oxide or sludge. Moisture can be detected by the crackle test.

If the oil is dark brown it may contain dissolved asphaltenes; if green the presence of copper salts is indicated, and it may be expected that further deterioration will be rapid.

An acrid smell is often indicative of volatile acids which can cause corrosion and which may render the oil unsuitable for treatment on site. A petrol-like or acetylene odour may indicate a low flashpoint due to a fault or some other cause.

Tests are laid down in BS 5730 to determine free water, acidity, electric strength, sludge, flashpoint and resistivity and suggested minimum safe limits are indicated. Frequency of testing is also indicated which varies from 6 months up to 2 years depending on the test concerned and the environmental conditions. It is important that the oil level is maintained.

At regular intervals it is strongly recommended that breathers and breather pipes are checked. If a silica-gel dehydrator is fitted then the condition of the oil bath and crystals should be checked. An oil-immersed transformer should never be allowed to breathe through silica-gel crystals supersaturated in moisture. Breathers should be changed, or re-activated, when two thirds of the silica-gel crystals have changed colour from blue to pink.

The Code also recommends that the protective paint finish of the tank is checked for damage and that there are no leaks.

If oil temperature and winding temperature indicators are fitted operation of the switches should be checked. Gas and oil relays and pressure-relief devices fitted with switches should also be checked occasionally for ease of movement.

Fire-resistant liquid-filled transformers

For reasons stated earlier, askarels are being replaced in transformers and so the maintenance of these units will not be discussed. If an askarel-filled transformer is found to be leaking, advice should be sought from the manufacturer as to how the situation is to be dealt with. It is important that askarel-type materials are not allowed to contaminate the environment around the transformer, or to enter any water drainage systems.

Of the available range of fire-resistant liquids, silicone fluid and synthetic esters (Midel) are the most commonly used types. These materials possess superior thermal stability to mineral oils and will consequently degrade at a slower rate. This implies a reduction in the amount of maintenance required by transformers filled with these fluids, but the recommendations made by manufacturers with respect to maintenance procedures are to be followed. IEC Specifications 836−1 and 837−1 give general information relating to the use of silicone fluids and synthetic esters (Midel) for electrical purposes.

In the event of silicone or synthetic ester fluids requiring to be cleaned, similar equipment to that used for the treatment of mineral oil may be used. It is important that synthetic fluids are not cross-contaminated, or contaminated with mineral oil. A very low level of mineral oil contamination in a high fire-point fluid will significantly reduce the fire-point. It is preferable that separate pumping and handling equipment are used for different fluids. Silicone fluids have minimal lubricating properties and care must be taken in selecting the appropriate pumping equipment to handle these materials. Solid contaminants may be removed from these fluids by cartridge filters or by filter presses, moisture being extracted by conventional degassing equipment.

Table 13.10 details physical and electrical properties of silicone and synthetic ester fluids that can be checked as part of maintenance operations. Transformers containing these fluids are normally of hermetically sealed construction, thus contamination by air-borne pollutants is prevented and the amount of maintenance required is reduced. As with mineral-oil-filled transformers, synthetic liquid cooled types should be checked for leaks and to ensure that deterioration of the paint finish has not occurred.

Dry-type transformers

The most important aspect of maintenance is that of ensuring that all cooling air ducts are kept free of foreign objects and contaminating dirt. A vacuum cleaner can be used for this purpose. Any blockage can cause the winding temperature to rise to levels that could affect the performance and life characteristics of the insulation. If a transformer is taken out of service for any length of time, and heaters have not been fitted, it is a wise precaution to dry out the unit by the means suggested earlier under 'Commissioning'. This is particularly true of the

Table 13.10 Recommended maintenance tests for high fire-point fluids.

Test	Acceptable values		Unacceptable values indicate
	Synthetic ester (Midel)	Silicone fluid	
Minimum			
Visual	Clear straw-coloured liquid, free of particles	Crystal clear, free of particles	Particulates, free water colour change
Odour	Faintly sweet	Odourless	Arced or burned insulation, volatile contamination (solvent pcb)
Dielectric breakdown	50 kV new	50 kV new (first test)	Particulates, contamination, water
BS 148: 1984 test cell	25 kV in transformer	25 kV in transformer	
Additional			
Water content	200 ppm maximum	100 ppm maximum	Excessive water
Viscosity	95 cSt at 20°C	50 cSt at 25°C	Fluid degradation, contamination
Fire-point	310°C minimum	350°C minimum	Contamination of fluid by volatile materials
Permittivity	3.2 at 20°C	2.7 at 25°C	Polar/ionic contamination

Class C or H designs. Before connecting any supply to the transformer for this purpose the terminations at both ends of the h.v. windings, the pressure rings and the insulators should be dried with a cloth. The l.v. winding can then be short-circuited and a low current passed through the h.v. winding. The whole drying-out operation should not take more than 5 hours. Alternatively, fan heaters can be employed to circulate warm air through the windings.

In the event of a minor winding fault, where the damage is restricted, it is possible to repair both conventional dry-type and cast-resin transformers on site. Major electrical breakdowns will require replacement of the winding subject to fault. This may be accomplished on site with a cast-resin type of transformer, but it is usual to return a conventional dry-type transformer to the factory, or suitably equipped workshop. Any repairs are best carried out by the manufacturer of the transformer, or by a reputable repair organization.

Future development

The development of distribution transformers tends to be an evolutionary process, rather than a revolutionary one. Most developments are aimed at reducing the manufacturing costs, as well as improving the overall efficiency, without reducing the inherent reliability of the modern distribution transformer. Improvement in the performance of the magnetic materials used for transformer core laminations has led to a reduction in the no-load losses and this trend will continue as the cost of energy increases. Advances in coil manufacturing methods have enabled the production of windings possessing superior short-circuit strength in conjunction with compact dimensions. The introduction of high fire-point liquids, and the various types of dry-type transformer, have given the potential purchaser the opportunity to choose a suitable design for almost any situation. It is anticipated that these types of transformer will undergo further development in order to improve their cost-effectiveness. The use of synthetic cooling fluids at elevated temperatures, in conjunction with suitable insulating materials, will produce dimensional and cost reductions in this type of transformer. The manufacturers of insulation materials also improve their products and these improved materials are then incorporated into distribution transformers by manufacturers as part of the evolutionary development.

CHAPTER 14

Switchgear

A. Headley, BSc, PhD, CEng, MIEE
(Reyrolle Ltd)

R.W. Blower, BSc(Eng), FIEE

DEFINITIONS

Switchgear

A general term covering switching devices and their combination with associated control, measuring, protective and regulating equipment, and also assemblies of such devices and equipment with associated interconnections, accessories, enclosures and supporting structures, intended in principle for use in connection with generation, transmission, distribution and conversion of electric power.

Metal enclosed switchgear
Switchgear assemblies with an external metal enclosure intended to be earthed, and complete except for external connections.

Metal-clad switchgear
Metal enclosed switchgear in which components are arranged in separate compartments with metal enclosures intended to be earthed. There will be separate compartments at least for the following components: each main switching device; components connected to one side of a main switching device, e.g. feeder circuit; and components connected to the other side of the main switching device, e.g. busbars.

In the UK, circuit-breaker switchgear is expected to be metal-clad, which is an extension of metal-enclosed through the use of internal metal partitions. In addition, since the majority of faults on distribution switchgear tend to be cable faults, at least in origin, there is a demand for the cable compartment to be closed off from the circuit compartment, either by a metal partition or, if clearances require it, by an insulating partition.

Circuit-breaker
A mechanical switching device, capable of making, carrying and breaking currents under normal circuit conditions, and also making, carrying for a specified time and breaking currents under specified abnormal circuit conditions such as those of short-circuit. Almost everywhere in the world except the UK requires that the 'short time rating' be for one second. Due to the number of protection stages and

the use of many older style relays, faults in the UK can exist for longer than this if there is a failure of the first stage protection. For this reason, most UK switchgear has a three second short time rating.

Indoor circuit-breaker
A circuit-breaker designed solely for installation within a building or other housing, where the circuit-breaker is protected against wind, rain, snow, abnormal dirt deposits, abnormal condensation, ice and hoar frost.

Outdoor circuit-breaker
A circuit-breaker suitable for installation in the open air, i.e. capable of withstanding rain, snow, dirt deposits, condensation, ice and hoar frost.

Switch

A mechanical switching device capable of making, carrying and breaking currents under normal circuit conditions, which may include specified operating overload conditions, and also carrying for a specified time (three seconds in the UK) currents under specified abnormal circuit conditions such as those of short-circuit. It must also be capable of making, but not breaking, short-circuit currents.

Disconnector (isolator)

A mechanical switching device which provides in the open position an isolating distance in accordance with specified requirements.

A disconnector is capable of opening and closing a circuit when negligible current is broken or made, or when no significant change in voltage across the terminals of each of the poles of the disconnector occurs. It is also capable of carrying currents under normal circuit conditions and carrying for a specified time currents under abnormal conditions such as those of short-circuit.

Note: Negligible currents imply currents such as the charging current of busbars, bushings, etc. and a current not exceeding 0.5 A is deemed to be a negligible current for the purpose of this definition.

Summary

As will be seen from the above definitions, the most versatile switching device is the circuit-breaker, as this is the only equipment capable of interrupting short-circuit currents and then restoring supply a number of times without requiring maintenance or the replacement of any parts. The least versatile device is the disconnector, which is only capable of carrying short-circuit currents and has to be operated off-load.

Because of the need to interrupt short-circuit currents, the design and mode of operation of a circuit-breaker tends to be dominated by this requirement and it is usually categorized by the arc extinguishing medium used for this purpose. In this chapter we consider only a.c. circuit-breakers, as the interruption of d.c. is a very limited specialist requirement.

CIRCUIT-BREAKING

An essential element in the operation of any a.c. circuit-breaker is the electric arc, which permits the current in the circuit to continue flowing after the contacts have parted, until a suitable current zero occurs. This is illustrated in Fig. 14.1. As will be described below, the sudden cessation of current flow at any time, other than very close to a natural current zero, has undesirable consequences in all normal distribution systems, so the existence of the arc as a natural commutating device is a very important factor in the operation of a.c. circuit-breakers.

An ideal circuit-breaker is one which acts as a perfect conductor until current zero is reached, at which point it becomes a perfect insulator. As no practical circuit-breaker meets this condition, the result is modified to a greater or lesser degree by the circuit-breaker characteristics. The objective of the circuit-breaker designer is to create the necessary conditions to sweep away the ionization products in the contact gap at current zero and replace them by a medium which will withstand the application of a very rapidly rising voltage of considerable amplitude, the transient recovery voltage. This is shown at H in Fig. 14.1.

MEDIUM VOLTAGE SWITCHGEAR

The term *medium voltage* is usually used to describe switchgear rated from 3.6 kV to 52 kV. In the UK it is general practice for m.v. distribution switchgear manufacturers to supply completely factory-assembled switchgear. This means switchgear equipment containing busbars and all circuit components up to the cable terminations, including any switching devices (circuit-breakers, switches, etc.) and their isolating means. It is, therefore, almost equally general for the switchgear to be type tested in its completely assembled form. This is not always the case for l.v. switchgear, since a thriving business exists in this area for the manufacture of l.v. switchboards incorporating circuit-breakers and other l.v. switching devices made by a small number of specialist manufacturers. Also m.v. circuit-breaker switchgear is designed principally for use with separately protective relays and therefore has a short time rating, usually equal to the breaking current rating for three seconds, unless where very heavy fault levels exist, the operating conditions permit a one second short time rating. The switch or circuit-breaker also has to be able to close fully against a short-circuit making current peak equal to 2.5 times the rated breaking current, and to be able to interrupt that breaking current with a degree of asymmetry which will depend on the opening time of the circuit-breaker being considered. With some of the modern high speed circuit-breakers using vacuum or sulphur hexafluoride (SF_6), this time can be quite short and the d.c. component correspondingly large. A figure of 35% d.c. component is not uncommon under these circumstances.

All circuit-breakers have to be able to close and open satisfactorily under all conditions of service. In particular they must interrupt all currents from zero to full rated breaking current with all possible combinations of power factor, current asymmetry and recovery voltage that occur. Figure 14.1 illustrates the typical form of the current and recovery voltage which occurs when a fault current is

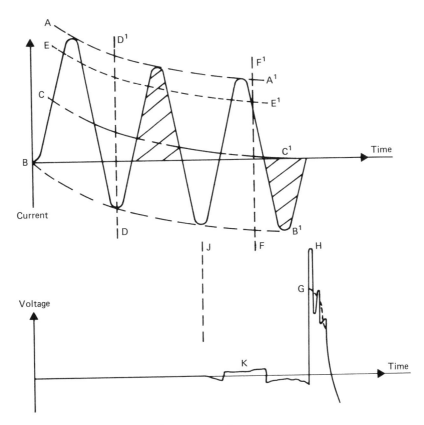

Fig. 14.1 Typical fault current and recovery voltage diagram.

AA^1 and BB^1	envelope of current wave
CC^1	displacement of current wave zero line
DD^1	typical contact separation point for asymmetrical breaking duty
EE^1	rms value of symmetrical current wave measured from CC^1
FF^1	typical contact separation point for symmetrical breaking duty
G	peak value of 50 Hz recovery voltage
H	peak value of high frequency transient recovery voltage
J	assumed contact separation point to illustrate typical voltage across circuit-breaker contact gap
K	arcing voltage.

interrupted. The asymmetry which is illustrated in this single-phase example arises as a function of the instantaneous value of the system voltage when the fault occurs, and the power factor of the system voltage under fault conditions. This reduces from the usual operating figure in excess of 0.8 to a much lower figure such as 0.1−0.3. This creates much more severe conditions for arc interruption as the current zero now occurs at a time when the supply voltage is closer to its peak. The high frequency transient recovery voltage is a function of the system inductance and capacitance.

In addition to the fault condition, circuit-breakers and switches often need to be able to switch low currents of a highly inductive nature (low lagging power factor) or of a capacitive nature (low leading power factor). Special conditions attend these operations.

Low inductive currents

As illustrated in Fig. 14.2, the available arc extinguishing effort may tend to force the current to zero prematurely. When this happens energy is trapped in the inductance of the load and subsequently transferred into capacitive energy stored in the leakage capacitance of the system. This process continues in an oscillatory manner as part of an augmented transient recovery voltage and eventually dies away as the energy is dissipated as heat in the resistive circuit elements. Some of the energy (maybe as much as 30%) is also lost by iron losses if the load has an iron core as is usually the case, since these conditions normally arise when

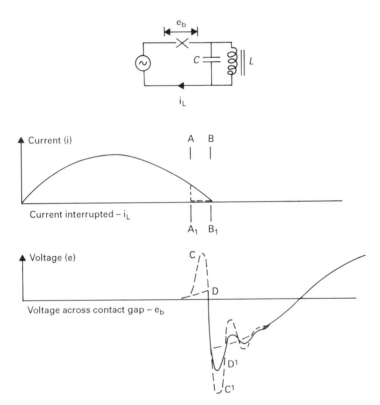

Fig. 14.2 Conditions occurring when interrupting small inductive currents ($\cos \theta \simeq \theta$ lagging).

L   inductive load
C   leakage capacitance
AA_1 current interruption point with 'current-chopping'
BB_1 current interruption at natural current zero
CC^1 recovery voltage with 'current-chopping'
DD^1 recovery voltage with interruption at natural zero.

switching transformers and motors on no-load. It is a necessary part of the circuit-breakers designer's job to ensure that the degree of 'current-chopping' (as this phenomenon is often called), is not allowed to reach a level capable of generating dangerous overvoltages.

Capacitive currents

Conditions here are again special, because a d.c. charge is trapped in the capacitance once the current flow ceases, and the contact gap is quickly stressed to a value of double the peak phase voltage as the polarity of the a.c. recovery voltage changes. This is illustrated in Fig. 14.3, and again the circuit-breaker or switch

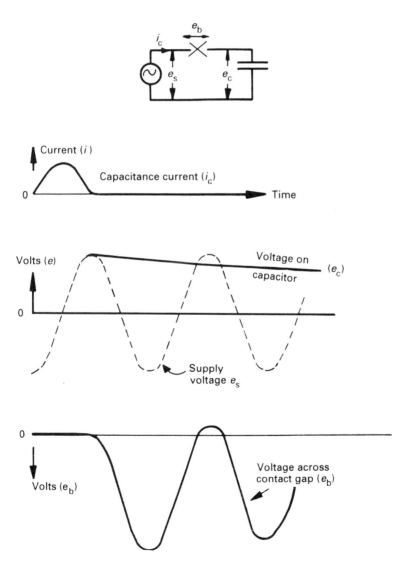

Fig. 14.3 Conditions occurring when interrupting capacitance currents ($\cos \theta \approx \theta$ leading).

designer has to ensure that the contact gap does not break down under these conditions, however early the initial clearance occurs, as this could lead to the build-up of even higher voltage stresses.

TYPES OF SWITCHING DEVICE

Circuit-breakers

An important feature in considering the application of any form of circuit-breaker is the behaviour of its operating mechanism. Because of its protective duty, it is essential that a circuit-breaker is always in a condition to open even if the power supply to the mechanism closing device has been interrupted. This is achieved by biasing the circuit-breaker towards the open position by springs in various ways. The circuit-breaker is then held in the closed position by some form of catch which can either be released manually or, more usually, by an electrical solenoid or trip coil.

Because circuit-breakers could at any time find themselves being closed onto a faulty network, it is forbidden for them to be closed by direct manual means. Therefore all circuit-breaker closing mechanisms require some form of stored energy to operate them. The two most common devices in use have either a solenoid and armature providing the necessary closing force, or springs which may be either hand-charged or charged by an electric motor. The spring close mechanism has the advantage that it can always be operated even if the source of supply has been disconnected. The solenoid requires a heavy duty d.c. source, usually a secondary battery. Where such a source exists for other purposes, such as emergency lighting, then the solenoid operating mechanism has certain advantages.

The vast majority of modern circuit-breakers are supplied with either hand- or motor-charged spring mechanisms. The latter is increasingly demanded for remote operation applications. At the distribution voltages considered here, up to 11 kV, by far the most common arc extinguishing media used until recently were oil, and air at atmospheric pressure. For particular heavy duty there have been a limited number of circuit-breakers made using compressed air as the interrupting medium, but this was more usually confined to designs of circuit-breaker for use at transmission voltages. Today, vacuum and SF_6 are the dominant interrupting media in circuit-breakers.

Oil circuit-breakers

The oil used in oil circuit-breakers complies with BS 148 and is a hydrocarbon oil of fairly low viscosity and good insulating properties. When the contacts part considerable heat is generated and this not only vaporizes the oil but disassociates it into its hydrogen and carbon constituents. The hydrogen is then thermally ionized, which generates the electrons and positive ions that carry the current across the space between the contacts in the form of an electric arc. To control the flow of gases in the arc region the contacts are normally enclosed by an arc control device (Fig. 14.4). The intense heating of the hydrogen gas and the dissociation of the oil generates pressure which is utilized within the arc control device to improve the efficiency of operation.

Fig. 14.4 Contacts and arc control device of an oil circuit-breaker.

During the arcing period the presence of the arc tends to prevent the exhaust of gases from the arc control device through its side vents, but as the current reaches current zero these gases are released and sideways displacement of the ionization products occurs due to the high pressure gas which surrounds the arcing zone. The replacement of the ionization in the contact gap by clean hydrogen ensures a dielectric strength sufficient to withstand the rapidly rising transient recovery voltage. In an oil circuit-breaker of this type the effort required to extinguish the arc increases as the current rises, but so does the energy injected into the electric arc. Consequently, the extinguishing effort rises to match the increasing fault current.

The arc voltage (see Fig. 14.1) is important for two basic reasons. First, the arc voltage controls the amount of energy being generated in the arc and this has an important effect on the mechanical design of the enclosure. Second, the arc voltage plays a part in modifying some of the electrical parameters concerned with the circuit-breaking operation, such as the power factor and the high frequency transient recovery voltage, by adding resistance to the circuit.

Bulk oil circuit-breakers

There are very many bulk oil circuit-breakers in service in the UK, but fewer and fewer are being ordered today as the new technologies of vacuum and SF_6 continually increase their share of the market. Since both the oil and the gases released during arcing are flammable, there is a desire to use safer designs today, even though the safety record of the oil circuit-breaker has been exemplary. The longer life without maintenance of the new technologies is another attractive feature. Much the same arguments apply to the continental minimum oil designs which are also being replaced by oil-less circuit-breakers. New designs of oil circuit-breakers are extremely rare and supply to existing designs relates to special local conditions (e.g. manufacturing and maintenance technology).

Magnetic air circuit-breaker

The basic principle of the magnetic air circuit-breaker consists of so manipulating the arc that a very high arc voltage is generated, eventually exceeding the supply voltage. Under these conditions current flow cannot be maintained and the arc is extinguished.

This is usually achieved either by forcing the arc to extend itself close to solid materials which extract heat from the arc, or by breaking the arc up into a series of arcs in which case the anode and cathode voltage drops are added to the total arc voltage to assist in achieving the objective. Designs exist where both methods are used in combination. A typical air circuit-breaker is shown diagrammatically in Fig. 14.5 and it will be seen that the arc extension means is contained in the arc chute. The arc is encouraged to enter this device by the arc runners to which the arc roots transfer when the contacts open. Magnetic circuits are provided to generate a field within the arc chute which will cause the arc to move into the plates of that chute. A typical arrangement has moulded plates with ribs which are arranged so that the ribs on one of the parallel plates interlock with the ribs on the other plate gradually causing the arc to take a more and more serpentine path as it is driven deeper into the arc chute.

Figure 14.6 illustrates the arcing and extinction process and shows how the arc voltage rises to quite high values during this process. It should be compared with Fig. 14.1 which shows the same condition for the oil circuit-breaker.

At very low currents of the order of 100 A the magnetic field may be inadequate to cause the necessary arc entry into the chute. To assist this operation at such low currents it is usual to fit an air 'puffer' comprising a piston operating within a cylinder connected to a nozzle under the contacts. This gives the necessary impetus to the low current arc to ensure its early extinction.

Magnetic air circuit-breakers were particularly advantageous for applications where frequent switching took place, because of the long life of the interruption unit (i.e. between maintenance). It was typical for this technology to be used for power station auxiliary supply switchgear.

Vacuum circuit-breakers now find favour in this application because of a lower cost interruption element.

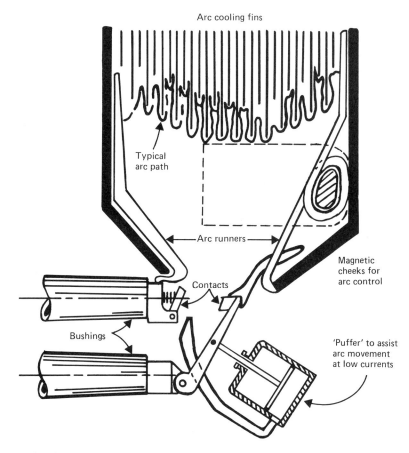

Fig. 14.5 Sketch to illustrate the principle of magnetic air circuit-breaker.

Vacuum circuit-breaker

In the vacuum circuit-breaker the contacts and arc control devices of the traditional circuit-breaker are replaced by a vacuum interrupter. The vacuum interrupter is a sealed vessel with insulating walls which contains two contacts. One of the contacts is fixed to one of the end shields and the other contact is free to travel in an axial direction; the vacuum is maintained by metallic bellows connected between the moving contact and the other end shield. Inside the vacuum interrupter are metallic screens between the contacts and the inside walls of the insulating tube. The performance of the vacuum interrupter is dependent upon three factors: the existence of a sufficiently hard vacuum within the device; the selection of a suitable contact material for the contacts; and the provision of some form of magnetic control for the arc.

The vacuum circuit-breaker has a contact gap of the order of 10 mm for voltage ratings up to 12 kV. Therefore the necessary operating power is much reduced from that needed in the traditional designs considered earlier. When operating within its designed breaking current limit the contact gap in the vacuum interrupter

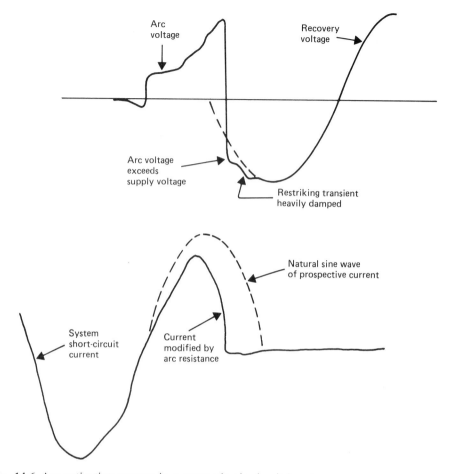

Fig. 14.6 Arc extinction process in a magnetic air circuit-breaker.

recovers its full dielectric strength within one microsecond of current zero. This makes for an extremely efficient interrupting device. With the high contact opening speed and short travel the contacts are usually fully open in 10 ms, leading to maximum three-phase arcing times of the order of 13 ms. Due to the small contact gap and the physical characteristics of the vacuum arc, the arc voltage is very low, leading to much lower energy generation than exists in oil or air circuit-breakers.

A modern 12 kV vacuum circuit-breaker equipment is shown in Fig. 14.7a.

Sulphur hexafluoride circuit-breakers

SF_6 is a heavy, chemically inert, non-toxic and non-flammable gas, which is odourless and colourless. Its outstanding insulating and arc extinguishing properties have made it the automatic choice for transmission applications to the highest voltages (52–800 kV). At atmospheric pressure its dielectric strength is between

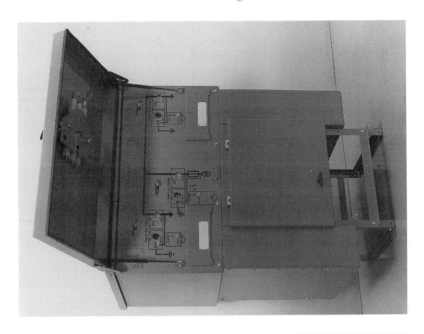

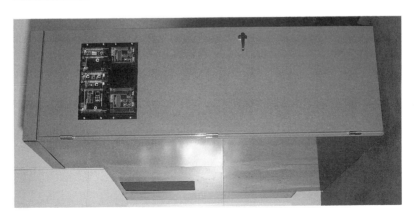

Fig. 14.7 (a) A modern 12 kV vacuum circuit-breaker equipment *(left & centre)*.
(b) A modern 'ring main unit' equipment using SF$_6$ technology *(right)*.

two and three times that of air, while at an absolute pressure of about three bars it equals the dielectric strength of insulating oil. The superior arc quenching ability of this gas is partly attributable to the fact that it is electronegative, which means that its molecules rapidly absorb the free electrons in the arc path between the circuit-breaker contacts to form negatively charged ions, which are relatively ineffective as charge carriers due to their much greater mass compared with that of the free electrons. This electron gathering action results in a rapid build up of dielectric strength at current zero.

In order to interrupt fault currents of the magnitudes found in today's distribution systems it is necessary to cause significant relative movement between the arc and the gas. In typical distribution circuit-breakers of the ratings being considered here, this can be achieved in two or three ways. Until comparatively recently the most used technique was to have a piston connected to the moving contact, which generated a blast of gas through a nozzle surrounding the contact gap.

In this type of gas-blast circuit-breaker, usually referred to as a 'puffer' circuit-breaker, the arc extinguishing energy is supplied from the operating mechanism. At the highest ratings this makes for a powerful mechanism, but this technique still has to be used for the highest ratings. It is in fact the same technique that is used in very high voltage circuit-breakers, which have very high energies indeed. This type of SF_6 circuit-breaker has ratings up to 50 kA at 12 kV and is suitable for use in power stations and large industrial sites, where this sort of fault level can arise.

Two other techniques have now come on the scene for use at more moderate ratings. The first, now well introduced into the market, is the rotating arc design. This is suitable for use up to about 25 kA fault level, and employs a magnetic field to create the rotation of the arc between suitable electrodes and hence produce the relative motion required for arc extinction. The movement of the contact and the subsequent path of the arc is used to bring a coil circuit into play, so that the coil is only in circuit during the arc extinction process. Several versions of this technique exist, due to individual manufacturers needing to avoid the restrictions imposed by the other manufacturers' patents. In all cases, however, the basic principle is followed, of establishing a magnetic field in the arc region which causes the arc to rotate at high speed between two electrodes.

As in all cases where the arc extinguishing effort is derived from the breaking current, the designer has to ensure a correct balance between the current to be interrupted and the effort available. At the low-current end, the field may have little effect on the arc, so the principal arc extinguishing effort comes from the properties of the gas. The efficiency in this region can be adjusted by varying the gas pressure. For this reason, the standing gas pressure in a rotating arc SF_6 circuit-breaker tends to be higher than for the puffer type, where the extinction energy is supplied mechanically.

At the limiting breaking current, the arc rotates so fast that the gas into which it is travelling has not had time to fully de-ionize. This effect can be adjusted by the dimensions of the contact structure and the number of turns in the magnetizing coil. It can be seen that these adjustments at the ends of the current scale can be interactive, which makes for an interesting design exercise.

The other recent development in arc extinction technique uses the thermal energy of the arc to create an increase in gas pressure. Typically, the contacts part in a chamber of which the only opening is blocked by the moving contact at the early part of the opening travel. Thus the pressure inside the chamber rises until the tip of the moving contact passes through the opening and the gas flow is released. A strong deionizing action is created along the arc path, causing interruption at the next current zero. An arcing contact of small diameter is better for this type of interruption, so the normal current carrying is via a set of parallel main contacts.

This type of circuit-breaker is usually limited to voltages up to about 15 kV and breaking currents up to 20 kA.

In addition to the virtual elimination of fire risk in the new technologies of vacuum and SF_6 circuit-breakers, these devices are sealed-for-life equipments, where it is not expected to make repairs or to replace the contact system during a normal service life. This makes the effectiveness of the sealing just about the most important aspect of performance. The vacuum interrupter is usually sealed by a brazing or welding technique, the most difficult part being the seal between the metal parts and the insulating enclosure. Many years of experience show that this is very reliable.

One of the original 'advantages' claimed for the SF_6 circuit-breaker was the facility that a pressurized device gave for the checking of its contents for leakage. Today, the emphasis on a sealed-for-life enclosure leads to efforts to omit pressure measuring devices so as to reduce possible leakage sites.

Switches

In many distribution systems there exist situations where a simpler and cheaper switching device than the circuit-breaker would be adequate for control purposes. The switch does everything that the circuit-breaker can do except interrupt fault currents. The breaking capacity is limiting to rated load current at a power factor of 0.7.

Until recent years switches used oil or air as the insulating medium. In the case of the oil switch, the oil also served as the interrupting medium, but with air insulated switches some help was required for the interruption of the load currents. Various techniques were used from air 'puffers' to gas producing plastics materials.

Now, this field has also been changed through the introduction of SF_6 gas and vacuum. On the whole, the vacuum interrupter is too expensive for use when the objective is a cheap device, and at present there are very few vacuum switches on the market. On the other hand SF_6 lends itself to very simple arc control devices when limited to load currents, and a large number of manufacturers have introduced this type of equipment.

In application, the protection of the circuit controlled by the switch relies on some other fault interrupting device, usually either an upstream circuit-breaker or a high-voltage fuse. It is common to incorporate high-voltage fuses with the load-break switch to give a protective equipment particularly suitable for connecting distribution transformers to the network. Many thousands of such fuse-switches are installed and working very well in this application. In the UK these were all oil switches until the late 1980s when a number of competing designs of SF_6 switches appeared on the scene. The associated fuse switches almost always had the fuses immersed in the oil. This gave two advantages, firstly by making the insulation of the fuses and their supports independent of the ambient conditions, and secondly by giving good cooling to the fuse, enabling high current ratings to be achieved. Close discrimination with the low-voltage fuses on the transformer secondary was also facilitated.

The introduction of gas-filled switches meant that the fuses had to be in external chambers filled with ambient air, to make it possible to change them conveniently after operation. This usually led to them being in insulating housings to achieve the required dielectric integrity. As high voltage fuses have a relatively high resistance, this enclosure by insulation created the need to de-rate the fuse to avoid premature operation and nuisance blowing. Hence the change to gas-filled switches with their advantages has also had an influence on maximum fuse ratings and in some cases consequent discrimination.

These issues have led to the review of the concept of the protection of transformers 'teed-off' a ring main distribution system, and in some cases involving high ratings or switching automation schemes a circuit-breaking element has replaced the fuse switch. In particular, the SF_6 rotating arc circuit-breaking element with its low energy demand mechanism has found favour.

However, the 'ring main unit' concept of two switches and a fuse switch still remains the dominant equipment worldwide.

A modern 'ring main unit' equipment using SF_6 technology is shown in Fig. 14.7b.

COMPARISON OF CIRCUIT-BREAKER TYPES

Table 14.1 compares some of the main characteristics of the five different types of h.v. circuit-breakers discussed. The oil and air circuit-breakers are now in the minority for new business, but there are still many thousands of them in service, so they are kept in the tables for comparison purposes, since they are well known to most switchgear users. Most users accept that the two new technologies are generally equal in performance, but for certain applications there are differences in behaviour that may make the choice of one preferable to the other. These differences are mentioned in the table.

By 1995, the supply of new circuit-breakers became dominated by those using the technologies of vacuum and SF_6.

SPECIFICATION AND TESTING

The general characteristics of all forms of switchgear are guided by the existence of a considerable number of international and national standards. These are under continuous review and changes are always being considered in the light of changes to the technology of switching devices and to the techniques of control and protection.

Table 14.1 Comparison of h.v. switchgear characteristics with various interrupting media.

Characteristic	I Air-break	II Bulk oil	III Vacuum	IV Sulphur hexafluoride
(1) Switching small inductive currents	At low currents the air circuit-breaker has a gentle extinguishing action and arcs for several half-cycles. This results in negligible 'current-chopping' and therefore negligible voltage surges.	Oil is a good insulator and arc extinction is more efficient than in air circuit-breaker giving short arc duration and a greater degree of 'current-chopping'. The current chopped is less than 10 A and gives rise to measurable overvoltages in highly inductive circuits. The value of these overvoltages is not sufficient to cause damage to system insulation.	The vacuum circuit-breaker practically always interrupts at the first current zero, irrespective of current value. As the arc plasma is formed from metallic vapour derived from the contact surfaces, the stability of the arc at low values of current is a function of the contact material. The contact material in the interrupters made in the UK restricts current-chopping to values below 7 A, which is no worse than in oil circuit-breakers.	Tests on the different forms of SF_6 circuit-breaker indicate that the degree of current-chopping depends on the amount of energy available for arc extinction. Thus the puffer types have a performance very similar to the copper–chrome vacuum interrupters, while the rotating arc and self-expansion varieties have very low values of chopping current.
(2) Switching capacitance currents	For the same reason that air circuit-breakers are gentle at extinguishing currents in inductive circuits they are prone to restrike when interrupting capacitance currents and have a very limited capacity for such duty.	Bulk oil circuit-breakers sometimes employing two breaks per phase at 12 kV have a build-up of dielectric strength across each pole sufficient to ensure restrike-free capacitance current interruption.	The dielectric recovery of the vacuum gap is extremely fast, giving restrike-free interruption of capacitive currents up to the full load current rating.	With the electronegative characteristic the contact gap de-ionises quickly and restrike-free capacitance switching is assured.

Table 14.1 (*contd*)

Characteristic	I Air-break	II Bulk oil	III Vacuum	IV Sulphur hexafluoride
(3) Mechanical behaviour	The international standards require 2000 no-load operations to be performed without attention and with negligible wear. Regular lubrication at this sort of operational interval is envisaged in the design.		The short travel and low energy requirements assist the designer to construct a robust mechanism capable of matching the maintenance-free life of the vacuum interrupter. At least 10000 maintenance-free operations are usual.	By their nature, puffer-type SF₆ circuit-breakers need a more powerful mechanism than the other types, particularly when the breaking current rating is high. However, heavy duty circuit-breakers do not usually need extended mechanical endurance since normal service requirements avoid frequent operation. The types that use the arc energy or magnetic arc rotation for arc extinction have mechanism energy requirements similar to vacuum circuit-breakers. In either case the design is intended for long periods of service without maintenance.
(4) Disturbance during fault operation (a) Pressure development	The rapid establishment of a heavy current arc in the arc chute generates high	The dissociation of oil into hydrogen and hydrocarbons by the arc current generates very high pressure	The metallic vapour produced during arcing increases the vapour density in the contact region synchronously with	In all cases, there will be a small local pressure rise in the contact and arcing zone, and the enclosure is designed with an adequate

pressure and shock waves which have to be withstood by the mechanical construction. This leads to large and heavy arc chutes which are accordingly expensive.	within the arc control device, and this contributes to arc extinction. In the bulk oil circuit-breaker a proportion of this pressure is transmitted to the metal tank but the existence of an adequate air cushion in the top plate of the breaker helps to keep the tank pressure to modest proportions.	the current. There is no general pressure rise within the interrupter, the envelope of which only has to withstand external atmospheric pressure.	margin of strength to allow for this, as well as the effect of maximum ambient temperatures and solar radiation, if applicable.
(b) Exhaust gas emission Large quantities of hot ionized air are discharged from the arc chutes, creating a need for insulating cooling and ducting arrangements for safe discharge.	Modest quantities of exhaust gases are discharged from the oil circuit-breaker after passing through a baffle compartment within the top-plate which both cools the gases and separates the gas from the oil.	The vacuum interrupter is permanently sealed and all metal vapour produced during arcing condenses immediately. There is no emission of any sort.	The breaker is totally sealed so there is no gas emission. Some of the gas is dissociated into other SF compound plus free sulphur. These other gaseous products are absorbed by special filters inside the circuit-breaker.
(c) Mechanical reaction on foundations Very heavy	Heavy	Negligible	Slight
(d) Noise generation Heavy	Moderate	Negligible	Slight
(5) Fire risk While no oil is used and no flammable gases are emitted, the hot exhaust gases	The use of oil as the interrupting medium and the emission of flammable gases	Fire risk is negligible as no flammable materials are used and no gases of any sort are emitted.	As for vacuum switchgear.

Table 14.1 *(contd)*

Characteristic	I Air-break	II Bulk oil	III Vacuum	IV Sulphur hexafluoride
	produced during fault interruption constitute a minor fire risk.	(hydrogen, acetylene, methane, etc.) during operation must constitute a fire risk. In practice, reputable designs rarely give rise to any fire unless some serious malfunction occurs. The quantity of oil is not a factor either in the liability to fire or in the amount of damage created in the event of fire. In environments where fire could have serious consequences, enclosed substations with fire precautions may be necessary.		
(6) Maintenance requirements (a) Routine		BS 6626 gives general guidance on this subject and the manufacturer's handbook is required by that document to give detailed guidance on all the recommended maintenance procedures. When the circuit-breaker life exceeds 20 years or so, it makes sound sense to reduce the maintenance intervals, since the probability of maloperation must then be increasing.		Vacuum and SF₆ are sealed circuit-breakers and not intended for site overhaul. BS 6626 recognizes this, and divides the overall task of maintenance into four regions. Inspection and servicing involve no dismantling and combine a visual inspection, reading of any diagnositc aids to contact condition that may be provided and lubrication. The frequency is determined by the user through experience, based on what is found during each

	inspection. The other two branches of maintenance, examination and overhaul, involve dismantling and the comparative rarity of these operations makes it sensible to involve the manufacturer, who has the tools, workshop conditions and know-how to perform this work expeditiously. Neither of these technologies should require this work during a lifetime of service in the usual distribution system.		
(b) Post-fault	It is normally recommended that maintenance after operation on fault is undertaken at the earliest opportunity so that the condition can be restored to the usual standard of security.	No post-fault maintenance is required. The system described above should be followed, unless it is known that very substantial number of fault interruptions have been undertaken. Then the manufacturer's diagnostic procedures should be followed to determine the internal condition of the circuit-breaker.	It is usually in the industrial environment that the infrequent maintenance is of greatest advantage. The mechanical requirements should be studied if frequent operation is required particularly in the case of puffer circuit-breakers of heavy fault rating.
(7) Suitability for industrial systems	Industrial systems often result in onerous environments for the switchgear installation and add frequent operating duty as well. The comments above together with the operating duty require frequent maintenance attention, particularly in respect of insulation surfaces.	Very suitable for industrial applications, the only drawback being the need for oil and contact changes when the operating duty is severe. Oil maintenance is more frequent for small oil volume units.	Industrial situations with frequent operation provide the application where the advantages of vacuum switching are more apparent. The annual costs of vacuum switchgear in this environment are markedly less than the traditional types.
(8) Operational facilities (a) Integral fault-making earthing facilities	Except on heavy and expensive UK equipment designed mainly for power-station auxiliaries, air-	The vertical withdrawal used with bulk ocbs makes it simple to provide safe, integral, fault-making	Some versions are provided with transfer earthing facilities, at least for the circuit side. This may be combined with an earthing switch, either integral or portable, for busbar earthing. Fixed circuit-breaker designs have built-in switches for earthing and testing.

Table 14.1 *(contd)*

Characteristic	I Air-break	II Bulk oil	III Vacuum	IV Sulphur hexafluoride
	break switchgear rarely has this feature. When required separate earthing trucks are used.	earthing by circuit-breaker transfer.		
(b) Injection test facilities	Require the removal of the circuit-breaker and then test sticks can be inserted into the isolating sockets.		On fixed circuit-breaker versions, separate test orifices are provided enabling test sticks to be inserted while the circuit is earthed as above. Other types as for columns I and II.	As for columns I or II.
(9) Factors in substation design	Substation width is dictated by the depth of the switchgear unit augmented by an access passage for cable termination at the rear of the units and a wide aisle in front of the switchboard to provide space for circuit-breaker withdrawal and maintenance. The cost of the roof structure increases more than linearly with the span so this dimension plays a large part in determining costs. Switchgear types imposing dynamic loads on the floor during operation require expensive, rigid foundation arrangements. With switchgear having a fire risk, if there is a serious possibility of extensive consequential damage it may be decided to install fire-fighting equipment, e.g. CO_2 or other inert gas or sprinklers. Even if the risk is not considered great enough to warrant fire-fighting equipment, the larger switchboards are often sub-divided by		With the fixed circuit-breaker designs no maintenance or withdrawal space is needed so the substation depth can be minimized. No fire walls or fire-fighting equipment are required, floor loadings are light and the unit width is small so that substations can be smaller and lighter in construction than for traditional switchgear types, significantly reducing building costs. When withdrawable circuit-breakers are employed a deeper	SF$_6$ switchgear usually has withdrawable circuit-breakers and the substation needs to leave space for this purpose, but as fire risk is negligible, fire walls and fire-fighting measures are unnecessary, leading to simple and compact substations.

fire walls, built across the substation, to reduce the likely damage in the event of fire. The substation length is a function of the width of each switchgear unit plus the space taken up by any fire walls and associated busbar trunking. Because of the use of oil as the main insulation medium in the bulk circuit-breakers this type would normally be narrower than either air-break or small oil volume for comparable ratings [see (2) above].

substation will be needed but savings in fire walls and fire-fighting equipment will still be made, thus contributing to more economical substation design.

(10) Special factors

Air-break switchgear can be made with very heavy fault ratings and normal current ratings. As it is generally considered to have less fire risk than any circuit-breaker it is frequently chosen for installations where these factors outweigh a higher first cost. The frequency of maintenance is likely to be less than that for oil circuit-breakers of comparable rating, but the high first cost and high substation costs together with the high cost of spare parts detract from this advantage.

Bulk oil switchgear has a long history of satisfactory performance under arduous conditions. Maintenance is frequent when service conditions call for a large number of operations per month but this is easy to do and costs of spares are reasonable. It is compact, which makes for smaller, and hence cheaper, substations than air-break or small oil volume.

In many cases the vacuum switchgear substation can be sited where other types of switchgear would be unwelcome and this can usually lead to other economies, e.g. in cable costs. Running costs of vacuum switchgear show substantial savings over other types, so the 'life-cycle' cost of a vacuum switchgear installation is usually significantly less than other types even when the effects of economic inflation are ignored.

SF_6 distribution circuit-breakers are 'sealed-for-life' devices and similar to vacuum equipment in that they lend themselves to use in environments where oil-filled apparatus would be unwelcome. This can make for significant savings in installation costs.

British Standards tend to be closely related to the corresponding IEC documents because of a process of harmonization agreed by the European Union (the body responsible for the electrical standardization process in Europe is CENELEC). In practice there can be differences, often arising because of differences in the dates of issue of documents and continuous development to meet technological needs, but also to reflect different local technical or legal requirements in Britain. When a British Standard departs from its corresponding IEC Standard a listing of differences is included in the Standard. These standards lay down a number of criteria for the rating of switchgear and the conditions of use for which it is intended. Also, a considerable amount of space is given to the subject of testing.

It is obviously impractical to subject every switching device produced to a series of tests simulating all the likely switching conditions that it might meet in service, in order to prove to the purchaser that it meets its specification. Therefore the quality control policy is followed of subjecting a sample piece of equipment (very often a prototype) to a comprehensive series of major type tests, and then to carry out routine tests on each production equipment of that type to ensure that it matches the type tested article in all essential characteristics.

The major type tests are: short-circuit proving up to full-rated breaking current at rated voltage; high-voltage (h.v.) tests; temperature-rise tests at rated normal current; and mechanical endurance tests.

HIGH-VOLTAGE CIRCUIT-BREAKER SWITCHBOARDS

As mentioned earlier, typical British distribution switchgear units are metal-clad, in that they have separate compartments for the circuit-breaker, the busbars and the circuit components (Fig. 14.8).

The busbars are customarily insulated copper bars and mounted on connections which terminate in an insulated housing with plugs onto which the withdrawable circuit-breaker is engaged when in the service location. All these components are in a metal compartment.

The circuit side also has an insulated housing with plugs for the circuit-breaker bushing contacts, and connects these contacts through a bushing, on which the protective current transformers are mounted, and then through a partition into the cable termination compartment.

The two insulated housings described above open into the circuit-breaker compartment, the openings being protected by metal shutters when the circuit-breaker is not plugged in. This circuit-breaker compartment may have a door which encloses it and gives access to the circuit-breaker for operational purposes. In other designs, the front plate of the circuit-breaker carriage serves this purpose.

A further compartment, usually mounted above the circuit-breaker compartment, contains the auxiliary low-voltage components, such as relays, control switches, meters, instruments and fuses. Those components requiring access for reading, resetting or operation are usually mounted on the front panel of this compartment.

Typical metal-enclosed switchgear units incorporating vacuum and SF_6 circuit-breakers are shown in Figs 14.7 and 14.9–14.11.

The cable compartment was traditionally designed for the termination of paper

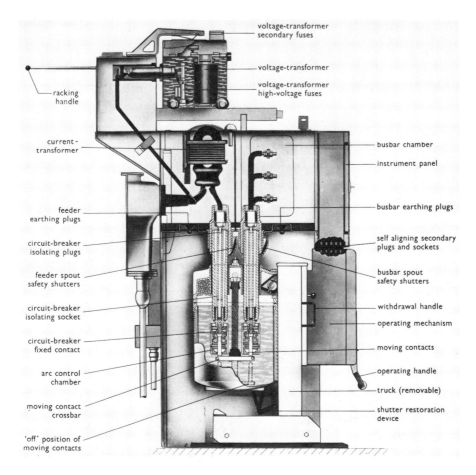

Fig. 14.8 Typical UK switchgear unit with vertically isolated withdrawable bulk oil circuit-breaker.

insulated cables, and made provision for filling with bitumastic compound as insulant for the completed termination. Nowadays there is increasing use of elastomeric cables terminated in air-filled cable compartments, or sometimes even paper insulated cables are terminated in this way. The introduction of heat shrinkable plastics, or elastic 'cold-shrink' sleevings, has facilitated the adoption of this technique. Some of these sleeves can be made semiconducting, which facilitates the grading of the electrical stress along insulation between the cable core and the sheath at the termination.

Another technique for the termination in air of polymeric cables involves the use of proprietary cable connectors. These have two forms, one with an inner and outer semiconducting sheath which allows for the earthing of the outer coating, and which uses termination components that also deal with the problem of stress grading of the exposed cable insulation. The other type is unscreened and just takes care of the right-angle joint which occurs in the majority of switchgear cable

Fig. 14.9 Typical vacuum switchgear unit with withdrawable circuit-breaker.

compartments, and provides the demountable jointing feature. The final step of the cable preparation is the completion with the proprietary sleeves described above.

The cable compartment is often the chosen location for the connections to the voltage transformer, where this is required. In most parts of the world these are required to be fuse-protected, which brings in its train a requirement for the provision of facilities for safely changing these fuses with the circuit energized.

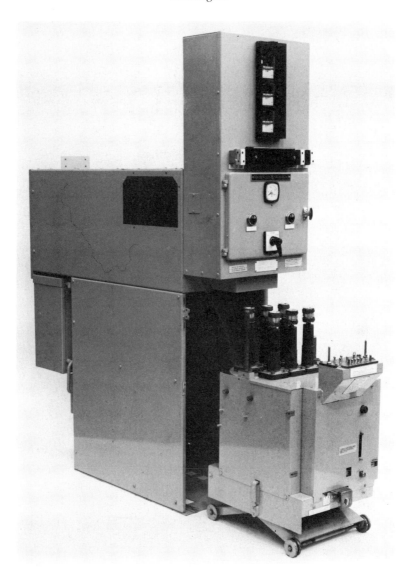

Fig. 14.10 Typical SF$_6$ switchgear unit.

As the switchgear usually forms the most convenient point at which access can be gained to the main conductors of a distribution network, it is usual to provide safe facilities for making connections to those main conductors. When the circuit-breaker is withdrawable, then access to either the busbars or circuit side conductors can be obtained through the orifices into which the circuit-breaker is normally plugged. When the circuit-breaker is of the fixed type then separate orifices for testing and access to circuit conductors are provided.

Fig. 14.11 SF$_6$ switchgear unit with horizontal withdrawal.

Interlocking and padlocking

In order to ensure safe operation of switchgear, particularly when obtaining access to conductors for such operations as cable fault detection, phasing out, testing of cables, etc., it is usual to provide mechanical interlocks or padlocks to control access. The first interlocking requirement on all forms of withdrawal switchgear is to ensure that the circuit-breaker cannot be withdrawn or plugged in while it is closed. Generally, metallic shutters are provided, which automatically cover the plugging orifices when withdrawable circuit-breakers are removed from the switch cubicle. Similarly, facilities are made available so that these shutters can be padlocked in the closed position to maintain the full safety of the metal-enclosed equipment.

Operation and maintenance

The most common position of the circuit-breaker is in the normal service location so that when it is closed it connects the circuit to the switchboard busbars. It is usual for metal-enclosed switchgear to incorporate secondary connections between the cubicle and the circuit-breaker which are automatically made when the circuit-breaker is racked into the service location. To ensure correct operation in service the interlock used to prevent the withdrawal of a closed circuit-breaker must be in the correct position for circuit-breaker operation.

Earthing

It is standard British practice for earths to be applied to distribution systems by using a proven fault-making device such as the circuit-breaker itself. The application of an earth to any system is normally controlled through a permit-to-work system. This ensures that all the necessary switching operations to make the relevant circuit dead have been carried out before the system is earthed.

With vertically isolated switchgear it is convenient to use the facility of moving the circuit-breaker into and out of the unit to provide earthing locations. These are illustrated in the case of a typical bulk oil switchgear unit in Fig. 14.8. This shows typically the disposition of the VTs, CTs, instrument panel, etc.

The selection of the correct location in which to raise the circuit-breaker into either the service, circuit earth or busbar earth position is under padlock control.

Other methods in use for providing earthing facilities with proven fault-making capability are the provision of extension plugs for the circuit-breaker itself so that it can be used in the service location with an earth connection to the other side of the circuit-breaker. A further way is the provision of a special fault-making earthing switch which is interlocked with the circuit-breaker so that the circuit-breaker must be removed from service before the switch is used. Finally, the most expensive way of achieving fault-making earthing is by the provision of an alternative device which replaces the circuit-breaker and its carriage and is so designed as to contain a fault-making earth switch, usually with facilities for carrying out tests as well.

Maintenance

The maintenance of electrical switchgear is covered in detail by two British standards: BS 6423 for switchgear up to 650 V; and BS 6626 for switchgear above 650 V and up to and including 36 kV. There is a third for switchgear above 36 kV, but that is outside the scope of this book.

These standards contain a substantial section on safety precautions and on the use of permit-to-work systems for ensuring safe access to the plant to be maintained.

The maintenance itself, for the high-voltage equipment is divided into four sections: servicing; inspection; examination; and overhaul. BS 6626 differentiates for the first time between the oil and air types of circuit-breaker which require regular examination and overhaul, and vacuum and SF_6 circuit-breakers which usually do not.

This standard also makes it clear that it is the manufacturer's responsibility to inform the user when and how to maintain switchgear, and contains an appendix listing the required content of the manufacturer's handbook for this purpose.

As the names imply, the servicing and inspection procedures involve a visual study of the switchgear and could include an operational check to ensure that all is in working order. The servicing includes such items as the lubrication of operating mechanisms and possibly cleaning of accessible parts. Neither operation is supposed to need the use of any tools, although it will probably be necessary to take the circuit-breaker out of service to obtain the required access.

Examination means the dismantling of certain parts of the equipment to give access to internal components. Usually this would follow an inspection that had given rise to doubts concerning the condition of such components.

Overhaul means the replacement or refurbishment of components which have been shown by inspection or examination to require such attention.

With vacuum and SF_6 equipment, the manufacturer is strongly urged to provide facilities and advice for diagnostic testing techniques which will indicate the need for examination. As mentioned earlier, these types of sealed-for-life switching devices should require no internal maintenance over periods of the order of 20 years.

ERECTION OF SWITCHGEAR

The switchgear cannot always be assembled in the factory and shipped to site as a complete plant, so it is usually necessary to pack the equipment in its individual cubicles, ship to site and eventually erect it. Erection includes the setting in place of these cubicles, their interconnection both mechanically and electrically, and the fitting of any extra relays together with the power and control cables. The chambers may have to be filled with insulating media of different types.

Once the equipment has been erected it needs to be commissioned, which is defined as the work of testing and finally placing in service of the installed apparatus. Before being put into service tests must be carried out on the complete equipment to prove that it meets the required specification.

Storage

It is best to deliver the switchgear cubicles at a time convenient for their immediate erection. However, it sometimes happens that the site works are delayed, in which case the switchgear must be stored carefully until it is needed.

On delivery it is important to check that all items are present and correct in accordance with the delivery note and then to store these components carefully to ensure that no parts go astray. At all times, care must be taken to maintain the condition of the stored items; precautions must be taken to ensure that the equipment is properly stacked, for example, and that it is kept clean and dry. If possible, arrangements should be made to keep the building temperature in excess of the dew point so that condensation on the equipment is prevented and thereby the plant is protected from corrosion.

Erection

Before commencing erection it must be ensured that the workforce is in possession of all the drawings and instructions required to erect the equipment, which are normally provided by the manufacturer. The substation in which the switchgear is to be erected should be as clean and dry as possible and all debris should have been cleared away.

During erection particular attention should be paid to a number of points:

- Dirt and debris should be excluded from partially erected cubicles.
- All openings that are not in immediate use should be blanked off or covered by clean sheets.
- All electrical insulation should be kept clean and dry by being kept covered and, if necessary, heated.
- Materials which have been issued from stores and not yet used on erection should be stored safely and tidily.
- When handling the cubicles and major components care should be taken to observe the correct lifting arrangements and to make certain that slings are attached to the manufacturer's designated lifting points. This ensures that no parts are subjected to undue strains or sudden stresses which could result in disturbed settings or other damage.

Foundations

The successful erection and operation of the switchboard depends very largely on the accuracy of the foundations. The most useful form of fixing medium for the type of switchboard considered here is some form of proprietary channel embedded in the floor, containing captive adjustable nuts. These channels should be assembled truly parallel and level and should project slightly above the surrounding floor. This is so that the switchgear can be clamped to the channels themselves which are then known to be at the correct level. Less commonly nowadays, holes may be cast in the floor in which suitable foundation bolts can be grouted after the switchgear has been erected carefully.

General assembly

It is normal to commence the assembly of a distribution switchboard from the centre unit. This minimizes any effect of a build-up of errors as the equipment is erected. Any inter-unit tie-bolts are loosely positioned as erection proceeds, and all units lined up onto the centre unit. Plumb lines should be used on each unit to ensure good alignment. When all units have been placed in position and checked for alignment, the inter-unit bolts may be tightened and checks may be made that any withdrawable portions can be entered or withdrawn smoothly. Instrument panels and any other component details should then be added to the switchboard.

Busbar and circuit chamber assembly

The actual section of the busbars to be used should be quoted on the manufacturer's drawings and care must be taken to ensure that the appropriate section is used for a particular unit.

Some general guidelines on preparation of contact surfaces are given below, but it cannot be overstressed that at all times the manufacturer's instructions must be followed.

Contact surfaces must be cleaned carefully before assembly, using fine emery cloth where the surface of the contact is of natural finish, and using a proprietary silver polish if the surfaces are plated. Where the connections have threads these should be cleaned with a fine scratch brush. After cleaning, the joint faces must be wiped with a clean lint-free cloth to remove any dust, and the joints assembled as soon as possible.

Normally, the lower busbar is connected up first, with the joints insulated as required, ensuring that a good electrical connection is made between all copper faces and that all securing nuts are tight. The next bar to be assembled will be the centre one and, finally, the top one.

The same treatment of joint faces must be carried out on all other connection joints within the circuit chamber, such as connections for VTs.

The manufacturer's instructions should be checked carefully to see whether any or all of the joints that have been made on site are required to be insulated. If this is the case then the components needed should have been supplied with the equipment and the instructions given by the manufacturer should be followed carefully, particularly in respect of any recommended safety precautions, with special attention to the chemical reagents that may be used.

Cabling

The preparation of the cables for connection to the switchgear is a specialist occupation and an experienced jointer should be employed for this purpose. The cable has to be laid carefully to avoid tight bends and the length to the cable lug measured very carefully to avoid any stress on the cable once the joint has finally been made. Minimum bending radii for cables are specified in BS 6480.

Insulation of the exposed cable and of the joint is frequently provided by means of heat shrinkable plastics sleeves. These should be applied and terminated in accordance with the manufacturer's specific instructions. Where proprietary cable connectors are supplied for the termination of the cables, apply these with careful reference to the termination maker's instructions, or employ a specialist contractor. Where crimping is employed for the attachment of lugs or thimbles to the cable core, it is vitally important only to use the type and size specified by the maker for the core size in use.

Occasionally, paper cables are still terminated in compound-filled boxes and where this is the case care should be taken to ensure that the compound used is continually stirred during the melting period, to avoid excessive heating and possible burning of the compound at the bottom of the boiler. Care must be taken to ensure cleanliness of the cable box and all utensils associated with this operation. The cable box itself, and any buckets or ladles which are used in the operation, must be pre-heated before use and the compound at the appropriate temperature should be poured slowly to avoid splashing and the inclusion of air bubbles. When the cable box is filled to its indicated level, the compound should be allowed to cool slowly, avoiding draughts. If the compound falls below the indicated level on cooling, it should be topped up while still warm to ensure a good bond between the main mass and the topping layer.

Earthing

Each unit forming part of any installation must be earthed by means of a copper strip fastened by a bolt or stud connection. This earth strip must be continuous and connected to all units and fastened to the main earth. All joints must be cleaned and treated as busbar joints. Special attention should be paid to the earthing of cable sheaths at cable glands.

Oil filling

Where chambers have to be filled with insulating oil on site it should first be checked that the oil supplied is in accordance with the manufacturer's recommended grade. All electrical oils must conform with BS 148 and the oils must be tested for electrical breakdown before filling, as described in BS 148 and BS 5730. The equipment must be filled to the indicated level, and it is advisable to fill VTs after they have been positioned on the switchgear. The indicated level is usually based on an ambient temperature of 15°C, and if the actual site temperature differs significantly from this then due allowance should be made.

Gas filling

Gas-filled circuit-breakers or switchgear compartments are usually shipped filled. Where the design pressure is relatively high the manufacturer may avoid shipping problems by filling only to just above atmospheric pressure and topping up to the recommended filling pressure on site. In this connection it is necessary to understand the two defined pressure levels used in the standards for gas-filled switchgear. These are the *filling pressure*, which is the pressure to which the enclosure is filled before being put into service, and the *minimum functional pressure*, which is the lowest pressure at which the switchgear retains all its ratings. It is important to determine the filling pressure recommended by the manufacturer for the particular switchgear being erected, so that it can be checked to ensure that it is filled to this level before commissioning. In some cases it may be recommended to take a sample of the gas for test before energizing the switchgear. If this is the case, then the manufacturer's instructions must be very carefully followed, to avoid loss or contamination of the gas. BS 5209, *Guide for checking SF$_6$ gas*, is also helpful.

Vacuum interrupters

It is often recommended that the integrity of the vacuum in the interrupters be checked before putting into service to ensure that no damage has been done during the shipping process. The manufacturer's handbook should be consulted for guidance on this point, since it usually involves applying a high voltage across the open contacts of the (or each) interrupter. Precautions should be taken to safeguard personnel from the possible generation of X-rays during this procedure. The manufacturer will have given guidance on this point also. Generally it is possible to apply this voltage without exposing the vacuum interrupter, so that the metal enclosure can act as a screen. In such a case there will be no possibility of radiation, external to the enclosure.

Small wiring

It is usually necessary to complete small wiring connections between the instrument panels on adjacent cubicles and to connect external multicore cables to the control cable terminating boxes, which are usually mounted at the rear of the switchgear units. If it is not already contained on the wiring diagram, a useful aid to ensure correct connection for multicore cables is to list the cores in order and record the appropriate connection to be made to each core at each end of the cable.

Final inspection

After the switchboard has been finally erected, all securing bolts tightened, all busbars and other connections completed, all insulation finished, cables connected and small wiring completed, a final inspection should be made. The following is a typical, but by no means exhaustive, check list.

(1) Before fitting covers, all chambers should be checked for complete cleanliness and the absence of all foreign matter.
(2) Once the covers have been fitted, it should be checked that all fixing screws are in place, and secure.
(3) All labels, where required, should be fitted and visible.
(4) Any mechanical interlocks should be checked; safety shutters, where fitted, must be checked to be operative.
(5) All withdrawable items should be proved to be capable of extraction and isolation as required.
(6) All fuses and links of the correct current rating should be inserted into the appropriate holders.
(7) All exposed insulation surfaces must be clean and dry.
(8) A final check should be made for continuity of earthing.
(9) All tools used in the erection of the switchboard should be carefully accounted for.

ELECTRICAL TESTING AND COMMISSIONING

Routine factory tests will have been carried out during manufacture in order to check design criteria and the maintenance of the supplier's standards. These tests provide a valuable reference when any query is made concerning the apparatus during its commissioning tests and operational life. Commissioning tests are conducted on site after installation of the equipment. These are to ensure that the apparatus will perform its duties in service, that interconnection with other equipment is correct and to provide data for future maintenance and service work. The conducting of these tests also provides valuable training for the purchaser's operating staff. Normally, such tests are carried out once only, after installation, prior to putting the apparatus into service. The most common form of commissioning test is to simulate operating conditions using portable voltage and current sources. A test log should be maintained listing all tests carried out, the results and objectives.

Typical commissioning tests include: visual checks, earth impedance measurements, insulation resistance checks, current and voltage transformer checks, portable primary current injection testing, circuit-breaker operation checks, control and scheme tests, secondary current injection testing and load testing.

Visual checks

The object of such a check is firstly to ensure the mechanical integrity of the equipment. Then all electrical connections made on site should be checked to the appropriate diagram to ensure their correctness. Check that all fuses and links are in place and have been fitted with the appropriate fuse wire or cartridge.

Insulation resistance checks

Insulation resistance measurements on all small wiring should be carried out at 500 V d.c. This test is to ensure that such wiring is in good condition. Any apparatus which may be damaged by application of such a voltage, e.g. static protective devices, should have the appropriate terminals short-circuited for the duration of this test. When measuring the insulation resistance to earth of the individual circuit, all other circuits should be normal, i.e. earth links closed and d.c. circuits normal. This ensures that the insulation of the particular circuit is satisfactory both to earth and to all other circuits.

It is impractical to give a definitive lower limit for the resistance measurement from this test. Such readings are dependent on the cable length, the size of cores and the number of parallel paths. As a guide, any reading of less than 1 MΩ at 500 V d.c. should be investigated.

Current and voltage transformer checks

Normally the VT and CT wiring will have been completed at the manufacturer's works and only if this wiring has been disturbed is it necessary to carry out any tests on site. Should this be the case then one essential test to be carried out is to ensure that the CT secondary winding has been connected correctly in relation to the primary winding. Should this have been reversed in any phase then the operation of the protection will be seriously affected.

All CTs have some means of identifying the primary and secondary terminals: common practice is to identify as P1–P2 (primary 1 and 2 S1–S2 (secondary 1 and 2) as recommended by BS 7626, with test windings identified as T1–T2.

The following test is recommended to prove these relative polarities (Fig. 14.12). A low reading d.c. voltmeter is connected across the CT secondary winding terminals and a battery across the primary. If relative polarities are correct, on closing the circuit a positive 'flick' is observed on the voltmeter, and

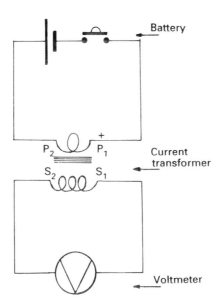

Fig. 14.12 Circuit required to test for CT terminal polarity.

on opening the circuit a negative 'flick' is seen. Where CTs are mounted in transformer bushings it is necessary to short-circuit the main transformer l.v. winding, thus reducing the equivalent impedance of the transformer to give a good deflection of the instrument needle. Voltage transformers may be checked in similar fashion.

Proving of protection

The operation of the various forms of protection which can be fitted to distribution switchgear is covered in Chapter 22. Suffice it to say here that the operation can be simulated either by injecting heavy currents through the CT primaries or by the injection of suitably chosen currents into the CT secondary circuits. In some instances the CTs may be fitted with a special tertiary winding intended for test purposes in this way. Whichever method is adopted, the intent is to inject currents that represent fault conditions of various types, and to check that the relevant relays operate correctly and that any auxiliary relays that are called upon to perform do so also in a correct manner.

To adjust the current settings it is normal to use a variable autotransformer to provide the supply. As the waveshape of the test current can have an influence on the behaviour of the relay it is recommended that as much series resistance as possible should be included in the injection circuit.

High-voltage tests

If called for in the contract documents h.v. tests can now be applied to the insulation. The test values should be those laid down in the appropriate standard for site testing and should be applied by an a.c.h.v. test set or, particularly if the cables have been connected, at equivalent d.c. test set. Where vacuum circuit-breakers are in use it is usually recommended that a commissioning voltage test be applied across the open interrupter as a check that the vacuum is still in perfect condition.

Circuit-breaker operation tests

Following the completion of insulation testing the circuit-breaker should be checked for operation. These checks should cover tripping and closing operations from both local and remote positions, and should preferably be carried out at minimum and maximum supply voltages. It should be checked that correct indication of the closed and open position is given by both the mechanical and any electrical indicators that are fitted. The correct operation of the interlocks should be checked. Not only should one ensure that the permitted operations can be undertaken but that the interlocks prevent the carrying out of prohibited operations as intended.

Before carrying out these tests care must be taken to ensure that the circuit is isolated from any source of power supply.

Load testing

Once all the checks on the equipment have been carried out, the operational tests are satisfactory, the voltage tests passed, and all the functional tests on relays, etc. completed to the customer's satisfaction, the equipment can be energized. Once the circuit is on-load it can be checked that all the instruments and indicators are reading correctly.

If test blocks are available or the relays have test plug connections it may be worthwhile doing a final check to see that the CT polarity is correct and the instrumentation is giving correct readings.

CHAPTER 15

Rotating Machines

D.B. Manning, CEng, MIEE
(Mawdsley's Ltd)

Electricity is a particularly attractive form of energy in that it can easily be produced, transmitted and converted into some other form. The commonest form of energy into which electricity is converted is mechanical driving energy and more than 60% of electrical energy produced is utilized in this way. Conversion of energy from its electrical to mechanical form is achieved using electric motors. The vast bulk of industrial electric motors is used to drive pumps, fans and compressors. All industrial installations, whether manufacturing units or complex process plants, include electric motors often of many types, sizes and voltages to suit a wide variety of applications.

Often much abused, electric motors are expected to perform efficiently and reliably for many years with the minimum of maintenance. In relation to the cost of the energy they convert they are extremely inexpensive. Typically, the electrical energy costs of supplying a motor can equal the prime cost of the motor in only two months of operation. Installation, cabling and control gear costs will perhaps double or even treble this period but this is still a small cost in the normal expected machine life. Many motors supplied over 50 years ago are still performing well in plants around the world.

Therefore, selection of the right type of machine for the application and duty and correct installation and maintenance of the machine is as important as the machine design itself.

This chapter concentrates on motors but does make reference where appropriate to alternators. Similarly, reference is limited to machines up to 11 kV and 10 MW.

MOTOR TYPES

All motors fall into one of two classes, those suitable for use on a.c. supply and those suitable for d.c. systems.

Alternating current motors

There is a large range of a.c. motors available, the most widely used being the cage design.

Cage motors

Three-phase cage rotor induction motors are available in ratings up to 10 MW. Essentially simple and robust in construction, the ubiquitous cage motor has a distributed stator winding which establishes a rotating field and a rotor comprising copper or aluminium bars welded or brazed at each end to a short-circuiting ring. It is the 'cage' appearance of the rotor which gives rise to the name by which the machines are known. In smaller motors the cage may be cast in aluminium alloy such as LM6.

The rate at which the field rotates is determined by the number of poles on the distributed stator winding and on the supply frequency. The more common rates or synchronous speeds are given in Table 15.1. The synchronous speed N_s is given by

$$N_s = \frac{f}{P} \times 60 \, \text{rev/min}$$

where f is the supply frequency (Hz) and p is the number of pole pairs.

Cage induction motors do not run at a synchronous speed because the torque they produce depends upon the speed difference or slip between the speed of the rotating field (synchronous speed) and the lower speed of the rotor. The slip however is quite small, can be predetermined within limits by motor design, and will be about 5% on a small 10 kW motor reducing to less than 1% on motors above the 100 kW.

The speed/torque relationship of Fig. 15.1 is typical in shape for a cage motor started direct-on-line. The motor operates at the point where its speed/torque curve intercepts the load characteristic at speed N. Slip rev/min = $N_s - N$ (where N_s is the synchronous speed and N is the rotor speed). The vertical difference between motor and load speed/torque characteristics is the accelerating torque. Clearly therefore the motor torque available must always exceed the load requirement up to the operating speed if the load is ever to reach full speed.

Table 15.1 Synchronous speeds.

No. of poles	Synchronous speed (rev/min)	
	50 Hz	60 Hz
2	3000	3600
4	1500	1800
6	1000	1200
8	750	900
10	600	720
12	500	600
14	428	514
16	375	450
18	333	400
20	300	360

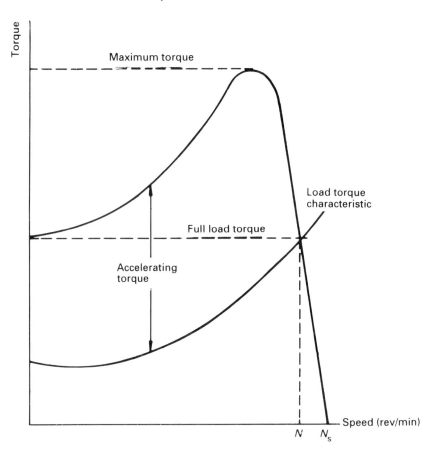

Fig. 15.1 Cage motor speed/torque curve.

The induction motor designer can influence the shape of the speed/torque characteristic to achieve say a higher torque at start. He must however strike a balance between conflicting requirements. With a cage motor it is not possible to insert additional rotor resistance. A compromise then has to be made between high starting torque (and high rotor copper loss and on-load slip) and higher efficiency low slip operation.

Where high starting torque and low on-load slip are necessary a double cage rotor may be used. This has two separate cages – one close to the surface of the rotor producing its maximum torque near standstill and the other 'inner' cage which has a lower resistance (and higher reactance) for maximum torque closer to synchronous speed. The speed/torque curve of a double cage motor is the sum of the speed/torque curves of each of the two cages, see Fig. 15.2.

Induction motors may be designed to operate at more than one discrete speed by pole changing. That is by having tappings in the stator winding to allow

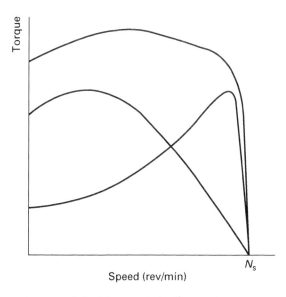

Fig. 15.2 Torque/speed curve of double cage induction motor.

selection of say four- or eight-pole speeds. Alternatively the stator may have two separate windings to permit the motor to run at say four- or six-pole speeds. Pole changing complicates the control gear which sees the motor as two separate differently rated machines.

Three or four different pole numbers can be accommodated in one motor but this is not common bearing in mind the fact that continuously or infinitely variable speed drives are so freely available and are relatively inexpensive.

Starting currents

These are quite high and are up to eight times full load current for machines up to 30 kW and up to six times for larger motors. Except at the extremities of supply lines the system capacity is normally adequate to supply these currents without too much voltage drop. Where the starting currents are likely to be an embarrassment special high resistance or deep slot rotor bars can be used which reduce the starting current to below five times full load current. These are by definition special machines and more costly so that reduced voltage starting is a more usual method of restricting the starting current (Chapter 17).

Whether star-delta or auto-transformer reduced voltage starting is employed it is important to remember that while the starting current will be proportional to the applied voltage, the starting torque will theoretically be proportional to the square of the applied voltage, see Table 15.2. In practice, the effect of stray losses reduces further the available starting torque at reduced voltage.

If reduced voltage starting is used then a careful check must be made to ensure that accelerating torque is available.

Table 15.2 Nominal starting currents and torques.

Starter type	% applied voltage at start	% full load current at start	% full load torque at start
Direct-on-line	100	700	120
Star-delta	58	400	40
Auto-transformer	80	560	77
Auto-transformer	70	490	59
Auto-transformer	50	350	30

Soft starters are becoming a common method of starting induction motors. They are thyristor-controlled a.c. voltage regulators which ramp up the voltage applied to the motor during starting. The ramp rate or starting time can be adjusted whilst at the same time a current-limiting circuit limits the maximum current drawn by the motor during starting.

As with other forms of reduced-voltage starting the starting current is restricted at the expense of starting torque. Soft starters therefore are particularly useful for motors driving centrifugal loads and, because of their infinitely variable voltage range between zero and system voltage, for minimizing the starting current depending upon the motor speed torque characteristic and load requirements.

Slip-ring induction motors
Much less widely used are slip-ring induction motors where the cage rotor is replaced by a distributed wound rotor, the three-phase rotor winding being connected to slip rings mounted on the shaft. Except during starting the slip rings are short-circuited and the rotor behaves exactly as a cage rotor. The advantage of the wound rotor slip-ring machine lies in the possibility of inserting external rotor circuit resistances during the starting period.

The rotor control gear can be arranged to be switched sequentially, automatically reducing the values of resistance in the rotor circuit as the motor runs up to speed. A family of speed/torque characteristics, one for each resistance value, is generated as shown in Fig. 15.3. Note that changing the value of externally connected rotor resistance does not change the value of maximum torque – only the speed or slip at which it occurs.

The number of resistance steps can be chosen appropriate to the starting torque requirements and starting current limitations. The external resistors usually take the form of metallic grids and if appropriately rated, allow the motor to run for long periods at reduced speeds. Very large slip-ring motors may still from time to time employ liquid rotor resistors for starting. These are better able to dissipate the heat generated and the speed variation during start up is stepless. The high losses appearing as heat in the rotor circuit resistances prevent this system being more widely used for economic continuous reduced or variable speed running.

The principal disadvantages of the wound rotor motor compared with the cage

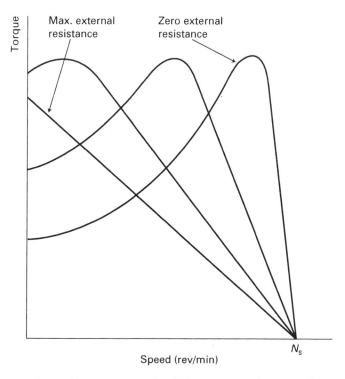

Fig. 15.3 Family of speed/torque curves for different external rotor resistances on a slip-ring induction motor.

machine are higher prime cost of motor and starting equipment and increased maintenance due to the slip rings and brushes.

Synchronous machines

Synchronous machines are most commonly encountered as alternators in the range up to 5 MVA as standby power generators or even as main base load generators in remote areas. As motors, synchronous machines above 5 MVA are used mostly where their ability to generate reactive power and so improve the power factor of the installation, coupled with their slightly higher efficiency, offset their higher cost when compared with the cage motor.

Synchronous machines are almost always brushless. The increased availability in recent years of thyristors and diodes has enabled design engineers to dispense with slip rings and to opt for static excitation and brushless systems.

The construction of the majority of brushless synchronous machines comprises a stator with a distributed winding and a rotor with salient pole field windings. These are supplied through a rotating diode rectifier assembly fed by a shaft mounted exciter as shown in Fig. 15.4. For economic reasons, the exciter for larger power ratings is itself controlled using a pilot exciter or static excitation

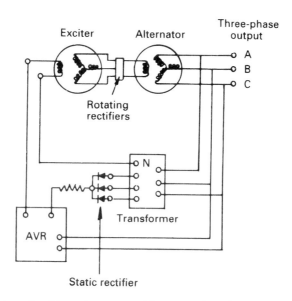

Fig. 15.4 Three-phase brushless alternator with automatic voltage regulator.

regulator. Although the machine runs at synchronous speed without any slip the rotor lags behind the rotating field by a variable angle, known as the load angle. This is approximately proportional to the load imposed upon the motor. The starting of synchronous motors depends upon the size and rating. Smaller machines, say up to 10 MW, are usually started as cage motors utilizing pole face damping windings or linked solid pole faces as the cage. Above that rating they may employ pony motors or even variable frequency power sources to accelerate from rest to almost synchronous speed.

High-speed (two-pole) synchronous machines almost certainly have cylindrical rotors due to the problems of centrifugal forces on salient poles although, in general, turbo-type machines are available only at powers in excess of 10 MW.

Synchronous reluctance motors
The majority of fixed-speed drive applications tolerate the slip rev/min associated with induction motors. Nevertheless some applications do require a precise known drive speed. Synchronous reluctance motors, available up to about 250 kW, fulfil this need. An example of their application is as a drive machine on a buffer or uninterruptible power supply where the output frequency of the alternator must remain the same as the input frequency to the motor alternator set.

Essentially the synchronous reluctance motor has a stator with a distributed winding for the appropriate number of poles or speed. A shaped laminated rotor with a cage winding follows the rotating field at exactly the same speed but with a variable load angle. A typical machine is shown in Fig. 15.5.

Started as cage induction motors, they draw large starting currents and run at relatively low power factor. On the other hand, they are extremely robust and inexpensive.

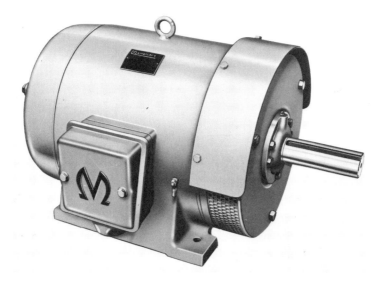

Fig. 15.5 A typical Mawdsley synchronous reluctance motor rated at 15 kW and 1500 rev/min.

Direct current motors

Far from being old fashioned, d.c. motors continue to be widely used even though few factories have a d.c. power distribution system. Once again it is the thyristor regulator which, in conjunction with one or more d.c. motors, provides a very formidable and flexible variable speed drive system. Direct current motors have apparently changed little over the years and still retain a salient pole stator to accommodate the field or excitation windings and interpoles. The rotor is a distributed winding in slots soldered, brazed or welded to the segments of a commutator.

In fact the construction has changed to recognize the particular characteristics of thyristor regulators. Modern d.c. motors have laminated poles and even laminated yokes. The laminated construction helps to improve commutation by allowing the magnetic circuit to respond more quickly to flux changes occasioned by thyristor regulators. Square frame designs on the market have much improved power/weight ratios together with other advantages, Fig. 15.6.

One outstanding advantage of d.c. over a.c. motors is the electrical accessibility of the field winding and the possibility therefore to influence the excitation by field forcing to achieve very fast response drives.

The speed and torque characteristics of a d.c. motor are very easily controlled. In general, for constant flux in a shunt-wound separately-excited machine its speed and output power is approximately proportional to the applied armature voltage. (The speed is actually proportional to the back e.m.f. which is less than the applied voltage by the amount of the armature voltage drop due to its resistance.)

Fig. 15.6 A Mawdsley's square frame industrial d.c. motor rated at 250 kW with forced ventilation.

For constant armature voltage its speed and output power are inversely proportional to field voltage (flux). The speed range over which the motor operates by field voltage control is known as the shunt range, see Fig. 15.7.

Most d.c. motors used are separately-excited shunt-wound machines with interpoles. Figure 15.8 shows the connection of such a motor and includes a series compounding winding which may sometimes be fitted to produce a drooping speed torque characteristic.

To avoid excessive flux distortion and maximize output the motors have interpoles as well as main poles. The interpoles carry the armature current and influence the flux as the load and hence armature current changes to minimize the effect of armature reaction. They significantly reduce the flux appearing in the regions between the main poles where the brushes are located to reduce sparking between brushes and commutator.

Figure 15.9 illustrates the section of a four-pole square frame d.c. motor. Tucking the interpoles up into the corners gives maximum utilization of field space and allows the use of a larger diameter armature than would be the case with a conventional round frame four-pole motor.

On larger d.c. motors, say above 500 kW, it is common to achieve the flux necessary with minimum distortion by employing a six-pole construction when the square frame gives way to a conventional round shape.

Certainly on these larger motors, and often at much smaller ratings, pole face compensating windings will be incorporated. As their name suggests, they are accommodated in slots in the main pole face and are closed coils. They produce a flux under the main poles to cancel out the effect of the flux produced by the

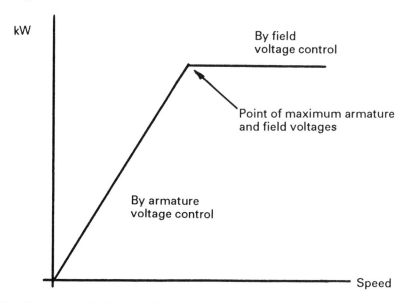

Fig. 15.7 Speed control of separately excited d.c. motor.

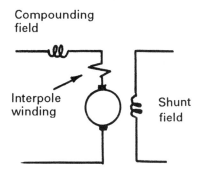

Fig. 15.8 Separately excited shunt-wound d.c. motor.

current in the armature conductors. Sudden changes in load cause corresponding flux distortions due to armature reaction. By cancelling these out the compensating windings prevent large voltage surges in the armature conductors which might otherwise result in flashover.

Separately-excited shunt motors have a fairly flat speed/torque curve for constant excitation (Fig. 15.10) although some highly stressed motors will have a characteristic which rises with load. This is not necessarily disadvantageous since almost all d.c. motors form part of a closed loop speed-control system with tachogenerator feedback. Providing that the maximum speed can be achieved off-load with maximum applied armature voltage and with a cold shunt field the speed-control system is capable of regulating the motor speed over the load range by armature voltage and in some cases by additional field voltage control.

Compounding, that is the provision of an armature current-carrying winding on

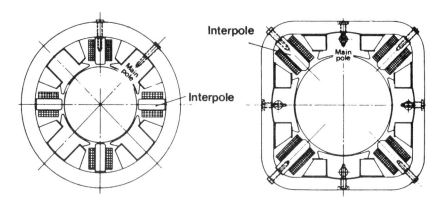

Fig. 15.9 Round and square frame d.c. motors of the same height.

and assisting the main pole winding, has the effect of increasing the flux and reducing the speed with increasing load, resulting in a more stable drooping speed/torque curve (Fig. 15.10). This can be useful and assists load sharing if several motors are to operate in parallel from a single thyristor regulator. If, however, a compound motor is to be reversible the compound winding connections must also be reversible by external control gear, so that its flux is always assisting that of the main pole.

Series wound d.c. motors derive their excitation only from a field system connected in series with the armature (see Fig. 15.11). This means that at no-load

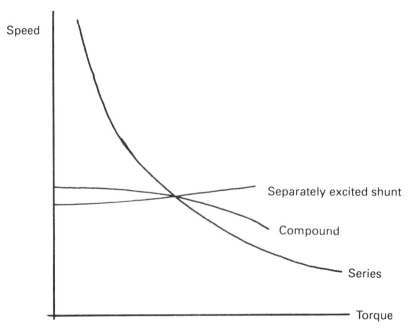

Fig. 15.10 Comparison of d.c. motor speed/torque characteristics.

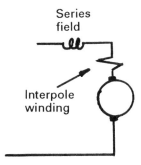

Fig. 15.11 Series-wound d.c. motor.

the flux is weak and the motor tends to run fast. As the load increases so does the flux with large speed reduction and high torque capability. For this reason series-wound d.c. motors are used only where they cannot run unloaded and are found particularly in applications such as traction.

Motor generator sets

Motors are invariably used to convert electrical to mechanical energy and vice versa in the case of generators. Motor generator sets, on the other hand, are used to convert one form of electrical energy to another.

This may be d.c. to a.c. such as in traction auxiliaries applications (Fig. 15.12) or when battery power is required in an a.c. system (e.g. uninterruptible power supplies). Battery charging, welding and Ward Leonard drive systems are typical applications for a.c. to d.c. converters.

Both d.c. to d.c. and a.c. to a.c. sets may be required, the former for voltage raising, lowering or regulation of a d.c. power source, and the latter for frequency changers. Machines changing 50 to 60 Hz and 60 to 50 Hz allow equipment from one part of the world to be used in another and for shore-to-ship power supplies.

Fig. 15.12 Unit construction motor alternator set. Input 600 V d.c., output 6 kVA, 240 V, single-phase 50 Hz.

A wide range of computer, telecommunications, radar and military equipment operates at 400 Hz and therefore demands motor generator sets for frequency changing.

Machines operating at 50/50 Hz and 60/60 Hz are commonly used as buffer sets or load isolation sets in order to protect sensitive loads from mains-borne spikes, noise and brown-outs. They will also prevent confidential computer data entering the supply system in the form of noise where it could be received and decoded by unauthorized personnel. Owing to their rotational inertia, the sets will provide ride-through for several cycles.

Two individual appropriate machines can be coupled together and mounted on a bedplate to form a motor generator set but often weight and size restrictions encourage the dedicated design of two machines in a single frame supported between two bearings. Figures 15.12 and 15.13 illustrate two examples of unit construction sets of this nature.

Conversion of one form of electrical energy to another can often be achieved using static equipment but the rotating motor generator set is still generally much less expensive without being larger or heavier and offers three major advantages:

(1) *Ride-through* The inherent inertia of a rotating system gives it stored energy and the ability to continue delivering power, albeit with some loss of speed and output frequency, during short duration losses of input supply.
(2) *Load isolation* No electrical coupling can occur between the load and the supply. The only connection between the two is a mechanical one in the shaft between the two machines of the set.
(3) *Short-circuit current maintenance* Rotating machines have the ability to deliver up to five times full load current. Thus if a short-circuit fault develops in the load the set will continue to feed the fault with sufficient capacity to cause the short-circuit protection to operate. In the case of uninterruptible power

Fig. 15.13 250 kVA 50/400 Hz frequency converter.

supplies static equipment would transfer to bypass, if it is available, to provide the energy to clear the fault. When the load requires a frequency different from the supply (or bypass) this transfer option is not available to static equipment.

VARIABLE-SPEED DRIVES

A variable-speed drive comprises a motor and some form of speed regulator. Increasing energy costs are resulting in a widening acceptance of variable speed drives. Particularly on centrifugal pump and fan drive applications, large energy cost savings are possible by adopting variable-speed drives in preference to throttling by valves or dampers, Fig. 15.14. The energy cost savings can quickly repay the capital cost of the speed control equipment.

Most pumps and compressors have a centrifugal characteristic. In other words, whilst volume is proportional to speed and pressure to the square of the speed, the power requirements are proportional to the cube of the speed. Centrifugal loads are thus normally associated with a short speed range and derive large benefits from a variable speed drive.

In the example given in Fig. 15.14 a pump running at 100% speed is delivering 40 m³/min and operates at point A where its head/volume characteristic, 1,

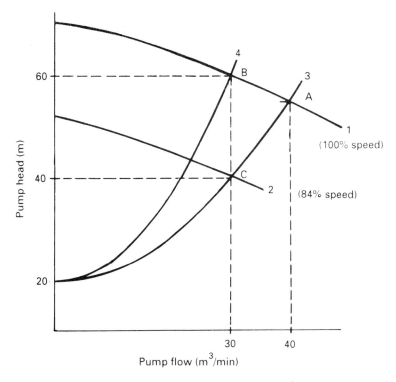

Fig. 15.14 Variable-speed drives provide significant energy savings.

intercepts the system characteristic, 3, and is driven by a 400 kW motor. The pump flow is to be reduced to 30 m³/min. Two possibilities exist:

(1) The pump flow may be restricted or throttled which produces a new system characteristic, 4, and the pump then runs at intercept B.
(2) The pump speed may be reduced by 16% which gives a new pump head/volume characteristic, 2, and the pump then operates at intercept C.

The power absorbed by the pump at B is approximately 330 kW and at point C, 220 kW. Thus in this example the power saving possible by pump speed adjustment rather than by throttling is 110 kW.

For process and process line control, for example in continuous production processes such as rubber, plastics, paper and metals, variable speed drives are essential. The ability to control motor speeds means that other process variables can be accommodated and that product quality and production rates can be optimized.

Alternating current variable speed drives

There are a number of a.c. variable speed drives available, the choice often being dictated by the requirements of the application. Gaining in popularity are variable speed inverter-controlled cage motors, especially for single motor pumps and fan drives having centrifugal load characteristics.

Variable frequency inverters

Apart from single motor drives, variable frequency inverters have for some time been employed on multiple motor drives such as steel mill run-out table applications. The advantage here is that several usually identical motors can be supplied in parallel from a single inverter in order that they all run at the same speed at any time and the speed of the group can be controlled as a whole.

Thyristor regulators are presently in use which provide controlled adjustable frequency outputs up to 15 MW. Depending on rating, the regulators may comprise one of two predominant systems:

(1) *Fixed voltage d.c. link*   The link is supplied from a free running diode bridge which in turn feeds a thyristor or transistor chopper unit. The chopper output is a series of pulses which build up an alternating current, Fig. 15.15. This is known as a pulse width modulated (pwm) system in which the voltage output is controlled normally in proportion to frequency output so that the motors they supply receive constant flux conditions.
(2) *Variable voltage d.c. link*   These are current or voltage fed inverters which have a conventional thyristor regulator as the input. The chopper unit is more simple, its output frequency being determined by the d.c. link voltage. Known as quasi square-wave inverters they also maintain constant output voltage to frequency ratio and commutate or switch only at output frequency rate, see Fig. 15.16.

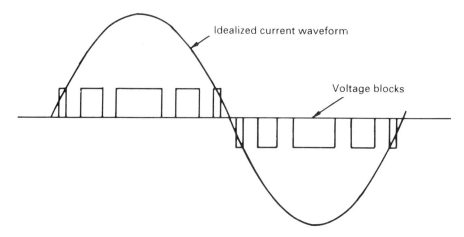

Fig. 15.15 Output waveforms of pulse width modulated (pwm) inverter.

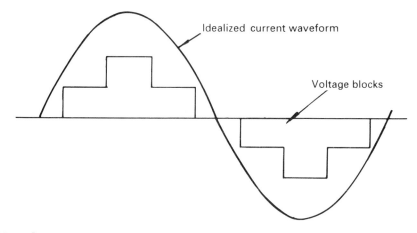

Fig. 15.16 Output waveforms of quasi square-wave inverter.

Force commutated variable frequency inverters are used to control the speed of cage motors up to 800 kW. Above 800 kW the machines used are synchronous motors because they are able to provide the reactive kVA necessary for commutation and so avoid the components and complexity associated with forced commutation. The inductance of the drive motor distributed stator windings has a profound smoothing effect upon the current in the stator. The current thus approximates closely to a sine wave. Obviously any harmonic currents present produce heat without contributing to the output torque of the motor. In general, the harmonic currents present require derating of both cage and synchronous motors by between 10 and 15%.

Additional motor derating is probably necessary if they are to run below normal full speed for extended periods unless the cooling system is independent of speed.

Static Kramer drives

Based upon the slip-ring induction motor, the static Kramer drive relies on extracting energy from the rotor and recovering it either electrically or mechanically. The rotor energy at slip frequency is rectified and fed via a line commutated inverter back into the a.c. supply, Fig. 15.17. A transformer is usually necessary to match the recovered and supply voltages. Alternatively, the rectified rotor energy and supply a d.c. motor mounted on the a.c. motor shaft, Fig. 15.18.

The speed control is stepless and is from sub-synchronous speeds down to zero. However, the rating of the energy recovery equipment is proportional to the speed range and the system is therefore economical only for short speed range applications such as pump drives. The ratings of static Kramer drives can be anything between 50 and 5000 kW.

Static Scherbuis drives

When the energy recovery equipment associated with static Kramer drives is arranged to inject power into the slip-ring induction motor rotor, as well as to extract it, then the drive is known as a static Scherbuis system and although more complex than the Kramer, it is capable of speeds above synchronous.

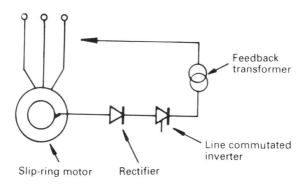

Fig. 15.17 Electrical slip energy recovery of static Kramer drive.

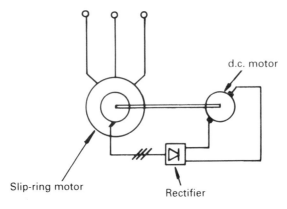

Fig. 15.18 Mechanical slip energy recovery.

Alternating current commutator motors

The rotating brush arm Schrage and regulated NS motors are Scherbuis drives from which the static Scherbuis has been developed.

All the a.c. drives described except the inverter-fed cage motors are not now so common owing largely to their costs and maintenance requirements.

Eddy current couplings

The attraction of eddy current couplings is their ability, in conjunction with conventional cage motors, to provide a variable speed output by control of the coupling excitation. Subsynchronous speeds only are possible since they operate effectively as slipping clutches. They also at reduced speeds dissipate energy in the form of heat and are usually limited to ratings below 15 kW. Much larger couplings are available up to several hundred kW but these require water cooling to dissipate the heat generated and are therefore not popular or economical.

Direct current variable speed drives

Variable speed d.c. drives have been widely used for many years, in particular for process line speed control. They provide excellent dynamic performance, wide speed range and have always proved to be very reliable in service. Planned maintenance techniques have successfully increased mean times between failure due to the fundamentally simple commutator and brush assembly, to periods much longer than associated control equipment.

The original Ward Leonard systems were superseded by magnetic amplifiers and mercury arc rectifiers and then, in the early 1960s, by thyristor regulators for the speed control of d.c. motors.

Switched reluctance drives

This type of drive uses a reluctance motor which has salient poles on both rotor and stator. The distributed three-phase winding on the stator is replaced by a number of salient pole d.c. field windings. The number of poles, as in the case of d.c. motors, does not determine motor speed. The rotor is built up of castellated laminations and carries no winding at all.

Unidirectional current pulses are applied sequentially to the field poles at a rate determined by the required speed and rotor position. Control of the pulse timing is derived from a position transducer mounted on the motor shaft. By appropriate pulse timing either positive or negative torques can be achieved and full four quandrant control is available over a wide speed range and both constant torque or constant power characteristics can be provided. The motor and its regulator are each dedicated and can only be used together as a drive system.

Thyristor controlled d.c. drives

The development of thyristors having increasing current carrying capacity and reliability permits d.c. motors to retain predominance as the most popular form of variable speed drive. Almost all present designs are for fully controlled thyristor regulators.

Anti-parallel connected suppressed half or circulating current regulators in conjunction with d.c. motors allow full four quadrant control. Armature control provides a d.c. motor with constant torque characteristics and field control provides constant power characteristics. Combined armature and field control is common on winder and coiler drives.

The thyristor controlled d.c. motor is usually fitted with a tachogenerator which provides a speed related signal for comparison with the reference signal. It is the error or difference between these signals which advances or retards the firing angle of the thyristors in the regulator to correct the d.c. voltage and hence speed of the motor, Figs 15.19 and 15.20. In addition to, or instead of using the motor speed, other parameters such as current, load sharing, web or strip tension and level may be employed.

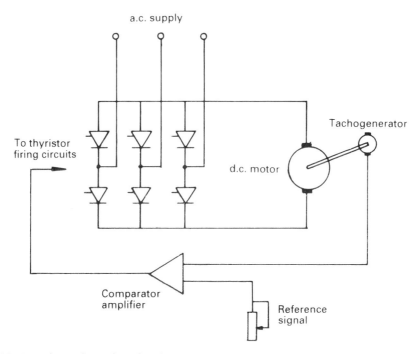

Fig. 15.19 A three-phase six-pulse thyristor bridge and d.c. motor.

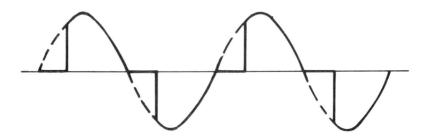

Fig. 15.20 Delayed firing of thyristor reduces mean d.c. voltage.

MOTOR APPLICATION

For reliable performance, the correct motor must be selected for the application. Torque and starting requirements can easily be identified. The choice of motor type is largely determined by rating and whether fixed or variable speed is necessary.

Voltages

On a.c. motors, the supply voltage is determined by motor rating. It is difficult to lay down hard and fast rules when relating voltages to motor outputs. Up to 150 kW it is usually l.v. There is a grey area between 150 and 750 kW when the supply voltage can be from 380 V to 6.6 kV. Higher voltages of around 11 kV would be expected on ratings above 750 kW.

Mounting

Mechanically, the type of load determines whether the motor shaft is to be horizontal or vertical and the particular mounting necessary. The mounting designations are covered in BS 4999 and IEC 34−7; some of the more common ones are shown in Fig. 15.21.

Bearings

The types of bearing incorporated depend on rotational speeds and radial and axial shaft loadings. Designers pay particular attention to bearings to maintain the highest level of operational reliability. Modern anti-friction bearings, of the ball

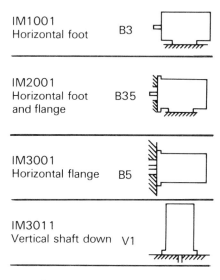

Fig. 15.21 Four common mounting arrangements.

and roller type, are selected to give an L10 life of 100 000 hours. This means that, given lubrication in accordance with manufacturer's recommendations, fewer than 10% of the bearings will fail within the 100 000 hour life. Re-greasing of bearings should be at intervals of about 4000 running hours, unless they are of the 'sealed-for-life' type commonly used on machines rated below 40 kW.

The larger and high speed motors incorporate oil lubricated sleeve bearings. Typical threshold ratings for sleeve bearings are for two-pole machines, 2000 kW and for lower speed machines, 5000 kW. Oil lubrication may be of the oil bath type or, for higher bearing loadings, forced oil circulation.

Enclosure

A motor installed outside in a coastal situation must have a different form of enclosure from that of a motor installed down a coal mine or in a computer room.

The degrees of protection to which motors are constructed are specified in National and European standards such as BS 4999: Part 105 and IEC 34–5. For a clean environment an 'open' machine may be perfectly acceptable. Screen protected, drip proof, splashroof and totally enclosed machines are commonplace.

The degrees of protection, or IP numbers most commonly met are:

IP 21   Screen protected against solids greater than 15 mm and against vertically falling dripping water.
IP 22   Protected against solids greater than 15 mm and against dripping water falling at any angle up to 15 degrees from the vertical.
IP 23   Protected against solids greater than 15 mm and against water falling as a spray at any angle up to 60 degrees from the vertical.
IP 44   Protected against solids greater than 1 mm and against water splashing from any direction.
IP 54   Protected against dust and against water splashing from any direction.
IP 55   Protected against dust and against water jets from any direction.

The letter 'W' inserted between IP and the designation numbers indicates that the machine is also weather protected to limit the ingress of rain, snow and airborne particles.

Ventilation and cooling

Some heat is produced in all electrical machines as a result of copper, iron, friction and windage losses. The main loss and therefore greatest heat produced is in the windings as a consequence of copper losses. These equate to the I^2R in the windings. Improved winding and slot insulation allow motors to run hotter without necessarily shortening the machine life. A hotter machine provides greater kW/unit volume, i.e. smaller machine for a given output. Smaller motors are less able in general to dissipate the heat produced and often incorporate some external method of heat removal. The accepted forms of cooling are specified in BS 4999: Part 106 and, IEC 34–6. Some of the methods most often encountered are indicated below.

Self-ventilating (IC 01)

A fan is mounted on the rotor shaft which draws surrounding air through the motor from one end to the other and discharges it to the atmosphere. This method cannot be used in a dirty or moist environment without adversely affecting machine life.

Force ventilated (IC 06)

An independent separately driven fan is mounted on the machine which forces air through the machine. The separately driven fan can be fitted with filters to remove dirt in the air, although if the environment is dirty the filters would quickly become choked and the motor overheat. Where filters are fitted it is usual to provide protection in the form of a differential pressure switch to measure the pressure drop across the filters and to shut down the plant, or at least raise an alarm when the pressure drop becomes excessive.

Force ventilation is used in particular for variable speed drives where a shaft driven fan becomes ineffective at reduced speeds.

Totally enclosed fan cooled (IC 0141)

Here the motor is totally enclosed and the fan mounted on a shaft extension external to the machine. It draws air from the surrounding atmosphere and a simple cowl directs the air over the surface of the motor along its length. The body of the motor normally incorporates cooling fins to increase the surface area exposed to the cooling air. An internal shaft driven fan is fitted to move the air inside the motor to minimize hot spots in the windings.

This method of cooling is particularly common on cage induction motors up to about 500 kW which are produced in vast quantities.

Closed air circuit, air cooled (IC 0161)

Usually encountered on machines above 500 kW which are required to operate in dirty or difficult environments, the motor has a shaft driven fan which circulates cooling air through the machine and around a closed circuit. An air-to-air heat exchanger, itself cooled by a separate fan, shaft or more usually separately driven, extracts the heat from the closed air circuit. This cooling system is usually referred to as CACA.

Closed air circuit, water cooled (ICW 37A81)

In this case the heat exchanger of the CACA machine is replaced by an air-to-water heat exchanger. The material from which the water tubes are made depends upon water quality. This method of cooling is more efficient than the air-to-air exchanger so that, for the same output, a closed air circuit water cooled (CACW) motor tends to be smaller than one which is CACA. However, water is not always available and when it is the attendant pipework, valves, pumps and installation costs do not always favour CACW.

In order to extract the heat within the motor, air is forced through axial holes in the rotor core and often also through radial slots in the rotor. Design of the machine cooling is exacting, requiring that temperatures within the machine are kept as even as possible from one part to another so avoiding hot spots. Keeping the temperatures even increases the utilization of material and minimizes size and cost. Typically the temperature variation from one part of the winding to another is about 15°C. It is of course the highest temperature which determines the motor rating.

Noise

The ventilation system is a major contributor to the noise generated by a motor. For effective cooling the air has to be turbulent within the motor and heat exchanger with a good scrubbing action. This applies also to totally enclosed fan-cooled machines which generate additional noise due to the air being directed by a cowl and rudimentary fan.

Larger and more modern motors often incorporate sound-absorbing material into their construction.

A two-pole cage motor with CACA ventilation may produce a noise level of 95 dBA if not acoustically treated while a lower speed CACW motor with acoustic treatment may produce only 75 dBA. Because the dB scale is logarithmic a noise level of 95 dBA is ten times greater than one of 75 dBA.

Noise levels in industry are becoming important considerations and the subject of various codes and standards such as BS 4999: Part 109. This is due to the permanent damage to hearing resulting from excessive exposure to high noise levels.

Insulation

Improvements in insulation materials and techniques permit higher temperatures. Insulation materials do not suddenly break down at some threshold temperature, they gradually age and deteriorate. The temperatures permitted by British (BS 2757: 1986) and European standards for each class of insulation are such as to give an insulation system life of 20 000 running hours. This equates to, say, 15 years at 6 hours per day, 5 days per week, 45 weeks per year. In practice the life is much longer, because a machine is unlikely to operate always at its maximum temperature rise in maximum ambient temperature. Exceeding the allowed temperature reduces the insulation life by about half for each additional 10°C. Conversely, the insulation life is doubled for each 10°C working temperature below the stated limit. The absolute temperature of the windings and insulation is the criterion but these are always expressed in terms of a temperature rise above an ambient temperature, normally of 40°C. It is rare in the UK and many other countries to find an ambient temperature of 40°C, even though the ambient is the air immediately surrounding the motor rather than the outside temperature. For water-cooled machines the accepted datum for water temperature is 30°C.

Insulation materials and in particular, insulation systems are classified by type. The most common classifications of insulation system in current use and as defined in BS 2757: 1986 (IEC 85: 1984) are:

Class B   Permitting a winding temperature rise of 80°C above a 40°C ambient, or 120°C maximum.

Class F   Permitting a winding temperature rise of 100°C above a 40°C ambient, or 140°C maximum.

Class H   Permitting a winding temperature rise of 110°C above a 40°C ambient, or 150°C maximum.

The winding temperature rises quoted in each case assume that hot spots in the winding may be 10°C hotter. For continuous operation it would be prudent and extend the machine life if the temperature rise was limited to a figure of 10°C or even 20°C lower than the maximum permitted by the class of insulation used.

Clearly the temperature rating of a motor is more of an art than a precise science and it may be perfectly acceptable to exceed allowable temperatures for short periods, such as during starting, providing that subsequent running is at a lower temperature level. The insulation life can then be estimated and provided that it is not less than 15 years there may be no need to use a larger and more expensive motor.

However, it is obviously essential to determine the frequency of starting, starting currents and other duty cycle data in order to size a machine correctly.

Winding temperature protection can be provided by motor makers in the form of thermostats or thermistors embedded in the stator windings. Unfortunately it is not possible to fit them in the rotor windings but the motor designer is aware of the temperature gradients in the machine and is able to select a protection operating temperature which is appropriate. On d.c. machines the temperature detectors can be fitted in the interpole windings. While not in the armature windings, they are at least in the armature circuit and do more accurately monitor the armature temperature. The protection is usually arranged to shut down the plant, although on some essential duties additional protection may be fitted which operates at a lower temperature in order to give an advanced warning of an impending shut down. At least the warning will give the operator an opportunity of initiating a controlled plant shutdown.

EFFICIENCY

Motor and generator manufacturers produce machines which perform reliably and well. Pressures exist however due to the consideration by many buyers only of first cost. These compel manufacturers to minimize the amount of active material (copper and iron) and to take advantage of the maximum temperature rise permitted by the class of insulation used. The cost of the machine is related to the amount of materials used whilst the buyer may be interested only in kW/rev/min or torque. The pressures are therefore to increase the power-to-weight ratio and significant advances have been made over the last 20 years. Usually these advances have been made at the expense of high efficiency.

With the increasing cost of electrical energy, and the fact that motors consume the value of their first cost in electrical energy in only a few weeks, there is good economic reason to consider the effect on running cost of higher efficiency.

A higher efficiency motor will cost more but not only will the premium quickly be recovered, the entire first cost of the motor can be recovered in energy savings in less than one year depending upon the motor type, loading and utilization.

To achieve higher machine efficiency the designer will incorporate more copper in the windings to reduce the I^2R copper loss. This may mean some increase in machine size to accommodate the additional copper.

Iron losses also will be reduced by utilizing more expensive lower loss and possibly thinner electrical sheet steel for the laminations and by specially treating the laminations to improve the stacking factor.

Fewer losses mean less heat generated and therefore less cooling with consequent reduction in windage losses.

Some companies already offer, at a premium, motors having an efficiency 2–3% higher than a normal commercially available machine. That is, efficiencies of 93% compared with 90% for 100 kW motors are readily available. One UK company is consistently achieving an efficiency of 97% on specialized 100 kW cage motors. The lower cooling demands of a high efficiency motor will also reduce the associated noise level.

STORAGE

Efforts should be made to minimize the deterioration of the machine during storage on site or in a warehouse. Dry conditions, protected from damp and condensation and from extremes of temperature, are obviously desirable. Ingress of dust to a machine on site should be avoided, especially of concrete dust which may be present if building and civil work are still going on. Machines should not be stored in an area which is subject to vibration because this could cause damage to the bearings. If some degree of vibration cannot be avoided then the shaft should be turned by hand every few weeks to load different parts of the bearings.

Where doubt exists about the dryness of the storage area and the machine is fitted with anti-condensation heaters it may well be worth considering using the heaters during storage. If the heaters are used for this purpose the safety aspects should be borne in mind and terminal box covers be replaced and warning notices attached. It is important to inspect the machine in storage every six months or so and to maintain protective coatings on bright parts which may be liable to rust.

INSTALLATION

Careless installation can affect the life and operation of a machine. Contractors and users are urged to follow the machine manufacturer's instructions. General considerations are indicated below.

Checking

Machine manufacturers are not always made aware of any need to pack machines specially for prolonged storage. Storage, for however long, may have been under unsuitable conditions and it is essential to check before installation that the motors have not deteriorated. While signs of rusting are obvious, the insulation resistance may have been affected by moisture. If, on checking, the insulation is found to be less than $1\,M\Omega$ the machine must be thoroughly dried out.

However careful the manufacturer may have been to pack the machine for difficult storage conditions, damage can occur in transit. The storage period may be considerably longer than was envisaged at the time the machine was dispatched. Close inspection of the machine prior to installation is therefore essential. Included in the check should be the nameplate data to ensure that it is appropriate for the actual site supply and other operating conditions. Machines should certainly not operate in flammable, dangerous or corrosive conditions unless they have been specifically designed and supplied for the purpose.

Any dust should be blown out of the machine and a detailed inspection, particularly of any machines fitted with commutator or slip rings, should be undertaken to ensure that no damage has resulted from exposure to chemical and corrosive fumes.

Bearing grease, except in the case of sealed-for-life bearings, must be checked and if necessary replaced if it has deteriorated during storage. Shaft clamps which may have been fitted to minimize the risk of brinelling during storage should also be removed.

Erection

The machine must be mounted on a solid and level foundation in order to avoid vibration. For a foot-mounted machine design the foundation is often a bedplate of fabricated steel upon a reinforced concrete plinth. Where this is the case the bedplate foundation bolts must be adequately proportioned and the bedplate packed so that when it is finally pulled down on to the plinth it is flat and correctly aligned. Only when flatness and alignment are established should the bedplate be grouted down to provide a solid homogeneous base for the machine.

The cost of manufacture and installation of bedplates is high and machines are now often bolted down directly on to a concrete foundation using foundation pads set into the concrete, Fig. 15.22. Foundation pads are usually cast iron with a top machined face and longitudinal tapped hole. The body is normally fluted for the best possible bond in the concrete. The pads are bolted on to the machine feet and the machine itself packed up on the concrete foundation for correct alignment. When the machine has been aligned the foundation pads are permanently grouted into pockets in the concrete. After final alignment and level have been satisfactorily established, dowel holes are drilled through diagonally opposite motor feet into the foundation pads and dowel pins inserted, Fig. 15.23. The use of dowel pins greatly facilitates any subsequent reinstallation of the machine on its foundation. Once the foundation pads have been permanently grouted in they behave and function in exactly the same way as a conventional fabricated or cast bedplate.

Minor alignment or height adjustments or a change of machine are accommodated by the use of shims under the machine feet. On metric machines, the height of the shaft centre line above the underside of the feet has been standardized and that dimension in millimetres incorporated into the frame size nomenclature. Thus a 400 frame size machine has a shaft centre line height of 400 mm. The only tolerance on this dimension is negative to allow one machine to be replaced by

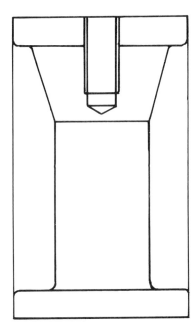

Fig. 15.22 Cast iron foundation pad.

another of the same shaft height dimension with only feet shim changes. Bedplate mounted machines are also dowelled after grouting and final machine alignment adjustments.

Large machines with pedestal mounted bearings are usually built upon and supplied with a cast iron subframe or bedplate for mounting directly on to prepared reinforced concrete foundations.

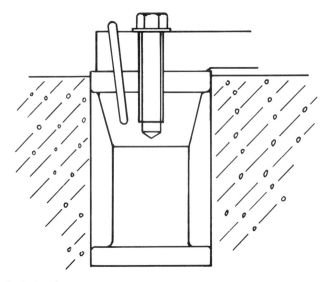

Fig. 15.23 Dowel pin in place.

Flange mounted or vertical machines are usually erected on a specially designed subframe or machined facing which itself forms the baseplate or foundation. Vertical motors are often mounted on fabricated skirts, especially for pump drives, and the skirts are treated as bedplates described above.

Alignment

Accurate alignment is an essential requirement if damage to bearings and flexible couplings is to be avoided. Alignment between an electrical machine and its prime mover or driven machine has to be established before the coupling may be connected.

The faces of the coupling must be parallel and any separation dimension between the faces established in accordance with the coupling manufacturer's recommendation. The final coupling alignment would normally be established and checked using a dial gauge, Fig. 15.24. If the machine is to be coupled to an internal combustion engine it is necessary to remove engine spark plugs or injectors to facilitate alignment so that the set may be turned easily by hand.

Two-bearing electrical machines are usually coupled to their prime mover or driven machine by means of a flexible coupling. The purpose of a flexible coupling is not to permit any misalignment but to minimize transmission of bearing shock loadings.

Single-bearing electrical machines are coupled to their prime mover or driven machine using a solidly bolted coupling. They cannot be connected by flexible couplings because these are not designed for, or capable of, supporting the downward thrust due to the weight of the machine rotor.

Mechanical erection is completed when the coupling halves are connected but some further checks are necessary before power can be applied. These should include ensuring that cooling air flows are not restricted by obstructions to air intakes or exhausts. Insufficient space between air intakes and adjacent walls can cause overheating. Ensure that covers have been replaced and that any doors

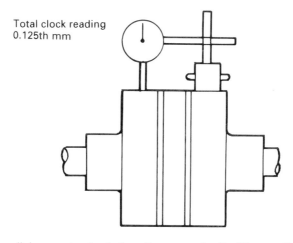

Total clock reading
0.125th mm

Fig. 15.24 Using a dial gauge to check the alignment of a flexible coupling.

which have to remain open while the machine is running are clearly marked. Check that auxiliary equipment such as blowers, tachogenerators, external coolers, filters, bearing vibration or temperature sensors and bearing oil circulation equipment has been satisfactorily fitted.

On machines fitted with air/water heat exchangers the water flow rate must be measured by means of a separate flow meter to ensure that it is in accordance with the flow rate specified on the rating plate of the machine. If a separate flow meter is not available the flow rate can be calculated by measuring the volume of water leaving the outlet over a set period of time.

Cabling and protection

Power and control cables to any machine should be of adequate current-carrying capacity and voltage grade. Specially rated cables are necessary for 400 Hz and other high frequencies. Effective and positive termination of cables onto machine terminals is necessary for good and lasting electrical contact. The machine is provided with earthing studs and it is essential that a proper earth is made in accordance with local regulations prevailing on site. Special regulations apply to increased safety Ex(e) and flameproof or explosion proof Ex(d) machines. These require special attention to earthing and to the use only of copper cables.

The power supply to which the machine is connected must have an appropriate fault clearance level in relation to the fuses and other protection devices being used in the system. Thermal protection relays are necessary to protect the machine against over-temperature resulting from overcurrent. However, excessive temperature may also result from obstruction of the cooling medium or operation at an excessively high ambient temperature. An overcurrent protection device will not of course recognize this. A protection device which directly monitors winding temperature, such as thermistors or thermostats if fitted, detects high temperatures whatever the cause. Current-dependent thermal protection relays must be provided in addition to any winding temperature detectors.

Current relays should be selected to provide thermal protection under short-circuit, single-phasing, locked rotor or other fault conditions. The protection device should isolate the machine from the supply and, for an Ex(e) machine, within the t_e temperature-rise time of the appropriate ignition group.

The t_e time is the time in seconds taken for the hottest surface of the machine and its windings to reach the ignition group temperature under stall conditions after it has reached its steady state full load temperature in an ambient temperature of 40°C.

COMMISSIONING

Once the machine has been installed and cabled up correctly, a further inspection should be made to ensure that bearings are appropriately lubricated, cooling systems functioning properly and that cooling air inlets or exhausts are not obstructed in any way. Power may be applied to any separately driven ventilating fan in order to confirm that it is rotating in the correct direction. This is normally indicated by an arrow on the fan casing. Providing that the fan motor has been connected with the correct phase rotation as specified in the connection diagrams this test should be just a formality but nevertheless an important formality, because wrong fan rotation will almost certainly significantly reduce its effectiveness with consequent overheating of the main machine. This overheating would not be detected by any overcurrent relay as already stated and damage to the main machine windings could result.

Direction of rotation should similarly be checked on the main machine itself. Again the correct direction is indicated or specified on the nameplate or elsewhere. Often machines, particularly the larger ones, are designed for rotation in only one direction. Reversing the direction of rotation therefore may require some mechanical alteration, albeit only a change to the shaft driven fan. Unidirectional fans are more common on larger motors because they are both more efficient and less noisy.

Alternators which have been in storage for some time may have insufficient residual magnetism to allow proper excitation to build up. Larger alternators having a permanent magnet pilot exciter should not suffer this problem. Some manufacturers provide the normal exciter with pole face permanent magnets embedded to eliminate the possibility of failure to excite.

When failure to excite is experienced and a maximum voltage regulator setting is not successful and any rotating diodes have been examined for correct functioning, flashing of the exciter field winding is necessary. This can be achieved by running the alternator up to speed and by touching the leads from a 6 or 12 V battery across the exciter field terminals in the correct polarity.

Checking of a rotating rectifier system requires each diode in turn being removed from its mounting plate and its forward and reverse resistances being measured. The forward resistance should be low and its reverse resistance high. If both resistances are approximately equal a failed diode is indicated and this should be replaced by an identical unit.

When refitting a diode a silicon heat sink compound should be lightly smeared on the diode seating to improve heat transfer and inhibit corrosion. A torque spanner to the correct setting must always be used when tightening semiconductors to their mounting plates or heat sinks.

Insulation resistance tests can be made on a brushless alternator using a 500 V or 1000 V instrument but not on the rectifiers themselves. Similarly, high voltage flash tests should not be attempted unless the rectifiers have first been short-circuited and any auxiliary equipment such as the voltage regulator disconnected.

Once the preliminary tests have been completed and the machine is running and loaded it is desirable to check for vibration and to monitor and log all meter readings and speed.

MAINTENANCE

Rotating electrical machines are inherently robust and reliable and require little maintenance. Generally, maintenance involves maintaining standards of cleanliness and regular inspection. A machine which is running within its design parameters is likely to need very little attention. The areas to be given most attention are, obviously enough, the bearing and, if fitted, the brushgear and associated slip rings or commutator.

Bearings

Large machines which are fitted with anti-friction ball and roller bearings normally have bearings which require re-lubrication. Grease is forced into the bearing through grease nipples while the machine is running. A grease gun may be used or sometimes bearings are fitted with Stauffer greasing facilities. In both cases, air bubbles in the grease should be avoided if lubrication is to be effective and the grease must be of the recommended type. Cleanliness is essential to avoid dirt and grit being forced into the bearings. Some bearings have grease escape valves to prevent overfilling the bearing and consequent overheating. The excess grease can be cleaned away periodically without adverse effect on horizontal shaft machines. On vertical shaft machines this excess grease is sometimes led to a container located below the bearing and this must be emptied at the recommended intervals to prevent grease entering the windings. The interval between re-lubrication depends upon rotational speed, bearing size and loading and whether the machine is horizontal or vertical. Typically, the period is 4000 running hours for a ball bearing and approximately half that period for a roller bearing of similar size. Specific recommendations will be given for each machine by the manufacturer but the period between re-lubrication should not exceed one year.

At longer intervals of about 8000–12 000 running hours the bearings should be inspected by removing the outboard bearing cap and noting any serious discolouration of the grease. If serious discolouration is evident, the bearing should be removed from the shaft, cleaned and examined more closely. Replacement of the bearing is indicated if it shows signs of blueing, cracking, brinelling or excessive wear.

Small machines are often fitted with sealed-for-life anti-friction bearings which do not require re-lubrication. Inspection of a sealed-for-life bearing is not possible so that where malfunction is suspected it should be replaced. Listening to a bearing through a sounding rod or stethoscope can give a useful indication of its condition but some experience is necessary before drawing too many conclusions. A high-pitched whistling sound may indicate defective lubrication while a low-pitched rumbling probably results from dirt or damage.

Removal of anti-friction bearings from a shaft involves some dismantling of the machine although it may not be necessary to remove the rotor from the stator. Bearings are an interference fit on to the shaft so that correct withdrawing tackle, which can apply even pressure round the bearing, should be used. Tapping or hammering of the outer race will almost certainly damage the track. Pre-heating of the new bearing to about 95°C in an oven assists in pressing it on to the shaft and a lightly tapped drift assists in pressing the bearing home squarely on to its seating. The drift should be applied to the inner race if the bearing is being fitted on to a shaft and to the outer race if into a housing.

Oil-lubricated sleeve bearings require an oil viscosity appropriate to the ambient temperature. The oil level in the bearings must be maintained and the oil replaced completely at regular intervals or at any sign of overheating.

Brushgear

Machines fitted with slip rings or a commutator have brushgear. To be effective the slip rings or commutator must be in good condition and free from surface defects. Slip rings which are badly scored need to be skimmed in a lathe. For removal of minor grooves a commutator stone can be used, rested on a brush arm as the rotor is slowly driven round. Finally, the surface can be polished using a piece of 00 grade glasspaper wrapped around the square end of a piece of wood as the rotor is turned by hand.

A similar technique can be applied to commutators but 'stoning' is a dangerous operation which should be carried out only by someone experienced in this work. A commutator which is in good condition should have developed a smooth brown surface patina. When this low friction patina is present the commutator surface should not be stoned or polished. Raised commutator bars or mica are indicated by heavy and uneven blackening and then skimming and undercutting are required. When the commutator is uniformly black, cleaning and polishing should be considered.

Longitudinal undercutting of the mica insulating between commutator segments is always necessary after skimming in a lathe, Fig. 15.25. Although the bars of the commutator should be chamfered lightly, the bar width must not be reduced.

It is sometimes impossible to achieve the brown surface patina which provides long brush life. The cause is usually light load running and insufficient brush current density. So often margins are built into the specified machine rating that light loadings are inevitable. However, where this is the case, it is worth considering the removal of some of the brushes. An equal number of brushes may be removed from the same track on each brush arm so increasing the current density of the remaining brushes. The machine manufacturer is always willing to give advice as to the number of brushes required per arm for the actual loadings on site.

Inspection of the brushgear will show up excessive or uneven brush wear and the presence of carbon dust. All carbon dust should be vacuumed out of the machine at each inspection. The brushes should be checked for wear and for freedom of movement in the brush holder. Although brushes may be allowed to wear down almost to the metal of the pigtail, it is worthwhile replacing them much earlier. If replaced when they are only reduced 50% in length the brushes are most unlikely to require attention, or cause an unscheduled plant stoppage, before the next routine inspection. Many d.c. motor manufacturers now offer

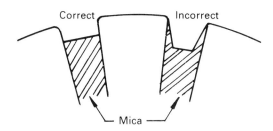

Fig. 15.25 Correct and incorrect undercutting of the mica.

brush wear detection facilities which provide early warning of advanced brush wear and reduce the possibility of an unscheduled stoppage for maintenance.

Coolers

After a period of time any heat exchanger whether air/air or air/water will have developed some dirt build-up which reduces the cooling effectiveness. Cleaning of the air passages, tubes and water tubes is therefore necessary from time to time. Filters fitted to force ventilating blowers need to be inspected and, if necessary cleaned or replaced depending upon the type.

If the machine is to be shut down in freezing conditions any water circuits should be drained. Condensation forms and accumulates in the air/water heat exchanger and could be drawn into the machine through the air passages and enter the windings. When a machine is to be shut down for any time the cooling water flow should also be stopped to limit the build-up of condensation. Even under normal operation this condensation can be significant and justify raising the water temperature in the cooler and reducing the ambient humidity to reduce the quantity of condensation formed.

Windings and insulation

A machine which has been out of service for a long period could have too low an insulation resistance to allow it to return to operation. If the resistance is less than $1\,M\Omega$ on an l.v. machine, or equivalent value on h.v. machines, the windings need to be dried out. The machine can be placed in an oven or dried using a hot air blower. In either case the temperature should not exceed 90°C or damage could result to rotating diodes for example. Initially, the insulation resistance will fall and then slowly increase to a maximum level. The resistance readings should be monitored as the drying out process proceeds and, to avoid misleading readings, the temperature should be kept as constant as possible.

Before taking any insulation resistance measurements, any diodes or capacitors should be disconnected or short-circuited. High-voltage insulation breakdown tests should be avoided as they have a progressive effect in causing deterioration of the winding insulation.

CHAPTER 16

HBC Fuses and Fusegear
in Low Voltage Systems

J. Feenan, MBE, CEng, FIMechE
*(Electrical Consultant and formerly Technical Director,
Fusegear Division, GEC Installation Equipment Ltd)*

A major change in the UK Wiring Regulations occurred in 1991 with the introduction of the 15th edition of the IEE Wiring Regulations. These regulations
differed considerably from the 14th and previous editions in that they were
aligned with the wiring regulations of the International Electrotechnical Commission
(IEC) which were more specific on many aspects of electrical protection including
overcurrent protection and protection against electric shock.

The 16th edition of the IEE Wiring Regulations, which contained amendments
arising from further agreements reached in CENELEC, was introduced in 1991
and this in turn was accepted by BSI as BS 7671 *Regulations for electrical
installations* in 1992.

When the first set of Wiring Regulations was issued in 1882 by the Society of
Telegraph Engineers and Electricians, mention was made of the fuse as an ideal
device for 'the protection of wires'. Over a century later the fuse, in the form of
the hbc fuse, is still a major protective device. In fact, BS 7671 gives precise
regulations regarding overcurrent protection in which the l.v. hbc fuse emerges as
a unique device for fault current protection as well as providing the necessary
overload protection to PVC insulated cables which are acknowledged to be the
most difficult to protect under overload conditions.

The design of the hbc fuse has progressed during the last 60 years until it has
become a highly sophisticated protective device with very precise characteristics.
The British and International Standards covering l.v. fuses have also progressed
during these years and in June 1981 agreement was finally reached in the committee
of the International Electrotechnical Commission dealing with l.v. fuses (SC32B),
for one set of time-current 'gates' within which the time-current characteristics of
all fuses for general purpose applications must fall. This decision was the culmi-
nation of many years of intense activity in IEC and it is significant to note that the
time/current characteristics of the modern British hbc fuse-link fall within these
agreed 'gates', thus indicating that British practice with regard to hbc fuse-link
design has been the correct one. It is therefore an opportune point in time to
consider the modern hbc fuse from the viewpoints of design and performance,
application in general purpose circuits and in motor circuits, and its effect on the
design of associated fuse switchgear and motor starter combinations.

417

HBC FUSELINKS DESIGN AND PERFORMANCE

The performance requirements of hbc fuse-links, for voltages up to 1000 V a.c. or 1500 V d.c. are covered by BS 88: 1988, *Low voltage fuses*. BS 88 defines a fuse as consisting of a fuse-link and fuse-holder. The fuse-holder consists of a fuse-base and in most cases a fuse-carrier. A diagrammatic representation of a typical fuse is shown in Fig. 16.1. It also defines three types of fuse-links, type gG for general purpose applications, and types gM and aM for motor circuit applications.

These symbols signify the breaking range and utilization category of the fuses. The symbol 'g' represents full range breaking capacity, and the symbol 'a' represents partial range breaking capacity. A gG fuse-link is one with a full range breaking capacity for general application, a gM fuse-link is one with a full range breaking capacity for the protection of motor circuits, and an aM fuse-link has a partial range breaking capacity for the protection of motor circuits.

The specification gives detailed requirements for each type, which are summarized below.

Type gG fuses

The minimum performance requirements which must be met by gG hbc fuses, complying with BS 88: 1988: Part 2, Sections 2.1 and 2.2 are:

(1) A breaking capacity of not less than 80 kA at 415 V a.c. 0.2 pf. Fuses to this standard can successfully interrupt any fault current from the minimum fusing current of the fuse up to the rated breaking capacity at the specific power factor.
(2) The temperature rise of the fuse when carrying rated current must not exceed 70°C at the fuse terminals. (Lower values are specified if the terminal is not silver plated.) In addition, maximum permitted power losses are specified for each fuse-link size and these are such that for the majority of fuse ratings the

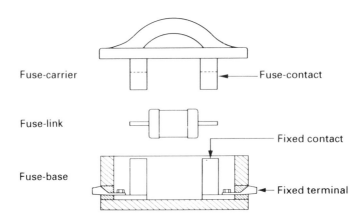

Fig. 16.1 Component part of a fuse.

terminal temperature rises attained are very much less than the permitted maximum.

(3) The ability to protect PVC-insulated cables from damage due to overload currents. The specification determines this by requiring a general purpose fuse-link (type gG), to have a fusing current not exceeding 160% of rated current (I_n) and a non-fusing current (I_{nf}) not less than the 125% I_n. It must also comply with the requirements specified in the conventional cable overload protection test now included in BS 88. Fuse-links meeting these requirements are deemed to provide overload protection to PVC insulated cables when the rated current of the fuse is equal to or less than the continuous rating of the cable for a given installation condition.

(4) The time−current characteristics of a fuse must fall within the specified time current zones given in the Standard. Here again, it should be noted that BS 88: Part 2, Section 2.2 contains much narrower time−current zones than the previous edition of BS 88: Part 2, and these zones fall within the time−current gates specified in BS 88: Part 1 for all general purpose fuses.

Figure 16.2 illustrates the test requirements of BS 88 and it can be seen that for verification of breaking capacity, the gG fuse-links are submitted to five test currents, I_1, I_2, I_3, I_4 and I_5, covering a range of prospective currents from a breaking capacity of 80 kA down to a current equal to 1.25I_f, (I_f is equal to 1.6I_n).

Type gM fuse-links

As these fuse-links are for the protection of motor circuits, they cannot in that role provide overload protection to the associated cables, which are usually rated on the basis of the full load current of the motor. Nevertheless the same tests for breaking capacity and compliance with the time−current zones, as that specified for gG fuse-links, are applicable to gM fuse-links. This ensures that the fuse-links will interrupt any current from rated breaking capacity down to minimum fusing current safely, and allows great flexibility of co-ordination with the associated overload devices.

Type aM fuse-links

These fuse-links by their designation are a back-up type, having a partial range breaking capacity and have a minimum breaking current approximately equal to 6.3I_n. The I_n of the aM fuse-link is not related to the rated current of the fuse, but signifies the rated current of the motor which it can protect. They do not provide overload protection and must be co-ordinated with a suitable device, which, whilst providing overload protection, ensures that this fuse is not subjected to low overcurrents below its minimum breaking current.

It should be noted that the reference to fusing factor, contained in all previous editions of BS 88, has now disappeared, and is no longer applicable.

The principal requirements are shown in Fig. 16.2. Proof of compliance with parts (1) and (2) of this figure necessitate testing at a high power test laboratory in closely controlled test conditions required by BS 88.

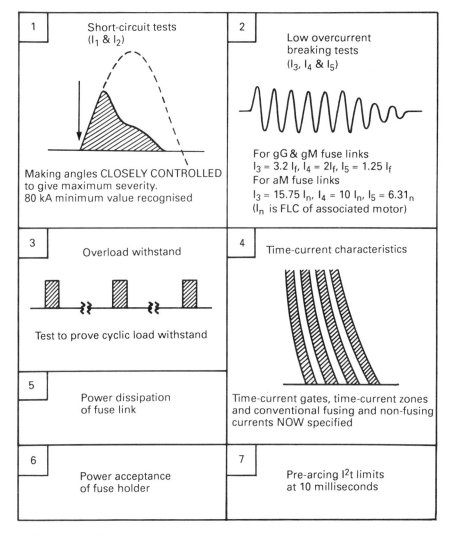

Fig. 16.2 BS 88: 1988, principal test requirements.

The high power laboratory is controlled by ASTA Certification Services, and an ASTA certificate of rating can be issued as proof of compliance with the breaking capacity clauses of the standard. Test conditions for breaking capacity are as indicated in Fig. 16.2, arranged to ensure that the fuse interrupts all possible combinations of voltage, current and inductance which would occur in practice on systems having fault levels up to the breaking capacity of the fuse.

A new requirement in the revised standard concerns verification of the pre-arcing I^2t limits of the fuse at a time of 10 ms. This is to ensure that the fuse will meet a discrimination ratio of 1.6:1, which is now also required by the revised standard.

The discrimination ratio of 1.6:1, which is based upon comparison of pre-arcing

I^2t values at 10 ms, has been included in the standard at the request of the IEC Wiring Regulations Committee, in order to ensure that with fuse-protected systems, an improved degree of discrimination can be provided.

When compared with the 1975 edition, BS 88: 1988 includes additional verification tests which are related to safety. These are:

- protection against electric shock;
- resistance to heat;
- mechanical strength;
- resistance to corrosion;
- resistance to abnormal heat and fire.

It also includes, with respect to fuse-holders, verification of the ability of contacts to perform satisfactorily if:

(1) left undisturbed for long periods, and
(2) after repeated engagement and disengagement.

These additional safety tests reflect the growing demands by inspection and test authorities in many countries, that any performance claims in a specification must have a test requirement which is deemed to verify that claim.

DESIGN OF CARTRIDGE FUSE-LINKS

Although the fuse-link complying with BS 88 has a similar outward appearance to a fuse-link complying with earlier editions of the British Standard, changes in element design and element configurations within the fuse-link have achieved significant improvements in breaking capacity and time−current characteristics within existing dimensions.

The modern cartridge fuse-link consists of a ceramic body, one or more fuse elements depending upon the current rating, an inert arc quenching filler such as granulated quartz, tin-plated copper or brass endcaps and, usually, tin-plated copper tag terminations. A typical fuse-link is shown in Fig. 16.3. The fuse elements used in modern l.v. fuse-links are made from a tape or strip. Different fuse manufacturers have different ideas on the precise form of the element but all of them employ a number of reduced sections along the length of the element in order to promote multiple arcing under short-circuit conditions (i.e. creating a number of arcs in series which burn simultaneously).

When arcing commences in a modern hbc fuse-link, a high transient pressure is set up within the fuse body due to the vaporization of parts of the element which can only expand to a limited extent in the interstices in the quartz filler. This high pressure, combined with the high temperature generated in the arc column, necessitates the use of special ceramic material for the body. The joints between endcaps and body must also be strong enough to withstand the stresses created. Granulated quartz is chosen as the filler because it is inert and has excellent heat-absorbing properties which make it an ideal arc quenching medium. Its purity and

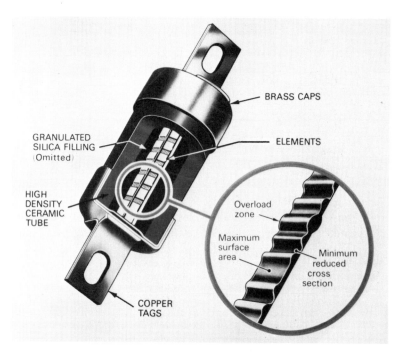

Fig. 16.3 Sectioned view of typical hbc fuse-link.

grain size are also very important factors in obtaining a high breaking capacity. The design of the fuse element controls both the mode of operation on short-circuit and the shape of the time−current characteristics of the fuse.

The number and lengths of the reduced sections of the element govern the arc voltage of the fuse. Arc voltage is the voltage which appears across the terminals of the fuse during the arcing time of the fuse-link and it is the resultant of four components:

(1) the voltage dropped across the arcs in series;
(2) the resistive voltage drop of the rest of the circuit due to the passage of a large fault current;
(3) the circuit e.m.f.;
(4) the inductive voltage, $L(\mathrm{d}i/\mathrm{d}t)$, produced in the circuit due to the dissipation of its inductive energy ($\frac{1}{2}Li^2$) within the fuse arcs.

Components (1) and (2) tend to reduce the current while (3) and (4) try to maintain its flow. The generation of an adequate arc voltage ensures that current ceases to rise after the commencement of arcing and the subsequent variation in arc voltage controls the rate at which the current reduces to zero.

A major factor affecting the arc voltage is the grain size of the quartz filler because this in turn can affect the pressure generated within the fuse body when the element vaporizes during the arcing period. An arc burning under pressure requires a higher voltage to maintain it than an arc in free air. It is therefore possible to design a fuse-link which will produce a predetermined maximum value of arc voltage. BS 88 stipulates an upper limit of 2.5 kV for the arc voltage produced by fuse-links rated up to 660 V but in a good design this value is never reached. The range of fuse-links using the type of element illustrated in Fig. 16.4 produces arc voltages in the region of 1 kV.

An indication of the speed of operation and current limiting ability of the hbc fuse is given in the oscillogram shown in Fig. 16.5. This is the record of a 400 A fuse interrupting a prospective current of 80 kA r.m.s. symmetrical at 415 V a.c. with a lagging power factor of 0.15. BS 88 specifies that the fault shall be initiated such that arcing commences in the fuse at approximately 65 degrees after voltage zero. This produces an asymmetrical fault current with a peak prospective value of 180 kA.

The limitation of thermal and electromagnetic stresses achieved by the fuse is better appreciated when it is remembered that both vary as the square of the current. An examination of the oscillogram serves to illustrate this point.

The peak or cut-off current permitted to flow by the 400 A fuse is 40 kA, i.e. 22% of the possible peak current. The electromagnetic force produced is therefore only about 5% of that which would otherwise have occurred. Similarly the thermal stress has been reduced to less than 1% of that which would have occurred if a circuit-breaker having a speed of operation of 0.02 s had been the protective device. The combination of current limitation and rapid operation enables this high degree of protection to be obtained. Figure 16.6 shows the cut-off currents of a range of fuses, having current ratings from 2 A to 800 A on a prospective current of 80 kA. This illustrates the reduction in electromagnetic stresses which can be achieved with hbc fuses.

Fig. 16.4 Fuse-element manufactured from silver strip.

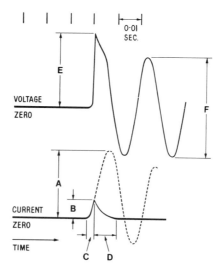

Fig. 16.5 Oscillogram of the operation of a 400 A 415 V fuse on a large fall circuit. Key: (A) Asymmetrical peak value of prospective current, 182.5 kA ≡ 81.1 kA r.m.s. sym; (B) cut-off current, 39.5 kA; (c) pre-arcing time, 1.6 ms; (D) arcing time, 3.4 ms; (E) arc voltage, 930 V; (F) recovery voltage, 461 V r.m.s.

On high short-circuit currents the energy required to melt a fuse is a minimum and constant value because there is no time for heat to be dissipated from it to its surroundings. The magnitude of this energy, known as the pre-arcing energy of the fuse, can be determined from the oscillograms of fuse operations on high short-circuit currents. Similarly the arc energy, which is the energy liberated in the fuse during its arcing time, can also be determined.

Unlike the pre-arcing energy, however, the arc energy can vary between wide limits depending upon the circuit conditions but the maximum value is obtained during the breaking capacity tests to BS 88. In effect the arc energy of a fuse is the energy stored in the circuit at the commencement of arcing which must be dissipated in the fuse before the circuit is finally opened. It must however be appreciated that the energy dissipated in the fuse is a function of $I^2t \times R_F$ where R_F is the resistance of the fuse during its pre-arcing and arcing periods.

When considering the short-circuit protection the fuse provides to equipment and cables in the circuit, the information of interest is the cut-off current and the I^2t value (A^2 s). The cut-off current, as stated earlier, is a direct indication of the limitation of electromagnetic stresses achieved by the fuse. The thermal stresses in the circuit can be expressed in terms of I^2t because this is the common factor between the fuse and other components in the circuit. Obviously in a cable the energy is a function of $I^2t \times R_C$ where R_C is the resistance of the cable.

It is common practice, however, for manufacturers and designers of associated apparatus such as fuse-switches, contactors, busbars and cables to express the

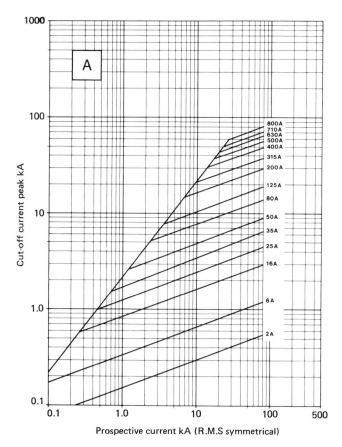

Fig. 16.6 Relationship between prospective current and cut-off current for a range of fuses.

short time thermal ratings of the equipment in terms of I^2t. This enables the manufacturers of such equipment to state in a simple manner the degree of thermal protection required by various items of equipment under short-circuit conditions. A fuse manufacturer can determine the I^2t of a fuse on large short-circuit currents from the relevant oscillograms and usually presents the minimum pre-arcing I^2t (i.e. the I^2t from the commencement of the fault until the arcing commences) and the maximum arcing I^2t (from commencement of arcing until circuit interruption) under worst fault conditions in the manner shown in Fig. 16.7. This shows the pre-arcing and total (pre-arcing plus arcing) I^2t values for a range of fuses from 2 A up to 800 A under short-circuit conditions on prospective currents up to 80 kA. Figures 16.6 and 16.7 represent a comprehensive picture of the short-circuit protection provided by this range of fuse-links.

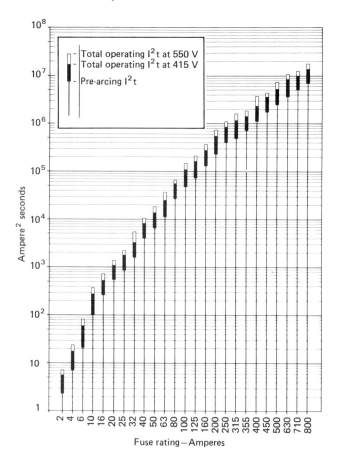

Fig. 16.7 I^2t characteristics for a range of fuses showing pre-arcing and total values.

OVERLOAD CHARACTERISTICS

The method of achieving compliance with the conventional fusing current mentioned earlier has traditionally been achieved by including an overload zone in the element. This is done by adding a precise amount of low melting point metal (tin) to the element, which is usually silver. Under overload conditions the temperature of the fuse element gradually increases until a temperature is reached at which the tin becomes molten and a metallurgical phenomenon, commonly known as 'M' effect, takes place. A eutectic alloy is formed at this temperature between the tin (melting temperature 230°C) and silver (melting temperature 860°C) and the alloy thus formed has the same melting temperature as that of tin (230°C). This ensures that the element will melt in the overload zone at the desired current of $1.6I_N$ without excessive temperatures being reached. All modern designs of cartridge fuse utilize this 'M' effect in one form or another.

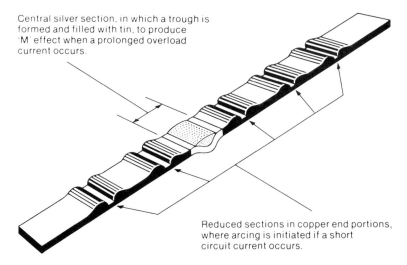

Central silver section, in which a trough is formed and filled with tin, to produce 'M' effect when a prolonged overload current occurs.

Reduced sections in copper end portions, where arcing is initiated if a short circuit current occurs.

Fig. 16.8 Silverbond element, showing that the 'M' effect zone is placed at a full section.

A recent trend, due to the high cost of silver, has been to utilize copper as an alternative material because the 'M' effect can be achieved between tin and copper although it occurs at a higher temperature (400°C), permitting higher temperatures to be reached by fuses under overload conditions which is not a desirable trend. A novel design of element which consists largely of copper, but still utilizes silver in the overload zone, is shown in Fig. 16.8. With this design the unequalled ability of the silver/tin eutectic is retained while economy is achieved in the use of copper for the majority of the element.

FUSE-HOLDER

The fuse-holder (i.e. the fuse-carrier and base) also plays an important part in the design of a cool running fuse. Robust contacts ensure a low temperature rise by helping to dissipate the heat from the fuse-link when carrying rated current. The method chosen for connecting the cable to the fuse terminals must be carefully considered to avoid the danger of a bad connection which could nullify an otherwise satisfactory design. Most of the front-connected fuse-holders to BS 88 employ tunnel-type terminals which are ideally suited for the connection of unprepared stranded copper conductors. This form of terminal is however not normally suitable for the direct connection of aluminium conductors. When aluminium conductors are used, it is preferable to terminate the conductor with a suitable cable lug and utilize a back-connected fuse-holder which provides a stud connection suitable for connecting the cable lug. High quality moulding materials are widely used in modern designs of fuse-carriers and bases which permit compact dimensions and extensive shrouding of current-carrying parts. These features are highlighted in Fig. 16.9 which illustrates a widely used design of fuse-holder.

Fig. 16.9 Sectioned fuse-holder showing how extensive shrouding of live parts is achieved.

APPLICATION OF HBC FUSES

BS 7671 clearly states the requirements which must be met when providing cables with both overload protection and short-circuit protection. These two terms are covered by a more general term, overcurrent protection. With regard to protection against overload, the Wiring Regulations state that the protective device must operate within a stated conventional time when carrying a current equal to $1.45I_Z$ (the current rating of the associated cable).

Type gG fuses to BS 88 provide the necessary degree of protection and both BS 7671 and the International Wiring Regulations recognize this fact. If the current rating of a gG fuse to BS 88 is not greater than I_Z, the rating of the associated cable, compliance with the requirements of BS 7671 for overload protection of cables is achieved. This rule confirms a practice which has existed in the UK for many years and it is interesting to note that this practice is now internationally accepted.

With regard to the protection of cables against short-circuit (the other form of overcurrent), the hbc fuse is indisputably the best device for the purposes, particularly if the fuse has a current rating not exceeding that of the cable. Even when the fuse has a greater rating than the cable it gives adequate short-circuit protection. The limiting parameter specified in BS 7671 with regard to the maximum size of fuse which can give short-circuit protection to a cable, depends on the minimum value of the short-circuit current which can occur, see Fig. 16.10.

There have previously never been any clear-cut guide lines for such a minimum value of short-circuit current and the only previous reference to a relationship between fuse and cable rating is given in Regulation A68 of the 14th edition of the IEE Wiring Regulations. Here permission was given for the use of a fuse of up to twice the rating of the associated cable in a motor circuit when overload protection is provided by the motor starter. In BS 7671 protection must be provided for cables for short-circuits of durations up to 5 s and Regulation 434–03–03 gives an adiabatic formula $t = k^2 S^2 / I^2$ (where S is the cross-sectional area of the conductor and k is a constant for each type of cable), which can be used to

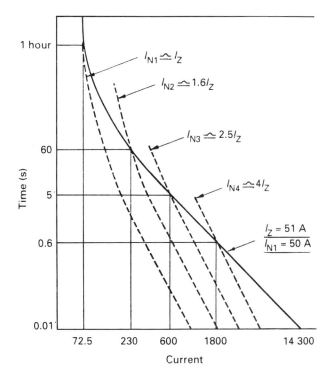

Fig. 16.10 Short-circuit protection of $10\,\text{mm}^2$ PVC-insulated cable by hbc fuses.

calculate the I^2t withstand capability of various types of cables at 5 s. It is therefore a relatively simple job to determine from the published information the maximum size of cable which can be protected by a given size of fuse on a short-circuit current of 5 s duration.

Table 16.1 shows, by utilizing the formula mentioned above, the appropriate maximum size of cable which can be protected against short-circuit in a motor circuit for both open and enclosed installation conditions. It can be seen that the old rule of twice the cable rating still applies in the majority of cases. An important point to appreciate is that if a fuse protects a cable under short-circuit conditions at 5 s, it will, because of its unique current limiting ability, protect that cable on all fault currents in excess of the 5 s value up to the breaking capacity rating of the fuse-link. This cannot be assumed for other types of protective devices such as circuit-breakers of the non-current limiting type where it is necessary to check whether protection is afforded at the maximum prospective fault current likely to occur at that point in the circuit, see Fig. 16.11. This illustrates how effectively the modern hbc fuse can provide overload and short-circuit protection in accordance with BS 7671 without sacrificing any of its ability to permit maximum utilization of the other components of the circuit it is protecting, e.g. cable, motor starter, etc.

Table 16.1 Short circuit protection of PVC-insulated cables in motor circuits.

Conductor size (mm²)	Maximum current rating (A) BS 7671 Table 4D1A		Maximum rated current of type gG fuse to BS 88: Part 2
	open conditions, Column 7	enclosed conditions, Column 5	
1.5	18	15.5	25
4	33	28	50
10	59	50	100
25	104	89	200
50	167	134	355
70	214	171	500
120	303	239	750

Application of Regulation 434−03−03 of BS 7671.

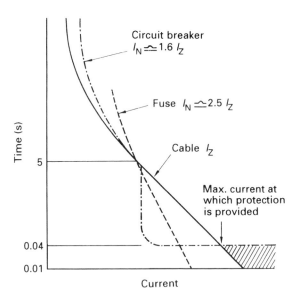

Fig. 16.11 Comparison of protection provided to PVC-insulated cable by hbc fuses and circuit-breakers.

DISCRIMINATION AND CO-ORDINATION

In any well-planned electrical system discrimination between overcurrent protective devices throughout the system is essential. Regulation 533−01−06 of BS 7671 states this as a requirement for compliance, to avoid danger and Regulation 314−01−02 draws attention to the need to consider the inconvenience likely to be caused.

Discrimination

Discrimination between any two devices in series is achieved when on the occurrence of a fault in the circuit protected by the minor device, only that device operates leaving the major device intact. Where hbc fuses are used as the protective device throughout the system the degree of discrimination which can be achieved is unequalled by any other form of protective device.

To achieve discrimination on the worst fault conditions, i.e. on high prospective fault currents, it must be ensured that the pre-arcing I^2t of the major fuse is in excess of the total I^2t of the minor fuse. The discrimination ratio of 1.6:1 mentioned earlier, is based upon comparison of pre-arcing I^2t at 10 ms. The assumption made is that in practice, particularly on final circuits, the fault conditions are such that the arcing I^2t can be ignored. Whilst this may well be true for such circuits, which are usually single-phase circuits where the prospective current is not very high, it is not a reasonable assumption for industrial applications, where high fault levels and low-power factors occur on three-phase systems. It is however agreed that the breaking capacity test conditions specified in BS 88 are unlikely to occur in practice. Thus the arcing I^2t values obtained on such tests indicated in Fig. 16.7, are unlikely to occur, even under worst practical fault conditions on 415 V a.c. three-phase systems. Tests on such circuits have established that a 1.6:1 ratio can be achieved with the modern design of fuse to BS 88: Part 2.

Co-ordination

In the modern electrical installation, there are a number of devices such as miniature circuit breakers (mcb) and contactors which although they are over-current protective devices, have a limited breaking capacity. In situations where the fault levels exceed their breaking capacity, it is necessary to provide back-up protection which will take over from the mcb or contactor on a fault current not exceeding its breaking capacity.

Such a back-up device must not only protect the circuit, but ideally it should not allow the mcb or contactor to be damaged by letting too much energy flow through them. This energy is related to the let through I^2t of the device, and the unique ability of the hbc fuse under fault conditions to keep the let-through I^2t to very low values, explains why it is the most effective means of back-up protection for these devices.

When selecting the fuse as a back-up device in a motor circuit, further objectives must be achieved. In addition to the ability to interrupt the fault current successfully and protect the associated contactor and cable under short-circuit fault conditions, it must be capable of withstanding the motor starting current for the starting period. The generally accepted method of determining the capability of a fuse to withstand motor starting conditions is to refer to the 10 s withstand current. Usually it is assumed that the starting current is approximately six times the motor full load, and that such a current would exist for up to 10 s.

One further aspect of correct protection of motor circuits has received considerable attention in recent times. This concerns the degree of co-ordination required between the back-up protective device and the starter. In BS 4941, on motor starters, there are three degrees of co-ordination specified: types 'a', 'b', and 'c'.

Types 'a' and 'b' permit varying degrees of damage to the starter, but type 'c' ensures no damage to the starter, particularly with respect to the characteristics of the overload relay. If, as is normal, the cable from the starter to the motor is chosen on the basis of the full load current of the motor, then the back-up device is only providing short-circuit protection to that cable. The overload relay of the starter is providing the desired overload protection to the cable as well as to the motor.

Thus in order to comply with the requirements of BS 7671 in such an installation, type 'c' co-ordination is necessary.

Tests have shown that the hbc fuse to BS 88 provides the most economic method of achieving this degree of protection. The modern motor starter is a very compact device and even when hbc fuses are used it is necessary to refer to the manufacturer of the starter to ensure that the correct fuse is chosen.

A more recent specification BS EN60947−4−1: 1992, *low voltage switchgear and controlgear − PT4 contactors and motor starters − Section 1 Electromechanical contactors and motor starters* has been issued to replace BS 4941. It is effective from 1 December 1997 and until then both specifications will run concurrently.

In the new specification type 'a' co-ordination disappears and type 'b' and 'c' co-ordination are replaced by types '1' and '2' respectively. Tests have indicated that the hbc fuse still plays a unique role in achieving type '2' co-ordination to this new specification, a necessary requirement for compliance with BS 7671.

Type gM motor circuit fuse-links

This type of fuse-link provides the necessary degree of back-up protection for motor circuits, but in a smaller physical package than that of the equivalent current rating of standard fuse-link. Its inclusion in BS 88 recognizes the fact that, when a fuse is used as back-up protection to another protective device (motor starter) the same degree of short-circuit protection is provided by either a standard fuse-link or the physically smaller motor circuit fuse-link of the appropriate current rating. For example, the 32M63 fuse-link in the A2 dimensions to BS 88 provides the same back-up protection as the 63 A fuse-link in the larger A3 dimensions for the same voltage rating and breaking capacity.

The dual rating or the gM fuse-link is characterized by two current values. The first value I_n denotes both the rated current of the fuse-link and the rated current of the fuse-holder. The second value I_{ch} denotes the time−current characteristic of the fuse-link. Therefore the 32M63 A fuse-link has a continuous rating of 32 A because of the limitation of the fuse-holder or fuse-switch in which it is installed, and has the same time−current characteristics as that of the standard 63 A fuse-link. This type of fuse-link permits full utilization of the make−break capabilities of the modern fuse switch. For example, a 200 A fuse-switch with a 200M315 fuse-link fitted can be used on a motor circuit having a load current up to approximately 200 A, whereas with a standard 200 A fuse-link fitted it would be limited to a circuit of 125 A, because of the reduced motor starting capability of the 200 A fuse-link. As will be discussed later, the make−break performance of the 200 A fuse-switch is such that it can safely break the stalled motor current of a motor, having a 200 A full load current. Thus, the motor circuit fuse-link has been an important contribution to the achievement of the compact dimensions of a modern fuse-switch.

It is important to note that the motor circuit fuse-link type gM is a general-purpose type and is tested to the same breaking capacity requirements as that of the gG type. The reduction in equipment dimensions which can be achieved by using motor circuit fuse-links is appreciable and has now been internationally recognized.

Type aM motor circuit fuse-links

This type of fuse-link which is of continental European origin, has considerably different characteristics from those of the standardised time−current zones, and as stated earlier, the current rating assigned to such a fuse-link is related to the full load current of the motor it is designed to protect rather than to the fuse-link itself. Because the minimum breaking current is a significant multiple of current rating, it is essential where such fuse-links are used, that its characteristics are co-ordinated with an overload device which will ensure that both the circuit and the aM fuse-link are protected against overload conditions. This fuse-link cannot safely interrupt overcurrents below its specified minimum breaking current.

FUSES IN HIGH AMBIENT TEMPERATURES

A fuse can carry its rated current continuously in ambient temperatures up to 35°C but if it is required to carry its rated current at higher ambient temperatures or in enclosures where the inside air temperature is more than 15°C in excess of this value, it is necessary to apply a derating factor, and most manufacturers give such information in their publications. If however a fuse is providing back-up protection in a motor circuit employing direct-on-line starti1.g, derating is not usually required at these higher temperatures because in such an application the fuse is only carrying approximately 50% of its rated current producing only 25% of the temperature-rise normally achieved at rated current. Therefore unless the local air temperature is excessively high (above 60°C) there is no need to derate in such motor circuit applications.

PROTECTION AGAINST ELECTRIC SHOCK

The biggest difference between the 14th and earlier editions of the IEE Wiring Regulations and BS 7671 concerns the rules which are included in BS 7671 regarding protection against electric shock. As fuses figure significantly in certain measures of protection, it is opportune to comment on those aspects of BS 7671 where fuses can be used to good effect in achieving protection against electric shock. The most popular measure of protection against indirect contact (contact with exposed conductive parts made live by a fault) is 'earthed equipotential bonding and automatic disconnection of supply'. The Regulations in BS 7671, notably Regulations 413−02−02 and a number of Regulations in Section 547, give details regarding the steps to be taken to achieve the necessary earthed equipotential bonding which is the first part of the requirement. The second part, automatic disconnection of supply, is covered by Regulations 413−02−04 to 413−02−26 specifying the requirements for (a) final circuits supplying socket outlets and (b) final circuits supplying only fixed equipment.

Regulations 413−02−10 and 413−02−14 give specific rules for the protection of circuits having a nominal voltage to earth of 240 V r.m.s. a.c. when protection is provided by means of a fuse. They state that in order to comply with the requirements for electric shock the earth fault loop impedance Z_S shall not exceed values given in Table 41B1 for final circuits supplying socket outlets, where the disconnecting time of the protective device under fault conditions must not exceed 0.4 s, and Table 41D for final circuits supplying fixed equipment, where the appropriate disconnection time shall not exceed 5 s. Among the devices mentioned in the tables are cartridge fuses to BS 88: Part 2, BS 1361 and BS 1362. The earth fault loop impedances given against each rating of fuse in each table have been determined from the time−current characteristics of these fuses and Appendix 3 of the Regulations shows the upper limit of the time−current zone for appropriate ratings of fuse-link.

It follows from these disconnecting time requirements and those mentioned earlier for short-circuit protection, that the current to cause operation of the fuse in 5 s is now of great importance and BS 88: 1988: Part 1 includes maximum values for the 5 s operating current for all rated currents of general purpose fuses. Furthermore, in order to provide information on currents causing the fuses to operate in 0.4 s the time−current characteristics must also comply with minimum and maximum values of current at 0.1 s.

Typical requirements are shown in Table 16.2 and Fig. 16.12 illustrates the time−current 'gates' which are included in BS 88: Part 2.

Another important requirement of BS 7671 is the need for circuits of an electrical installation to have a voltage drop not exceeding 4% of the system voltage under normal service conditions. It has been established during consideration of this particular problem, that in the vast majority of cases compliance with the

Table 16.2 Gates for 'gG' and 'gM' fuses, extract from Table III of BS 88: Part 1. Limits of pre-arcing time of fuse-links.

I_n of gG and I_{ch} of gM (A)	I_{min} (10 s) (A)	I_{max} (5 s) (A)	I_{min} (0.1 s) (A)	I_{max} (0.1 s) (A)
16	33	65	85	150
50	125	250	350	610
100	290	580	820	1450
200	610	1250	1910	3420
400	1420	2840	4500	8060
800	3060	7000	10 600	19 000

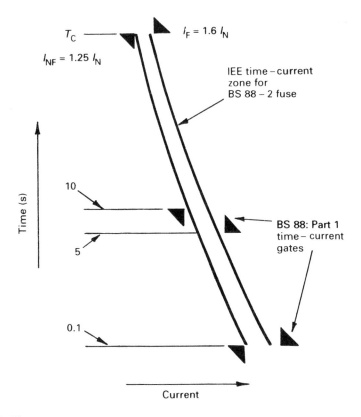

Fig. 16.12 Time–current gates in BS 88: Part 1.

4% voltage drop limit automatically ensures, in a TN system (an earthed neutral system), compliance with the values of Z_S given in the appropriate tables. In fact it can be shown that where a fuse is giving complete overcurrent protection to a circuit supplying fixed equipment installed in accordance with the requirements of BS 7671, the disconnecting time in the event of a negligible fault to earth is considerably less than 5 s and in the vast majority of applications less than 1 s. It can also be seen from the values of earth loop impedance given in Table 41B1 (of the Regulations) that in TN systems the hbc fuse is suitable for the protection of all socket-outlet circuits except socket outlets supplying equipment outside the equipotential bonding zone. Here the Regulations make the fitting of a residual current device (rcd) of 30 mA sensitivity a mandatory requirement and forbid the use of an overcurrent protective device for protection against electric shock for these particular socket outlets. It can still provide overcurrent protection to this circuit and back-up protection to the rcd.

In addition to the foregoing, the energy-limiting ability of the hbc fuse can be used to good effect in achieving economies in the size of the circuit protective conductor. Regulation 543–01–01 states two methods of selecting the size of this conductor. The first method is by calculation of the cross-sectional area using an adiabatic equation $S = \sqrt{(I^2 t / k^2)}$ which is applicable for disconnecting times not exceeding 5 s, and the second method relates the cross-sectional area of the circuit protective conductor to the associated phase conductor, the details of which are given in Table 54G of Regulation 543−01−04.

If the adiabatic equation is used, which takes into account the I^2t let through by the fuse, a much smaller size of protective conductor can be used, particularly when account is taken of the foregoing statement that in the vast majority of applications the fuse disconnects in times less than 1 s. The other important point is that, if protection is afforded to the protective conductor by a fuse under these conditions, it is assured on higher fault currents.

Even in motor circuits where the fuse is providing back-up protection and is therefore usually about twice the rating of the phase conductor, it can be shown that the use of the adiabatic formula results in smaller sizes of protective conductor than those specified in Table 54G.

One of the difficulties in applying the adiabatic formula is that it requires a knowledge of Z_e which is the value of earth loop impedance external to the installation. Where this value is known it is a simple matter to use the adiabatic formula. In those situations where the precise value of Z_e is not known, it can be shown that even where the value of Z_e is as high as 50% of Z_S, appreciable economies can be gained by using the adiabatic formula.

The foregoing confirms that on fuse-protected installations on TN systems existing practice usually achieves compliance with the new requirements for protection against electric shock by equipotential bonding and automatic disconnection of supply. This is applicable for fixed circuits, where the fuse is either giving complete overcurrent protection to a circuit or back-up protection to motor circuits, and also to circuits feeding socket outlets, except those socket outlets feeding equipment outside the equipotential bonded zone.

DOMESTIC FUSE APPLICATIONS

The fuse-links used in domestic installations are those complying with BS 1361: 1986 and BS 1362: 1986. Their characteristics are included in BS 7671 together with those of semi-enclosed fuses to BS 3036.

Fuse-links to BS 1362 for use in the 13 A plug are well established and widely used both in the UK and overseas, and the type II fuse-links of BS 1361 are the standard house service cut-out fuse-links used by the Electricity Association (EA). The BS 1361 fuse-link type I range (5 A to 45 A), provides a non-interchangeable range of fuse-links for use in domestic consumer units. They were initially launched in 1947 but due to lack of availability of replacement fuse-links, and the lack of effort in educating the general public, who had previously only been used to semi-enclosed fuses, the system was abused and the anticipated growth of this type of consumer unit was never achieved. It is interesting to note, however, that the fuse-links to BS 1361 which are used in domestic consumer units, have a breaking capacity of 16.5 kA which is the maximum fault level specified for domestic installations in the UK and discriminates with the 100 A house service cut-out fuse.

Furthermore, the BS 1362 fuse-link which is fitted in the BS 1363 plug is apparently used without any noticeable abuse, because the general public is now well educated in the use of cartridge fuse-links. The requirements of BS 7671 may well see an increase in the use of domestic consumer units incorporating cartridge fuse-links.

With the introduction of BS 7671, which required the determination of the fault level at the terminals of an installation, the EA declared that the fault level at the terminals of a 240 V single-phase supply in a domestic installation, could reach 16 kA. This posed an immediate problem for consumer units containing semi-enclosed fuse-links or mcbs, because their breaking capacities are considerably less than 16.5 kA at 240 V a.c.

Furthermore, the EA requirements at that time were that the final circuit protective devices should discriminate with the house service cut-out fuse. The impossible and impractical situation, created by the EA statement, has now been resolved by the EA permitting co-ordination between the cut-out fuse and downstream devices, and the modification of the time−current characteristics of the type II fuse-links of BS 1361. These two changes, coupled with an introduction of a test circuit representing typical domestic installations into BS 5486: Part 13, *Consumer units*, has solved the problem. It is now relatively easy to demonstrate that the combination of the house service cut-out fuse with its modified characteristics, with semi-enclosed fuses or mcbs of modest breaking capacity can satisfy the requirements of BS 7671 with regard to short-circuit protection.

The dimensions and performance characteristics of the BS 1361 and BS 1362 fuse-links are included in the requirements of IEC 269: Part 3, and the pending IEC 269: Part 3-1, covering fuse-links for domestic and similar purposes. When this work is complete in IEC, it is intended to introduce these requirements as BS 88: Part 3, and Part 3-1, respectively, to replace BS 1361 and BS 1362.

SEMICONDUCTOR FUSE-LINKS

BS 88: Part 4 covers semiconductor fuse-links. The basic requirements for such fuse-links (breaking capacity, etc.) are similar to those specified for industrial fuses. Because the main attributes of semiconductor fuse-links are their very low values of let-through I^2t and cut-off current, there are additional tests which must be made to ensure that the manufacturer's statements regarding such information as pre-arcing I^2t, total I^2t, cut-off current, arc voltage, etc. are verified.

The application of such fuse-links requires a knowledge of the semiconductor device and the type of circuit to be protected. Because of the precise nature of the application, more information is provided with regard to the performance of semiconductor fuse-links than for the conventional industrial or domestic types. These include variation of pre-arcing I^2t with prospective current, variation of total I^2t with voltage and power factor and variation of arc voltage with applied voltage. This information is required in order to optimize the capability of the semiconductor devices.

FUSE-LINKS FOR ELECTRICITY AUTHORITY NETWORKS

BS 88: Part 5 covers the range of l.v. fuse-links used by the supply authorities. As these fuse-links are dimensionally standardized, the specification includes these dimensions together with standardized time−current characteristics. The breaking capacity of these fuse-links is 80 kA at 415 V a.c. and the method of short-circuit testing is similar to that for industrial fuse-links although the time−current characteristics are somewhat faster at the short time end of the characteristic than the equivalent rating of fuse-links to BS 88: Part 2. Electricity Authority Standard EA 12-8 also contains these characteristics as well as information on the co-ordination achieved between these fuse-links and the h.v. fuse-links used in the 11 kV and 6.6 kV ring-main units.

A novel feature is the arrangement of the fuse tags. The fuse-links are accommodated in fuse-carriers which engage with the fixed contacts of the supply authority feeder pillar by means of a wedge action.

COMPACT FUSES TO BS 88: PART 6

With the international agreement now reached regarding standardized fuse characteristics, an innovation which has been introduced into the UK is a range of compact fuse-links and fuse-holders for applications up to 415 V a.c. for use in industrial and commercial applications.

This range of fuses extends up to ratings of 63 A for applications on both 240 V a.c. and 415 V a.c. circuits with a breaking capacity up to 80 kA. The characteristics align with the requirements of BS 88: Part 1 and they provide an economically attractive alternative to fuses to BS 88: Part 2 and to mcbs, for commercial and light industrial applications.

Figure 16.13 illustrates the dimensional benefits obtained from these compact designs.

Consideration is being given to the introduction of a larger size of fuse-link and fuse-holder with a current rating of 125 A. This will form part of a revision to BS 88 in the near future.

FUSE SWITCHGEAR

The combination of a switch and a fuse as a means of (a) providing short-circuit protection, (b) the switching of loads (both steady and fluctuating) and (c) means of isolation, dates back as far as the introduction of the fuse itself, but obviously the design of the modern types of such combinations meets the most onerous service conditions. These combinations fall into two categories, switch-fuses and the fuse-switches, and BS 5419 covers both types. However as BS 5419 is a reproduction of IEC 408 entitled *Low-voltage air-break switches, air-break disconnectors, air-break switch-disconnectors and fuse combination units*, it can be appreciated from such a cumbersome title that in IEC there was a definite need to differentiate between the switching function and the disconnection function in

Fig. 16.13 Dimensional comparison of fuses to BS 88: Parts 2 and 6.

such combinations. It is therefore worth examining the relevant definitions from these documents.

- *Switch (mechanical)*   A mechanical switching device capable of making, carrying and breaking currents under normal circuit conditions which may include specified operating overload conditions and also carrying, for a specified time, currents under specified abnormal circuit conditions such as those of short-circuit. It should be noted that a switch may also be capable of making but not breaking short-circuit currents.
- *Disconnector (isolator)*   A mechanical switching device which for reasons of safety provides in the open position an isolating distance in accordance with specified requirements. A disconnector is capable of opening and closing a circuit when either negligible current is broken or made or when no significant change in the voltage across the terminals of each of the poles of the disconnector occurs. It is also capable of carrying currents under normal circuit conditions and carrying for a specified time currents under abnormal conditions such as those of short-circuit.
- *Switch-disconnector (switch-isolator)*   A switch which in the open position satisfies the isolating requirements specified for a disconnector.
- *Switch-fuse*   A switch in which one or more poles have a fuse in series in a composite unit.
- *Disconnector-fuse*   A disconnector in which one or more poles have a fuse in series in a composite unit.

- *Fuse-switch* A switch in which a fuse-link or a fuse-carrier with fuse-link forms the moving contact of the switch.
- *Fuse-disconnector (fuse-isolator)* A disconnector in which a fuse-link or a fuse-carrier with a fuse-link forms the moving contact of the disconnector.

All of the foregoing combinations of a fuse and switch or disconnector are covered by another definition:

- *Fuse-combination unit* A combination of a switch, a disconnector or a switch disconnector and one or more fuses in a composite unit made by the manufacturer or in accordance with his instructions.

A number of important points emerge from these definitions. There is a definite distinction drawn between a device which performs the functional switching, i.e. making and breaking the circuit, and one which provides isolation. The earlier types of fuse-combination units probably only provided overcurrent protection by the fuse combined with an isolating function by the associated 'switch' but in more recent years, certainly in the UK, the popular combinations of either switch-fuse or fuse-switch have provided all of the functions mentioned earlier and the UK user has come to expect that such combinations provide him with the isolating function.

Discussion in IEC has revealed that this assumption is not necessarily valid with respect to equipment made internationally, because there are devices available, which, while performing the switching function, do not have the necessary requirements for satisfying the safety aspects of an isolating function. This is highlighted in BS 7671 in which Chapter 46 deals with isolation and switching, Section 476 of Chapter 47 describes the applications measures for isolation and switching, and Section 537 of Chapter 53 specifies the type of devices to be used.

Chapter 46 considers separately the functions of isolation, switching off for mechanical maintenance, and emergency switching. Section 537 states that for isolators the isolating distance between contacts or other means of isolation when in the open position shall not be less than those specified for isolators according to BS 5419 and that the position of the contacts or other means of isolation shall either be externally visible or clearly and reliably indicated. An indication of the isolated position shall occur only when the specified isolating distance has been attained in each pole. The same requirements are specified for devices for switching off for mechanical maintenance. It is also necessary in order to comply with the requirements of BS 7671 that such devices be provided with means for padlocking in the OFF position.

Section 463, *Emergency switching*, requires that for every part of an installation which it may be necessary to disconnect rapidly from the supply in order to prevent or remove a hazard, a means of emergency switching shall be provided.

Regulation 537−04−01 states that the device shall be capable of cutting off the full load current on the relevant part of the installation. Where appropriate due account shall be taken of stalled motor conditions.

Regulation 476−01−01 includes a statement that when more than one of these functions, i.e. isolation, or switching off for mechanical maintenance, are to be performed by a common device, the arrangement and characteristics of the device shall satisfy all requirements of these Regulations for the various functions concerned.

The modern combinations of fuse and switch are ideal for this purpose. The most popular arrangement is undoubtedly the fuse-switch although the switch-fuse is popular in circuits of small current rating. The test requirements of BS 5419 for these combinations are very rigorous. They include making on to a fault test where the switch shall be capable of making on to prospective r.m.s. currents as high as 80 kA, the peak current and energy let-through being limited by the fuse-link of maximum rating which can be associated with the switch. The switch itself must be able to withstand twenty times its thermal rating for 1 s. With regard to the overload make−break performance, there are various categories of duty, the most onerous of which is AC23, and this requires the switch to make and break up to ten times rated current at a power factor of 0.35 and 110% rated voltage. There is also a requirement for mechanical and electrical endurance where the switch has successfully to withstand up to 10 000 make−break operations with 5% of the operations being made at rated current and a power factor of 0.65.

There are many modern combinations of fuse and switch which can readily provide this performance up to 660 V a.c., having very compact dimensions. The full capability of the overload make−break performance of the fuse-switch can be utilized when motor circuit fuse-links are fitted. It can also provide all of the requirements for isolation by giving clear indication of the position of the contacts, adequate isolating distance and a padlocking-off facility. They are therefore ideal for use either adjacent to the equipment they are protecting or in a switchboard supplying a number of circuits. Figure 16.14 illustrates a very popular arrangement.

A new specification has recently been issued as part of a suite of specifications covering low-voltage switchgear and controlgear. BS EN60947−3: 1992 covering switches, disconnectors switch-disconnectors and fuse-combination units will run in parallel with BS 5419 until April 1998 when BS 5419 will be withdrawn.

Manufacturers have this period of time to re-test equipment to the new requirements. These new requirements are more onerous than the BS 5419 requirements with specific regard to motor switching duty. This arises from a detailed study of this duty by the IEC and CENELEC technical committees. It is further evidence of the essential role which fuse switches continue to play in modern electrical installations.

THERMAL RATINGS

One of the problems which has been encountered with fuse-switches is the fact that both the IEC and British Standards permit the thermal rating to be determined without an enclosure. This is known as a conventional temperature-rise test and can produce misleading results, particularly with the larger size of fuse-switch. British practice has always been based upon thermal tests conducted using the minimum size of enclosure normally associated with the fuse-switch and recent amendments of the IEC Specification make it mandatory for the manufacturer of the fuse-switch to state the enclosed current rating for his product. As

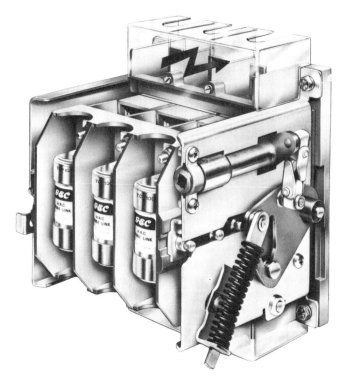

Fig. 16.14 100 A fuse-switch to BS 5419 suitable for use on system voltages up to 660 V a.c.

this has been common practice in Britain the amendment will be welcomed by the user as it eliminates a misunderstanding which could create overheating problems in service. There is some safeguard in the requirements of BS 5486, *Low-voltage switchgear and controlgear assemblies*, where it is necessary for the manufacturer to ensure when designing the switchboard that no item of included equipment is subjected to a condition which could result in higher temperature rises than those specified in its relevant specification. The above-mentioned amendment requiring a statement of the enclosed rating further emphasizes the need for close attention to this particular aspect.

The fact that such a compact relatively simple device as the modern fuse-switch can offer such a range of protection firmly establishes its future role in the design, installation and utilization of electrical systems.

CHAPTER 17
Motor Control Gear

C. Copestake

Revised by T. Fairhall, IEng, MIEIE
(Allenwest-Brentford)

Motor control gear and associated circuitry are fundamental parts of electrical distribution and utilization systems. Motor control gear, in its simplest form may be a direct on-line starter consisting of a switching device (contactor) and tripping device (overload relay) mounted within an appropriate enclosure.

Motor control gear has two main functions. It must make and break the connections between the motor and the supply and it must also automatically disconnect the motor (a) in the event of excessive current being passed which would overheat and ultimately destroy the windings, (b) if the supply fails, or (c) in combination with its external short-circuit protective devices (fuses or circuit-breakers), in the event of an earth fault.

In addition to the main functions described above, motor control gear may also be required to: limit the starting current to a value consistent with the requirements of the drive; apply braking when rapid stopping is required, reverse rotation or vary speed as necessary; and carry out automatic operating cycles or sequences.

It follows that the control gear must be appropriately selected to suit the characteristics of the motor and the drive.

Control devices for manual operation are normally pushbuttons or selector switches, alternatively automatic control of a starter may be achieved by use of a float switch, pressure switch, thermostat or other similar device; these are usually covered by the generic term 'pilot switch' on starter diagrams.

Actuation of the contactor (switching device) may be by a number of different methods, the most usual being electromagnetic, where the energization of a coil and magnetic system provides the force to close the spring-loaded contacts which switch the motor current. Although some hand-operated starters are available where finger pressure on the start pushbutton directly closes the switching device and others were available where the changeover from 'start' to 'run' was carried out manually. It should be remembered that these starters could not be remotely controlled, although in some instances the option of a remote stop was offered, also they may not provide no-volt protection so were unsuitable for machine-tool applications.

In order to provide a 'no-volt' release function and be able to provide remote control, an electromagnetic contactor is used in the majority of industrial applications.

Before considering in detail the different types of motor starter available and the criteria for choice, it is worthwhile examining the major components utilized in motor control.

CONTACTOR

The most widespread switching device used in a starter is the a.c. air-break contactor which consists of contact assemblies actuated by electromagnetic action. An operating coil is enclosed by the magnetic yoke, and when energized attracts an armature to which is attached a set of moving contacts which make with a set of stationary contacts. The rating of the contactor depends on the size, shape and material of the contacts and on the efficiency of the arc extinction method used. Modern contactors use a silver alloy contact tip, normally silver–cadmium oxide or silver–tin oxide alloy attached to a brass or copper backing strip. The choice of tip material is critical, and is normally established after many type tests. The contact materials are selected primarily for their weld and erosion resistance – the suitability of the material chosen is then confirmed by a sequence of type tests and special tests, such as contact life tests.

The silver alloy typically contains 10–12% cadmium oxide or tin oxide; the use of silver–tin alloys is now the first choice at the design stage as the cadmium can require special precautions during manufacture, and ultimately in the disposal of the contacts.

Modern contactors use double-break main contacts, usually of butt-type design, with either circular or rectangular contact tips. It must always be remembered that the attachment to the backing strip requires a silver enriched backing – this means that older maintenance techniques such as filing of contact faces actually reduce contact service life and will eventually expose the silver backing; this may permit contact welding.

The method of arc control is also critical in determining the performance of the contactor; generally smaller a.c. contactors of up to 22 kW rating do not require complex arc chamber design – the combination of the natural current zeros of the a.c. supply and the 'stretching' of the arc by the opening of the contacts gives adequate performance.

Larger sizes of contactor usually require the use of cooling devices inside the arc chamber to assist with arc extinction. These may take the form of cooling plates or shrouds which simply enclose the contacts or an array of de-ion plates similar to those in a circuit breaker. The cooling plates or de-ion plates do not need to be selected for their electrical conducting properties; in most cases the material used should have a relatively high melting point and mild steel can be used.

In all cases the objective is to put the arc out within typically 10–20 ms when breaking currents of eight times the rated AC3 value under type test conditions, and 5–10 ms in normal service. The relatively fast arc extinction is a major consideration where long service life (termed durability in British and International standards) is a design objective; an AC3 contact life of 1−2 million operations can be achieved with modern designs.

Another choice the designer must make is that of insulating materials. These not only act as a mounting base but as the mechanical slides and guides and also the wall of the arc chamber within the contactor. A rule of thumb is that most moulded components in contact with current-carrying parts will be produced from thermoset materials, usually a polyester glass material stable up to at least 160°C, but the use of recently developed high temperature thermoplastic materials is increasing. The use of asbestos has now been phased out by most manufacturers with replacement materials being adopted for the continued production of existing designs.

The contactor is magnetically held closed by maintaining the current flow through the coil. If the voltage to the coil fails or falls below a defined level, the contactor opens, thus disconnecting the motor from the supply. The coil must be constantly energized in order to keep the contactor closed. Alternatively, contactors can be fitted with a mechanical latch which does not require continuous energization.

In order to provide quiet operation, a.c. magnets are fitted with shading rings which normally consist of a single short-circuited loop of copper alloy. They fulfil the basic function of providing a secondary magnetic flux to prevent the magnetic pole faces parting and reclosing when the flux produced by the operating coil passes through zero. If a shading ring should be omitted when reassembling some types of contactor after maintenance, the omission will be obvious as the magnet will 'hum' loudly when energized. This same situation may also arise if a shading ring should break in service, although this is unlikely as modern contactors have captive shading rings of proven design.

A further advantage of well-designed shading rings is the reduction or elimination of magnet bounce which contributes to contact bounce during closing of the contactor; if a contactor has severe magnet bounce short contact life in service will result.

The remaining design parameters such as the ability of terminals to accept cables, service temperature rise limits, the ability to switch currents under defined conditions and the performance under short circuit conditions are now determined by British and International standards, the existing standards for low voltage motor control gear.

IEC 292: BS 4941, *Low voltage motor controlgear for industrial use*: Part 1, *Direct on line starters*. (Parts 2 to 4 covered specific requirements for reduced voltage and rheostatic starters).

IEC 158: BS 5424, *Low voltage motor controlgear for industrial use*: Part 1, *contactors for voltages up to and including 1000 V ac and 1200 V dc*.

These have now been replaced by a comprehensive document IEC 947 published in the UK under CENELEC rules as BS EN60947.

IEC 947 and BS EN60947 are published in several parts which are to be read in conjunction. In the case of the devices under consideration in this chapter, the applicable parts of the documents are:

IEC 947: BS EN60947, *Low voltage switchgear and controlgear*: Part 1, *General rules* (this contains requirements common to other low voltage products such as switches and circuit breakers) and Part 4, Section 1, *Electromechanical contactors and starters*.

CONTACTOR SELECTION

Contactors for use in direct on-line starters are normally selected by their AC3 rating, that is the switching on of a cage type induction motor and switching off the supply to the motor after the motor has run up to full speed. The other most common utilization categories are AC4, the switching on and off of a cage type induction motor before it has run up to full speed, sometimes termed 'inching' or 'jogging' the drive, and category AC2, the switching of the stator supply to a wound rotor motor where the starter circuit automatically inserts resistance into the rotor circuit for each start. Figure 17.1 shows a star-delta starter.

The catalogued ratings are published on the basis of known service conditions typically:

- Ambient temperature outside the starter enclosure between $-5°$ and $35°C$ average (with a maximum not exceeding $40°C$).
- Rate of operation to be specified by the manufacturer, typically 120 starts per hour.
- The duty cycle or utilization category, see Table 17.1 for typical values of current to be switched during type testing and in service, e.g. is the starter a reversing type which must reverse a motor which is already run up to speed, or does the reversing only occur after allowing the motor to come to rest? In the latter case only AC3 rated contactors could be used.
- The acceleration time of the drive, the ability of the contactor to carry the starting current must be considered. AC3 rated contactors are able to carry eight times rated current for 10 s (this applies to ratings of up to 630 A, above this the value is six times rated current).
- Any special contact life requirement.
- The type of short-circuit protective device to be installed in series with the supply to the starter, and the classification of type of protection to be obtained.
- Any special co-ordination requirement, e.g. are there any residual current devices, which on detecting a fault condition, may attempt to open the contactor on a current in excess of its breaking capacity.
- Any special requirement to attach cables of other than copper cored, PVC or rubber insulated types to the contactor terminals, e.g. the use of some types of high temperature insulation such as XLPE allow the cable to run much hotter than the manufacturer expected. The normal practice is to design for use with 70°C cable so the possible economies of using XLPE may be not be exploited as the core temperature may be as high as 250°C, consequently the cabling would act as a heat source, not a heat sink.

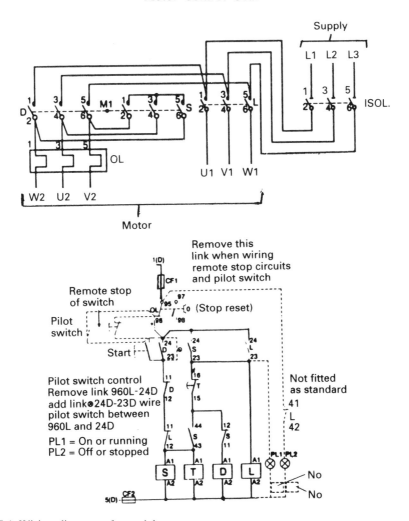

Fig. 17.1 Wiring diagram of star-delta starter.

Any of the above may need special consideration and may necessitate the selection of a contactor of a higher AC3 rating than catalogued values would initially suggest.

Final or run contactors used in auto-transformer starters, Fig. 17.2, should be selected using AC3 ratings; star and intermediate contactors should be selected in accordance with catalogue recommendations.

For stator-rotor starters, the stator contactors should be selected using AC2 ratings. Rotor contactor ratings are normally specified by the contactor manufacturer as enhanced ratings based on them being in circuit only during starting.

Another important consideration when selecting contactors is to ensure that the proposed cable conductor can be accommodated by the contactor terminals. Most manufacturers supply this information in their catalogue together with the differ-

Table 17.1   Contactor utilization categories.

| | Current as multiples of operational current (I_c) | | | |
| | Normal operation | | Proving operation | |
Utilization category	make	break	make	break
AC1 Non-inductive or slightly inductive loads such as furnaces and heating loads	1	1	1.5	1.5
AC2 Starting of slip ring motors. Plugging with rotor resistance in circuit	2.5	2.5	4	4
AC3 Starting of cage motors, switching of motors during running	6	1*	10 (8)	8 (6)
AC4 Starting of cage motors, plugging inching	6	6	12 (10)	10 (8)

All tests carried out at supply voltage for normal operation except as indicated below and at 105% voltage for proving operations. For full details and power factors, etc. refer to specification IEC 947/BS EN60947.
Values in brackets applied to type testing to IEC 158 BS 5424, for devices of I_c in excess of 100 A.
* At 17% supply voltage.

ent types of termination that are available. A range of contactors is illustrated in Fig. 17.3.

PRODUCT STANDARDS FOR CONTACTORS

Previously, contactors have been marketed as meeting IEC 158, BS 5424, or even BS 775. The current product standard as mentioned previously is IEC 947 published in the UK as BS EN60947. These standards all laid down general design and performance requirements for the products and included various design or 'type' tests to verify the ability of a device to perform the duty for which it is sold. The test currents and typical service switching currents for industrial motor control applications are summarized in Table 17.1. It should be noted that IEC 158 and BS 5424 allowed contactors of current ratings in excess of 100 A to be tested making eight times the rated current and breaking six times the rated current in the case of utilization category AC3.

An individual sample contactor must be able to complete a test sequence in

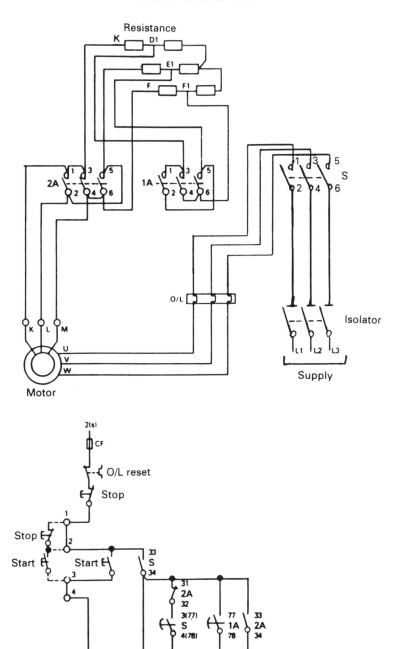

Fig. 17.2 Simplified wiring diagram of stator-rotor (or rotor resistance) starter.

Fig. 17.3 Typical range of contactors covering three-phase AC3 ratings 4–400 kW.

order to confirm its rating. For a contactor intended to be used in an AC3 application, the test sequence would consist of:

- 50 operations making ten times the rated current, 25 operations to be performed with the coil energized at 85% of rated voltage followed by 25 operations with the coil energized at 110% rated voltage. Followed by:
- 50 operations making and breaking eight times the rated current at 105% of rated voltage, followed by:
- 6000 operations making and breaking two times the rated current, followed by a dielectric test.

Any failure of the test item during this test sequence would mean that the complete test sequence must be repeated on a new test sample.

The test currents, voltages, power factors and time intervals between test operations together with the test sequence requirements for other utilization categories are specified in IEC 947 or BS EN60947 which should be consulted for full details of test requirements.

Vacuum contactors

Modern power switching equipment usually includes a double-break air-break contactor. An alternative to this is the vacuum contactor, which is frequently used below 1000 V where the switching duty is arduous and is very widely specified for 3.3 kV and above, as it is more compact than an air-break device of equivalent rating.

Figure 17.4 shows a typical unit, and it is constructed with three vacuum bottles and a d.c. magnet on the reverse side. The construction of the bottle, Fig. 17.5,

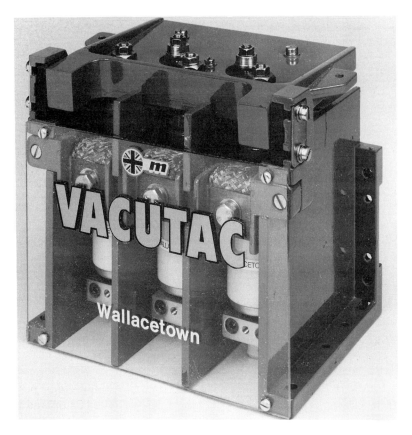

Fig. 17.4 Typical compact vacuum contactor.

consists of a ceramic envelope with a metal top and bottom. The bottom is the fixed contact and the top the moving contact with bellows attached. There are collector shields for any splutter which may arise. The contact tips, shown hatched, contain annular rings of impurity, the introduction of this impurity reducing the current chop level. Early vacuum contactors gave a fairly high current chop level of 1–2 A. This high current chop level generated h.v. spikes and surges, so certain impurities were introduced into the contacts which allowed the current chop level to be brought down to about 0.5 A. This is very important, and the reason for this in a vacuum contactor of the hard vacuum type is that when the contacts start to part, and despite the hard vacuum, an arc is formed. This arc is stretched by the operation of the moving contact to create an envelope containing a cloud of ionized gas between the two contacts. In due course that envelope is stretched, and through a magnetic and thermal effect conducts until such point that it snaps and the electrons are then diffused on to the cold face of the anode. Once the ionized gas path snaps, the current flow stops very quickly without persisting until a current zero, consequently current chopping is severe. By introducing an impurity on to the contact face when the arc is formed, the cloud of gas

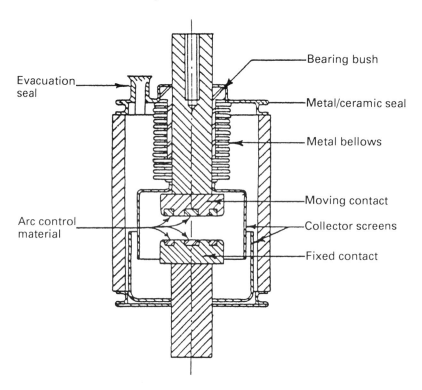

Fig. 17.5 Section of vacuum contactor bottle.

is thickened by the impurity and allows the conduction of current to carry on almost down to current zero. It is only when current zero is approached that the arc extinguishes and a very low current chop level is attained. This is the principle which has been used in the design of vacuum contactors for some time now, and avoids the fitting of any form of suppression devices.

The control circuit is slightly different from the normal air-break contactor circuit, in so much that there is a relay to drive the contactor, and the contactor is operated through a tap-changing system. To attain the required pick-up and drop-off characteristics, the contactor is run at a lower voltage than the pick-up voltage. The driving relay has contacts in both the input and output circuit of the bridge rectifier, to give faster drop-off characteristics, otherwise free-wheeling effects within the circuit itself would reduce the opening speed of the electro-magnet.

It is not possible to design a contactor where every contact opens at the same time. When the first contact opens, leaving the other two contacts a single-phase circuit, one of these will then open the circuit completely. The chances of a failure of the vacuum contactor are remote but contamination is one of the big problems, and for safety reasons a loss-of-vacuum detector or a resistance leakage detector is used, i.e. a common unit monitors the three outgoing lines and feeds into the detector unit, which latches and trips the circuit-breaker supplying the contactor.

A leakage of approximately $100\,k\Omega$ will cause the unit to operate and trip the back-up protection.

The vacuum contactor gives long life with minimal maintenance. The only preventive maintenance required is to check the setting of the device. There are no contacts to replace.

OVERLOAD PROTECTION

Together with the contactor(s) the second major component of any motor starter is the overload relay. The widespread use of CMR motors having a limited overload capacity make selection of the correct type of overload relay a major consideration for a motor starter.

The types of relay available are: thermal, electronic, magnetic, and thermistor. The characteristics of thermal, electronic and magnetic relays normally comply with the requirements of BS 4941, IEC 292 also with IEC 947 and BS EN60947. The performance criteria are not the subject of an individual product standard but are included in the 'motor starters' standards.

Thermal relays

A thermal relay normally consists of three bimetallic elements, one in each phase, each of which is heated by the motor current. The method of heating the bimetals may be direct, where the bimetal element carries the motor current or indirect where the motor current is carried by a heater element and the heat transferred to the bimetal through an insulating sleeve or small air gap.

Ambient temperature compensation is provided by a fourth bimetal, usually having the same deflection sensitivity as those which are heated, placed in air inside the relay body and adjusted so that the same deflection of the heated bimetals is required to cause a trip whether the starter is in a high or low ambient temperature. It also compensates for the temperature rise of the air inside the starter case and means that the user does not have to calculate derating factors and set the overloads accordingly.

Most modern thermal overload relays incorporate protection against loss of a phase, a faster trip being obtained by the use of a differential trip mechanism which trips as the de-energized bimetal cools and moves back to its rest position. An inherent feature of all thermal relays is a thermal memory which provides shorter tripping times when an overload is applied following a period of normal running than the tripping time starting from cold. A typical time−current characteristic curve is shown in Fig. 17.6 and a typical relay is shown in Fig. 17.7.

A second function of this thermal memory is that the relay cannot be instantly reset upon removal of the current; a cooling period must always be allowed to elapse. This delay in resetting is typically 3−5 minutes.

Thermal overload relays are usually suitable for operation by current transformers, as specified by the manufacturer; these enable isolation from the incoming supply voltage, operation over a wider current range and some modification of the

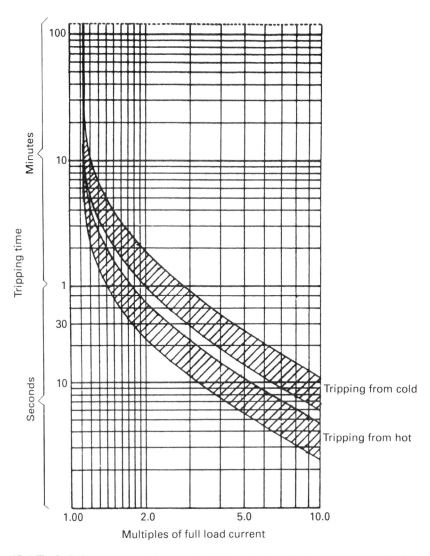

Fig. 17.6 Typical time–current characteristic for tripping of thermal overload relay.

time–current characteristic, for example to increase the run-up time available without the overload relay tripping.

Electronic relays

Electronic overload relays offer several advantages over conventional thermal or magnetic overload relays, including:

- faster responses to three-phase overloads;
- faster response to phase loss;

Fig. 17.7 Typical small thermal overload relay covering FLC range of 0.14−40 A direct connected.

- trip time−current settings are user adjustable;
- other features such as visual indication of overload current are available on mid-priced units.

A further advantage of electronic overload relays is that they dissipate less power than an equivalent thermal or magnetic relay so there will be a lower temperature rise in a starter enclosure where an electronic relay is fitted.

Magnetic relays

A magnetic relay employs a coil of wire in each pole of the motor supply; when current is passed through these coils the resulting magnetic field attracts plungers which when lifted actuate a trip switch. In order to cater for the starting current of a motor a restraint device such as an oil dashpot is provided. When installing a magnetic overload relay, care should be taken that the dashpots are filled to the recommended level with the fluid provided by the manufacturer. If nuisance tripping should occur in practice, the manufacturer should be consulted with regard to the use of an alternative fluid in the dashpots. In general vegetable or mineral oils should not be used, silicone fluids only are recommended. It should also be remembered that magnetic relays lack a differential trip mechanism and so will not provide such a high level of protection against loss of a phase as an electronic or thermal relay. Magnetic overload relays do not have a thermal memory, so may be reset immediately following a trip. This is an advantage in cranes and lifts where a minimum delay before restarting is required. It is essential that regular maintenance checks of dashpot fluid level and cleanliness are made.

Thermistor relays

Thermistors are used to detect the temperature of the motor windings. A thermistor is a thermally sensitive semiconductor resistor, which exhibits a significant change in electrical resistance with a change in temperature. A detector monitors the sudden increase in resistance of the thermistor at the critical temperature. Thermistors must be fitted by the machine manufacturer who knows the positions where hot spots occur on the machine.

It should be remembered that thermistor protection has the following disadvantages.

(1) Special motors required, thermistors cannot be retrofitted.
(2) Additional control cables required between starter and motor.
(3) Limited protection for cables between starter and motor. They do not have close excess current protection as with a current sensing overload relay.

POPULAR STARTERS

A knowledge of starter types is required to enable one to select the correct unit for a specific application. Choice of the right starter is important because failure to do this can drastically reduce the working life of a drive and result in un-economical operation. A modern range of starters is shown in Fig. 17.8.

Direct-on-line starter

The simplest and least expensive form of starting is direct-on-line (dol) and this is generally the standard choice for motors up to 7.5 kW. Subject to supply authority agreement, larger motors may be dol started.

Fig. 17.8 Typical range of small starters in deep drawn steel enclosures to IP 55.

The switch-on current is quite high, up to eight to ten times the motor full load current, falling off as the motor accelerates to normal running speed. As well as the current, the initial torque is also high, of the order of 125% of the full load torque. This high level of starting torque can produce rapid acceleration and may be detrimental with belt or chain driven loads or if starting is frequent.

Star-delta starter

If direct switching is not allowed, a method of reduced voltage starting may be used in order to limit the motor starting current. The most popular method of reduced voltage starting is star-delta. The starting current is approximately one third the value of dol starting current (about twice full load current) and the starting torque is one third of that with dol starting. A six-terminal motor must be used to enable the winding connections to be changed during the starting sequence.

The connection of the overload relay in a star-delta starter is important. Consider the relationship of currents during starting and running of a typical motor, having a full load current of say 10 A and taking twice full load current (flc) in star at start.

An overload relay connected in the line and selected on a basis of 10 A flc would experience a current of 20 A or 200% during starting. However, the motor windings are rated at only 5.8 A and are subjected to a current of around 350%. It follows that for correct motor protection during starting the overload should be connected in series with the motor windings during both starting and running.

Occasionally, tripping times may prove to be too short at start with the overload relay phase connected as recommended above; reverting to line connected may solve this problem. Any deviation from the theoretically correct method of protection should only be undertaken if it has been established that the motor can withstand the more onerous conditions placed on it. If very long accelerating times are experienced, the use of saturable current transformers may be necessary. As an alternative, the overload relay can be connected in series with the delta contactor, but this will provide no protection during starting. This disadvantage can be overcome by fitting a second overload relay in series with the line contactor, but rated at the higher current.

Auto-transformer starter

Another reduced voltage motor starting method is that of using an auto-transformer, which has significant benefits but is relatively expensive. It is particularly useful where progressive control of accelerating torque is required, for example with large fans or mixers, and is often used with submersible borehole pumps.

The starting voltage at the motor terminals (and therefore the starting torque) can be selected to suit the load by means of tappings on the auto-transformer – the usual values being 50, 65 and 80%. By using the Korndorfer system, the motor is not disconnected from the supply during the transition period in which the auto-transformer winding acts as a choke, thus eliminating the surge currents that would otherwise occur.

The above starting methods are summarized in Table 17.2 and it will be seen that the reduced voltage starters provide a reduction in starting torque.

Table 17.2 Summary of starting methods for cage motors.

Method	Starting torque % of dol value	Starting current % of dol value	Nature of start	Adjustment of torque	Advantage	Disadvantage
Direct-on-line	100	100	Quick starting at max. torque	Fixed (max.)	Cheapest, simplest, most reliable	Heavy current, heavy torque
Star-delta	33⅓	33⅓	Light starting heavy running torque	Fixed	Low cost and simple, reduced starting current	Fixed torque, six terminal motors
Auto-transformer (Korndorfer) *Tap*: 50% 65% 80%	25 42 64	25 42 64	Light Normal Heavy	Tappings selected to suit load	Smooth starting, transient peaks reduced	Costly

Stator-rotor starter

The slip-ring motor has the advantage that, while developing approximately the same starting torque as an equivalent size cage induction motor, the starting current is much lower. This type of motor is therefore largely used for applications where starting current restriction is essential and a relatively long time is taken to reach full speed due to the high inertia of the load. A stator-rotor starter is used only with a slip-ring motor. With this method the stator is switched direct to the full supply voltage but the current drawn from the supply is limited to typically between 125 and 200% of full load current by external resistors connected to the rotor windings. In addition to limiting the starting current, the resistors improve the power factor so that the current values given above produce starting torques of between 100 and 175% of full load value. As the motor accelerates, resistor sections are progressively cut out by rotor contactors closing in sequence under time control. The more rotor contactors, the lower the demand on the power supply and the smoother the acceleration.

Variable speed drives

In many industrial processes it is necessary to vary the speed of the machine depending upon the conditions prevailing at a specific time. Traditionally for low-power machines this has been achieved by running the mains connected motor at a fixed speed and interposing some device, such as slipping clutch, variable belt, or gearbox between the motor and the load, or for higher kW drives by schemes such as Ward Leonard. This can be achieved by using a variable speed drive which can provide an optimum process speed, and may introduce an element of energy saving.

The more widespread types of drive are a.c. and d.c. thyristor systems. The most usual application of thyristors is in a.c. drives where the thyristor provides a variable conduction angle which enables the amplitude of the output to be controlled between a maximum and zero. Thyristors can be connected in various converter configurations depending on the requirements of the load and electrical network.

When connected to an electrical network, a thyristor drive equipment loads the system in a manner that is different from conventional loads such as motors and lighting circuits. This difference is primarily associated with the following two characteristics of these converters, which both contribute to a distorted load current waveform:

(1) high reactive power consumption;
(2) harmonic currents arising owing to the non-linear impedance characteristic of the converter; these may give rise to electromagnetic interference problems in service.

With large converter power ratings both the nature of the supply network and the duty of the converter must be thoroughly studied so that the most suitable type of power factor correction or harmonic filtering can be selected. It should also be remembered that fuse selection for thyristor drives is more critical than contactor switchgear and that only fast acting 'semiconductor' fuses or other fuses specifically approved by the drive manufacturer can be used. For a fuller discussion of variable speed drives, see Chapter 15.

ISOLATION

It is a requirement of the Electricity At Work Regulations that equipment must be effectively isolated from its power supply before any maintenance, fault finding or modification is made.

The regulations also contain the 'essential requirement' that precautions are taken to prevent accidental reconnection of the supply while work is in progress. This is normally achieved by locking off the isolating device of the equipment receiving attention; it must be remembered that it is not simply a matter of attaching a padlock to the isolator handle – the locking must be verified by attempting to turn the isolator on with the lock in place. Where a number of persons are engaged in the work, each must hold their own padlock and if necessary an adapter fitted to accept the appropriate number of padlocks simultaneously.

The isolating device within a starter may typically be a triple-pole switch, a moulded case circuit-breaker, a fuse switch combination or an air circuit-breaker.

The most common device is the air break triple-pole switch which will be mechanically interlocked with the front door or cover of the starter. Where an isolating device is not fitted within the starter, the Regulations may be satisfied by locking out the switching devices feeding the starter. It is important to ensure that in addition to the isolation and locking off of the main power supply to the starter, consideration is given to any remote control, signalling or sequencing connections between the starter and any external equipment. Modern starters will incorporate auxiliary poles on the isolating switch to remove any potentially dangerous voltages from the control circuit at the same time as the main circuit, but it is important to verify the lack of power by means of a voltmeter prior to actually touching the internal components. If it is necessary to work on an existing starter of an older design, it may be that the differing regulations in force at the time of installation did not require the same level of isolation, so extra care should be taken prior to working.

It is current practice to fit shrouding to IP 20 minimum on all incoming terminals on isolators, and possibly over the main poles on contactors, overload relays and fuse-holders.

A further difference in design practice is that of the door interlock function between the isolator handle and door opening – the established practice in all cases is that the door cannot be opened with the isolator in the on position, but subject to the age of the equipment, it may be possible to insert the padlock in the isolator handle and lock the handle in this position but still open the starter door; modern practice is for the locking off to also lock the starter door closed.

Motor control centres

Modern installations and process plant require many motor starters and it is therefore often convenient to group these together in a single cubicle switchboard known as a motor control centre (mcc), typically shown in Fig. 17.9.

Motor control centres will have been type tested by the manufacturer to confirm compliance with BS 5486, IEC 439 and EN60439.

The general arrangement of most mccs consists of a horizontal air-insulated, copper (or possibly aluminium) busbar system with vertical droppers to the various components. Cable entry may be available from the top, or bottom with front and rear access.

Incoming supply cables may be connected to the busbar system via air circuit-breakers, moulded case circuit-breakers or switch-fuses. In association with any of these devices, mccs sometimes include bus-section switchgear, interlocking and metering systems to meet specific customer requirements.

The starter units are contained in standard sized cubicles which may be of a fixed pattern, for larger drives or withdrawable for drives up to around 160 kW.

A withdrawable starter has the advantage of being readily removed from service and replaced by another starter of suitable type and rating, thus ensuring minimum shutdown time on vital drives in the event of a failure. A further feature of the withdrawable starter is that the make up of the mcc (depending on the design) can be readily modified to suit changing process requirements.

Fig. 17.9 Motor control centres form a compact means of housing many starters.

Short circuit co-ordination

The behaviour of a combination of short-circuit protective device (SCPD) and starter under short-circuit conditions were defined in BS 4941 and IEC 292.

Three categories of damage to starter were accepted:

Type 'a' Starter interior may need total replacement after a fault.
Type 'b' The overload relay may be affected by the fault current and need replacement.
Type 'c' No damage to starter components, starter to be re-usable after the occurrence of a short circuit.

For all categories of pass the case of the starter must not become live during the fault, i.e. must not present an electrical shock hazard to an operator, and the enclosure must be intact, i.e. displacement or opening of doors or covers is not permitted. For types 'b' and 'c', the contacts of the contactor(s) can be lightly welded, provided the welds can be broken.

It is therefore essential that after a short-circuit the installation is inspected and that the mechanical movement of all switching devices is checked with the supply isolated prior to replacing fuses or reclosing the mccb feeding the starter.

Most manufacturers have carried out tests to demonstrate type 'c' co-ordination of starters with their recommended fuses or circuit breakers; in order to maintain the level of protection the recommended fuse ratings or mccb setting should not be exceeded without the manufacturer's approval.

The same restrictions on selection of short circuit protective device also apply to types 1 and 2 co-ordination to IEC 947 and BS EN60947. These standards have revised the test method to a two part short-circuit test sequence where the performance of, typically, the starter and fuse combination are checked at two fault levels referred to test currents r and q. In common with the other tests specified in these standards the short-circuit tests must be performed as a sequence; all parts of the test sequence must be 'passed' for certification to be possible.

The values of test current are derived from a table in the case of test current r, the selection is based on the rated current of the starter being tested. As an example, devices of rated currents up to and including 16 A are tested at a prospective fault current of 1 kA and then the test repeated at test current q which is the full fault level; the value is typically 50 kA, but actually subject to agreement between the manufacturer and user.

The starter must still satisfy the same operator safety criteria as before. The standards now give more detailed guidance on the evaluation of the test samples after the tests, and the post-test checks to demonstrate the suitability of the devices for further service also apply.

The same pass or fail criteria generally as in the old types 'a' and 'c' co-ordination still apply; as an approximation type 1 is equivalent to type 'a' and type 2 is equivalent to type 'c'. It should be noted that a type 'b' pass is no longer recognized, this is now a form of type 1 pass. The possibility of contacts tack welding during these tests is still recognized, so the recommendation to examine contactors after any operation of main fuses or tripping of main circuit breakers must be observed.

CHAPTER 18

Lighting

R.L.C. Tate, FCIBS

Revised by H.R. King BA (Hons)
(Thom Lighting Ltd)

THE NATURE OF LIGHT

The electromagnetic spectrum

Light is a form of electromagnetic radiation. It is basically the same thing as the radiations used in radio and television, as radiant heat and as ultraviolet radiation and the still shorter X-rays, gamma rays, etc. Visible light is radiation in that part of the spectrum between 380 and 760 nm, to which the human eye is sensitive. A nanometre is a wavelength of one millionth (10^{-6}) of a millimetre. Within these limits, differences of wavelength produce the effect of colour, blue light being at the short-wave and red at the long-wave ends of the visible spectrum. Because the human eye is more sensitive to the yellow and green light in the middle of the spectrum, more power must be expended to produce the same effect from colours at the ends of it. This is why the monochromatic low-pressure sodium lamps which emit all their visible energy in two narrow bands in the yellow region are more efficient in terms of light output than 'Northlight' fluorescent tubes that imitate natural daylight pretty closely and emit approximately equal packets of energy in each spectral band.

The areas of radiation on each side of the visible spectrum are important to the lamp maker. Wavelengths shorter than those of violet light are designated 'ultra-violet' or UV and have the property of exciting fluorescence in certain phosphors. That is, when irradiated with UV, they themselves produce visible light. Infrared radiation, at the other end of the spectrum, produces heating effects and may be considered simply as radiant heat. Where, as in the case of a filament lamp, light is produced by heating a coil of wire to incandescence, far more infrared radiation is produced than visible light, and this is radiated and may be reflected with the light.

Reflection

Light is reflected from a polished surface at the same angle to the normal that it strikes the surface. A truly matt surface scatters light in all directions, while a semi-matt surface behaves in a manner in between the two. It must not be forgotten that all reflectors absorb some of the light that falls upon them, transforming it into heat. Even super-purity polished aluminium absorbs about 5% of the light, while a white-painted semi-matt reflector will absorb as much as 20% or more.

Concentrating specular reflectors are usually parabolic or elliptical in cross-section. If a compact light-source is placed at the focal point of a parabola, its light will be reflected as a parallel beam, that from a source placed at one focus of an ellipse will be reflected through the second focus, while a hemispherical reflector will reflect light back through its own centre. These principles are made use of in the design of various types of spotlights, film projectors and high-bay industrial reflectors, Fig. 18.1.

Refraction

When light passes from one clear translucent medium to another, e.g. from air to water, or air to glass, it is 'bent' or refracted at the surface. This can be observed by dipping a straight rod into water, which will appear to be bent at the point of entry. This property is made use of in lenses and refracting panels in lighting fittings. Again, it is important to remember that some light will be lost in passing through the medium and some will be reflected from its upper surface, to a greater or lesser degree according to the angle at which it strikes that surface.

Diffusion

If light passes through a surface that is partially opaque, it will be scattered or diffused in all directions. Typical diffusers are sand-blasted or acid-etched glass, flashed opal glass or opal acrylic sheet. The distance the light-source is mounted from the diffuser, the opacity of the latter and the nature of the background or reflecting surface will determine whether the light-source appears as a bright spot of light or its light is evenly diffused over the surface. Light losses in a 'near perfect' diffuser are greater than those in a partial diffuser or a prismatic panel.

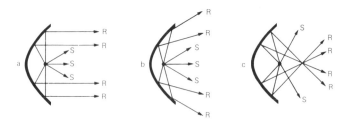

Fig. 18.1 Concentrating specular reflectors are used in the design of various types of spotlights, film projectors and high-bay industrial reflectors. Light (R) reflected from a reflector of parabolic cross-section will produce a parallel beam when the source is at the focus of the parabola (a), a divergent beam if behind the focal position (b), and a crossing beam if it is in front of it (c). Spill light (S) is not controlled by the reflector.

UNITS OF LIGHT MEASUREMENT

The unit of luminous intensity is the candela. This is the amount of light emitted in a given direction by a source of one candle power. From it is derived the lumen, the unit of light flux. This is quite simply defined as the amount of light contained in one steradian from a source of one candela at its focus, Fig. 18.2. The light output of all electric lamps is measured in lumens and their luminous efficiency (efficacy) is expressed in lumens per watt.

The unit of illuminance (measured illumination) is the lux. This is the illumination produced over an area of one square metre by one lumen. The 'foot-candle' or lumen per square foot has long been an obsolete term in the UK, but is still used in America.

Measured brightness is termed 'luminance' and should not be confused with 'illuminance'. Its units are the candela per square metre and the apostilb, the lumens emitted by a luminous surface of one square metre. Imperial units, now obsolete, but still occasionally quoted, are the candela per square foot and the foot-lambert.

Two other terms that are easily confused are 'luminance' and 'luminosity'. The first is measured brightness expressed in apostilbs or candelas per square metre, the second the apparent brightness as seen by the eye. A simple example is the appearance of motor-car headlamps by day and by night. Their luminance is the same in both conditions but their luminosity is far greater at night than when they are seen in daylight.

ELECTRIC LAMPS

Lamps produce light in three ways: by incandescence, by the excitation of metallic vapours in an electrical discharge, and by fluorescence initiated by a discharge in mercury vapour causing radiation in the ultraviolet region of the spectrum.

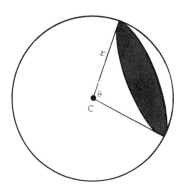

Fig. 18.2 The relationship between the candela and the lumen. If the shaded area $= r^2$ and a source of one candela is at the centre C, the light contained within the solid angle θ is one lumen.

All modern filament lamps consist basically of a coil of tungsten wire enclosed in a glass bulb generally filled with an inert gas. Tungsten is used because of its high melting point; most filaments operate at temperatures in the region of 2700°C. At this temperature the surface of the filament evaporates slowly, until a point is reached where it cannot support its own weight and fractures. The rate of evaporation is controlled by the gas filling, usually a mixture of argon and nitrogen, which in general service (gls) lamps exerts a pressure of about half an atmosphere on the surface of the filament. Evaporated tungsten settles on the inner surface of the glass bulb, discolouring it and reducing the amount of light emitted by the lamp throughout its life. The filament is coiled to reduce heat losses by convection currents in the gas; coiling substantially reduces the area exposed to them.

In the case of all standard mains voltage lamps made to BS 161: 1976, the life of the filament is 1000 h. Some manufacturers make a 'double-life' lamp in which the filament is run at a slightly reduced temperature, giving a 2000 h life at the cost of reduced light output. Lamps made for special purposes, such as projector lamps often have a shorter life.

Reflector lamps

Blown glass reflector lamps were first introduced in the late 1940s and are now made in sizes ranging from 40 to 150 W. Pressed glass lamps with a more accurately profiled reflecting surface followed in the mid-1950s. The greater efficiency of the reflector allowed the filaments to be slightly underrun without loss of light in the beam so that these lamps have a rated life of 2000 h. Lamps are available with a dichroic reflecting surface which reduces radiant heat in the beam, but care should be exercised not to mount them in enclosed luminaires, which would overheat.

Tungsten-halogen lamps

The use of a small quartz bulb allows the gas pressure to be substantially increased, thus lengthening the life of the lamp by reducing the rate of evaporation of the filament. Heavy blackening of the bulb by the tungsten deposited on it is avoided by the application of the tungsten-halogen principle. The addition of a small amount of a halogen, usually iodine, to the gas filling results in an unstable compound being formed with the evaporated tungsten, which is maintained at temperatures between about 200 and 2500°C. As the bulb wall temperature is in excess of 250°C, no tungsten is deposited on it. The tungsten halide is carried into the filament by the convection currents in the gas, where it separates into its original components, the tungsten being deposited on the cooler parts of the filament and the halogen being released to repeat the cycle. Note that the longer life and higher light-output of these lamps is entirely dependent on the increase of gas pressure and has nothing to do with the redeposition of the tungsten on the filament.

Mains voltage tungsten halogen lamps have from double to four times the life of GLS lamps of equivalent type and give up to 20% more light.

Recently the lighting industry has developed the miniature low voltage (12 V) lamp. Such lamps are used in capsule form, or more commonly are set into a metal or dichroic reflector. The combination of a tiny tungsten-halogen lamp set into a dichroic-coated mirror (diameter is only 50 mm for the 50 W rating) gives great optical benefits, particularly for the display world. The reflector produces a precise, uniform beam resulting in intense, efficient light of high colour rendering and good appearance. The dichroic mirror allows most of the heat to pass backward whilst reflecting a relatively cool beam of light forward. This resulting 'cool-lightstream' is most important in displays as a wide range of merchandise may suffer from heat. Owing to the efficiency, a 50 W lamp compares with a 150 W PAR 38 spot and thus leading retailers are now utilizing the benefits. Low-voltage tungsten-halogen lamps require a transformer to operate.

Discharge lamps

When an arc is struck in a gas or metallic vapour it radiates energy in characteristic wavebands. For example, neon gives red light, sodium yellow and mercury vapour four distinct lines in the visible and two in the ultraviolet region of the spectrum.

All modern discharge lamps operate in a translucent enclosure (Fig. 18.3), containing the appropriate metals or metal halides, the initial discharge is usually struck in argon or neon. As the metal or metal halide evaporates, it takes over the discharge from the starter gas and emits light at its characteristic wavelengths.

Because more light and less heat is radiated by these lamps, they are more efficient in terms of lumens per watt than filament lamps, but where a line spectrum is emitted there is a marked distortion of colours seen under their light.

Low-pressure sodium lamps

These are the most efficient lamps in terms of lumens per watt, because the monochromatic yellow light they produce is in the area near the peak of the eye

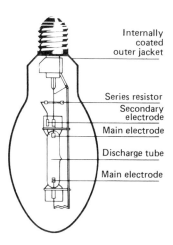

Fig. 18.3 A typical discharge lamp, in this case an MBF lamp. All modern lamps operate in a translucent enclosure containing the appropriate metals or metal halides.

sensitivity curve. Because only yellow light is emitted, however, objects are seen in tones of black and yellow only and colours cannot be perceived. They are extensively used for highway lighting and in situations where no colour discrimination is necessary.

High-pressure sodium lamps

If the internal vapour pressure of a sodium vapour lamp is increased, the sodium spectrum will broaden to include colours on both sides of the original yellow lines. Although this will result in a slight reduction of efficacy, it allows colours to be discriminated. High-pressure sodium lamps are rated in sizes from 1 kW down to 50 W and have taken over from mercury discharge lamps in side streets, industrial and commercial applications. The efficacy of these lamps is in the order of 100 lm/W compared with 150 lm/W for low-pressure sodium and 45 lm/W for mercury lamps. Most manufacturers have increased their high-pressure sodium lamp offers with increased output and improved colour versions, the latter opening up applications into commercial interiors.

Mercury vapour lamps

Mercury vapour lamps emit a considerable amount of energy in two wavebands in the UV region, the proportion of visible to invisible radiation being closely related to the gas pressure in the discharge tube. This is used to excite fluorescence in phosphors coating the inside of the outer bulb to add some of the missing red to their spectrum.

Metal halide lamps

Metal halide lamps use mercury vapour and the halides of a number of chemically active metals which are released as the temperature in the arc tube increases, producing a virtually continuous spectrum in which colours can readily be distinguished and seen with very little distortion. Their main advantage is that they provide a compact source of light with good colour-rendering properties at efficacies in the order of 80 lm/W. They are now available in lower wattages, typically 150 W, 70 W and even 35 W, and can thus provide high-quality white light for commercial interiors or for exterior floodlighting.

Compact-source iodide lamps

Two types of metal halide lamps are specially made for floodlighting and are extensively used for lighting football and sports stadia. One is a double-ended tubular lamp, with no outer jacket, designed to be used in an enclosed parabolic floodlighting projector and the other a very compact discharge capsule enclosed in a parabolic sealed glass reflector. This lamp has superseded the giant carbon arc projectors used in television studios and, because of its light weight and small size, is specially suited to outside broadcasts. It is available in a 'hot restrike' version.

The main characteristics of discharge lamps are summarized in Table 18.1.

Table 18.1 Basic characteristics of discharge lamps.

Lamp type	Lumens per watt (initial)	Colour appearance	Colour rendering	Power range (w)	Average life (h)
Mercury	50	Bluish white	Poor	50–1000	16 000–24 000
Metal halide double ended compact	80	White	Very good	70–250	6000
Metal halide double ended	80	White	Very good	750–2000	6000
Metal halide single ended compact	80	White	Very good	70–150	6000
Metal halide eliptical coated	80	White	Very good	250–1000	14 000
Metal halide tubular clear (horz mounted)	85	White	Very good	250–400	6000
Low-pressure sodium	140	Orange	None	18–180	16 000
Low-pressure sodium economy	180	Orange	None	26–131	16 000
High-pressure sodium	96	Yellow	Very poor	50–1000	24 000
High-pressure sodium increased output	100	Yellow	Very poor	50–400	24 000
High-pressure sodium improved colour	90	Golden	Good	150–400	12 000–16 000

Run-up time

All the discharge lamps so far referred to have a 'run-up' period varying from a few minutes to a quarter of an hour. They also require time for the vapour to cool and its pressure to drop before they will restrike after they have been extinguished, except for low-pressure sodium lamps which restrike immediately current is restored. High-pressure sodium lamps recover a great deal more quickly than other types.

Fluorescent tubes

A fluorescent tube is a low-pressure mercury discharge lamp. At this pressure, the mercury arc emits very little visible light and a considerable amount of UV radiation in two distinct wavebands, one in the harmless near-visible region and the other in the shorter-wave therapeutic area. This emission is used to excite fluorescence in the phosphors coating the inner surface of the tube, which may be up to 2400 mm long and either 38 mm (T12), 26 mm (T8) or 16 mm (T5) in diameter according to the power rating.

For the reasons already explained, lamps such as the 'colour-matching' or 'Northlight' type, which give a close approximation to north-sky daylight, are inherently less efficient than 'white', 'warm-white' or 'natural' tubes, with less satisfactory colour-rendering properties. Data for the majority of types of tube are shown in Table 18.2.

Two recent developments are the introduction of high efficacy fluorescent tubes utilizing the rare earths developed for colour television as well as the conventional halophosphates, and krypton gas to facilitate starting. Because the light from the phosphors is produced mainly in the three areas of the spectrum, blue, red and green, which combine to give the effect of white light, instead of having a virtually continuous spectrum like the conventional type of tube, they may occasionally give rise to the effect of 'metametric mismatching' of colours. In situations where accurate colour-matching is essential, therefore, the standard halophosphate tubes should be used. They must always be operated on a starter-switch or electronic circuit and consume fractionally less current than their conventional equivalents when operated on the same chokes.

In 1981 compact fluorescent lamps, such as GE's 2D and Philip's PL and SL, were introduced as an energy-saving replacement for the gls lamp. The 2D lamp is a 520 mm long by 13 mm outside diameter tube bent into a tight, double 'D' shape and supported by a central mounting plate which incorporates the starter. Consuming only 21 circuit watts, the 16 W 2D lamp has a light output approaching that of a 100 W lamp, but saves around 80% of the energy. Substantial savings are possible, especially where lights are in continuous use. But energy saving isn't its only feature. It has a life eight times longer than a filament lamp, is comparatively cool running and is similar in colour appearance to an incandescent lamp, Table 18.3. Improvements are growing all the time – there is now a family: 16 W, 28 W and 38 W.

While breaking away from the 2D shape, but still utilizing the principle, the two limb and four limb lamps use a 'U' shape and range from 7 W to 55 W. This

Table 18.2 Initial lumens for fluorescent tubes.

Length (mm) / Watts / Tube-Dia.	CCT	Ra	2400 / 125 / T12	2400 / 100 / T12	1800 / 75/85 / T12	1800 / 70 / T8	1500 / 65/80 / T12	1500 / 58 / T8	1500 / 50 / T8	1200 / 40 / T12	1200 / 36 / T8	900 / 30 / T8	600 / 20 / T12	600 / 18 / T8	450 / 15 / T8
White	3450	54	9500	8600	5850	5800	4600	4800	3800	3050	3000	2300	1225	1225	950
Warm-white	2950	51	9500	8600	5850	5800	4600	4800	3800	3050	3000	2300	1225	1255	950
Cool-white	4200	58	9300	8450	5700	5700	4500	4700	–	3000	3000	2250	1200	1200	900
Natural	4000	73	7700	6850	4650	–	3775	–	–	2375	–	–	–	–	650
Kolor-rite	4000	89	–	–	3900	–	3300	–	–	2000	–	–	850	–	–
Northlight	6500	93	6000	5500	3700	–	3150	–	–	2000	–	1500	850	–	650
Polylux	2700	80	–	–	–	–	–	5400	–	–	3450	2500	–	1450	1050
Polylux	3000	80	–	9400	–	–	–	5400	–	–	3450	2500	–	1450	1050
Polylux	3400	80	–	9400	–	–	–	5400	–	–	3450	2500	1450	1450	–
Polylux	4000	80	10900	9400	6700	–	5400	–	–	3450	3450	2500	1450	1450	1050

Polylux is GE's trade name for high-efficiency, high-colour rendering tri-phosphor tubes.

Table 18.3 Initial lumens for a selection of compact fluorescent lamps.

Type	Lumens (initial)	Rated life (h)
10 W 2D	650	8000
16 W 2D	1050	8000
28 W 2D	2050	10 000
38 W 2D	2850	10 000
18 W 2 limb	1250	10 000
24 W 2 limb	1800	10 000
36 W 2 limb	2900	10 000
40 W 2 limb	3500	10 000

compact fluorescent now extends applications in commercial areas, typically used in 600 mm × 600 mm ceiling modules with high-frequency control gear.

CONTROL GEAR AND STARTING

All arc lamps require a current limiting device to prevent them taking more and more current until they destroy themselves. In most cases a starting device is also necessary to strike the arc.

A high-pressure mercury lamp is operated in series with a choke having a laminated iron core. The starting device consists of a secondary electrode within the arc tube, connected through a resistor to one electrode and in close proximity to the other. When the lamp is switched on an arc is struck between the adjacent electrodes, ionizing the argon filling of the arc tube so that the main discharge is struck between the cathodes. The secondary electrode then ceases to function because of the resistor in series with it.

Metal halide and high-pressure sodium lamps require an external ignitor in addition to the choke, although some manufacturers sell SON lamps with an internal starter like that of a mercury lamp, so that they can be operated from a choke designed for approximately the same sized mercury lamp. The separate ignitor, however, is more reliable. It must be placed as close to the lamp as possible to avoid voltage drop in the lead to the lamp; the position of the choke is not critical.

Fluorescent tubes, being essentially low-pressure mercury lamps, also need a current limiting and starting device. The former usually takes the form of a choke, but occasionally a resistor is used. Some of the earlier domestic luminaires made use of a filament lamp wired in series with the tube and there are luminaires housing two 20 W 600 mm tubes controlled by a resistance wire available for domestic use. It will be appreciated that such devices largely offset the advantage of the low current consumption of the tubes. A starter switch, wired in parallel with the tube, is commonly installed, but transformer starters are also used.

In all cases where an inductive current limiting device is used a capacitor is included in the circuit for power factor correction. This is usually housed in the luminaire with the rest of the controlgear, but bulk power factor correction at a central point is occasionally used. Starters are always housed in the luminaire, but in some cases, as for example street-lighting lanterns and floodlights using metal halide lamps, the choke may be mounted elsewhere. This type of choke is very bulky and heavy and for street-lighting lanterns is often housed at the foot of the column.

Since 1985 a new era is taking place in the controlgear used to start fluorescent lamps – the use of electronics. The high-frequency electronic ballast provides instant, flicker-free starting for single and twin standard fluorescent lamps up to 1800 mm in length. It takes advantage of a characteristic of fluorescent lamps whereby greater efficacy is obtained at high frequency, typically 32 kHz. The overall achievement in a suitable luminaire is an energy reduction of 20−30% while still maintaining the lighting level. This is due to improved lamp efficacy at high frequency operation and reduced circuit power losses as the new solid-state unit contains no conventional copper windings. The quality benefits are no flicker, instant start, no stroboscopic effects and silent operation. In addition, the ballast is a complete gear unit requiring no additional power factor correction capacitor or starting device thus allowing fittings to be lighter in weight and more compact.

LUMINAIRES (LIGHTING FITTINGS)

Most of the light from a bare lamp is likely to be wasted; in addition, a bare unshielded lamp gives rise to glare which can hinder vision, so the use of a luminaire to house the lamp(s) and direct or diffuse its light is essential. The detailed design of luminaires is outside the scope of this book, but some knowledge of the principles involved is needed if satisfactory lighting schemes are to be produced. Luminaires may be classified by their light distributing qualities and their choice governed by the quality of light required. It may also be influenced by special safety considerations, such as where lamps are used in explosive or corrosive atmospheres or are liable to mechanical damage or excessive vibration.

There is an abundance of luminaire and control gear terms in common use. The following list does not attempt to provide precise definitions, but rather to suggest the most common meaning and where appropriate to point out the areas of potential confusion.

Air terminal devices   The means by which air is supplied or extracted from a room.
Ballast   Control gear inserted between the mains supply and one or more discharge lamps, which by means of inductance, capacitance or resistance, singly or in combination serves mainly to limit the current of the lamp(s) to the required value.
Barn doors   Four independently hinged flaps that serve to cut off spill light for spotlights.
Batten   A fluorescent luminaire with no attachments; just the control gear channel and a bare tube.

Bulkhead   A compact enclosed luminaire designed for horizontal or vertical mounting outdoors or indoors.

Ceiling system   A grid and panel system providing a false ceiling to incorporate lighting and other building services.

Choke   A simple low power factor inductive ballast.

Columns   Poles for mounting road lanterns or floodlights.

Controller   A clear plastic or glass with prism shapes in the surface, the overall shape may be flat or formed.

Diffuser   An attachment intended to reduce the lamp brightness by spreading the brightness over the surface of the diffuser.

Downlighters   This term is frequently misused to refer to any luminaire which produces no upward light, whereas the object of a downlighter is to provide illumination without any apparent source of light. They are therefore, extremely 'low brightness' luminaires referring particularly to small recessed, semi-recessed or surface 'can' luminaires.

Emergency lights   Lighting provided for use when the mains lighting fails for whatever reason. There are two types: (1) escape lighting and (2) standby lighting.

Escape lighting   See *Emergency lights*.

Eyeball   A recessed adjustable display spotlight.

Flameproof   An enclosure capable of withstanding the pressure of an internal explosion and preventing transmission of the explosion to gases and vapours outside the luminaire.

Floodlight   General term applied to exterior fittings housing all types of lamps and producing beams from very narrow to very wide. Floodlights usually employ a specular parabolic reflector and their size is determined by the light source employed and distribution required. The term also applies to interior fittings providing a fairly wide beam which could not be described as a 'spotlight'.

High bay   As the name implies, these are for use when mounting heights of around 8–10 m or above are encountered. They have a controlled light distribution to ensure that as much light as possible reaches the working plane. Although primarily intended for industrial areas they also find applications in lofty shopping areas and exhibition halls.

Hood   A baffle projecting forward from the top of a floodlight or display spotlight to improve the cut-off of spill light above the horizontal to reduce glare and light pollution.

Ignitor   A starting device, intended to generate voltage pulses to start discharge lamps, which does not provide for the pre-heating of electrodes.

Inverter   Electronic device for operating discharge lamps (usually fluorescent) on a d.c. supply obtained either from batteries or generators.

Lantern   Apart from reproduction period lighting units, this is an alternative term for a luminaire usually restricted to roadlighting or stage lighting equipment.

Louvres   Vertical or angled fins of metal, plastic or wood arranged at right angles to linear lamps or in two directions. The object is to increase the cut-off of the luminaire and so reduce glare from critical angles. Louvres can also form ceiling panels with lighting equipment placed in the void above.

Low bay   Luminaires housing high-pressure discharge lamps (usually mounted horizontally) to provide a wide distribution with good cut off at mounting heights around 4−8 m. Apart from industrial applications, these are used in many sports halls and public concourses.

Low brightness   Term usually applied to commercial fluorescent luminaires where, by the use of louvres, the brightness is limited. These are particularly appropriate for offices with VDTs.

Mast   Mounting columns higher than 12−15 m are generally referred to as masts and are used to support floodlighting for large areas.

Optic   The reflector and/or refractor system providing the light control for the luminaire.

Post top   Road or amenity lighting luminaire which mounts directly on to the top of a column without a bracket or out-reach arm.

Profile spotlight   Luminaire with focusable front (objective) lens producing a hard edge beam. By a set of adjustable shutters or masks at the gate of the projector, the beam can be shaped or 'gobos' (cut-out metal slides) can be used to project patterns or pictures.

Projector   Term for floodlight or profile spotlight.

Proof   Applied to all types of luminaires which have a higher degree of protection from the ingress of solids and liquids than standard interior lighting luminaires.

Recessed   The luminaire mounting arrangement where the whole or part (semi-recessed) of the luminaire body is set into a ceiling or wall or floor surface.

Refractor   Clear glass or plastic, panel or bowl where an array of prisms is designed to re-direct the light of the lamp into the required distribution.

Semi-recessed   See *Recessed*.

Spine   See *Batten*.

Spotlight   An adjustable interior luminaire with controlled beam using a reflector lamp or optical system.

Stand by lighting   See *Emergency lights*.

Starter   A starting device usually for fluorescent tubes which provides the necessary pre-heating of the electrodes and, in combination with the series impedance of the ballast, causes a surge in the voltage applied to the lamp.

Tower   Lattice, steel or concrete structure from 15 m to 50 m to support area floodlighting equipment.

Track system   A linear busbar system providing one to three main circuits or a low voltage supply to which display lighting can be connected and disconnected at will along the length of the system. The luminaires must be fitted with an adapter to suit the particular track system in use.

Transformer   A wire-wound or electronic unit which steps-up or down its supply voltage. Often fitted integrally into low voltage spotlights.

Troffer   Fluorescent or discharge luminaire designed to recess into suspended ceilings and fit the module size of the ceiling. See *Recessed*.

Trunking   Apart from standard wireway systems in ceilings and floor, trunking associated specifically with lighting usually provides mechanical fixings for the luminaires as well as electrical connection.

Uplights   Indirect lighting system where task illuminance is provided by lighting reflected from the ceiling. The luminaires can either be floor standing, wall mounted or suspended and can be arranged to provide either general or localized lighting.

Visor   Clear or diffused glass or plastic, bowl or flat panel closing the mouth of a luminance.

Wall washer   Interior floodlight or spotlight with asymmetric distribution intended to provide uniform lighting of walls from a close offset distance.

Wellglass   These consist of a lamp surrounded by an enclosure of transparent or translucent glass or plastic. They are usually proof luminaires and often used outdoors fixed to a bracket.

Zone 0, 1 and 2 fittings   Classification of various hazardous areas.

Lighting safety

The following provides explanations of some of the most common classifications applied to luminaires.

Class I   luminaires in this class are electrically insulated and provided with connection with earth. Exposed metal parts that could become live in the event of basic insulation failure are protected by earthing.

Class II   luminaires are designed and constructed so that protection against electric shock does not rely on basic insulation only. This can be achieved by means of reinforced or double insulation. No provision for earthing is provided.

Class III   protection against electric shock relies on supply at safety extra-low voltage (SELV) and in which voltages higher than those of SELV are not generated. (max. 50 V a.c.r.m.s.).

F Mark   luminaires suitable for mounting on normally combustible surfaces (ignition temperature at least 200°C) are marked with the 'F' symbol.

Ta Classification   this denotes the maximum ambient temperature that the luminaire is suitable for use in max. 25°C ambients.

British Standards

Lighting products should be designed and manufactured to the standards given in Table 18.4. European equivalents are also listed.

Wherever possible, luminaires should be purchased that are marked with a third party approval, such as the BSI Kitemark. Recently the ENEC mark has been introduced following European harmonization of standards. The ENEC mark indicates that a luminaire is suitable for use throughout Europe and that all of the most onerous special national conditions of EN60 598 have been complied with.

Ingress protection

The ingress protection (IP) code denotes the protection against dust, solid objects and moisture provided by the luminaire enclosure. If no code is marked, the luminaire is deemed to be IP20. The first digit of code denotes protection against

Table 18.4 Standards for lighting products.

Subject	British Standard	European standard
Luminaires – general types	BS 4533: 102.1	EN60 598 2.1
Luminaires – recessed	BS 4533: 102.2	EN60 598 2.2
Luminaires – roadlighting	BS 4533: 102.3	EN60 598 2.3
Luminaires – floodlights	BS 4533: 102.4	EN60 598 2.5
Luminaires – with transformers	BS 4533: 102.6	EN60 598 2.6
Luminaires – air-handling	BS 4533: 102.19	EN60 598 2.19
Luminaires – emergency	BS 4533: 102.22	EN60 598 2.22
Luminaires – track	BS 4533: 102.57	EN60 570
Radio interference	BS 5394	EN55 015
Emergency lighting	BS 5266	Under discussion

dust and solid objects, the second against moisture, for example IP65 is a dust-tight and jet-proof luminaire.

OUTDOOR LIGHTING EQUIPMENT

Floodlights

Until quite recently, the use of high-powered tungsten filament or discharge lamps in outdoor floodlighting equipment led to very bulky equipment. The diameter of a floodlight for a 1 kW incandescent or a 400 W mercury, metal halide or high-pressure sodium lamp is about 600 mm and its weight, exclusive of control gear, about 13 kg.

The introduction of tungsten halogen lamps has resulted in two 'families' of floodlights both much smaller than the conventional types. One, originally produced under the name 'Sunflood' by Thorn, and extensively copied, is a very simple open housing with a parabolic reflector for the linear type of lamp and is much used in car parks and domestic situations. The other, an enclosed floodlight, is found in a number of industrial and commercial situations, especially where an immediate flood of light is needed. Narrow-beam floodlights for single-ended mains voltage tungsten halogen lamps are also available.

The use of linear high-pressure sodium and metal halide lamps in similar floodlights followed almost immediately and this type of projector soon became more popular for such applications as the floodlighting of minor football fields and sports grounds, including trotting tracks, and for the lighting of the yards and loading bays of industrial premises. In addition to the longer life and higher light output of these sources, the type of floodlight described has much less weight and windage than the conventional types; for example a 1000 W fitting of this type weighs approximately 9.5 kg and is approximately 450 mm square.

Specialized floodlights for football stadia

Floodlights using metal halide (MBIL) linear lamps specially designed for football stadia in which matches are covered by coloured TV are usually intended for comparatively low mounting heights, on the roofs of the stadia or just above them. The 1500 W type is very compact and weighs only 6.5 kg. It may be fitted with an internal baffle to prevent direct light from the arc tube dazzling spectators on the opposite side of the pitch. All the weights quoted are exclusive of control gear, which is housed separately from the floodlight.

An even more specialized floodlight, mounted on high masts for lighting stadia and also for general television work incorporates the 1000 W CSI lamp which is itself housed in a PAR 64 reflector. This housing is designed to take either the ordinary CSI lamp or the hot restrike type and sufficient ventilation is provided to take care of the heat passing through the dichroic reflector used with the hot restrike lamp. The ignitor is mounted on the lamp housing, but the chokes and power factor correction capacitors are separately housed.

FLOODLIGHTING CALCULATIONS

Floodlights mounted on poles are extensively used for lighting areas such as loading docks, parking lots and so forth. The calculation of the illuminance is not difficult, but it varies from the method used in the interior of buildings in certain respects and other terms are introduced.

Beam angle

This is the width or spread of the beam and is defined as the total angle over which the intensity drops to a fixed percentage, 50%, 10% and 1% of its peak value. A single figure is given for a floodlight with a symmetrical distribution, two for a unit with an asymmetric distribution.

Beam flux and beam factor

Beam flux is the total flux contained within the beam angle to 10% of peak intensity and is usually expressed as the proportion of lamp lumens contained within the beam (beam factor). It is usually presented in tabular form by the manufacturers.

Instead of the usual 'utilization factor' a 'waste-light factor' multiplied by the 'beam factor' is used. The waste light factor is usually taken as about 0.9, but in awkward shaped areas may be as high as 0.5. It is influenced by the width of the floodlight beam.

Illuminance diagrams

A number of manufacturers publish illuminance diagrams which provide an easy method of calculating the illuminance over an area from a specific floodlight at a known mounting height and aiming angle. These can be superimposed to calculate the illuminance from an array of floodlights. A typical diagram is shown in Fig. 18.4. Conversion factors are used to adapt the figure for different mounting heights.

NEW AREA FLOOD

Description General area floodlight
Luminaire maintenance category
Catalogue number   Lamp   Data code
AFBS 250.T   250W SONXL-T   R0002274

Isolux diagram in lux / 1000 lm

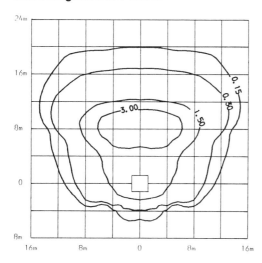

Fixed attitude - top horizontal
Mounting height 8m

Fig. 18.4 Data sheet for a typical floodlight showing area illuminance.

Zonal flux diagrams

Another method, which is very adaptable is the use of 'zonal flux diagrams', a typical example of which is shown in Fig. 18.5. It is divided through the plane of symmetry with isocandela lines plotted on the angular grid and figures denoting the flux per 1000 lamp lumens in each angular zone on the right. It can either be used to calculate point values of illuminance or the flux intercepted by the area to be lit, which can be drawn as an overlay using the same angular scale. Elevation of the floodlight can be shown by moving the diagram up and down the vertical grid line, but a fresh diagram must be drawn for adjustments in azimuth. The beam factor of the floodlight is expressed by the dotted line on the diagram. The beam lumens per 1000 lamp lumens is twice the value of the flux values enclosed within it.

Street-lighting lanterns

The basic principle of highway lighting is to achieve a bright road surface against which vehicles can be seen in silhouette. The non-specialist lighting engineer can usually ignore highway lighting techniques since he is only likely to be concerned

NEW AREA FLOOD

Description General area floodlight
Luminaire maintenance category

Catalogue number	Lamp	Data code
AFBS 250.T	250W SONXL-T	R0002274

Beam data

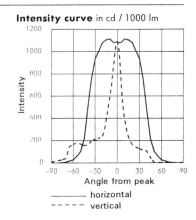

$\Theta = 21°$

Peak intensity (I) cd / klm			1110
Beam factor to 10% peak (I)			0.74
Beam angle to 10% of peak (I)	Horizontal	2 x 50°	
	Vertical	36°/66°	
Beam angle to 50% of peak (I)	Horizontal	2 x 39°	
	Vertical	7°/13°	
Beam angle to 1% of peak (I)	Horizontal	2 x 67°	
	Vertical	49°/89°	

Isocandela and zonal flux diagram

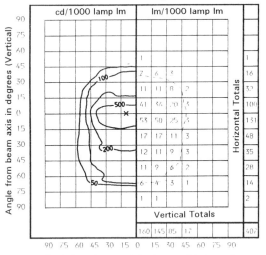

Angle from beam axis in degrees (Horizontal)

Intensity curve in cd / 1000 lm

X-indicates the position of peak intensity. The pecked line shows 10% peak contour. The direction of the peak intensity in the vertical plane is 21 degrees above the normal to the front face of the floodlight.

Fig. 18.5 Data sheet for a typical floodlight showing angular beam features.

with lighting subsidiary road and parking areas within a factory complex or similar location. Recommendations for exterior lighting are given in BS 5489, *Road lighting and the CIBSE lighting guide: Outdoor environment*. A number of lanterns mounted on top of a pole can be suitable for this purpose. They are usually designed to take any suitable light-source and the control gear may be mounted in the housing or in a chamber at the foot of the pole.

LIGHTING DESIGN FOR INTERIORS

Recommended task illuminance

The amount of light needed to perform a visual task satisfactorily varies according to the size of the object being handled and the contrast between it and the background, either in terms of brightness or of colour or both. For example fine assembly, as in watch-making may require an illuminance of about 1000 lux. Insufficient light may lead to eye-strain and slow down production, causing mistakes to be made or faulty workmanship to be overlooked.

CIBSE Code

The CIBSE in its Code for Interior Lighting give recommendations for good lighting practice, Table 18.5. Having 'sufficient' light is only one aspect of many. Illumination values are, in the Code, recommended as maintained illuminance, at a value below which the average illuminance should not fall during the life of the installation. This can be fulfilled by carrying out an appropriate maintenance

Table 18.5 Recommended maintained illuminance (CIBSE Code 1994).

Task group and typical task or interior	Standard maintained illuminance (lux)
Storage areas and plant rooms with no continuous work	150
Casual work	200
Rough work; rough machining and assembly	300
Routine work; offices, control rooms, medium machining and assembly	500
Demanding work; deep plan, drawing or business offices; machine offices, inspection of medium machining	750
Fine work; colour discrimination, textile processing, fine machining and assembly	1000
Very fine work; hand engraving, inspection of fine machinery or assembly	1500
Minute work inspection of very fine assembly	2000

Note: Adjustments to increase or reduce these values may be justified on the basis of task difficulty, task duration and where error can have serious consequences.

regime on the lamps, luminaire and room. The Code is an essential tool for the lighting engineer and indeed, for anyone concerned with lighting design.

Photometric data

To plan a lighting installation to give a specified illuminance on a working surface or plane, the engineer must know how to use photometric data published by reputable luminaire manufacturers and be familiar with such terms as 'light-output ratio' (LOR) of luminaires. The significance of the utilization factor in calculating the illuminance from an array of luminaires by the 'lumen method' of design must also be understood. Such matters as the relationship between the spacing to mounting-height ratio of luminaires, the evenness of the illumination on the task area and discomfort glare from the luminaires, must also be borne in mind.

Light-output ratio

The LOR is the proportion of the total lumen output of the lamp that is emitted from the luminaire. It must be understood that all reflecting and diffusing media absorb a certain amount of light when controlling light from the lamp(s).

The LOR of a luminaire is of little use by itself. For example, a bare fluorescent single 'batten' may have an LOR as high as 0.95, but in most cases will be less effective overall visually than one fitted with a reflector with a total LOR of 0.84. It is normal to specify upward and downward LOR. For example a bare tube might have an upward LOR of 0.29 and downward of 0.66, and the reflector unit 0.02 upward but 0.82 downward, an improvement where light delivered on the working plane and control of glare from the luminaires is concerned. This is especially important where luminaires are mounted under a glass or dark-coloured roof, a common situation in industrial interiors.

Luminous intensity and polar curves

Most manufacturers publish luminous intensity tables, giving the intensity in candelas per 1000 lamp lumens on a single vertical plane for symmetrical (tungsten or discharge lamp luminaires) or on a transverse and axial plane for fluorescent luminaires and those using linear sources. Most of the lighting data for the luminaires are derived from these measurements, although, in some cases more than two planes are measured.

From the table of luminous intensities the 'polar curve' for the luminaire can be derived, Fig. 18.6. Polar curves are a useful indication of the general light distribution of a luminaire. The point where the horizontal and vertical line crosses corresponds to the optical centre of the luminaire. The direct distance from this point to the curve shown is proportional to the luminous intensity at that angle. The left-hand side of the curve corresponds to the transverse plane (T) and the right-hand side of the curve corresponds to the axial plane (A). The tabular values of intensity are valuable if accurate calculations of the illuminance at a given point are required. It must be clearly understood that the area enclosed in a

CHALICE HI POWER

Description : Architectural lighting data presentation

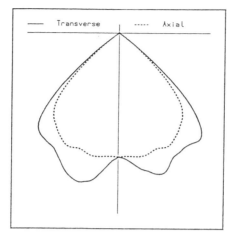

Polar Curve

Fig. 18.6 Polar curves are a useful indication of the general light distribution of a
luminaire. This is shown as axial and transverse where appropriate.

polar curve does not give an indication of the amount of light emitted by
the luminaire in that direction. Figure 18.7 shows that although the downward
component of the polar curve illustrated covers a greater area than the horizontal
one, the actual lumens contained in the latter, which represents a circular zone of
light around the luminaire, is considerably greater.

Utilization factors and room index

Where a general area has to be lit fairly uniformly over the working plane, as
opposed to situations where a considerable variation in illuminance may be
permissible or desirable, it is usual to use the 'lumen method of design'. In order

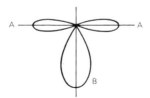

Fig. 18.7 A polar diagram must not be misread, in that the amount of lumens contained in
the horizontal plane may be greater than in the vertical plane:
(A) all light in equatorial area; (B) all light in polar area. Actual lumens
delivered at A is considerably in excess of B.

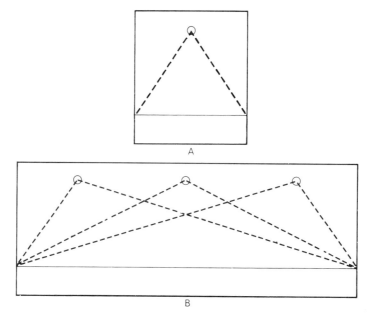

Fig. 18.8 Room dimensions affect illumination of room surfaces.

to do this, a 'room index' based on the proportions of the room has to be used.

Figure 18.8 makes it clear that in a tall, narrow room more light is absorbed by the walls, assuming the same mounting height for the luminaires than in one of more generous proportions. It must also be borne in mind that a glass roof can be considered as a black area, reflecting no appreciable light into the room and that uncurtained windows must be regarded as absorbing areas too.

The formula for obtaining the room index is:

$$k = \frac{L\ W}{H_m\ (L + W)}$$

where L　= length of the room (m)
　　　　W　= width of the room (m)
　　　　H_m = mounting height of luminaires above the working plane (m).

The utilization factor can then be obtained by reference to the photometric tables published by the manufacturer using the room index and an average room reflectance.

Spacing to mounting-height ratios

Where an array of luminaires is designed to give an even illuminance on the 'working plane', sometimes assumed to be 0.85 m above the floor, the recommended spacing to mounting-height ratio (SHRMAX) shown in the data must not be exceeded. This is S/H_m where S is the distance between fittings and H_m their mounting height above the working plane. Too wide a spacing will result in a serious falling off in illuminance between the luminaires.

In practice, the number and spacing of the luminaires are often determined by the ceiling height, the beam structure, or modular size of a suspended ceiling. The recommended spacing can be exceeded if the luminaires are mounted above work-benches or rows of desks, but this presupposes a permanent layout which may not be considered when the lighting scheme is designed.

Discomfort glare

Luminaires in a large room can be a serious source of direct glare, especially if they contain fluorescent tubes and are mounted at right angles to the principal line of view. Glare from luminaires can cause discomfort to people using and working in an interior. This is known as discomfort glare. The CIBSE code gives limits to a glare index. Luminaire data often includes uncorrected glare indices for the luminaire which after correction can be compared with the limiting value to establish whether the luminaire is a suitable choice.

Designing an installation

Lumen method of design

For hand calculations the lumen method of design is the easiest available and is always recommended where the precise use of a room and dispositions of work-benches or desks are not known. It provides a general illuminance over the whole of the working area and is generally employed in planning the lighting of speculative office blocks or factories where future requirements are not known.

After the task illuminance has been settled and the type of lamp and luminaires decided upon, the method starts with a simple calculation of the number of luminaires required, based on the formula:

$$N = \frac{E \times A}{n \times UF \times F \times MF}$$

where F = initial lumen output of the lamp or tube(s)
E = average maintained illuminance (lx)
A = area to be lit in (m²)
UF = utilization factor
MF = the maintenance factor
N = number of luminaires required
n = number of lamps per luminaire.

With the introduction of maintained illuminance in the 1994 CIBSE Interior Lighting Code the maintenance factor is derived from four factors associated with the lamp, luminaire and room surfaces. For a comprehensive explanation of the method it is suggested that the Code is referred to.

The utilization factor is derived from the photometric data provided by the luminaire manufacturer and a luminaire maintenance category may be provided. Lamp data would be sourced from lamp manufacturers.

The value of N seldom comes out as a whole number and may produce difficulties in the arrangement of the luminaires. In such a case, the smallest whole number of these that can be installed at adequate spacing can be calculated from the formula:

$$\frac{L \times W}{MS \times MS}$$

where L = length of the room (m)
$\quad\quad W$ = width of the room (m)
$\quad\quad MS$ = maximum permitted spacing between luminaires.

Further modifications may be imposed by the beam structure of the room and by obstructions in the space. The spacing to mounting-height ratio of the luminaires should not be exceeded.

The final scheme must be checked for glare index, and, if it falls outside the recommended limits, a type of luminaire with a lower glare index must be substituted and the calculations repeated.

Low energy installations

The need to conserve electrical energy has led to a reconsideration of lighting methods in offices and workshops. For many years draughtsmen have tended to prefer localized lighting on their drawing-boards, often using the heat from the filament lamps to dry the ink on their drawings. But it has long been the practice to provide a high level of uniform lighting at the working plane of offices and factories by means of a general lighting system. This has been much assisted by the use of fluorescent tubes.

Today, although general lighting of such places must still be provided, if only to avoid heavy shadows under desks or benches with consequent untidiness, and to facilitate movement, there is a tendency to keep it at a fairly low level and supplement it with local lighting at the work-point.

Considerable impetus has been given to this by the introduction of smaller diameter high-efficacy fluorescent tubes and to a lesser degree, by the use of tungsten halogen lamps. Both have led to the design of small, compact reflectors, taking up comparatively little space on a desk or above a work-bench.

The calculation of a general lighting system has already been described but that of the lighting from localized luminaires is less obvious. A useful rough guide to the illuminance that can be provided by such a light-source is the 'halving law', in which the intensity in the desired direction is divided by twice the distance in metres from the object to be lighted. The inverse square law really only applies to point sources and consequently may be used with compact filament or discharge lamps. A more accurate method of calculation of the illuminance from line sources is the use of aspect factors described in the CIBSE Report No 11. The aspect factors of fluorescent luminaires are usually shown in the photometric data published by the manufacturers. If they are not, one assumes a LOR of 50% from the luminaire and calculates the lumens falling on the area to be lighted according to the mounting height of the luminaire. Most localized luminaires are designed to be mounted about 600 mm above the working plane. Note that the local luminaire is usually mounted on the left-hand side of the desk to reduce the possibility of reflected glare, a common hazard with this type of lighting.

Lighting for visual display terminal (VDT) areas

Many people spend long periods of time using computers and display screens. The provision of suitable conditions for these users is a matter of good sense and a legal requirement as the EC Directive (90/270/EEC) for display screen equipment is now law. Also, the Health and Safety Executive Regulations 1992 are in force. These make it a legal obligation for employers to fulfil minimum requirements to provide suitable and adequate conditions for users of workstations with VDTs. The regulations came into force on 1 January 1993 and apply now to all new or modified workstations. They also require that all existing workstations comply before 1997.

These regulations are deemed to have been met, as far as lighting is concerned, if the advice in CIBSE Lighting Guide 3 (LG3) is followed. Options include direct lighting (provided by an overhead lighting system where the brightness or luminance of equipment is limited above critical angles to the downward vertical), indirect lighting (uplighting) or a combination of both. Correctly designed, uplighting is suitable for VDT applications, but direct downlighting fluorescent fittings are divided into three categories. Each category has a limiting angle and above this the luminance must not exceed $200\,cd/m^2$ (candelas m^2) according to the severity of the VDT task.

Luminaires conforming to Category 1, 2 or 3 must comply to this luminance limit at angles above 55 degrees, 65 degrees and 75 degrees respectively to the downward vertical. Category 1 is for the most stringent situations and Category 3 the least. Category 1 is not better that 3 as it depends upon the application. Category 1 areas are rare and for situations in which the visual displays are particularly difficult to read and the work is demanding, such as air traffic control.

Category 2 is the most common VDT situation, where conventional displays are used for long periods of time and the work can be demanding. An order entry room would be a typical example.

Category 3 contains the applications where the visual displays are easy to read or the work is less demanding such as an airline booking desk.

With downlighting, particularly with Category 1 or 2 luminaires, attention must be paid to avoiding a gloomy interior appearance that can result from low vertical surface luminance. This is usually achieved by lighting the surrounding wall areas.

EMERGENCY LIGHTING

Of all the uses to which lighting design can be put, the exercise of its skills in the provision of suitable and efficient emergency lighting is perhaps the most important, as ultimately good emergency lighting can save lives.

But why should anyone bother to spend money on a lighting system that hopefully will never be used? Well, apart from the moral obligation that we have to the occupants of a building to ensure their safety, The Fire Precautions Act, The Health and Safety at Work Act and Building Regulations of 1992 make it obligatory to provide suitable means of escape during an emergency and lighting is obviously an essential part of this requirement.

But what do we mean by 'emergency lighting'? There are two types of emergency lighting: escape lighting and standby lighting. Escape lighting is provided to ensure the safe and effective evacuation of the building and has special requirements. Standby lighting is provided to enable some activities to continue for a period after the mains lighting fails, for example in a hospital operating theatre. Standby lighting can be regarded as a special form of conventional lighting and dealt with accordingly.

There are two main types of systems available for powering emergency lighting: self-contained or central. For the former, in the event of an emergency, the luminaires operate from their own batteries. Each luminaire is fully equipped with battery, charger, charge indicator and changeover device. This can be within the luminaire or in a separate unit less than one metre from the luminaire. The normal unswitched lighting circuit, which charges the batteries and provides mains failure detection, is the only extra connection required for a self-contained emergency luminaire. With central systems, the power is provided by batteries or generators and the output is distributed through sub-circuits to feed a number of slave luminaires. These systems require comparatively large battery/generator rooms. Sub-circuit monitoring is used to protect against local failure and high integrity wiring must be used between power source and luminaire. Due allowance must be made for voltage drops.

Types of emergency lighting luminaire are classified according to modes of operation and duration. The battery is always being charged when the mains supply is present.

Non-maintained   In this mode the lamp(s) is only lit when the mains fails and is operated by an emergency power source.

Maintained   In this mode the lamp(s) is lit at all times. It is powered by the mains supply under normal conditions and under emergency conditions operates from an emergency power source.

Combined/sustained   This is a variant of the maintained luminaire in which one lamp is powered by the mains supply during normal conditions. A second lamp only operates under emergency conditions or by an emergency power source. This type of luminaire provides light at all times.

Duration is the period that the emergency luminaire remains lit to the minimum design output, after a mains failure. Various premises require different durations, as set out in BS 5266, from 1−3 h.

Before we can begin to design an emergency lighting installation we must establish what set of standards are in force. A great deal of discussion and speculation has recently been taking place regarding the forthcoming harmonized European standard for emergency lighting by CEN and CENELEC working groups. The standard may demand brighter illumination in a wide range of buildings and could also apply to existing premises and installations. Emergency lighting could be effected in the following ways:

Escape route lighting   The proposed standard calls for 1 lux minimum along the centre line of an escape route with a potential hazard. The UK may however negotiate to retain 0.2 lux for specified escape routes without hazards.

Anti-panic area lighting   The proposed minimum lighting level for these open areas (halls, shops and offices) with undefined escape routes is 0.5 lux. The standard will be significantly simplified to reduce design and measurement complexity.

High risk task area lighting   This new category, which is common practice in Germany, aims to ensure the safety of people carrying out potentially dangerous tasks, for example moving machinery and acid baths; and where there is a restricted escape route such as a turnstile. The European norm will require emergency lighting equivalent to the greater of 15 lux or 10% of normal lighting levels within a quarter of a second, thus allowing for a controlled shutdown to take place.

Signage   Standards for signage will follow ISO 6309 Euro legends in green/white colour. These will gradually replace the BS 5499 word/pictograms and older wording signs. Remember it is not permitted to mix signs of different formats in an installation.

But, what advice should be given to lighting designers currently dealing with emergency lighting schemes? Well, the UK has yet to determine the full implementation of the new proposals and to decide whether the standards will apply retrospectively (requiring all existing schemes to be refurbished). Additionally, implementation time scales are estimated to be around 2 years. Current schemes therefore should be designed to the present emergency lighting standards and Codes of Practice, which are: BS 5266: Part 1 (1988), CP 1007 (1955), for cinemas and BS 5499: Parts 1, 2 and 3 for fire safety signs. Details of these standards and design and installation methods are covered by CIBSE Technical Memorandum 12. Key points to remember are:

Exits   All exits used in an emergency and dedicated emergency exits must have signs that are visible at all material times. Where there is no direct sight of an exit, direction signs should be positioned to help those not familiar with the building find the nearest exit. Additionally, lighting outside the building should be adequate for safe evacuation and dispersal.

Escape route illumination   Corridors and gangways are classified as clearly defined escape routes. For these areas the horizontal illuminance at floor on the centre line of a clearly defined route should not be less than 0.2 lux and 50% of the route width up to 2 m wide should be lit to a minimum of 0.1 lux. Wider routes should be treated as several 2 m bands.

Large open areas   Offices, supermarkets, dining halls, conference rooms, laboratories, multipurpose rooms. These places will not have defined routes and the layout of furnishing may change from time to time. The average horizontal illuminance over the whole area on an obstructed floor should be not less than 1 lux with a uniformity of 0.025.

Signs   These may be externally or internally illuminated. The mounting height of exit signs should be between 2 and 2.5 m above the floor. Exit signs with the running man motif should be used in new buildings. When considering a building which already has some exit signs, the colour, format and style should be consistent with those already installed.

CHAPTER 19

Mains Cables

G.A. Bowie, DFH, CEng, FIEE

S.E. Philbrick, CEng, MIEE

Revised by T.L. Journeaux
(Engineering Manager, Pirelli Cables Ltd)

Mains cables are used to supply electricity to complete electrical installations and as the electrical loadings of such cables must take account of the total loading of the electrical installation, mains cables are generally of larger sizes than the wiring cables dealt with in Chapter 20.

For most installations mains cables operate on low voltage, that is about 440 V, but for very large installations, particularly where the site is extensive and large loads are carried over long distances, mains cables operating at higher voltages are used. In these cases the h.v. source might be provided from the public electricity supply or from a private generating station on the site. In some industries, such as paper-making, large quantities of process steam are needed and it is often convenient for the company to generate electricity at the same time as producing steam.

This chapter deals with systems for operating at voltages up to and including 11 kV and requiring conductor sizes larger than about 25 mm^2.

CABLE SPECIFICATIONS

All cables are fundamentally similar in that they contain conductors for carrying current, insulation for surrounding the conductors and some form of overall covering having metallic and non-metallic components to provide mechanical and possible corrosion protection to ensure that the insulation may continue to operate satisfactorily throughout the life of the cable once the cable has been installed.

The majority of mains cables used in UK industrial installations comply with British Standards Specifications (BS) whilst those intended for use by the regional electricity companies (RECs) are either covered by BSs or by standards issued by the electricity supply industry (ESI), Electricity Association (EA) or individual RECs.

Table 19.1 lists the more important BS types dealt with in this chapter. All these standards make reference to other BSs for the various materials and components used in the construction of the cables.

Table 19.1 British Standard types of cables used for mains cabling.

Specification	Title	Voltage ranges for mains uses
BS 4553	PVC-insulated split concentric cables with copper conductors for electricity supply	600/1000
BS 5467	Armoured cables with thermosetting insulation for electricity supply	600/1000 1900/3300
BS 5593	Impregnated paper-insulated cables with aluminium sheath/neutral conductor and three shaped solid aluminium phase conductors (CONSAC), 600/1000 V for electricity supply	
BS 6346	PVC-insulated cables for electricity supply	600/1000 1900/3300
BS 6480	Impregnated paper-insulated lead or lead alloy sheathed electric cables of rated voltages up to and including 33 kV	600/1000 1900/3300 3300/3300 3800/6600 6600/6600 6350/11 000 8700/11 000
BS 6622	Cables with extruded cross-linked polyethylene or ethylene propylene rubber insulation for rated voltages from 3800/11 000 up to 1900/33 000 V	3800/6600 6350/11 000
BS 6724	Armoured cables for electricity supply having thermosetting insulation with low emission of smoke and corrosive gases when affected by fire	600/1000 1900/11 300

British standards for electric cables are expressed in metric terms and wherever possible they align with applicable internationally agreed standards. Such international standardization includes harmonization documents agreed by the European Committee for Electrotechnical Standardization (CENELEC) and publications prepared by the International Electrotechnical Commission (IEC). UK manufacturers, users and the electricity supply industry take an active part in the work of international standardization.

Following the publication of the EEC Public Procurement Directive (PPD), all cables falling under its requirements have to be in accordance with harmonized standards. Thus from 1995 onwards, a new series of BSs based on the PPD HDs will be issued covering l.v. and m.v. cables used for electricity distribution. These will replace the existing standards but will not introduce significant technical changes.

CABLE CONDUCTORS

Conductors for all voltage ranges covered may be of copper or aluminium. The supply industry prefer aluminium for economic reasons in the context of large quantities of cable. Although aluminium cables are lighter for equal current rating, the copper conductor is smaller because of its better conductivity and is preferred by the contracting industry because of its easier termination.

BS 6360, *Conductors in insulated cables and cords*, specifies the sizes and constructions of standard conductors used in a wide variety of cables but the individual cable specifications lay down particular requirements for each type of cable.

Copper conductors

Sizes larger than $25\,mm^2$ are normally only of stranded construction. They are made up from a number of smaller circular copper wires twisted together. The wires are in the annealed condition to provide adequate flexibility to the cable to allow for handling during installation.

Most stranded conductors are compacted by various means using wires of the same or different size. The compacting process can be applied entirely to the outer layer or partially to all layers. A smaller conductor diameter results from this – approximately 8% reduction. The removal of the interstices between wires in the outer layer reduces the penetration of material when an extruded covering is applied as an insulation.

For multicore power cables, except in the case of smaller conductor sizes, stranded copper conductors are shaped to result in the overall cable dimensions being reduced, thus saving on materials and weight. Figure 19.1 illustrates how cable dimensions are reduced by the use of shaped conductors.

Copper conductors are occasionally composed of tinned wires, the thin coating of tin on each wire serving as a barrier to prevent chemical interaction between certain constituents of elastomeric insulation materials and the copper of the conductor. Tinned copper conductors have a slightly higher electrical resistance than plain copper wires.

Aluminium conductors

Solid circular aluminium conductors, because of the rather more ductile nature of aluminium, can be used in sizes considerably greater than with solid copper but in sizes larger than $300\,mm^2$. The circular cross-section is made up from four solid segments, as shown in Fig. 19.2. This form of construction provides a conductor which can be bent by normal cable handling techniques and is also within the handling capabilities of cable manufacturing plant.

Where greater ease of bending is required use is made of stranded circular aluminium conductors which are generally available in sizes up to $1000\,mm^2$.

Solid shaped aluminium conductors provide a compact form of conductor for multicore cables and are generally used in larger sizes up to and including $300\,mm^2$.

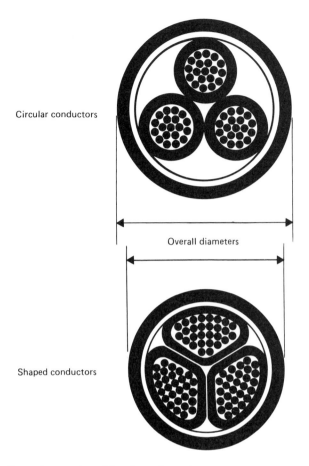

Circular conductors

Overall diameters

Shaped conductors

Fig. 19.1 Cable dimensions reduced by shaped conductors.

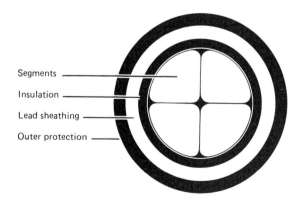

Segments

Insulation

Lead sheathing

Outer protection

Fig. 19.2 Large solid circular aluminium conductor made from four segments.

Stranded shaped aluminium conductors provide rather more flexibility for easier handling than solid shapes but produce a cable having slightly larger dimensions because small spaces exist between the wires of the stranded construction.

While copper conductors for mains cables are produced from annealed copper wire it is normal for aluminium conductors to have their individual wires in the semi-hard condition before stranding. While fully annealed aluminium provides good conductivity it is very soft and can create problems at conductor terminations of the mechanical clamping type.

INSULATION

Conductor insulating materials used in cables operating at system voltages up to and including 11 kV include impregnated paper, thermosetting synthetic insulation and thermoplastic synthetic insulation.

Impregnated paper-insulation

Cables with impregnated paper have been used successfully since towards the end of the 19th century and this form of insulation can provide a cable service life of upwards of 50 years, some of the earliest power cables still being in service. BS 6480 is the most widely used standard for paper-insulated cables for industrial use and gives constructional details and test requirements for a wide range of cables.

Cable conductors are lapped with layers of thin narrow paper tapes to the required total radial thickness. Cables for 11 kV operation have two layers of semiconducting carbon paper tape applied first to the conductor; this is known as conductor screening and its purpose is to reduce electrical stresses at the interface between the outside surface of the conductor and the insulation.

Multicore belted cables have their cores laid up together, any spaces between the cores being filled with paper or jute strings, with the whole assembly being lapped with further paper tapes to provide the belt insulation. Cables for 11 kV operation include a semiconducting carbon paper tape in the outer layer of belt papers for the purpose of providing an equipotential screen for controlling electrical stresses and minimizing the effect of any gap occurring beneath the lead sheath during load cycling.

Single-core cables for 11 kV operation have the outer surface of the core insulation lapped with a screening tape, which may be semiconducting carbon paper or metallized paper, or even thin copper or aluminium tape. This screening is for the purpose of controlling electrical stress distribution. Three-core 11 kV cables can be of the 'screened' type where each individual insulated core is screened with metallized paper or non-ferrous metal tape and the whole assembly of laid-up three cores is lapped with a binder of fabric tape incorporating interwoven copper wire in order to ensure electrical contact between the screens of all cores and the overall metal sheath. These copper-woven fabric tapes are also applied over the screening of large single-core screened cables for the same purpose of maintaining good electrical contact between screen and metal sheath.

The paper tape insulation is dried under vacuum and heat to remove moisture and is then impregnated with a special insulating compound based upon oil. Conventional impregnating compounds were based upon mineral oil containing about 20% of rosin, the rosin making the compound viscous. Cables impregnated with such compounds are entirely suitable in situations where the cable is laid in virtually horizontal runs but they have the disadvantage of tending to drain to lower points in runs where cables are laid vertically or on gradients. This results in higher parts of the cable becoming partially drained of impregnated compound with a resultant lowering of dielectric strength. Also, oil at lower levels tends to leak out of terminations or, at worst, can burst the metal sheath at the bottom of long vertical runs.

The most satisfactory solution for overcoming the drainage problems is to employ compounds composed of mixtures of oils and microcrystalline waxes with other substances which result in impregnating compounds which are virtually solid at the normal operating temperatures of cables, yet are sufficiently fluid at the impregnating temperature during manufacture to allow the paper tapes to become properly saturated. Cables having this type of insulating compound are known as mass-impregnated non-draining (MIND) types. While such cables present no drainage problems over their range of normal operating temperatures it has to be appreciated that at excessive temperatures due to overloads the compound may become sufficiently fluid to drain from higher to lower levels. Virtually all mains cables with impregnated paper-insulation are of the MIND type.

Impregnated paper-insulation must be enclosed within a metal sheathing to ensure that the insulation remains free from moisture. Ends must also be sealed from the atmosphere with a metallic or adherent synthetic cap.

Thermosetting synthetic insulation

These insulations include ethylene propylene rubbers (EPR) and cross-linked polyethylene (XLPE). Such materials are applied to cable conductors by extrusion techniques while in a hot thermoplastic state and once on the conductor the materials are subjected to a curing process which causes cross-linking of the molecules, resulting in a tough elastic material which does not soften again over the normal range of operating temperatures of the cable conductor.

BS 6899, *Rubber insulation and sheath of electric cables*, includes requirements applicable to ordinary EPR and a tougher material known as HEPR which is suitable for use in cables having conductor continuous operating temperatures of 90°C and under short circuit conditions 250°C.

Ethylene propylene rubbers have excellent electrical properties; their good ozone resistance making them suitable for high voltages. Like natural rubber, the material burns and is not particularly oil resistant, therefore the insulated cable cores must be provided with some form of external protection.

BS 6899 also covers requirements for XLPE for use in cables having conductor continuous operating temperatures up to 90°C. Before it is cross-linked, ordinary polyethylene has a somewhat sharp melting point of around 110°C so one of the important Standards tests ensures that the material removed from the finished cable is properly cross-linked and is thus capable of withstanding high temperatures resulting from short-circuit conditions where the cable conductors might reach about 250°C for short periods. The requirement for EPR and XLPE for m.v. applications are similar and are given in BS 6622.

Ethylene propylene rubbers and XLPE will burn without releasing significant quantities of corrosive gases. When judged in terms of their oxygen index, EPR will burn more slowly than XLPE.

For low voltages, both insulants will survive in wet conditions but EPR has the better potential for survival at higher voltages should moisture come into contact with either insulation.

Using the continuous vulcanization (CV) process the insulation is applied by extrusion which is followed by the passage of the insulated core through the section of the manufacturing machine where the material is subjected to the cross-linking, or curing, operation. Afterwards the insulated cores are assembled into complete cables of the required design.

As the CV method of manufacture is a long length process, an alternative, more flexible approach is to use the 'Silane' process where the cable is first extruded and then cross-linked in a controlled ambient whilst still on the drum, which is then used for the assembly of the cores into complete cables. This method is generally used for l.v. cables.

For m.v. cables it is necessary for the electrical stresses at the inside surface of the insulation adjacent to the conductor and at the outside surface of the core to be carefully controlled so that no points of concentrated high stresses develop during the operation of the cable. This screening is achieved by the use of layers of semiconducting materials immediately over the conductor and over the outside surface of the insulated core, both layers being in intimate contact with the insulation. Positive and efficient screening is achieved by extruding a thin layer of semiconducting rubber-like material over the conductor and also over the outside surface of the core. These materials are applied on the same machine and at the same time as the main insulation is extruded on to the cable conductor.

Thermoplastic synthetic insulation

For l.v. power cables polyvinylchloride (PVC) was the most widely used thermo-plastic insulation; polyethylene being used infrequently and then only for high voltages. As already stated, polyethylene has a sharp melting point around 100°C and is thus prone to cause failures where cable conductors exceed this temperature due to overloads or short-circuits.

PVC compounds are composed of mixtures of the basic resin and plasticisers together with small amounts of other compounding ingredients, such as fillers, stabilizers, lubricants and pigments. The plastics technologist, by selection of varying quantities of different ingredients, can produce a very wide range of PVC compounds having an extremely wide range of physical and electrical properties but BS 6746, *PVC insulation and sheath of electric cables*, closely specifies the requirements for compounds used in cables.

PVC insulating compounds have excellent electrical properties suited to power cables used up to and including 3.3 kV, but at higher voltages the material is not so suitable, mainly because it has a relatively high dielectric constant and dielectric losses become high. One electrical characteristic of PVC which is often not appreciated is the fact that its insulation resistance varies quite considerably with temperature. At 70°C a PVC-insulated cable has an insulation resistance some 700 to 1000 times lower than the value at 20°C.

Being a thermoplastic material, PVC is harder at lower temperatures and becomes progressively softer with increasing temperature. Unless PVC compounds are specially formulated for low temperatures (it is possible to produce compounds which can be bent at −40°C) it is usual to limit the temperature at which PVC cables are bent during installation to about 0°C otherwise there is a risk of the PVC compound cracking.

The grades of PVC used for mains cables are designed for conductor continuous operating temperatures of 70°C and under short-circuit conditions the conductor temperature should never exceed 160°C, otherwise the core insulation would flow and the cable would become permanently damaged.

When PVC is excessively heated the polymer suffers chemical degradation and at continuous temperatures above about 115°C serious chemical break-down begins to occur with corrosive hydrochloric acid forming in association with damp conditions. Standard PVC compounds are self-extinguishing when the source of ignition is removed but the material will continue to burn when subjected to fire from adjacent burning material and toxic and corrosive gases are produced. Special PVC compounds are available which have less tendency to burn and which produce lower levels of corrosive gases.

While polyethylene is little used for power cables it may be noted that it burns exceedingly well and, while burning, has the undesirable characteristics of dripping burning molten material which can easily cause spread of fire. By incorporating special additives into polyethylene the material can be made more flame retardant, but these additives seriously affect the electrical properties, resulting in an insulation not much better than PVC.

PVC insulation is applied to cable conductors by means of extrusion at elevated temperature and upon leaving the extrusion die the hot material in a soft plastic state is run through cooling water so that by the time the insulated core is wound on to the collection drum the PVC has become sufficiently hard.

For reasons associated with the burning performance referred to above and the higher operating and short circuit temperatures associated with XLPE and EPR, PVC is losing ground to XLPE as an l.v. insulation.

CABLE CONSTRUCTION

The manufacture of cable is carried out under carefully controlled conditions which include quality control procedures on the individual constituents. All manufacturing lengths are subjected to the electrical tests required by the specification.

Impregnated paper-insulated cables with lead sheathing

Paper-insulated cables to BS 6480 have a continuous sheathing of lead or lead alloy applied over the assembled cores. Lead alloys have their compositions specified in BS 801, *Lead and lead alloy sheaths of electric cables*, and for unarmoured cables Alloy E or Alloy B are generally used. Both alloys are more resistant than lead to the effects of vibration, Alloy B being used in cases where severe conditions of vibration are to be experienced. It may be noted that lead can develop a crystalline structure when subjected to vibration and cracks can then develop through the wall of the sheath on the cable with the ultimate result that moisture or water can enter and cause electrical breakdown of the insulation. Armoured cables often have lead or Alloy E.

The lead or alloy sheathing is generally provided with some form of outer protection. Single-core and unarmoured multicore cables are generally provided with an oversheath of PVC or medium density polythene (MDPe) but the majority of multicore cables are of the armoured type. Armour, for mechanical protection of the lead sheath, can be in the form of double steel tape armour or wire armour. Steel tape armour provides a certain degree of mechanical protection against sharp objects penetrating the cable from outside and consists of two helical lappings of tape, applied with gaps between convolutions, the outer tape covering centrally the gaps in the underlying tape. The tapes are applied with gaps to allow bending of the cable.

For protection giving longitudinal mechanical strength and a certain degree of protection from penetrating objects, single-wire armouring is used, consisting of a helical layer of galvanized steel wires. For additional longitudinal strength, for example for cables which need to be suspended vertically, double-wire armouring, consisting of two layers of galvanized steel wires, is applied, the outer layer of wires being applied in the reverse direction from the inner layer. A separator of waterproof compounded textile tape is applied between the two layers of wires to allow the wires a certain free movement in relation to one another when the cable is bent.

Armoured cables have a bedding applied first to the lead or lead alloy sheath and this may consist of lappings of compounded paper and textile tapes, all adequately coated with waterproof compound, or if the cable is to have an overall extruded PVC on MDPe sheath the bedding may also consist of an extruded covering.

Unless cables are to be left with the wire armour in the bright condition, it is usual for the armouring to have an overall protection in the form of serving applied in the form of compounded textile tapes or as an extruded sheathing.

If single-core lead sheathed cables are required to be armoured for mechanical reasons, the armouring must be composed of wires of non-magnetic material, such as aluminium alloy, otherwise considerable electrical losses and heating effects occur due to magnetic hysteresis in steel wires with alternating current loads of more than about 50 A.

Impregnated paper-insulated cables with aluminium sheathing

Instead of a lead or lead alloy sheath for providing the moisture-proof covering for protecting the impregnated paper-insulation an alternative design uses aluminium sheathing. Aluminium sheathing is mechanically tougher than lead therefore aluminium sheathed cables generally do not require any further armouring for mechanical protection.

Aluminium sheathing is also more rigid than lead sheathing and large diameter cables having smooth aluminium sheaths are more difficult to handle during laying, especially at bends. To overcome this problem corrugated sheathing is used, the annular corrugations being formed after the sheathing has been extruded on to the cable cores in smooth form. The result is a cable having equivalent bending performance to that of lead covered wire armoured cable. Figure 19.3 illustrates the design of a PICAS (paper insulated corrugated aluminium sheath) cable to ESI Standard 09−12.

Should the corrugated sheath become damaged there could be the possibility of moisture movement from the local damage moving along under the sheath, resulting in a long length of cable becoming unserviceable, so to avoid this possibility the annular corrugations are substantially filled with a suitable compound.

Insulated cores

Aluminium sheath

Bitumen compound

PVC sheath

Fig. 19.3 6350/11 000 V PICAS paper-insulated cable with corrugated aluminium sheath (*Pirelli Cables Ltd*).

The quantity of compound has to be carefully controlled so that when the cable is operating at its full load temperature the internal pressure within the sheath due to thermal expansion does not create undue pressure.

Aluminium sheathing, whether smooth or corrugated, must be adequately protected from corrosion when laid in the ground and the usual protection consists of a good coating of bitumen compound applied directly to the aluminium sheath followed by an extrusion of PVC.

The above types of aluminium sheathed cables are generally used only on 11 kV systems by regional electricity companies who have the necessary skills for installation of such cables.

CONSAC cables

Because aluminium sheathing provides a good electrical conductor there is one type of cable which makes use of this sheathing as the neutral conductor. This is CONSAC cable in accordance with BS 5593 or ESI 09−8 which has been used by some UK electricity boards for l.v. mains. The word CONSAC is derived from 'concentric aluminium cable'. Figure 19.4 shows the general construction which consists of three shaped aluminium conductors for the three phases, each paper-insulated and then having an overall belt, the whole insulation being impregnated with non-draining compound, then encased in a smooth aluminium sheath which is liberally coated with bitumen compound and finally sheathed overall with PVC. As with all aluminium sheathed cables, the bitumen compound shall adhere to both sheath and PVC.

CONSAC cables are used in systems where it is permitted to employ combined protective and neutral conductors (PEN). One advantage claimed for CONSAC cables over other PEN types having extruded synthetic insulation is that should the aluminium sheath become corroded there is every possibility that the cable insulation would fail because of ingress of moisture through the corroded sheath and the circuit protection would isolate the cable long before the neutral conductor became completely corroded away with the loss of the earthing connection. Other types of PEN cables could lose their neutral and earth continuity yet would still provide the phase voltage and thus could present a hazard to the electricity consumer.

CONSAC cables can be made in sizes up to and including $300 \, \text{mm}^2$ with 100% neutral conductor.

Thermosetting elastomeric insulated cables

Cables with EPR and XLPE insulation are gradually tending to replace impregnated paper-insulated cables because of easier handling and terminating techniques.

A typical 11 kV three-core cable is shown in Fig. 19.5. This has heavily compacted stranded aluminium conductors with semiconducting screening and extruded EPR insulation. The insulated cores are covered with semiconducting screening material which are in turn lapped with metal tapes in order to ensure longitudinal electrical continuity. The three screened cores are laid-up together, filled circular, then wire armoured and sheathed overall with PVC. This, and many other designs, are covered in BS 6622.

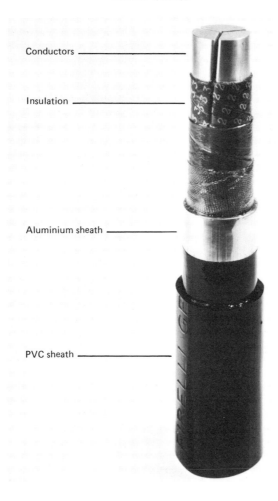

Conductors

Insulation

Aluminium sheath

PVC sheath

Fig. 19.4 600/1000 V CONSAC cable (*Pirelli Cables Ltd*).

BS 5467 provides requirements for wire armoured cables having EPR or XLPE insulation for operating at voltages up to 3.3 kV. The specification covers a wide range of conductor sizes in both copper and aluminium conductors. As well as wire armouring composed of circular wires the specification also includes aluminium strip armour consisting of a layer of strips of fairly hard aluminium. The bedding under the armour and overall sheath are of a grade of PVC suitable for the 90°C operating temperature.

BS 6724 gives requirements for a range of cables similar to those in BS 5467, using the same insulation materials but special sheathing materials such that the cables have low emission of smoke and corrosive gases when affected by fire. Modern low smoke sheathing materials can have mechanical properties similar to PVC and so the cables may be handled and installed in a similar manner to their pre-sheathed equivalents. However, because of differences in properties of sheathing compounds from one supplier to another, the advice of the cable manufacturer should be sought before general use of such cables.

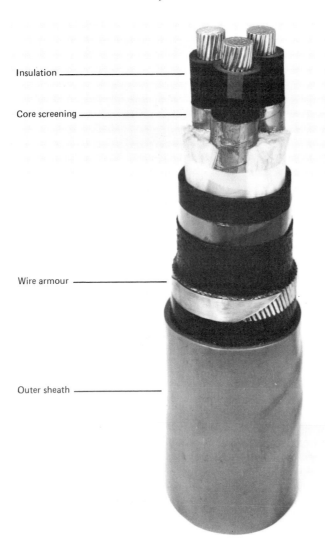

Insulation

Core screening

Wire armour

Outer sheath

Fig. 19.5 Elastomeric cable for 6350/11 000 V operation (*Pirelli Cables Ltd*).

Sheathing having low emission of smoke and corrosive gases can also be applied to cables generally in accordance with BS 6622.

Waveconal (aluminium waveform) cables

As an alternative to CONSAC cables extensive use is made in electricity distribution at low voltages of Waveconal cables to ESI Standard 09−9 which have solid shaped aluminium conductors insulated with XLPE. The three-phase cores are laid up together and lapped with an open spiral of clear plastic tape. The assembled cores are then covered with an extrusion of rubber-like material which acts as a

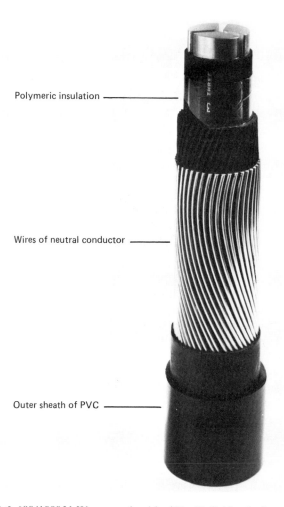

Polymeric insulation ─────────

Wires of neutral conductor ───────

Outer sheath of PVC ────────

Fig. 19.6 600/1000 V Waveconal cable (*Pirelli Cables Ltd*).

bedding for the concentric neutral conductor and the same material also completely covers the wires of the neutral conductor.

The aluminium wires forming the neutral conductor, which may be seen in Fig. 19.6, are applied in an oscillatory manner instead of in a helical form and this enables the cores to be exposed when making tee-joints under live conditions without the need to cut the neutral wires, and thus break the return path of the combined protective and neutral conductor of a TN-C-S system.

The aluminium neutral wires are carefully segregated so that they are individually surrounded by the rubber-like filling material in order to prevent moisture transference around the neutral in the event of mechanical damage.

The cable is finally sheathed with PVC to provide mechanical protection to the rubber-like covering around the neutral wires.

A similar construction of cable (copper waveform) is available which uses copper wires forming the neutral conductor and is particularly used where there are concerns of possible corrosion of aluminium wires.

Cables with synthetic thermoplastic insulation

Many low-voltage cables in and around buildings have PVC insulation and BS 6346, *PVC-insulated cables for electricity supply*, fully specifies a large range of cables having both copper and aluminium conductors, with and without armouring.

For vertical risers and some submains, single-core unarmoured designs would be used but in general it is more normal to employ wire armoured or aluminium strip armoured cables for the mains supply where PVC cables are required.

All the cables in BS 6346 have their PVC insulation applied to the copper or aluminium conductors by extrusion; multicore cables with shaped conductors have their insulation applied in sectoral shape so that the complete cable is compact and thus overall dimensions are kept to a minimum. Figure 19.7 shows examples of cross-sections of typical solid sector-shaped aluminium conductor cables.

Cables have a bedding for the armour of either an extruded covering of PVC or a lapping of two or more plastics tapes.

Armouring on copper conductor and some aluminium conductor cables consists of a conventional layer of galvanized steel wires, although single-core cables, requiring armouring, have wires of non-magnetic material if the cables are to be used in a.c. circuits.

Aluminium conductor cables can also have aluminium strip armouring, consisting of a single layer of aluminium strips of the sizes specified in BS 6346. This form of armouring is useful where low armour resistances are required; aluminium strip armouring has a much lower resistance than galvanized steel wire armouring. For smaller cables the strip armouring resistance is approximately half that of steel wire while for larger sizes the aluminium strip has only one quarter of the resistance.

All BS 6346 cables have an overall extruded sheathing of PVC.

Cables to the above specification are also available for operation on 3.3 kV circuits. Such cables are similar to those for use at l.v. but the thickness of core insulation is somewhat greater. At 3.3 kV there is no necessity for conductor or core screening as is required for higher voltage cables.

Concentric neutral-and-earth service cables having PVC insulation are made in a variety of combinations having copper or aluminium phase conductors and concentric neutral conductors of aluminium or copper wires. These cables are generally not used in sizes greater than $35 \, mm^2$ and can have one or three phase conductors. Details of these cables are given in ESI Standard 09−7.

Where a concentric cable is required with separate neutral and earth-continuity conductors, there is a design available in accordance with BS 4553, 600/1000 V *PVC-insulated single-phase split concentric cables with copper conductors for electricity supply*. These cables, in sizes up to and including $35 \, mm^2$, have the central circular stranded copper phase conductor insulated with PVC. The neutral conductor is made from a number of single copper wires, each covered by an extruded layer of PVC to a diameter the same as the wires forming the earth-continuity conductor.

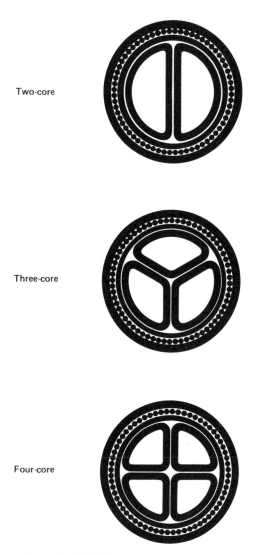

Two-core

Three-core

Four-core

Fig. 19.7 Cross-sections of typical 600/1000 V solid aluminium conductor PVC-insulated cables.

The earth-continuity conductor wires are bare and not individually covered with PVC.

The covered neutral wires and the bare earth-continuity wires are applied around the central phase core as a concentric layer, the neutral wires being together as a band and the earth-continuity wires as a separate band. PVC string separators, about the same diameter as the bare copper wires, are laid either side of the group of earth-continuity wires to separate the group from the covered wires forming the neutral conductor.

A binder of non-hygroscopic tape is applied immediately over the concentric layer and the whole assembly is sheathed with PVC.

There is now a marked preference for XLPE as an insulation instead of PVC for a wide range of l.v. cables. Whilst the dimensions and handling characteristics of both cables are similar, XLPE has a much better rating for continuous, emergency and short-circuit conditions. Even when these are not required, the margin of safety is enhanced compared with PVC.

INSTALLATION

General care in handling

Cables can be perfectly manufactured and meet all the required specification tests at the cable factory and then be easily ruined and made unfit for service if not handled with care during installation.

Unlike building wiring cables, mains cables are generally large and are on heavy drums, so in designing an installation careful thought must be given to the method by which the drum of cable is to be delivered to site and removed from the delivery vehicle. Dropping a drum of cable from a lorry tail board not only causes possible damage to the cable but can cause collapse of the wooden drum and subsequent severe damage to the cable when attempting to remove the tangled turns from the broken drum.

If drums of cables have to be rolled any distance over the ground the battens should be left in their fixed position so that some protection can be given to the outside turns on the drum. Many cables have been damaged by rolling over a removed batten on the ground with the sharp ends of the nails pointing upwards and penetrating the cable passing over it.

Where drums are rolled it is important that the direction of rotation should be in accordance with the arrow on the side of the drum. If this is not observed the turns of cable on the drum can work loose and very large longitudinal forces set up which either push the inner end out of the flange or, if it is rigidly secured, may damage the cable itself.

When pulling a cable off a drum a spindle of adequate size and strength should pass through the central holes in the flanges and proper drum jacks should be used to support the spindle. A person should be stationed at the drum position to stop the drum unwinding as soon as cable pulling is stopped, otherwise the drum can over-run and a damaging kink form in the cable when it becomes bent too sharply.

With long pulls, especially with heavy cable and also where straight-pulls are not possible, cable rollers should be placed at frequent intervals to support and guide the cable. Care should be taken to ensure that the cable runs in the centre of each roller and does not become wedged at roller edges.

Where cables are cut or where factory seals are removed during the laying operation the exposed ends should be quickly sealed. This applies not only to impregnated paper-insulated cables which are quickly damaged by ingress of moisture but to all types of insulation. If cut ends are left exposed to the elements, or worse, left in the ground, water can penetrate even along elastomeric and thermoplastic cables, leading to untold problems later during jointing or terminating operations.

Methods of installations

The method of installing a cable obviously depends upon where the cable has to run but the shortest route is not necessarily the most economic route. Rocky ground can be expensive to excavate and corrosive ground can call for expensive anti-corrosion finishes to the cable.

Conventional methods of installing mains cables are: laid direct in the ground; pulled through ducts in the ground; laid in troughs; or installed in the air on supports such as cable trays or cable ladders.

Laying cable directly in the ground involves expensive excavation and cables must be laid at a sufficient depth to ensure that no disturbance or damage to the cable can take place under normal circumstances. The soil immediately around the cable should contain no sharp stones or similar objects, ideally it should be sifted. In order to comply with the IEE Wiring Regulations (BS 7671) cables buried in the ground should be marked with tiles or by some other suitable means so that persons excavating the ground later have some warning of the presence of the cable before reaching it.

If several cables are laid side by side in the trench the spacing between them should be carefully maintained otherwise their current-carrying capacities may be seriously affected.

Cable trenches should be filled in again as quickly as possible after cables have been laid in order to reduce possible damage to the cable at the bottom of the trench.

Where cables pass under busy roads it is useful for ducts to be used, spare ways being left so that additional cables can be later drawn through without the need to excavate the road again.

Where cables are laid in troughs thought has to be given to the possible effects of adding more cables at a later date, as these can adversely affect the current ratings of all the cables.

Cables installed in air must be supported at intervals sufficiently close so that no mechanical strain is exerted on them. Cable manufacturers are able to give necessary guidance. Where cables are run on trays or cable ladders the positioning of the trays or ladders requires careful consideration as they are often used by other trades for walking upon to gain access to adjacent areas and the cables can become damaged.

The installation of single-core armoured cables requires care because generally there is a considerable current circulating in the armouring. The rating of the circuit depends on how the cables are laid, i.e. in trefoil, flat touching or flat separated. The use of two cables per phase is not unusual and these should be laid in accordance with the manufacturer's instructions to avoid overheating within a cable or at the terminations.

For all methods of installation, cables should not be bent in radii close to terminations less than those given in the manufacturer's recommendations but during the actual operation of pulling the cables into position every effort should be made to ensure that cables are only bent into somewhat large radii.

In cold weather conditions cables having PVC components should not be installed when the cable or ambient temperatures are at 0°C or lower. Installation should not commence until the temperature has been above 0°C for at least 24 h unless the cable can be stored in a warm atmosphere to ensure that it is well above freezing point and it can be taken out and laid quickly. In freezing conditions there is a chance that the PVC material may shatter. The same care must be taken with any cables having bituminous compounded coverings because these also can crack.

Precautions against spread of fire

In all installations precautions should be taken to ensure that cables cannot spread fire. Where cables pass through walls or partitions the holes should be made good with fire-resisting material.

Where vertical risers pass upwards through floors fire barriers must be fitted at each floor. Not only does this restrict spread of fire but alleviates a build-up of heat at the top of the cable run which would otherwise affect the current rating of the cable.

Where cables run together in large groups there is always a risk that one burning cable may ignite adjacent cables and eventually the whole group can be involved in a major cable fire. Special PVC compounds and compounds having low emission of smoke and corrosive gases, having better flame-retardant properties than standard grades, are available for cables in fire risk areas but wherever possible cables in large groups should be avoided.

Cables without continuous metal sheathing should not be run in ground liable to be saturated with hydrocarbons, such as from the petrochemical industry, because some liquids, while not attacking PVC and some other elastomers, are able to penetrate through cable coverings and insulation and flow along the cables into switchgear and apparatus to which the cables are connected, thus creating a dangerous fire and explosion risk. Where cables have to be run through such ground the cable design should include a lead sheathing as this is impermeable to such liquids. Precautions might have to be taken to protect the lead sheath from corrosion.

Problems with electric cables involved in fires have led to several improvements in cable design and their all-round fire performance. The first of these involved fire propagation to avoid the effects arising from bunches of cable installed close together. BS 4066: Part 3 specifies a test method for various levels of cable loading known as Categories NMV 7 (A), NMV 3.5 (B) and NMV 1.5 (C) and should be invoked where the possibility of fire propagation exists.

Secondly, the location of the fire source, escape and fire fighting are greatly hampered by smoke, and the presence of corrosive gases in the smoke can result in damage to electrical equipment and structures as well as being injurious to the health of personnel. Cables having low emission of smoke and corrosive gases when affected by fire are widely available and overcome these problems. Test methods for smoke emission and corrosive gas emission are given in BS 7622 and BS 6425 respectively.

JOINTING AND TERMINATING

Jointing and terminating mains cables of any type involves making adequate electrical connections to conductors, ensuring continued satisfactory functioning of the cable insulation and providing adequate electrical continuity to the metal sheath, armouring or other metallic covering.

Conductor connections

Any joints or terminations in conductors must be capable of carrying any load current to be expected during the life of the cable and also very heavy currents which may occur as a result of faults and short-circuits. All conductors expand when they become heated by the passage of currents and contract when the current is reduced or removed. This expansion and contraction is both longitudinal and radial, thus, cable conductors become longer when hot from the passage of the normal load current and much longer from the passage of short-circuit currents. On cooling, the conductor tries to return to its original length. These longitudinal movements can create considerable forces which must be taken into account in joint and termination design.

The radial expansion and contraction has to be accommodated in the device for making the electrical connection to the conductor; the electrical resistance of the connection must remain very low throughout the life of the connection and this requires that the conductor metal must remain in constant intimate contact with the connecting device at all times. The passage of current through any increased resistance will result in the generation of additional heat which may lead to the ultimate failure of the connection and its insulation.

For copper conductors soft-soldering into lugs or ferrules has been the traditional method for terminating or jointing for many years and is still considered the most reliable method when properly carried out but it involves the use of heat and certain skills. The alternative is to use crimping methods where lugs or ferrules having sufficient wall thickness are compressed, usually by hydraulic means, on to the conductor. Many designs of compression-type joints are available but the user should always first ensure that the one selected has been fully type tested under all conditions. The most reliable tools for undertaking the compression operation are those which cannot be removed from the joint until the pressure device has travelled the whole distance of its compression stroke. In all cases the sequence of operations recommended by the manufacturer must be followed.

Aluminium conductors can be soldered; there are satisfactory solders and fluxes available from a large number of sources, but such soldering requires more skill than for copper conductors. It is well known that aluminium develops a thin oxide layer on its surface which is difficult to break down and therefore special techniques are necessary in order to obtain satisfactory alloying between the aluminium and the solder. In the case of stranded conductors it is not only necessary to make a good electrical connection to the outside layer of wires but the inner layers in large strands must also be jointed so that they may carry their full share of the load current.

The jointing of solid aluminium conductors is more usefully undertaken by crimping methods, using specially designed hydraulically operated tools. Again, only systems which have satisfactorily withstood type tests, involving long-term cyclic testing should be used.

Mechanical connectors using a clamping screw or bolt are used for both copper and aluminium over a wide range of conductor sizes. The need for the application of a correct torque is particularly important for aluminium conductors and this has led to the development where the clamping screws are designed to shear at predetermined torque levels.

The much wider use of thermosetting insulants having much higher temperature capabilities is reducing the use of soldering techniques owing to their restricted temperature range.

Insulation at joints and terminations

Impregnated paper-insulated cables must have their insulation protected from ingress of moisture and conventional joints involve making good the insulation over the conductor joints by lapping with impregnated paper tapes then enclosing the whole joint in a metal sleeve which is plumbed or soldered to the ends of the existing metal sheathing, the whole sleeve then being filled carefully with special compound which is generally poured in hot. Conventional terminations are dealt with in a similar manner, the end of the cable being enclosed in a suitable case which is filled with special compound to maintain it air- and moisture-tight.

These conventional methods have withstood the test of time, but changes in switchgear and transformer cable boxes and changes in cable design have resulted in the rapid growth of 'dry' terminations and joints, particularly for m.v. cables. Terminations of paper-insulated cables are now 'sleeved' either with heat shrink or float-down techniques. These sleeves are fitted on to an extended crutch cap and, where the cable has a screened insulation, an additional stress control tube is also applied. Some designs incorporate a synthetic tape beneath the tube in order to provide extra protection against the ingress of moisture. The heat shrink materials must be adequately and evenly heated whilst the float-on technique involves no heat but uses a simple foot pump to expand the sleeving, Fig. 19.8.

These features can also be applied to straight-through and other joints for paper-insulated cable. Whilst acrylic or polyurethane resin is widely used, the more modern sleeved type of joints are becoming popular. Resin joints have a shelf-life limitation and may be affected by ambient temperature. While the principle is simple, the mixing of resin and hardener must be undertaken conscientiously and the pouring must ensure that no air bubbles exist as these become potential breakdown areas. Adhesion to the cable sheaths is essential to prevent the entry of moisture and, in addition with polymeric m.v. cables, good adhesion to the core insulation is essential to avoid introducing a breakdown path under the influence of lightning impulse voltages. Figure 19.9 shows a typical profile of a modern straight-through joint in an 11 kV polymeric insulated cable.

Low voltage polymeric cables having EPR, XLPE and PVC insulation can have their outer coverings just trimmed back for sufficient distance to ensure no surface tracking at terminations and there is no need for the ends of the cores to be enclosed in any sort of compound in normal dry situations. At joints for these l.v.

Fig. 19.8 Outdoor termination with 'float on' protection for 11 kV polymeric- and paper-insulated cables (*Pirelli Cables Ltd*).

cables it is necessary to seal the jointed cores from ingress of moisture from the ground in the case of buried joints and the usual manner is to enclose the whole joint in compound or resin. In pouring hot compound into PVC cable joints the temperature of the compound must not, of course, be so high as to cause softening of the insulation. As with paper-insulated cables, polyurethane and similar resins must be mixed and poured so that no air bubbles remain after the material has hardened.

Medium voltage polymeric cables normally have a screen around each core consisting of a non-metallic and a metallic part. The former is generally an extruded layer of semiconducting material which is designed to be stripped in the cold condition with the assistance of specialized tools which are commercially

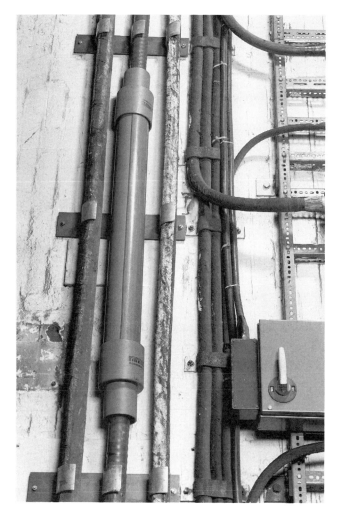

Fig. 19.9 Typical profile of modern straight through joint in 11 kV polymeric-insulated cables (*Pirelli Cables Ltd*).

available. Removal of the screen is a critical step in jointing, especially where a knife is used. It is useful to practise stripping the screen from a scrap end of the cable before actually jointing.

At the point where the screens are removed from the core exists an area of high electrical stress requiring special precautions. To overcome these problems, stress-relieving devices are applied. They may be in the form of a moulded cone, heat-shrinkable or cold-shrink tubes. These devices must be carefully fitted to ensure that no air pockets exist in the area of high stress.

Continuity of metal sheath or armouring

The metal sheath, armouring or other metallic covering of any cable must be capable of carrying fault currents which may flow through it and where such metal is relied upon as an earth-continuity conductor it must provide a low resistance path at all times. This means that at any joints or terminations the continuity must be efficiently maintained throughout the life of the installation.

Over the years conventional joints and terminations with paper insulation have proved satisfactory because the metal sheathing has been plumbed or soldered to whatever it is connected and armour wires have been adequately connected into massive metal joint box enclosures or glands. With the advent of more modern jointing techniques the continuity of metal sheathings and armouring has depended more upon various forms of metal clamps. Tests with overload and fault currents and with cyclic loadings, also under damp corrosive conditions, have been conducted on many new designs of jointing methods for metal sheaths and armour wires and many have been found to be unreliable, developing into high resistance connections with passage of time.

Before any jointing or terminating methods are selected for use in a particular cable installation the suppliers of the system should be requested to produce evidence to show that the technique for providing continuity has satisfactorily withstood appropriate tests when used in conjunction with the particular type of cable to be installed.

In general, for resin-type joints, armour wires are best connected from one side of the joint to the other by means of crimped joints, a bunch of wires from one side being carried over and joined by a crimped ferrule to a bunch of wires from the other side.

At terminations the wire armouring or metal strips of polymeric cables are most efficiently secured in glands to BS 6121, *Mechanical cable glands*. This standard covers a wide range of glands and the correct type for a particular cable must be selected. It is important that full details of the cable to be accommodated are stated at the time of ordering.

Copper wire screens are a common feature with m.v. extruded cables. When bonding them at joints or terminations, it is usual to twist them into a pigtail with a crimped connector, but in areas of high ambient temperature, this could become a hot spot unless additional copper wires are inserted.

In corrosive conditions all types of metal sheath, armour and metal strip connections should be well encased in suitable material to reduce risk of corrosion of the electrical connection.

CABLE RATINGS

The passage of current through a cable conductor raises its temperature. In the case of sustained current an equilibrium is eventually established and the heat generated is equal to the heat dissipated through the insulation, sheathing and other coverings and finally into the surrounding ground or air. The highest value of current a cable can carry continuously depends upon the highest safe operating temperature applicable to the insulation or other parts of the cable. Table 19.2 gives the maximum temperatures for continuous operation for a range of insulants and these values are used for the calculation of sustained current ratings.

Current ratings may be determined from practical field tests for any size and type of cable installed in a particular manner by making sufficient temperature

Table 19.2   Temperature limits of mains cable insulants.

Insulating material	Maximum conductor temperature (°C)
Impregnated paper below 6350/11 000 V	80*
Impregnated paper 6350/11 000 V other than belted	70[†]
Impregnated paper 6350/11 000 V belted	65[†]
Polyvinylchloride (PVC)	70
Hard ethylene propylene rubber (EPR)	90
Cross-linked polyethylene (XLPE)	90

* 60°C for unarmoured lead sheathed cables laid in ducts.
[†] 50°C for unarmoured lead sheathed cables laid in ducts.

measurements at appropriate values of current in association with relevant ambient temperature conditions. Alternatively, ratings may be calculated from first principles or may be more easily obtained from published tables.

Methods for calculating ratings have been standardized on a world-wide basis and are given in IEC Publication 287, *Calculation of the continuous current rating of cables (100% load factor)*. In order to calculate current ratings one must have detailed information of the cable construction as dimensions and properties of the components enter into the calculations.

The usual method is to obtain ratings from published tables which cover standard cable types and cater for a variety of installation conditions. For l.v. cables it will be found that the tables of current ratings included in the IEE Wiring Regulations relate generally to cables installed above ground and mostly in air. These ratings are based on an ambient air temperature of 30°C. Ratings for power cables are provided in more detail in the 69−30 series of reports issued by ERA Technology Ltd and cover installation methods for cables in the ground, in underground ducts and in air.

These ratings are tabulated for a standard ground temperature of 15°C and an ambient air temperature of 25°C; the different air temperature results in slightly different tabulated current ratings from the IEE values. Factors in both documents allow adjustments to be made for particular ambient temperatures.

Current rating tables are also included in catalogues of many cable makers.

In using published ratings tables it is most important to take cognisance of all the notes and footnotes in order that the correct factors are applied to the standard ratings to take account of the actual installation conditions. In particular, attention must be paid to the temperature of the surroundings in which the cable is to be installed, the precise method of installation and the effects of grouping.

If the temperature of the air or the ground surrounding the cable is higher than the standard value appropriate to the published rating the temperature-rise applicable to that rating would result in the cable conductor temperature exceeding the maximum permitted operating temperature for the conductor insulation. For this reason the rating must be reduced by the application of the appropriate factor. Only in cases where it can safely be assumed that the surrounding temperature will always be lower than the standard value is it possible to take advantage of a slight increase in rating.

The manner in which a cable is installed in relation to its surroundings and any other cables affects its ability to dissipate heat. For this reason the factors appropriate to such matters as depth of laying, grouping, type of soil, must be applied to the published ratings.

For cables laid in the ground the ERA reports provide tabulated ratings based upon a solid thermal resistivity of 1.2°C m/W which can be a fair average value for much of the UK but the type of soil affects its thermal resistivity considerably and it is strongly recommended that the information regarding different types of soil provided by ERA is studied. In some areas of the world very dry conditions for long periods can result in high values of soil resistivity which can reduce the safe ratings of cables.

Impregnated paper-insulated cables are well established and their current ratings in all sizes of copper and aluminium conductor up to high voltages are provided in ERA reports. Similar information relating to extruded dielectric cables is also contained in ERA reports.

Care should be taken when selecting the conductor size for medium voltage cables as the smallest possible size does not necessarily provide the lowest system cost. This is due to the cost of I^2R losses when they are used to evaluate a system cost.

TESTING AND FAULT FINDING

After mains cables have been installed, jointed and terminated they should be subjected to electrical tests to ensure that they have not become damaged and that joints and terminations have been properly made.

The essential test is a h.v. test which is more conveniently carried out with d.c. than a.c. as long lengths of cables having high capacitances would require an unacceptably large size of testing transformer for a.c. tests (due to the high charging currents involved).

Table 19.3 shows the values of applied voltages which should be increased gradually to the full value and maintained continuously for 15 minutes between conductors and between each conductor and sheath. No breakdown should occur.

The test voltages given are intended for cables immediately after installation

Table 19.3 Test voltages for cables after installation.

Cable voltage designation (V)	Test voltage d.c.	
	between conductors (V)	between all conductors and screen or sheath or armour (V)
600/1000	3 500	3 500
1900/3300	10 000	7 000
3800/6600	20 000	15 000
6300/11 000	34 000	25 000

and not for cables that have been in service. When testing is required for cables that have been in service, the cable manufacturers should be consulted for appropriate test conditions which will depend on the individual circumstances. A lower voltage should be applied taking in account the age, environment, history of breakdowns and purpose of the test.

If switchgear or other apparatus cannot be disconnected from the ends of the cable during the test, the value of test voltage must be the subject of agreement with all parties concerned.

During the testing period special precautions must be taken to keep all persons away from any exposed parts which may be charged at the high voltage.

If cables with special anti-corrosion sheathing have been used and it is thought that the sheathing could have become damaged during cable laying operations it is sometimes beneficial to apply a voltage test between the cable sheath and the ground before the cable is terminated. The value of test voltage should be no more than 4 kV, per mm of oversheath thickness with a maximum value of 10 kV applied for 1 minute.

If a cable should break down as a result of h.v. testing, the position of the breakdown must obviously be found so that a repair or replacement can be made. If the test set can provide sufficient power the fault can often develop into a low resistance path which may be located after the removal of the h.v. by conventional methods using resistance bridge techniques. More often than not, the fault remains of high resistance which can involve problems of location. Sometimes the position of the fault can easily be detected by ear if the test voltage is left connected, preferably at a lower value than the full test value.

Various makes of sophisticated fault location gear are available for finding high resistance faults and most cable manufacturers offer a fault location service.

One quite effective method is to charge a large capacitance by the h.v. test set through a resistance and then discharge it into the faulty cable core through a spark gap. The voltage and size of spark gap are arranged so that the capacitor discharges every few seconds. By walking along the cable the position of the fault can usually be heard as the insulation breaks down each time the capacitor discharges. A stethoscope, consisting of a microphone, amplifier and earphone, is useful for detecting the fault position in buried cables.

CHAPTER 20

Selection of Wiring Systems

R.G. Parr, ACGI, CEng, MIEE
(Consultant)

D.W.M. Latimer, MA(Cantab), CEng, MIEE
(Consulting Electrical Engineer)

INTRODUCTION

This chapter deals with the selection of wiring systems for voltages up to and including 1 kV. It provides a survey of the more frequently used systems and comments on the electrical and environmental duties for which each is suitable. Systems for alarm and control circuits are included.

BS 7671 Requirements for Electrical Installations (the IEE Wiring Regulations) requires at clause 130−02−03 that all conductors shall be of sufficient size and current-carrying capacity for the purpose for which they are intended. Clause 311−01−01 goes on to say that in determining the maximum demand of an installation diversity may be taken into account.

The starting point for selection is an assessment of the load, considering both the number, location and size of power consuming outlets, required to fulfil the purpose of the installation, and the total load to be taken from the supply. Guidance on the determination of maximum demand from connected loads in domestic and commercial installations is given in the IEE Guidance note No. 1 to BS 7671; the guidance given is of a general nature and experience of particular types of equipment and of installations may allow the designer to adopt different values. For industrial installations, methods used to establish demands differ with the type of industry and must be based mainly on past experience with particular industrial processes.

Once the expected maximum demand has been determined, a meeting with the supplier can decide the type of supply, earthing arrangements and the prospective fault-currents at the origin of the installation. This information will have an important bearing on the selection of wiring system(s) for the installation.

Division of the installation into circuits and the routing of each circuit is a matter which concerns both safety and convenience. Attention is drawn to the requirements of BS 7671: Part 3, *Assessment of general characteristics*, which are relevant at this stage.

Some types of circuits such as emergency lighting and fire alarms have particular requirements for wiring systems laid down in standards and are also required to be segregated.

517

When the number of circuits, loads, routes, lengths and environmental conditions have been decided, consideration can be given to selection of appropriate wiring systems.

Fuses or circuit-breaker ratings are initially selected on the basis of the load which the circuit is designed to carry. There are situations in which the fuse rating is selected on the basis, not of the load in the circuit, but of the necessity of ensuring that in the case of a short-circuit or earth fault, an upstream protective device does not operate before the device intended to protect the faulty circuit.

In the case of two fuses or two MCBs this can usually be achieved by ensuring that the rating of the upstream device is two or more rating steps higher than that of the down stream device.

In the case of a combination of MCBs and fuses the difference in ratings may need to be substantial. Figure 20.1 shows typical characteristics for a 63 A MCB and various BS 88 fuses and indicates that to be certain of discrimination it is necessary to select a 315 A or possibly a 400 A fuse upstream of the MCB.

While failure to discriminate is an inconvenience to the user (and to the electricity supplier if the upstream device is his cut out fuse) as far as the designer of the installation is concerned it is necessary to select the cable on the basis of the fuse rating to achieve discrimination and not the load.

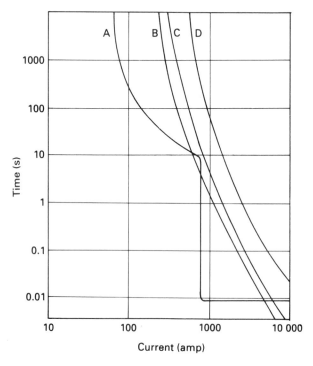

Fig. 20.1 Discrimination between a 63 A MCB and various BS 88 fuses: (A) 63 A MCB; (B) 250 A fuse; (C) 315 A fuse; (D) 400 A fuse.

WIRING SYSTEMS

A wiring system is defined in BS 7671 as 'an assembly made up of cable or busbars and parts which secure, and if necessary enclose, the cable or busbars'. A description of a wiring system includes the type of cable (or busbar) and how it is supported and protected. Most systems relate to cables and these are the types mainly dealt with here. Systems involving busbars are usually factory assembled items and are covered later in the chapter.

It is conventional to recognize three types of circuit; mains, sub-mains and final, with which experience tends to associate certain types of wiring system. A main circuit feeds a distribution board located at the centre of a group of loads or it may feed a single large load. Where the demand is high the main circuit may operate at a voltage higher than 1 kV. Conductors in such systems are usually in the form of cables, as described in Chapter 19.

In extensive installations it may be appropriate to divide the circuits further using submains which connect a main distribution board to smaller boards more convenient to load groups. In very large installations the main switch board may feed one or more section boards which in turn feed further distribution boards. Final or wiring circuits connect distribution boards to individual points of utilization.

For small installations, such as in most domestic and small commercial premises it is usual for a distribution board to be located at the supply intake, when it takes the form of a consumer unit, with final circuits going directly to points of utilization.

Some types of cable are suitable for use in either mains or wiring circuits (e.g. the 1 kV cables in BS 6346), while others are used almost exclusively in final distribution circuits and are regarded as wiring types. Popular examples of the latter are cables containing a protective conductor and the insulated conductors used with conduit or trunking. On the other hand mineral-insulated cable, while being generally regarded as a wiring type, can be used in the larger sizes for sub-mains circuits as may the larger sizes of BS 6004 cables run in trunking.

Cable types most commonly used for main circuits are the larger sizes of single or multi-core PVC insulated, wire-armoured and PVC sheathed (BS 6436) and single or multi-core elastomeric insulated, wire-armoured, covered with a synthetic material (BS 5467 and 6724).

Table 20.1 lists the types of cables available for wiring purposes and gives brief indications of the main features of their applications. More detail is given later and there are useful guides on the applications of the various types in the relevant British Standard specifications.

The British Standards for most cables reflect the requirements laid down in the European Standards. The process of achieving this throughout the Common Market is called harmonization and such cables are referred to as harmonized cables. Some cables used in the UK, such as flat twin and earth, are not included in this European Standard and are therefore not harmonized and may not be acceptable in other EU countries, even though they are fully recognized in the IEE Regulations.

Where a manufacturer wishes to have a cable of new and novel design recognized, he may have it assessed by the British Approvals Service for Electric Cables, who,

Table 20.1 Types of cable and flexibles.

BS no.	Insulation	Construction	Conductor	Application
6004	PVC	Unsheathed single-core	Copper, solid, stranded or flexible	In conduit, trunking or similar enclosures where mechanical protection is provided. Above ground. Cable should not be immersed in water. Not directly in plaster, etc.
6004	PVC	Sheathed single- and multicore	Copper, solid, stranded or flexible	On trays, ladder supports, cleats, wiring surface where mechanical damage is unlikely. Single-core cables without armour for a.c. circuits. In the ground if provided with additional mechanical protection. Can be embedded in plaster in defined zones see p. 555.
6346	PVC	Steel wire or aluminium strip armour with PVC covering, single- and multicore	Copper or aluminium, solid or stranded	No restrictions except in cases where severe mechanical impact is expected. Can be used directly in the ground without additional mechanical protection provided it has PVC covering. Single-core cables must have aluminium armour when used on a.c.
4553	PVC	Split concentric single-phase + earth	Copper, stranded	General use including underground without further protection. Should not be installed where mechanical damage is a likely hazard.

Code	Insulation type	Construction	Conductor	Remarks
6231	90°C PVC	Non-sheathed single-core	Copper	High temperature locations in control and equipment wiring, may be useful in restricted wiring applications.
7211	Thermosetting, low smoke and fire type	Non-sheathed single-core	Copper	Suitable for use in above-ground installations such as conduit and trunking or with similar protection.
5467	Elastomeric: cross-linked polyethylene or EP rubber	Armoured	Copper or aluminium, stranded	As for BS 6346 cables. Caution should be exercised where these higher temperature cables are run together with PVC-insulated cables.
6724	Elastomeric: cross-linked polyethylene or ethylene propylene rubber	Single- and multicore, armoured	Copper or aluminium	For installation in dry conditions where it is important to minimize the amount of smoke and corrosive gases in the event of fire.
6007	Elastomeric: vulcanized rubber, butyl or EP rubber, silicone rubber	Unarmoured single-core with various coverings. Multicore, flat or circular with elastomeric sheath	Copper, aluminium as an alternative, stranded or flexible	Butyl, EP rubber and silicone insulated cables used for high ambient temperatures. Braided coverings in dry situations only. Entry to luminaires. Sheathed cables resistant to some oils and solvents. Can be obtained with a lead sheath where pollution is severe.
6141	Rubber or PVC, capable of operating at high temperatures up to 150°C	Non-sheathed single-core cables. Sheathed two- and three-core flexible cords	Copper plain, tinned or plated, stranded or flexible, only small sizes	Zones having high ambient temperatures, such as in the vicinity of hot process plant, boiler houses. Non-sheathed versions must be protected mechanically by being enclosed.

Table 20.1 (*contd*).

BS no.	Insulation	Construction	Conductor	Application
				Wiring inside and into (high temperature tails) hot equipment such as luminaires and immersion heaters. Considerable variety of types, see BS for details.

6500 (flexible cords)	Rubber, PVC or glass fibre	Multicore, unsheathed and sheathed	Copper flexible or tinsel	Large variety of types of flexible cords with working temperature up to 150°C. Some types are similar to those in BS 6141.
				    Includes all general purpose flexibles and the high temperature varieties for use in or into hot equipment.
				    Rubber or PVC sheathed varieties can be used out of doors or for industrial applications. See extensive guide in Appendix of Specification.

7269	Thermosetting	Sheathed, two, three and four-core, plus circuit protective conductor	Copper	Primarily for use in alarm and emergency lighting circuits. For installation above ground including in enclosures.

| 6207 | Mineral | Plain copper sheathed or with additional PVC covering | Copper (solid), rigid | General wiring; larger sizes are used for rising mains and distribution circuits, especially where continuously high temperatures are expected. Excellent fire resistance (alarm and safety circuits). Looks well as surface wiring, not affected by most contaminants. Unlikely to need further mechanical protection. Limited overvoltage withstand capability, see p. 571. Excellent for use in very low temperatures. Weather resistance excellent. Limitations depend on seals at terminations and joints and on material used to insulate conductor tails. |

if they consider the cable is suitable, will issue a certificate to the effect that the cable, if used in the manner indicated in the certificate, will provide a degree of safety not less than that required by the IEE Regulations.

Such cables may be described as BASEC certified but do not totally comply with a British Standard for cables and may not be described as complying with the IEE Regulations. Their use must be recorded in the Completion Certificate.

CONSTRUCTION OF WIRING CABLES AND FLEXIBLES

When selecting a wiring cable there are three components that must be considered, conductor material, insulation and protective covering.

Conductors

Conductors of wiring cables are made of annealed copper, and for sizes of $16 \, mm^2$ and above, also of aluminium. Apart from solid, or single-wire conductors, which are available up to $2.5 \, mm^2$ in copper and for all sizes in aluminium, conductors are built up from several wires twisted together to impart various degrees of flexibility to the conductor. (*Note*: All sizes of conductor in mineral-insulated cables are of solid annealed copper.) The wires used in ordinary cables are of such a diameter that the cable is sufficiently pliable for convenient installation, but in no case should they be subjected to continuous movement. Excessive vibration can cause breakage of the wires, usually at or near terminations. Flexible conductors are built up of fine wires and are available for applications where continuous movement or vibration occurs. Where a cable route includes frequent bends or room for installation is limited, the flexibility of stranded conductors is an advantage.

Insulation

PVC compounds provide the most economical choice of insulating material for all normal uses and special grades are available where ambient temperatures exceed the normal limits. For much higher temperatures, materials such as ethylene propylene rubber (EP rubber), cross-linked polyethylene (XLPE) and mineral insulation (MI) are available. Where flexibility is important, the various types of rubber can be more appropriate.

Mechanical properties are usually the most important ones for insulation of cables up to 1 kV, and loss of strength and brittleness are the best signs to look for when judging the state of a cable. This applies also to the insulation on the tails of mineral-insulated cables, but not of course to the insulation of the cable itself.

The higher maximum temperatures permitted for elastomeric materials are of value where part of a cable route passes through a thermally adverse location (groups, thermal insulation, high ambient temperature) and it is undesirable to increase the conductor size over the whole circuit length. In addition, these materials offer the advantage of high permissible short-circuit temperature and are useful where high fault-currents are expected.

High temperature cables should not be grouped with others having lower permissible temperatures. Either the cable types should be grouped separately or care should be taken to confirm that the actual operating temperature of the whole group does not exceed that permissible for the lowest temperature material present.

It should be noted that terminals of some equipment may not be intended for use with conductors operating at temperatures higher than 70°C, although high temperature cables may be useful if not essential for connections to equipment which itself operates at a high temperature such as heating equipment, luminaire control gear or even electric motors.

It may not be possible to take full advantage of higher conductor temperatures because of the effect on conductor size of voltage drop requirements.

Sheaths and coverings

Cable and cord insulation have been designed and tested to provide adequate electrical strength and any additional coverings are required only in order to protect the insulation against environmental hazards. PVC insulation is tough enough to withstand pulling into conduits without further protection.

Coverings can consist of a simple sheath of PVC or rubber which provides good protection against abrasion and liquids and modest resistance to impact. Because some hydrocarbon solvents leach out the plasticiser of PVC compounds and creosote softens and destroys them it is best to avoid routes where such contaminants are likely. Otherwise a lead sheath is generally used, but the type of metal/polymeric laminate used on cables such as the Pirelli FP 200 could be suitable. Mineral-insulated cables have a solid-drawn copper sheath and are resistant to water and oils. However it is attacked by some chemicals and a covering of PVC is advisable.

Where a cable is likely to be subject to impact, or installation conditions are difficult, a cable having steel wire or aluminium strip armour incorporated in the protective coverings should be chosen. The cable can be used without further mechanical protection but it is general practice, particularly for outdoors or below ground use, to prevent corrosion of the armour by adding a further covering of PVC.

MI cable is not suitable for general use underground. In an exceptional case a short length may be so used, provided that it is protected against corrosion and mechanical damage. The PVC covering is not robust enough to withstand being pulled into a pipe or duct unless the run is short and there are no bends.

The remark above on the care necessary with groups containing cables of different types includes consideration of the temperature limits of sheath and covering materials, Cables designed for lower conductor operating temperatures are likely to be sheathed with low temperature material which might deteriorate at temperatures permissible in groups of high temperature cables.

Flat conductor cables

Flat conductor under-carpet cables (see BASEC Certificate No. 88101790) are a convenient way of providing supplies to socket outlets away from walls or partitions, without the necessity for cutting into the floor or the hazards of trailing flexibles. Manufacturers' purpose-made joint boxes and pedestals must be used and joints must be made strictly in accordance with the manufacturer's instructions using approved tools. Standard socket outlets can be accommodated in the pedestals. Special wall terminal boxes are used to provide connections to the fixed installation wiring. These cables have been approved for use only under carpet tiles.

Flat data and telephone cables with standard outlets are available to complete the office cabling needs.

This system is particularly useful for office refurbishment where limited floor/ceiling heights prohibit the installation of a false floor. In new buildings savings can be made in slab thicknesses and ceiling heights, reducing the weight and cost of a building.

CABLE ENCLOSURE AND SUPPORT SYSTEMS

Cable enclosure

Unsheathed cables are intended to be installed in a cable enclosure which provides protection against mechanical damage. Such enclosures are either conduit or trunking, although there are minor variations of each. Both are available manu-factured either from steel or plastic with some of the variants being manufactured from aluminium extrusions. Conduit and trunking should be manufactured to the relevant standard.

Steel conduit

This is the most common type of conduit for commercial and industrial applications and is electrically and mechanically continuous, with screw joints between lengths and into suitable boxes and fittings; conduit can be connected into trunking systems. The conduit itself must be earthed and therefore must be mechanically and electrically sound so that it can act as its own protective conductor; it can therefore be used as a circuit protective conductor.

The most common finishes are black enamelled or a variety of zinc coatings; either may need further protection against corrosion.

Conduits can be mounted on the surface of a building structure, screeded or plastered in, or cast into the structure of a building.

Guidance on the number of cables which may be drawn into a conduit is available from the IEE Guidance Note No. 1, *Selection and erection*, the require-ment of the Regulations merely being that 'other mechanical stresses' should not damage the cables. These stresses include those arising from drawing in; minor damage to cables frequently occurs during the drawing in process, particularly when additional cables are drawn into a conduit already containing cables.

Plastic conduit

Conduit is also available manufactured from PVC. It is cheaper than steel conduit and is easy to install. However, it has a number of disadvantages. First, it is not as strong as steel conduit, secondly there are upper and lower temperature limits within which it must be used and thirdly its high rate of expansion makes it necessary to fit expansion joints. These features make it necessary to give attention to mechanical damage, to the weight of the cables installed, their operating temperature and to fixing distances.

Flexible conduit

For connections between fixed installations and equipment which either vibrates or which is required to be moved from time to time (for belt adjustment) flexible conduits can be used. They vary in construction from a simple helically wrapped galvanized steel strip to more elaborate arrangements of plastic linings and sheaths with either helical interlocking strip or plain steel strip or steel wire.

A circuit protective conductor must always be provided.

Trunking

Trunking is available manufactured out of folded sheet steel, extruded plastic or extruded aluminium. By definition, trunking has a detachable lid and is intended to have the cables laid in; it is not intended that cables should be drawn in.

Standard cable trunking is installed fixed to the structure of a building; it may be installed flush with a wall, but in all cases its removable cover makes for great flexibility in wiring. In long vertical lengths steps must be taken to provide arrangements at suitable intervals to relieve the cables of stress arising from their own weight.

If it is suitably electrically connected, it may be used as a protective conductor; the cross-sectional area of the interconnections is subject to the requirements of Chapter 54 of BS 7671.

Trunking is available divided into compartments to allow for the segregation of various categories of circuit. The same restrictions which apply to the use of PVC conduit also apply to PVC trunking. There are a number of variants on the trunking theme.

Skirting trunking

Skirting trunking which is available manufactured commonly out of steel, less commonly out of PVC or moulded plywood provides a ready means of both distributing cables around the perimeter of offices and also ready access to them. Accessories can be mounted on the surface.

Floor ducting and trunking

Where it is necessary to provide supplies in open plan offices, floor ducting is available; it can be provided with compartments for segregation purposes. Floor ducting and trunking are manufactured in sheet steel, plastic or mixed fibre and are cast into the floor finish screed, with interconnecting outlet boxes at intervals as required. Once cast in alterations cannot be made without difficulty.

Some floor trunking systems have a removable cover allowing access to the whole of the wiring system.

Service post and poles

Alternatively, electrical services can be brought down to user positions from above the working area by use of service poles or posts. These are essentially vertical self-supporting trunking, divided into compartments into which the various categories of circuits can be placed. They are commonly used in association with lighting trunking.

Lighting trunking

Lighting trunking is in many ways similar to standard trunking but incorporates flanges and other features enabling it to be used either in association with a false ceiling structure or as a structural part of the false ceiling. It is intended to provide accommodation for cable and to support luminaires. It is sometimes, but not commonly, available in compartmented form for segregation purposes.

Cable support systems

Cable support systems are used for the installation of sheathed or sheathed and armoured cables where the mechanical protection of a full enclosure is not required. Some slight mechanical protection is provided by tray or ladder racking which, if suitably electrically connected, may be used as a circuit protective conductor, making it possible to dispense with a protective conductor in sheathed cables.

However, it is not a requirement that either ladder racking or tray should be electrically continuous and where the cables have adequate intrinsic strength, elbows, tees, offsets, etc. can be dispensed with. BS 7671, *Requirements for electrical installations*, recognizes that sheathed cables, although not to be described as being double insulated, provide an equivalent degree of safety and therefore can be regarded as providing protection against direct and indirect contact, in which case isolated lengths of tray or ladder racking do not need to be earthed.

However, where there is an electrically continuous system throughout a building it must be bonded to the main earthing terminal.

Cable tray

Light gauge cable tray is most suitable for small cables, and a well-designed system will allow access to the cables so that redundant cables can be removed and cables added.

Attention must be paid, particularly in the case of plastic trays carrying cables operating at higher temperatures, to the span permitted for a given weight of cable which must not be exceeded otherwise sagging occurs.

Where larger numbers of cables or longer spans are necessary then heavy gauge return flange tray should be used.

Ladder racking

Ladder racking is used mainly in industrial premises for the support of larger cables over longer spans. Cables can be laid on the ladder racking or can be fixed with cable cleats. These are essential for vertical runs but for horizontal runs plastic cable ties are often sufficient to locate the cables.

Both tray and racking are commonly made of mild steel or aluminium. Aluminium does not require additional protection for normal environments; steel tray may be painted but it is more commonly galvanized. For severe environmental conditions

PVC coating, epoxy coating, nylon coating and other protective finishes are applied. In extreme cases stainless steel may be used.

TEMPERATURE LIMITS

Insulating materials must be used within certain temperature limits if a satisfactory performance and an adequate service life are to be obtained. Upper limits of temperature are determined by the thermal sensitivity of each material, the higher the temperature, the shorter the period for which the necessary electrical and mechanical properties are retained. The values in Table 20.2 are based on extensive experimental work and field experience and are recognized internationally.

Temperature limits for MI cables are the exception in Table 20.2 in that they are not based on insulation performance. Their limits are related to the temperature capability of the over-sheath material, or to a maximum cable surface temperature where they are exposed to touch, or can be in contact with thermally sensitive materials. Where neither of these limitations applies the temperature at which MI cables operate is not critical except that it must not exceed the maximum value set by the end seals or by the material used for insulating the tails.

CABLE RATINGS

The size of conductor used must be large enough for it to carry the expected load current without exceeding the temperature limit appropriate to the insulating material involved. Factors affecting the current that a cable can carry are discussed below.

Table 20.2 Temperature limits of wiring cable materials.

Insulating material	Maximum conductor temperature (°C)	
	continuous operation	short-circuits
Polyvinylchloride compounds (PVC)	70	160/140[†]
90°C PVC	90	160/140[†]
60°C rubber	60	200
85°C rubber	85	220
Ethylene propylene (EP) rubber	90	250
Silicone rubber	150	250
Cross-linked polyethylene (XLPE)	90	250
Mineral insulation (MI) (see text)	70	160
Glass fibre	185	*

* Refer to manufacturers.
[†] The second figure refers to conductor sizes larger than $300\,mm^2$

Sustained current carrying capacities

The highest value of current a cable can carry continuously without exceeding its maximum insulation temperature given in Table 20.2 is referred to as its current-carrying capacity and can be derived either by calculation or by experiment. This current is dependent on the rate at which heat can be dissipated from a cable. Clearly a cable mounted by itself on a wall can dissipate much more heat than one which is in a bunch of cables or is enclosed in conduit. As a consequence the rating depends on the installation conditions. UK cables are made to internationally accepted specifications and as a consequence their ratings are comparable with those used in other countries. Methods for calculating current-carrying capacities are given in IEC Standard 287 and ratings for the more frequently used types of cable and methods of installation are provided in IEC 364−5−523 and have been harmonized by CENELEC.

Tabulated ratings, such as those given in the IEE Regulations and in the 69−30 series of reports issued by ERA Technology Ltd., are calculated using methods given in IEC 287 and care is taken to ensure that, where appropriate, they harmonize with those issued by IEC and CENELEC. The major differences between the IEE ratings and those issued by ERA lies in the types of installation methods covered and the ambient temperatures for which the ratings are calculated. The 69−30 reports include ratings for cables laid in the ground. To keep the individual tables within reasonable limits of size, ratings are given for single-circuits only (i.e. two single-core cables or one twin cable in a single-phase or d.c. circuit and three or four single-core cables or one multicore cable in a three-phase circuit). Where more than one circuit is installed close together or in the same enclosure, thus forming a group, mutual heating occurs between the cables and their rating must be reduced by multiplying by an appropriate group factor.

Where cable current-carrying capacities other than those provided in BS 7671 are used, or where derating for groups has been calculated using methods other than those provided, it is advisable that this should be quoted on the IEE Completion Certificate, if the installation is to comply with the IEE Regulations.

The sustained current-carrying capacity of a cable is affected by its thermal environment. The most important environmental influences are ambient temperature, grouping with other cables, enclosure affecting heat dissipation and partial or total enclosure in thermal insulating material. A less frequent influence, but of equal importance, is that of solar radiation or radiation from nearby heated surfaces such as those associated with industrial processes. Although solar radiation is not usually expected inside a building, it should not be forgotten that cables running underneath roof lights are subject to considerable heating, both from direct radiation and also due to the significant increase in ambient air temperature. Comments on the influence of solar radiation are given in a later section.

The effects of the more common types of cable support or enclosure are dealt with by publishing tables of current-carrying capacity for each method of installation, e.g. in conduit, trunking (and its derivatives) and in thermally insulated walls. Such ratings are refered to as 'Reference ratings' because the capacities have been determined by extensive tests. Capacities for other methods of installation are then obtained by the application of theoretically or experimentally derived factors. Factors are also needed where the cable capacity is affected by the size of the enclosure or the thermal properties of the enclosing material (e.g. cables in thermal insulation).

Where environmental conditions differ from those assumed for the published tables of current-carrying capacities, appropriate factors should be applied to obtain effective capacities for the particular installation.

Ambient temperature

The rating of a cable is derived directly from the amount of heat it can liberate when it is running at its maximum operating temperature and dissipating that heat into surroundings which are at a stated ambient temperature. Given that the installation conditions are the same as those assumed for the determination of the ratings, there is no margin on the conductor temperature-rise. If anything changes to reduce the amount of heat which can be dissipated then the cable will become overheated and its expected life reduced. The most likely causes of overheating in practice are high ambient temperatures and installation of cables in groups.

Tables of cable ratings for wiring installations are generally based on an ambient temperature of 30°C. This is sufficiently high to include most general installations, so that it is possible in many cases to use the values in the tables directly without correction.

However, there are situations where the ambient temperature does exceed 30°C, such as in boiler houses, in the vicinity of industrial processes and in domestic airing cupboards. Unless there is an accepted value of ambient temperature for the particular situation, an estimate or a measurement must be made, if necessary referring to a similar type of installation. The measurement can be made using any suitable type of thermometer and appropriate instructions are given in the IEE Regulations. It is important to observe that the measurement must take into account all sources of heat, but should not be influenced by heat from the cable under consideration, that is if the cable is in place then it should not be loaded or the thermometer should be appropriately shielded.

Ambient temperature correction factors are given with tables of cable ratings and the correct rating is obtained by multiplying the tabulated rating by the appropriate correction factor. Where it is known that ambient temperatures will be consistently lower than 30°C it is possible to take advantage of this fact and to increase a rating. Here again appropriate factors are usually provided.

Groups of cables

Where more than one circuit or cable run together in conduit, trunking, or on the surface the mutual heating between them reduces their current-carrying capacities. For most arrangements of cables in general use factors are issued with the rating tables whereby the maximum current-carrying capacity of cables in groups can be calculated.

Appendix 4 of BS 7671, *Requirements for electrical installations*, provides such factors together with guidance and formulae for selecting suitable conductor sizes. A method for checking that the selected sizes are capable of carrying expected overloads is included.

Because they have to deal with the subject in a general way, these factors must be derived on the assumption that all the cables in a group are of the same size and are equally loaded. This situation does not often arise in practice, but it is usually assumed that the application of such factors to a practical group of mixed sizes of cables would be on the safe side. In fact, this is not true.

A simple example will illustrate the degree of overheating which can arise with the smaller conductor sizes. Consider a group of mixed sizes of cable which contains a small bundle of seven $1.5\,\mathrm{mm^2}$ cables and a $25\,\mathrm{mm^2}$ cable. The seven $1.5\,\mathrm{mm^2}$ cables occupy approximately the same space in the group as the $25\,\mathrm{mm^2}$ cable but, if rated using a common group reduction factor, could dissipate about 3.4 times as much heat. Overheating of smaller cables is accompanied by over-sizing of the larger ones.

Extensive work on this matter has resulted in the issue of several new ERA reports in the 69–30 series which provide ratings for cables in groups of mixed sizes carrying different loads. At the moment these reports deal with cables made to BS 6004 and BS 7211 installed in trunking, to BS 6346, BS 5467 and BS 6724 (LSF cables) installed in multilayer groups (groups where the cables are in contact with each other) on trays and ladder racks.

It is obvious that, because the ratings for cables in such groups are specific to the cable and load content of each group, rating factors and current-carrying capacities cannot be tabulated in the usual way. In fact a cable ceases to have a specific rating under these circumstances and, dependent on the size of the group and the loading of the other cables, its permissible load may be even greater than that for the cable in isolation. The method of conductor size selection given in the ERA reports provides for the adjustment of individual sizes (subject to other constraints such as voltage drop) so that the most economical mix of sizes can be selected.

Overcurrent protection is also influenced by the thermal effect of such groups and the ERA reports include formulae for checking that each conductor will meet the requirements of the IEE Regulations.

The ERA reports on methods for designing groups with mixed loads using the well documented and proven uniform heat generation (UHG) principle. Design based on UHG allocates correctly the permitted power losses between small and large cables, generally provides a more economical combination of cable sizes and may permit the addition of further circuits later without requiring a change in the original circuits.

The following descriptions of applications to groups in conduit, trunking and on a tray illustrate the method and its advantages. The procedures described are applicable to circuits which must comply with given voltage-drop requirements. If this is not the case reference should be made to the ERA reports.

Cables in conduit and trunking

The procedure starts by selecting conductor sizes to comply with voltage drop requirements. These sizes are minima and the size of a conductor will be increased only if it is thermally too small. An application of the UHG principle checks that individual cables are thermally adequate and that the total power loss is within the capability of the enclosure.

An optimization technique is described which augments the current-carrying capacity of certain circuits, thus improving the economy of the installation.

The procedure for conduit and trunking is outlined in Fig. 20.2 and examples are provided in Tables 20.3 and 20.4.

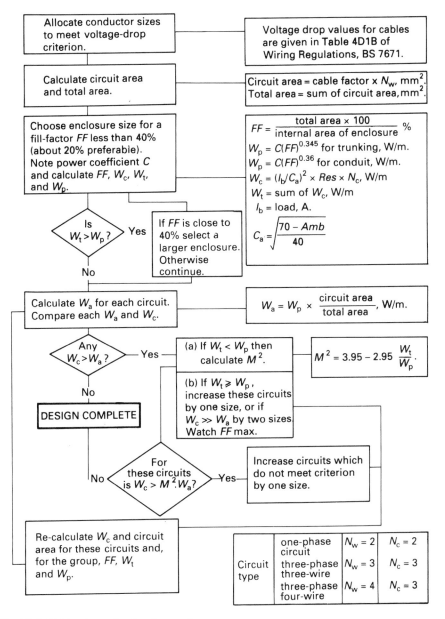

Fig. 20.2 Cable size selection for enclosures.

Table 20.3   Group of cables in conduit.

No.	Load (A)	Wires	Length (m)	Voltage drop size (mm²)	Circuit area (mm²)	Circuit loss, W_c (W/m)	Permitted loss, W_a (W/m)
1	25	4	50	6.0	91.6	6.84	7.85
2	35	3	40	6.0	68.7	13.41 *	5.89
3	10	2	30	2.5	27.8	1.80	2.38
4	6	4	30	1.5	38.4	1.57	3.29
				Totals:	223.5	(W_t) 23.62	(W_p) 19.41

Table 20.4   Group of cables in conduit.

No.	Size	Circuit area (mm²)	Circuit loss, W_c (W/m)	Permitted loss, W_a (W/m)	Conventional group factor size (mm²)
1	6.0	91.6	6.84	7.07	10.0
2	10.0	108.9	8.09	8.41	16.0
3	2.5	27.8	1.80	2.15	1.5
4	1.5	38.4	1.57	2.97	1.5
	Totals:	266.7	(W_t) 18.30	(W_p) 20.60	

The cable type used in the examples, is single-core PVC insulated to BS 6004, in steel conduit to BS 4568 or steel trunking to BS 4678. The relevant cable and trunking data required to effect the designs are given in Tables 20.6 and 20.7.

Example 1   Group of cables in conduit
Ambient temperature 30°C, $C_a = 1.00$.
(*Note*: For any other ambient temperature

$$C_a = \sqrt{\left(\frac{70 - Amb}{40}\right)}$$

where Amb = ambient temperature, °C.)
Voltage-drop limit = 6000 mV single-phase.
Suitable conduit size could be 32 mm, for which $C = 5.3$ W/m.
Conduit internal area = 615 mm².
Fill-factor = 223.5 × 100/615 = 36.8%.
Maximum permitted total power loss = 5.3 × 36.8$^{0.36}$ = 19.41 W/m.

The sizes determined by voltage drop are adequate, except for circuit No. 2, marked (*), where the power loss is greater than that permitted for uniform heating. Further, the total power loss W_t, is higher than the permissible value,

W_p. This would suggest that either larger cables or a larger size of conduit is needed.

In this example, the fill-factor is below the maximum recommended so that an increase in the size of one circuit could be acceptable. This would reduce the power loss and increase the permitted power capacity of the group.

Examine the effect of increasing the size of cables for circuit No. 2 by one size. The result is shown in Table 20.4.

All conductor sizes are now thermally adequate.

Comparison with the last column, which shows the sizes determined by conventional use of an across the board group rating factor, shows that the UHG procedure permits circuit Nos 1 and 2 to be wired with conductors one size smaller than that indicated by the conventional procedure. This is in line with the expected result from UHG ratings mentioned above.

Example 2   Group of cables in steel trunking

Ambient temperature 25°C, $C_a = 1.061$.
Voltage-drop limit = 6000 mV single-phase.
The first stages of selection are given in Table 20.5.
Suitable size of trunking would be 100 mm × 100 mm.
The internal area of which is 9448 mm².
The fill-factor, FF 1758 × 100/9448 = 18.6%.
For 110 × 100 trunking $C = 16.25$ W/m.
Maximum permitted power for trunking = $16.25 \times 18.6^{0.345}$ W/m.
$$= 44.59 \text{ W/m.}$$

The losses for circuits Nos 2, 7 and 9, marked (*), are too high. This could be put right by increasing the conductor sizes for those circuits. However, as the total

Table 20.5   Group of cables in trunking.

No.	Load	Wires	Length (m)	Voltage-drop size (mm²)	Circuit area (mm²)	Circuit loss, W_c (W/m)	Permitted loss, W_a (W/m)
1	5	2	100	4.0	36	0.24	0.91
2	16	2	50	6.0	46	1.66 *	1.17
3	32	2	100	25.0	150	1.64	3.80
4	50	4	100	25.0	300	6.00	7.60
5	30	3	100	16.0	150	3.36	3.80
6	50	4	150	35.0	380	4.33	9.63
7	75	3	100	35.0	285	9.75 *	7.22
8	5	3	100	2.5	42	0.60	1.06
9	15	3	100	6.0	69	2.19 *	1.75
10	50	4	100	25.0	300	6.00	7.60
				Totals:	1758	(W_t) 35.77	(W_p) 44.54

Table 20.6 Cable data for single-
core unsheathed PVC cable in
conduit or trunking.

Size (mm^2)	Cable factor (mm^2)	Resistance[†] (mV/A/m)
1	3	4
1.5	9.6	14.50
2.5	14.0	9.00
4.0	18.0	5.50
6.0	23.0	3.65
10.0	36.0	2.20
16.0	50.0	1.40
25.0	75.0	0.90
35.0	95.0	0.65
50.0	133.0	0.475
70.0	177.0	0.325

[†] *Note*: All values apply to one single-
core PVC cable to BS 6004, at 70°C.

power loss W_t is well within the maximum permitted value W_p it is feasible to try the optimizing procedure on these circuits.

$$\frac{W_t}{W_p} = \frac{35.77}{44.54} = 0.803$$

$$M^2 = 3.95 - 2.95 \times 0.803 = 1.58$$

For circuit Nos 2, 7 and 9, $W_a \times M^2$ is greater than W_c, hence these sizes can be retained.

The optimizing procedure is based on experimental work which showed that a limited number of circuits can dissipate a certain amount more than the average UHG power, provided that the overall losses of the group are less than the permitted maximum. It can be seen that, as the total power loss of the group increases towards its maximum permitted value, the optimizing factor M^2 decreases to unity.

The maximum permissible value for M^2 is 2.4, irrespective of the ratio W_t/W_p.

The optimization procedure is applicable only to groups in conduit or trunking.

Increasing the conductor sizes of circuit Nos 2, 7 and 9 would also provide a thermally acceptable design, but the cost would be higher.

Comparison with sizes derived by an across the board application of a conventional group rating factor, $Cg = 0.48$, would show, as before, that savings can be made on some of the larger conductors. Voltage-drop limitations tend to mask the comparison for smaller conductors but, where a direct comparison is available, the conventional group factor may lead to sizes which are thermally inadequate.

Table 20.7 Trunking and conduit data for groups of cables.

Nominal size (mm)	Internal area, A (mm^2)	Power coeff., C (W/m)
1	2	3
Trunking		
38 × 38	1296	8.15
50 × 38	1728	8.93
50 × 50	2304	9.90
75 × 50	3456	11.40
100 × 50	4646	12.66
75 × 75	5271	13.24
150 × 50	6948	14.58
100 × 75	7086	14.69
100 × 100	9448	16.25
150 × 75	10628	16.92
150 × 100	14308	18.76
150 × 150	21550	21.60
Conduit		
16	121	3.3
20	214	4.1
25	345	4.8
32	615	5.3

Groups on trays

The procedure, given in Fig. 20.3, is initially much the same as that for conduits and trunking, with the width and depth of the group replacing the size of the enclosure and the fill-factor. Because a group on a tray may be of such a width that heat losses from the edges do not affect the temperature rise at the centre, the concept of optimization to take advantage of under-utilization of the whole group is not applicable.

On the other hand, a shape factor C_s may be applied which corrects for the effect of added heat loss from the edges of a group. This tends to compensate, at least in part, for the rapid decrease in cable current-carrying capacity as the depth of a group increases. Further comment on this is made when discussing the example.

A further difference is introduced, due to the fact that groups on trays and racks are laid in an orderly manner so as to facilitate fixing with binders or clips. This usually results in an approximately rectangular cross-section for the group and the rectangular outline of the cables is the outline of the heat-dissipating surface of the group.

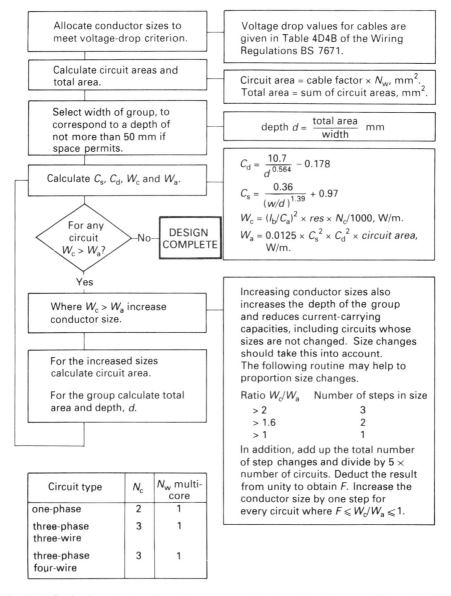

Fig. 20.3 Cable size selection for cables on trays (single and multi-core cables to BS 6346).

Under these circumstances it is convenient to assume that each cable occupies a square area with sides equal to the diameter of the cable. The total cross-sectional area of the group is then equal to the sum of these squares. The depth, calculated by dividing this area by the width, approximates to the visual depth. The unit system, used to calculate the area occupied by each circuit, differs therefore from that used for cables in conduit and trunking by being based on the square of the cable diameter and not on $\pi/4$ times that value.

Example 3   Group of SWA cables on a perforated steel tray
Ambient temperature 25°C, $C_a = 1$.
(*Note*: The reference ambient temperature for cables on trays is 25°C; this being a more usual temperature for industrial installations. For any other ambient temperature

$$C_a = \sqrt{\left(\frac{70 - Amb}{45}\right)}$$

where Amb = ambient temperature, °C.)
Voltage-drop limit = 6000 mV single-phase.
   For cable data see Table 20.8. The circuit schedule is given in Table 20.9.
A tray width of 450 mm has been chosen.
Depth of group is then $25916/450 = 57.6$ mm.

A wider tray would be desirable because the depth of the group would be less, and smaller conductors sizes would be needed. On the other hand the supporting steelwork would be more expensive and space available might be limited. A

Table 20.8   Armoured cable data for use with groups of cables on trays.

Size (mm²)	Resistance[†] (mV/A/m)	Cable area (diameter²) (mm²)		
		twin	three-core	four-core
1	2	3	4	5
1.5	14.50	137	151	169
2.5	9.00	172	185	210
4.0	5.50	228	250	317
6.0	3.65	272	324	369
10.0	2.200	404	449	520
16.0	1.400	480	534	671
25.0	0.875	511	605	751
35.0	0.625	600	724	906
50.0	0.490	751	906	1197
70.0	0.315	900	1170	1475
95.0	0.235	1204	1482	1892
120.0	0.190	1384	1714	2314
150.0	0.150	1640	2144	2746
185.0	0.125	2043	2520	3285
240.0	0.095	2500	3158	4109
300.0	0.0775	3003	3795	4956

[†] *Note*: Values apply to one-core of a multi-core cable, at 70°C.

Table 20.9 Group of three-phase, four-wire circuits on a perforated tray, brackets or ladder support.

No.	Load (A)	Length (m)	Voltage drop size (mm^2)	Circuit area (mm^2)	Circuit loss (W/m)	Permitted loss (W/m)
1	100	100	50	1197	13.95*	12.39
2	150	120	70	1475	21.26*	15.26
3	250	100	95	1892	44.06*	19.58
4	300	125	150	2746	40.50*	28.41
5	300	150	185	3295	33.75	34.09
6	125	200	95	1892	11.02	19.58
7	90	150	70	1475	7.65	15.26
8	100	150	70	1475	9.45	15.26
9	250	125	120	2314	35.63*	23.94
10	300	150	185	3295	33.75	34.09
11	250	200	240	4109	17.81	42.52
12	50	100	25	751	6.56	7.77
			Total:	25916		

smaller width would require very much larger conductors and would not be practicable except for installations where space is extremely limited.

Depth coefficient $C_d = 0.9098$,　　　$C_s = 1.0$.

Conductor size adjustment

Conductors Nos 1, 2, 3, 4 and 9 are too small and should be increased in size. While some conductors, No. 1 for example, need increasing by one size step, others, such as No. 3, will clearly need more than one size increment. It is useful to have a simple guide relating the value of W_c/W_a to the likely number of size increments.

Two features must be taken into account at this stage:

- As conductor, and hence cable, sizes are increased they occupy a greater area and the group depth for a fixed width increases. This in turn decreases the load carrying capacity of conductors (see the depth factor, C_d) so that a size increase greater than that originally thought suitable may become necessary.
- The increase in depth and reduction in current-carrying capacity affects all cables in the group, so that cables which were just adequate before are now too small.
- It is not feasible to provide a precise guide because the changes in size and the effect of group enlargement are unique to each load schedule. Further, it is difficult to test for an oversized conductor, so that it is preferable to underestimate the size increases needed and, if necessary, use a second stage of increases. The guide provided in Fig. 20.3, based on a study of groups of armoured multi-core cables, may be found helpful.

Application of this guide to the two last columns of Table 20.9 produces Table 20.10.

These increases are successful for circuit Nos 1 and 3 and almost so for circuit No. 2. The general effect of increasing the group depth has made circuit 12, previously adequate, now marginally too small.

A second stage, increasing circuit Nos 2, 4, 9 and 10 by one size, is successful for all circuits.

For the larger cables, which may be mounted on supports so that the cables are spaced apart to permit better heat dissipation, information for calculating rating factors is given in ERA Reports 74−27 and 74−28. Rating factors for certain specific cases have been derived from such information and are given in the IEE Regulations and in IEC 364−5-523. However, these factors are subject to the limitation that all the cables are presumed to be of the same size and equally loaded.

A point worth making here is that, except for MI cable, a wiring cable which is continuously loaded to even 10% more than its maximum permissible value has its expectation of life reduced by about 50%. Where a large number of simultaneously loaded cables must be accommodated it can be preferable to split them into several small groups so as to avoid the economic penalties of the low ratings required for large groups.

Overcurrent protection and selection of cable size

For circuits of practical length, voltage drop may be the decisive factor when choosing conductor sizes. This has been recognized in the previous section on the selection of conductor sizes for groups. The calculation of voltage drop is dealt with later in the chapter.

Table 20.10 Types of cables and flexibles, UHG increases in conductor size.

No.	First size (mm^2)	Second size (mm^2)	Circuit area (mm^2)	Circuit loss (W/m)	Permitted loss (W/m)
1	50	70	1475	9.45	12.16
2	70	95	1892	15.86*	15.60
3	95	185	3295	23.44	27.16
4	150	185	3295	33.75*	27.16
5	185	240	4109	25.65	33.87
6	95	95	1892	11.02	15.6
7	70	70	1475	7.66	12.16
8	70	70	1475	9.45	12.16
9	120	150	2746	28.13*	22.64
10	185	240	4109	25.65	33.87
11	240	240	4109	17.81	33.87
12	25	25	751	6.56*	6.19

One of the functions of circuit protection is to prevent currents greater than the rating of a cable flowing for a duration long enough to cause damage. There are two types of such overcurrents; those which arise in a healthy circuit because of an excessive load and those resulting from a faulty condition such as a short-circuit.

Overloads
A properly designed circuit will be capable of safely carrying the highest expected load. In general, however, there will be a small unquantifiable risk that, under exceptional and unpredictable circumstances, the design load could be exceeded. It is a general requirement, therefore, that all circuits (apart from a few exceptions noted in the IEE Regulations) must have overload protection, chosen so that an excess current cannot persist long enough to cause damage.

As a result of considerable experimental work and field experience it has been concluded that a PVC-insulated cable can safely withstand overload currents up to 1.45 times its continuous rating. However, it is essential to realize that such currents can only be tolerated if they occur for a strictly limited duration and only very occasionally, say not more often than a few times during the life of the installation. Most other cable materials can withstand this treatment, so the rule limiting overload to not more than 1.45 times the cable rating has been adopted generally for all wiring.

Circuit design must aim to avoid such overloads by taking into account the nature of the loads likely to be imposed on the circuit and their frequency. Some guidance on load diversity is provided in the IEE Regulations, but care must be taken to anticipate reasonable exceptions. The overload capability of cables is intended only to cover unforeseeable circumstances, generally arising from faulty conditions and not from a characteristic of the load. Overload protection should be carefully matched to the load, taking into account the magnitude and duration of any peaks.

For cases like starting currents and varying loads in motor circuits it is feasible to take advantage of the intermittent load capability of cables, where thermal inertia enables them to carry a current higher than their steady state rating for a short period. This example also illustrates the situation where the thermal effect of exceptional overload may be taken care of by protective features built into equipment associated with the cable, e.g. in the starter or a thermal trip in the motor.

Selection of an appropriate combination of protective device and cable is illustrated in Fig. 20.4. Having estimated the maximum sustained load (design value) I_B, a circuit protection device is chosen which has a rating of I_n, equal to or greater than I_B. A cable size is then chosen so that its rating, corrected for ambient temperature and for grouping, I_z is equal to or greater than I_B, and furthermore 1.45 times I_z is equal to or greater than the long duration operating current of the device, I_2. The cable will then be capable of carrying the design load safely and can also withstand infrequent overloads which will not be cut off by the protective device. This selection process is a requirement of Clause 433 of the IEE Regulations.

Recognizing the need for devices which can provide suitable overload protection,

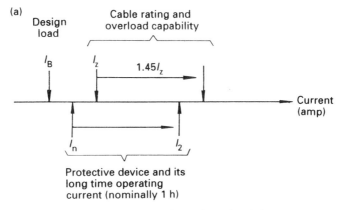

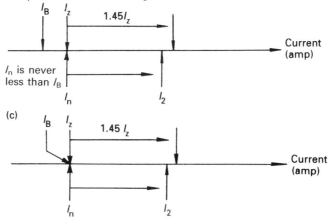

Fig. 20.4 Overload protection: (a) general situation; (b) situation possible with certain protective devices; (c) the load I_B can be as high as I_n and I_z.

fuses to BS 88: Part 2 or Part 6 or to BS 1361 and circuit-breakers to BS 3871: Part 1 or BS EN60898, Part 1 are made with characteristics such that under practical installation conditions their long duration operating currents are not greater than 1.45 times their rating I_n. This greatly simplifies the process of selection as it is then only necessary to follow the simple rule given in the IEE Regulations, that:

$$I_B \leqslant I_n \leqslant I_z$$

In the case of semi-enclosed fuses to BS 3036 the long duration operating current, I_2, is equal to $2I_n$. It is therefore necessary to match the cable with a fuse having a rating such that $2I_n < 1.45I_z$, or $I_n < 0.725I_z$.

As $I_B < I_n$, the above requirement effectively limits the circuit to 72.5% of the full thermal rating of the cable. It is convenient, when semi-enclosed fuses are frequently installed, to tabulate these reduced ratings, I_z (as was done in the 14th edition of the IEE Wiring Regulations for 'coarse protection') so that a direct comparison can be made with values of load current and fuse ratings. If this is done, then the simple relationship $I_B < I_n < I_z$ holds once more. For example, see Table 20.11.

Short-circuit

Cables are able to withstand quite high conductor temperatures providing they are of very short duration, and only occur very infrequently. The short-circuit conductor temperatures given in Table 20.2, which are based on both experimental evidence and field experience, can be used to determine the combinations of current and time which a conductor can carry without a significant reduction in operating life or in insulating capability, see Table 20.12. (Actually a quantity (I^2t) is used which is proportional to the energy liberated during the fault and hence to the conductor temperature-rise.) A similar relationship can be determined for the capability of protective conductors to carry earth-fault currents, see Tables 20.13 to 20.15, based on the maximum temperatures that cable insulation or covering materials can tolerate.

Tables 20.12 to 20.15 can be used to check whether a cable chosen to meet the requirements for voltage drop, current-carrying capacity and protection against electric shock will be adequate for short-circuit and earth-fault conditions using the following procedure. For each fault path and its appropriate conductors:

(1) Calculate the short-circuit and/or earth-fault currents:
   - for a live conductor the fault current is calculated using its resistance at a temperature midway between the maximum operating temperature and the maximum permissible final temperature (see IEE Regulations, Section 43);
   - for a protective conductor the fault current is calculated using its resistance at a temperature midway between its initial temperature and its final temperature (see IEE Regulations, Section 43).
   Further detail on calculation of short-circuit and earth-fault current can be found in the IEE Guidance Note No. 6, *Protection against overcurrent*, particularly Section 4, *Determination of fault current*.
(2) Determine the operating time from the characteristic of the appropriate protective device (time current characteristics of some devices are given in Appendix 3 of the IEE Regulations, or information may be obtained from manufacturers).
(3) Divide the fault-current capacity given for the cable in Tables 20.12 to 20.15 by the calculated fault current in kA.
(4) Square the ratio so obtained and compare it with the appropriate operating time.
(5) If the ratio is equal to or greater than the operating time of the protective device the conductor size is adequate. If not then a larger conductor is required or a protective device having a shorter operating time must be chosen.

Table 20.11 Single-phase current ratings for cables provided with coarse protection by fuses to BS 3036 (ratings adjusted to take account of protection see text on overload protection).
Single-core insulated non-armoured cables to BS 6004 or BS 6346.

Area (mm^2)	Non-sheathed		Sheathed
	2 cables enclosed in conduit in a thermally insulating wall (A)	2 cables enclosed in conduit or trunking on a wall (A)	2 cables clipped direct to a wall (A)
1	8	10	11
1.5	10.5	12.5	14.5
2.5	14	17.5	19.5
4	19	23	27
6	25	30	34
10	33	41	47
16	44	55	63

Two-core cable PVC insulated and sheathed. Clipped direct to a wall.

Area (mm^2)	Non-armoured to BS 6004 or BS 6346 (A)	Armoured to BS 6346 (A)
1	11	—
1.5	14	15
2.5	19.5	20
4	26	28
6	33	36
10	46	49
16	62	65

Light duty mineral-insulated, two-core cables, exposed to touch or having an overall covering of PVC clipped direct to a wall.

Area (mm^2)	Current (A)
1	13.5
1.5	16.5
2.5	22
4	29

Table 20.12   Short-circuit withstand of conductors and protective conductors incorporated in a cable (for initial and final temperatures see IEE Regulations).

Insulation size (mm²)	kA for 1 s				
	70°C PVC and MI	60°C rubber	85°C rubber	XLPE, EPR	silicone rubber
Copper conductors					
1	0.115	0.141	0.134	0.143	0.109
1.5	0.173	0.212	0.201	0.215	0.164
2.5	0.288	0.353	0.335	0.358	0.273
4	0.460	0.564	0.536	0.572	0.436
6	0.690	0.846	0.804	0.858	
10	1.15	1.41	1.34	1.43	
16	1.84	2.26	2.14	2.29	
25	2.88	3.53	3.35	3.58	
35	4.03	4.94	4.69	5.01	
Aluminium conductors					
16	1.22	1.49	1.42	1.50	
25	1.90	2.33	2.23	2.35	
35	2.66	3.26	3.12	3.29	
50	3.80	4.65	4.45	4.70	
70	5.32	6.51	6.23	6.58	
95	7.22	8.84	8.46	8.93	
120	9.12	11.2	10.7	11.3	
150	11.4	14.0	13.4	14.1	
185	14.1	17.2	16.5	17.4	
240	18.2	22.3	21.4	22.6	
300	22.8	27.9	26.7	28.2	
380	25.8	35.3	33.8	35.7	
480	32.6	44.6	42.7	45.1	
600	40.8	55.8	53.4	56.4	
740	50.3	68.8	65.9	69.6	
960	65.3	89.3	85.4	90.2	
1200	81.6	112	107	113	

Assessment of fault-current capacity of cable

Consider a 50 A/400 V motor circuit where the fault-current protection is provided by a fuse with a 100 A operating characteristic. The supply impedance at the distribution board is $0.025 + j0.03\,\Omega$.

The motor circuit consists of 50 m of 16 mm² three-core cable.

The resistance of 16 mm² cable at 70°C is 2.4 mV/A/m, see Table 4D4B of Appendix 4 of The Regulations.

Table 20.13 Earth-fault capacity for separate protective conductors (for initial and final temperatures in IEE Regulations Table 54B).

Insulation size (mm²)	kA for 1 s			
	70°C PVC	60°C rubber	85°C rubber	XLPE, EPR
Copper conductors				
1	0.143	0.159	0.166	0.176
1.5	0.215	0.239	0.249	0.264
2.5	0.358	0.398	0.415	0.440
4	0.572	0.636	0.664	0.704
6	0.858	0.954	0.996	1.060
10	1.43	1.59	1.66	1.76
16	2.29	2.54	2.66	2.82
25	3.58	3.98	4.15	4.40
35	5.01	5.57	5.81	6.16
50	7.15	7.95	8.30	8.80
70	10.0	11.1	11.6	12.3
95	13.6	15.1	15.8	16.7
120	17.2	19.1	19.9	21.1
150	21.5	23.9	14.9	26.4
185	26.5	29.4	30.7	32.6
240	34.3	38.2	39.8	42.2
300	42.9	47.7	49.8	52.8
400	57.2	63.6	66.4	70.4
500	71.5	79.5	83	88.0
630	90.1	100	105	111
800	114	127	133	141
1000	143	159	166	176
Aluminium conductors				
16	1.52	1.68	1.76	1.86
25	2.38	2.63	2.75	2.90
35	3.33	3.68	3.85	4.06
50	4.75	5.25	5.50	5.80
70	6.65	7.35	7.70	8.12
95	9.03	9.98	10.4	11.0
120	11.4	12.6	13.2	13.9
150	14.3	15.8	16.5	17.4
185	17.6	19.4	20.4	21.5
240	22.8	25.2	26.4	27.8
300	28.5	31.5	33.0	34.8
380	36.1	39.9	41.8	44.1
480	45.6	50.4	52.8	55.7
600	57.0	63.0	66.0	69.6
740	70.3	77.7	81.4	85.8
960	91.2	101	106	111
1200	114	126	132	139

Table 20.14 Earth-fault current withstand for steel wire armour on multicore BS 6346 cables.

Cond. size (mm^2)	Fault current withstand, kA for 1 s						
	Copper conductor cable				Aluminium conductor cable		
	2-core	3-core	4-core	4-core reduced neutral	2-core	3-core	4-core
1.5	0.66	0.70	0.75	–	–	–	–
2.5	0.75	0.84	0.88	–	–	–	–
4	0.92	1.01	1.54	–	–	–	–
6	1.06	1.58	1.76	–	–	–	–
10	1.80	1.94	2.16	–	–	–	–
16	2.02	2.20	3.17	–	1.85	2.02	2.90
25	2.64	2.90	3.34	3.34	2.38	2.73	3.08
35	2.91	3.26	3.70	3.61	2.55	2.99	3.43
50	3.26	3.70	5.37	4.14	2.90	3.43	4.97
70	3.70	5.24	6.07	5.94	3.26	4.97	5.63
95	5.37	6.07	7.04	6.91	4.80	5.63	6.47
120	5.76	6.60	9.68	9.46	–	6.07	8.84
150	6.34	9.28	10.60	10.30	–	8.40	9.68
185	8.84	10.12	11.66	11.44	–	9.46	10.78
240	9.90	11.44	13.16	12.72	–	10.56	12.06
300	11.00	12.72	14.65	14.21	–	11.66	13.38
400	12.28	14.04	20.55	19.89	–	–	–

In order to obtain the average value of the fault-current it is necessary to adjust the cable resistance from 70°C to its average temperature (115°C) during the fault. From Table 3 of the IEE *Guide on protection against overcurrent*, the factor to do this is 1.15.

The average cable resistance is then $1.15 \times 2.4 = 2.76$ mV/A/m.

For 50 m run the resistance is $50 \times 2.76/1000 = 0.138\,\Omega$.

This is the line to line value, for a three-phase fault the line to neutral value is required, or $0.138/\sqrt{3} = 0.0797\,\Omega$.

The total fault loop impedance is:

(supply) $0.025 \quad + \text{j}0.03$
(cable) $\quad 0.0797$
—————————————
Total $\quad 0.1047 + \text{j}0.03$
$\quad\quad = 0.1089\,\Omega$

The three-phase symmetrical fault current $= 230/0.1089 = 2112$ A.

From the characteristic of the 100 A fuse, the circuit will be interrupted within 0.03 s.

Table 20.15 Earth-fault current withstand for aluminium wire armour on single-core BS 6346 cables.

Copper conductor cables		Aluminium conductor cables	
Cond. size (mm^2)	Fault current withstand (kA) for 1 s	Cond. size mm^2	Fault current withstand (kA) for 1 s
50	3.16	50	2.84
70	3.56	70	3.24
95	4.05	95	3.65
120	5.83	120	5.35
150	6.48	150	5.83
185	7.13	185	6.32
240	7.94	240	7.13
300	8.34	300	7.78
400	12.39	380	11.18
500	13.69	480	12.15
630	14.99	600	13.20
800	21.06	740	18.23
1000	23.41	960	20.66
		1200	22.60

From Table 20.12, a 16 mm^2 copper conductor can withstand 1840 A for a duration of 1 s without exceeding 160°C.

At 2112 A the duration should be reduced to $(1840/2112)^2 = 0.76$ s. The fuse will operate well within this time limit.

In the case of circuits with separate devices providing protection against overload and fault current, in this example the overload trips of the motor starter and a fuse, it is important to check that the energy let through by the fault protecting device does not damage the overload device. For example, motor starters are not intended to interrupt fault currents, though they may be able to close on to currents of such magnitudes.

In the example given above the fault-current at the switch board, where the motor starter might be located, is $230/(0.025 + j0.03) = 5890$ A; the fuse would be expected to interrupt the current in no more than 0.004 s. The starter trips are unlikely to open the contacts as quickly as this and damage is unlikely. Even if the starter were to be located at the motor end of the cable, the speed of operation of the fuse is probably adequate. It is advisable to check with starter manufacturers whether the proposed fault-current protection is suitable.

Calculation of the maximum likely value (prospective value) of fault-current, both for short-circuits and, sometimes, for earth-faults is required in order to check that the fault-current protective device is capable of operating safely and reliably.

To calculate prospective fault-currents, details of the impedance at the supply terminals is needed; for the smaller installations the Electricity Association's Engineering Recommendations P23/1, P25 and P26 should be consulted. For the larger installation it is advisable also to consult the supplier.

Fault levels can be up to 30 kA at 11/15 kV and 50 kA on 400 V, so that the level should be checked before switchgear and cables are selected. Current limiting protective devices, such as high breaking capacity (HBC) fuses or current limiting breakers will limit the energy (I^2t) let through and prevent damage to switchgear and cables.

These figures assume that the cable is installed so that there is no significant crushing pressure on the insulation such as may arise under fixings or over the edges of objects.

Regulations remove the need to check the fault-current protection of conductors provided that the circuit protective device has adequate fault-current breaking capacity and complies with Section 433. In general, this means that calculation of fault-currents is required where short-circuit protection is provided by a device not complying with Section 433; such may be the case in motor circuits, and for earth faults.

Voltage drop

The energy loss involved in distributing electrical power takes the form of a reduction in voltage at the receiving end of each cable run. This reduction is dependent on the impedance of the cable, values of which are given in mV/A/m (i.e. mΩ/m) in the IEE Regulations or can be calculated. For cable sizes up to about 35 mm^2 the impedance is practically equal to the d.c. resistance of the conductors, but above this size, especially with single-core cables the inductance becomes increasingly important and at the largest sizes is the dominant component. As the inductance decreases only logarithmically with conductor diameter, it becomes more and more difficult at these large sizes to reduce the voltage drop by increasing the size of conductor and splitting the circuit into two or more independent parallel limbs may be necessary. Circuit inductance increases as conductors are separated so that from the point of view of voltage drop it is best to keep single-core cables as close together as possible.

Although it is general practice to use tabulated values of voltage drop directly without any adjustment, they are in reality vectorial values and in a.c. circuits neglecting the effect of their phase angles amounts to taking the worst case. Actual circuits may have a lower voltage drop than is given by such a simple approach. The effective voltage drop depends on both the power factor of the load and the phase angle of the cable impedance. The 16th edition provides a method for calculating values of voltage drop which takes both of these factors into account and the tables of cable impedance include both resistive and reactive components. Application of the method is straightforward for multicore cables and for single-core cables installed in trefoil.

Single-core cables in trefoil have slightly lower ratings than those laid in flat formation but have the advantage, which is important at the load currents for which single-core cables are commonly used, of presenting a lower impedance; in addition this impedance is balanced between the lines. As a general rule for 400 V circuits it is usually better to obtain a high current-carrying capacity by paralleling more than one run of cables in trefoil than to employ larger cables in flat formation.

For three-phase circuits having single-core cables installed in flat formation the apparent impedances of the three conductors are unequal and the impedance between the two outer conductors is tabulated, this being the one having the greatest magnitude. This will yield the highest value of voltage drop for all sizes of armoured cable with all values of load power factor. However, when using the voltage drop equations in Appendix 4 Section 7 it is not a matter of the largest magnitude of impedance voltage but of the largest component of that voltage in phase with the line-to-line load end voltage. For non-armoured cables the impedance across the outer conductors does not yield the highest voltage drop, as defined in Appendix 4, when the load power factor is higher than about 0.9 and may be seriously low for the larger single-core conductors. Under these conditions the highest voltage drop occurs between the centre and an outer conductor. The matter is somewhat complicated and designers of installations with very large unarmoured cables are recommended to consult a paper by A.H.M. Arnold in *JIEE*, Vol. 67, 1929, p. 90.

The unbalanced nature of the impedances of single-core cables in flat formation introduces considerable difficulties when cables are connected in parallel. Of all the available combinations of phase positions only a few offer the possibility of reasonable current sharing between cables in parallel in the same phase. With three cables per phase there is no arrangement of cables which will give equal current sharing; however, with a few arrangements balance within about 5% can be obtained.

It should be noted that the problem is not confined to the line conductors; neutral and protective conductors run with the phase conductors may be subject to considerable circulating currents. A summary of a few of the more suitable circuits is given in the IEE Guidance Note No. 6, *Protection against overcurrent*.

Especially with shorter runs, the arrangement at the terminations where cables break formation in order to connect to equipment terminals can seriously affect current balance between parallel conductors. Suitable terminal positions can reduce this difficulty and it may be advisable to use larger conductor sizes to provide for inevitable poor current sharing.

If materials which either surround or are close to cables (e.g. conduits, trunking, trays, structural items) are magnetic the effect on current ratings is small but the voltage drop can be increased. Unfortunately the general situation is too complicated to deal with in a simple manner as it is critically dependent on the disposition of the cables in relation to the steel. However, variation in cable disposition in conduits is limited and it is possible to tabulate values of voltage drop figures in rating tables for 'enclosed cables' which are based on experimental data. The impedance of multicore cables and of single-core cables up to and including $25\,mm^2$ is not affected to any practical extent when they are run in contact with steel.

Where single-core cables are used to feed large discharge lamp loads the neutral conductor carries a considerable third harmonic current which results in a significant voltage drop unless the neutral conductor is laid close to the phase conductors. The loaded neutral must then be taken into account when considering the effect of grouping on the rating of the cables.

FACTORS AFFECTING THE SELECTION OF TYPES OF CABLE

Cable selection is based on a considerable number of factors including regulations in force in the locality concerned, the environment and other matters discussed below.

Statutory regulations

Certain statutory regulations, referred to in the IEE Wiring Regulations, limit the range of cables from which the choice may be made, without removing the necessity for taking into account other factors such as environmental conditions.

In the UK the Electricity at Work Regulations 1989, which apply to all places of work, require that:

All conductors in a system which may give rise to danger shall either
(a) be suitably covered with insulating material and as necessary protected so as to prevent, so far as is reasonably practicable, danger; or
(b) have such precautions taken in respect of them (including, where appropriate, their being suitably placed) as will prevent, so far as is reasonably practicable, danger.

The Memorandum of Guidance produced by the Health and Safety executive refers the user to the IEE Wiring Regulations as giving guidance in installations up to 1000 V.

The guidance given in the Memorandum to the now repealed Factories Act (Electrical) Special Regulations 1908–1944 that in most cases 'protected' meant surrounded by earthed metal has been replaced by the offering of the use of steel trunking and conduit or the use of steel armoured cables as an example only. The Electricity at Work Regulations allow lightly protected cables, such as PVC/PVC to be used in situations where the risk of damage is low, but the burden of decision is upon the specifier.

Coal mines have very special requirements and are not considered here. The Miscellaneous Mines (Electricity) Regulations and the Quarry (Electricity) Regulations have been replaced by the Electricity at Work Regulations and an approved Code of Practice. This lays down that all cables operating at a voltage of 125 V a.c. to earth or between conductors should be entirely surrounded by earthed metal, which should have a conductance to earth of not less than half the conductance of the phase conductors. PVC/SWA/PVC cables are permitted to be used and are an excellent choice in such situations, but in the smaller sizes the armour may not have the necessary conductance. Cables with a copper strand in the armour may be specially ordered, if the quantity is sufficient. Otherwise a cable with an extra core may be used. In quarries, but not in mines, the lead sheath of a cable may be taken into account.

Flexes should have a protective conductor with the same conductance as the phase conductors and should be surrounded by earthed wire with a conductance of half that of the phase conductor. Smaller protective conductors are permitted if a suitable and reliable residual current device initiates disconnection.

The Cinematograph Regulations require cables for safety circuits, i.e. secondary lighting, emergency lighting and alarms to be in a separate enclosure from ordinary circuits and in particular from projection circuits. Cables must be enclosed in metal or other rigid non-ignitable enclosures or have a rigid metal sheath. It is doubtful whether PVC conduit complies with the former requirement; lead does not comply with the latter although copper does.

Before commencing the design of any installation which is covered by these regulations, it is recommended that copies be obtained from Her Majesty's Stationery Office and their contents studied in detail.

Environmental conditions

The relationship of the various environmental influences on the selection of cables and cords is given below.

Aesthetics
The aesthetic considerations governing the selection of cables are not really environmental conditions nor are they considered in the IEE Regulations unless it is under the heading of 'Workmanship'. Nevertheless, they are sometimes very important in the choice of a suitable cable, particularly in domestic and some commercial installations.

It is obvious that a flush system, i.e. one in which the cables are hidden beneath the plaster or within partitions, presents no difficulty; in situations where additions are to be made and decorations or wall finishes are not to be renewed, the problem is more acute. The choice falls on a cable which is small in diameter (for its current-carrying capacity) and which will dress well, i.e. is sufficiently easily bent to be accommodated, where necessary, to curves, etc. and yet which is stiff enough not to sag between clips. In these cases MI cable or Pirelli FP 200 or PX are likely choices, with PVC/PVC as a possibility. However, the advantage to be gained from the small diameter of MI cables may be offset where brass glands have to be used.

Temperature
The question of derating of cables for high ambient temperature is dealt with on p. 531. All cables are suitable for the normal range of temperatures, i.e. the IEC Classification AA5; it is the extremes of temperature which cause problems. All cables are suitable for handling and installation in the normal range of temperatures, from 5°C to 40°C, the average temperature over 24 hours not to exceed 35°C; this range is classified by the IEC as AA5. The problem is less serious when no movement is involved and specially formulated PVCs are available where the additional cost is considered worthwhile. PVC cables should not be installed during cold conditions; even a shock (such as may be caused by rough handling of a drum of cable) at low temperatures has been known to cause damage. It is best

to store cables at temperatures above 0°C for at least 24 hours prior to handling. It can be some time before heat penetrates thoroughly into the insulation, so that while a cable may appear externally to have survived handling when cold, it could be damaged internally. As far as flexibles are concerned, insulation and sheathing of rubber retain their flexibility to lower temperatures than those of PVC compounds.

Higher temperatures present different problems; the derating applicable to all cables when operating at high temperatures is prohibitive, although, when their seals are outside the hot zone, MI cables can operate at very elevated temperatures. Table 20.1 gives an indication of suitable cables for higher temperatures; for extreme temperatures where flexing is minimal, such as in luminaires, glass fibre insulation is suitable.

Fire, as an environmental condition is not considered by the IEE Regulations. However, circuits which are required to continue operating during a fire must rely on either MI cable which will continue to operate up to the melting point of the copper (this does not apply to the termination) or on Pirelli FP 200. The latter is destroyed by fire but the insulation is reduced to silica which still provides a degree of insulation, allowing the circuit to continue operating.

Presence of water

All rubber and PVC sheathed cables are reasonably resistant to the presence of water and it is their terminations which require protection. However, if continued operation in wet conditions is envisaged, it is necessary to use either unsheathed cables in a watertight enclosure or MI cable with a PVC oversheath or PVC/SWA/PVC cables. Cables with an aluminium content are not suitable.

Presence of dust and foreign bodies

Dust and foreign bodies, unless presenting hazards of corrosion or mechanical damage do not affect the choice of cable, although accumulations of dust may reduce the cable's current carrying capacity. The effect on cable choice and installation methods of stored materials (and dust arising therefrom) are considered on p. 558.

Presence of corrosive or polluting substances

There are substances which have a deleterious effect on steel, aluminium, copper, PVC and rubber, all of which are used in cable construction. The choice of a cable must depend upon the nature of the corrosive substance. Water, as a corrosive substance, is resisted by paints or galvanizing, provided that these are continuous and not liable to damage. For more actively corrosive agents, galvanizing may not be adequate and either black enamel conduit with protective grease tapes, or PVC may be necessary to provide protection.

The copper sheath of MI cables in damp and mildly corrosive situations will form a patina and corrosion will not proceed beyond that point, unless the patina is removed. However, the effect is to reduce the thickness of the copper sheath and so, if the sheath is relied upon as a protective conductor, the design calculations may be invalidated. For this reason, MICC cables used in damp and mildly corrosive conditions must be provided with a sheath of PVC (which is available in a variety of colours) and glands must be provided with PVC shrouds. Connection boxes should be made of plastics or of hot dip galvanized, cast or malleable, iron rather than zinc-electroplated mild steel.

Besides the more obvious sources of corrosive chemicals in cements and plastics, care should be taken in the selection of cables to be fixed to oak and similar woods, some kinds of stone if damp, and to metals which are dissimilar to the sheath. Organic liquids, particularly creosote or petroleum products can seriously damage PVC and rubber; cables insulated or sheathed with these materials should be routed so as to avoid contact with such substances. Where this is not possible cables with a lead sheath should be used.

Mechanical stresses

Nail and screw damage
No cable can resist penetration by nails and screws; even steel conduit can be penetrated. For this reason cables must be so located that penetration is unlikely. Regulations define zones in which cables must be run and in which screw and nail fixings should not be made. Alternatively cables must be protected by an earthed metallic sheath or armour or enclosed in earthed metallic conduit so that, if penetration occurs and the screw or nail makes contact with a live conductor, the circuit-breaker will trip or the fuse will blow.

Impact
Cables should be installed to minimize the possibility of impact but this is not always possible and so an assessment of the likelihood of impact must be made. In domestic and office installations, all cables and cable systems provide adequate resistance to impact, but at points of particular likelihood of damage (at skirting boards for instance) further protection should be provided for PVC/PVC cables.

Although MI cables can be shown to continue to operate satisfactorily when hammered flat this characteristic should not be relied upon where damage of this nature is thought likely to occur.

PVC/SWA/PVC cables provide an adequate degree of protection for most industrial applications, but do not have the same degree of resistance to impact as steel conduit. PVC cables in steel conduit provide a system which is most resistant to impact; any impact of such severity as to damage cables in conduit is likely to destroy the conduit system.

Flexible cables and cords have a very limited resistance to impact and should be so installed as to render impact unlikely; where such a possibility remains, flexible cables and cords with metallic braids are available.

Vibration
Vibration of the fixed installation of a building is unlikely except at final connections to motors or other equipment. While conduit cables themselves are reasonably resistant to vibration, particularly if stranded, steel conduit is not and is likely to fracture, particularly at box spouts. Patent flexible or semi-flexible conduit systems or PVC conduit do not suffer from this defect.

Copper work hardens and fractures under vibration; to guard against slight vibration a loop of cable adjacent to the equipment will be adequate but for installations where continuous vibration is experienced PVC/SWA/PVC cables are to be preferred. All flexibles except glass fibre types are resistant to vibration. Where there is sustained movement appropriate types of flexible cable must be used. However, for very occasional limited movement, for example when starting long running process plant mounted on anti-vibration mountings, extruded insulation armoured cables may be used provided that an adequate length of free cable is arranged to drop down vertically to the equipment. The weak point is likely to be where the cable enters into the equipment and well-proportioned armour clamps and glands are essential.

Animal damage

Animals, in relation to electrical installations fall into two classes, those which are intended to be present and those which it is feared may be present. The first class occur in agricultural installations and the IEE Regulations state that all fixed wiring systems shall be suitably protected from livestock.

The second category, of which rats and mice are the most usual, is vermin. Insects are not known to attack cables in the UK, but may be a problem elsewhere.

Rats have been known to chew lead, copper, PVC and even to attack wire armouring; to be absolutely certain of resisting attack, steel conduit or PVC/SWA/PVC cables should be selected. It must be said however that there is little or no evidence that rats or mice attack PVC conduit or trunking. This may be because of the different grades of PVC used for conduit and trunking as opposed to that used for cable sheaths.

Solar radiation

The ultraviolet component of solar radiation can have deleterious effects on PVC. Where cables are to be exposed to sunshine, a metallic sheathed cable, cables in steel conduit or cables with a black PVC oversheath should be used.

Figure 20.5 illustrates the reduction in rating due to direct solar radiation on armoured PVC-insulated cables installed in free air and lying at right angles to the sun's rays. The percentage reduction in rating is approximately proportional to the intensity of solar radiation. Local values of intensity depend on latitude and, above all, clarity of the atmosphere. The intensity at or above 3000 m height is about $1200\,W/m^2$. At sites with a clear atmosphere this is reduced to about $1000\,W/m^2$ at sea level at UK latitudes. In and around towns and industrial sites the intensity is reduced to about $800\,W/m^2$. If the cable rating is critical, it is advisable to seek guidance on local values of solar radiation intensity. IEC Publication 287 provides a method for calculating the effect of radiation. These figures apply to times of maximum radiation, i.e. around midday in the summer. If the cable load is greatest at other times of the day or year the effect of solar radiation is likely to be unimportant.

It is often overlooked that cables installed under transparent roofs can be subject to solar radiation. Although direct radiation received by the cable may be

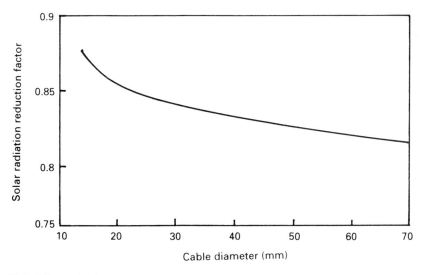

Fig. 20.5 Effect of solar radiation on rating of three-core BS 6346 armoured cable at a radiation intensity of $1000\,\text{W/m}^2$.

reduced by the roof material, there is an important increase in the effective ambient temperature, not only of the air in the roof space but of the general radiative ambient due to the surrounding surfaces of the building. Effective ambients could well reach 40°C, without taking into account the effect of heat from any processes in the building. Such ambients introduce a reduction in rating in addition to that due to direct solar radiation.

Utilization

The IEE Regulations include utilization as a class of external influence and list a number of categories which may influence the choice of cable.

Capability of persons
This would not seem at first glance to affect the choice of cable, but if extended to the action of persons, particularly of children, there is the problem of deliberate damage. No system can totally resist deliberate attack, but in places where it is envisaged that such attack may take place and where a surface installation is concerned, there is no doubt that the most resistant system is steel conduit and that PVC/PVC sheathed cables are the most vulnerable. MI cables have an increased resistance. The use of flexible cords should be avoided wherever possible.

Contact of persons with earth potential
All metallic sheathed cables propagate potentials whether earth or higher. The IEE Regulations seek to limit these potentials, particularly by equipotential bonding; in certain situations of high risk, such as bathrooms, extra supplementary bonding is required of all exposed conductive parts and extraneous conductive parts. Since the exposed conductive parts of an installation are primarily sources

of raised potentials it seems sensible in bathrooms and similar situations to use either PVC conduit or PVC sheathed cables. PVC sheathed MI and PVC/SWA/PVC cables require metallic terminating glands, although these can and should be either shrouded or contained within a non-conducting enclosure.

Conditions of evacuation
The only consideration likely to affect the choice of a cable has been dealt with under 'Fire'.

Nature of processed or stored materials
Processed or stored materials can produce three environments which may affect the choice of cable; these are explosive vapours and gases, dust and stacked materials.

Hazardous atmospheres
The selection and installation of cable systems in potentially flammable atmospheres is dealt with in CP5345. The requirements of this Code of Practice are over and above those of the IEE Regulations; the selection and sizing of cables must comply with the requirements of both documents.

Cables appropriate to hazardous atmospheres are single-core cables to BS 6004 when enclosed in suitable conduit systems, or MI cables, or PVC/SWA/PVC cables, or PVC/LS/SWA/PVC cables, provided that all these are terminated with suitable glands.

Dust
Dust is universally present, but in some installations it is a considerable problem. The explosion risk arising from dust is dealt with in BS 6467 which concerns itself with the design and installation of equipment for use in dusty atmospheres. All cable entries must be sealed, thus requiring gasketted conduit boxes and grometted glands. The standard offers little guidance to the cable installer, but then, cables themselves offer little hazard unless mechanically damaged. For this reason cables in conduit or armoured cables should be selected.

Accumulations of dust are a different matter. All factory occupiers are required under the Factories Act to keep their premises clean. The cabling designer and installer has, however, to decide whether this prescription will invariably mean that accumulations of dust will not occur. If dust does accumulate it can be of significant thickness (50 to 75 mm is not unknown).

Cables should therefore be so installed as to facilitate cleaning and to reduce the accumulation of dust. Where practicable, cable trays should be installed on edge or a screen should be provided to reduce the dust accumulation.

Accumulations of dust do not affect the selection of the type of cable, unless the dust has a deleterious effect on the sheath, but cable ratings are affected because of the thermal insulating effect of the dust. Damage to PVC insulation is likely and in the extreme with MICC cables temperatures sufficient to ignite the dust could be reached.

The precise effect of dust accumulation has not been studied, but the effect is not dissimilar to that of installing the cable in an insulating wall. The current-carrying capacities given by the Regulations for this situation are reduced from the clipped-direct value by approximately 30% and all cables including MICC should be derated by this amount in situations where dust is likely to accumulate. In any case, when dust is present, MICC cables should be selected on the basis of their exposed to touch ratings.

Stacked materials

Where the building or structure within which the cables are to be installed is intended for the storage of materials in stacks or heaps, cables should be installed outside of the zones in which the materials are likely to be kept. If this is not practicable, provision should be made so as to ensure that the circulation of air will be maintained around the cables.

Fires can be and are caused by surrounding loaded cables with stored materials; it is not possible to calculate a derating factor to allow for this possibility. Where personnel can climb on to stacked materials, any conduits should be installed in such a way as to ensure that they cannot be used as hand or foot holds.

Commercial considerations

Consideration of the environmental influences to which the installation, or any part of the installation may be exposed, may reduce the range of choice of suitable cables, but rarely does it eliminate choice entirely. From the remaining possible cables it is the duty of the designer to select the most economical type. Table 20.16 gives the relative costs per metre installed of various types of cable for a 5 m run of a single-phase 16 A circuit, protected by hbc fuses and run at normal working height on plaster, in a normal 25°C ambient. It includes the cost of necessary terminations.

To arrive at the figures in Table 20.16 the labour content was based on published data and the material at trade prices. The labour cost for Pirelli cables was based on MI costs. Table 20.16 is not, nor is it intended to be, definitive. It seems to indicate that for an installation in which the environmental influences are not severe, the aesthetic considerations low (i.e. the cable is concealed) and the customer is not prepared to pay for a long life or rewireability, PVC/PVC cables are the only choice, hence the widespread use in domestic and small offices. It indicates that a single circuit of cable in steel conduit is marginally more expensive than a two-core PVC/SWA/PVC. It also indicates that two circuits of cable in steel conduit are cheaper than two PVC/SWA/PVC cables.

Table 20.16 takes no account of the ability of conduit (and trunking) to span wider gaps than any other system, and its ability (in certain circumstances) to support luminaires. Nor does it take account of the neatness of MI cables or Pirelli cables, of the relatively shallow chases needed (if at all) for MI cable, and of its long life. It does not take account of the ability of PVC/SWA/PVC to stand a certain amount of vibration and movement. Neither does it take into account the third factor affecting the selection of cable, namely thermal considerations.

Thermal considerations

The choice of a cable to carry a given current is determined from the tables in Appendix 4 of the IEE Regulations applying where necessary the relevant cor-

Table 20.16 Relative cost of 5 m of cable to carry a 16 A single-phase load, including terminations where necessary.

Type of cable	Relative cost	Notes
1.5 mm² PVC twin + CPC*	1	Nailed clips
2 × 1.5 mm² singles in steel conduit	8.2	
2 × 2.5 mm² singles in steel conduit with two other circuits	3.0	
2 × 1.5 mm² singles + 1 mm CPC in PVC conduit	6.5	
Two-core 1 mm² light duty MI	5.2	Sheath used as CPC, screw on seals
One pair in a seven-core 2.5 mm² heavy-duty MI	3.2	Sheath used as CPC, screw on seals
Two-core 1 mm² light-duty MI/PVC	5.5	Sheath used as CPC, screw on seals
One pair in a seven-core 2.5 mm² heavy-duty MI/PVC	4.3	Sheath used as CPC, screw on seals
Two-core 1.5 mm² PVC/SWA/PVC	6.8	
Three-core 1.5 mm² Pirelli FP 200	3.7	

* CPC = circuit protective conductor.

rection factors for grouping and ambient temperature and the presence of thermal insulation. It should be noted that these factors are cumulative only where more than one condition calling for a factor is present; so that if a cable is grouped in one place, is in a high ambient in another and in thermal insulation at yet another, each factor is applied once and the cable is selected to meet the worst condition. For large cables jointing of different sizes of cable may be worthwhile; the cost of jointing must be assessed against savings in cable and installation costs.

A cable protected against overload is also protected against short-circuit, but the size of the protective conductor incorporated in a cable may not be adequate to withstand the thermal effect of the current flowing through it at the time of a fault.

Of the wiring systems under consideration the following form the protective conductors:

PVC/PVC cable	Copper protective conductor incorporated
PVC cables in metal conduit	Conduit
PVC cables in PVC conduit	Separate conductor drawn in (size to suit)
MI cable	Sheath
PVC/SWA/PVC cable	Armouring
Pirelli FP 200 PX cable	Copper protective conductor incorporated
Flexible cables and cords	Copper protective conductor incorporated

Tables 20.17 and 20.18 indicate that in the range of cable sizes considered, the protective conductor incorporated in a cable, be it a separate copper conductor, cable sheath or armouring, has a cross-sectional area which is adequate for the energy let-through demands of semi-enclosed fuses to BS 3036 which will also protect the circuit conductors against overload.

BS 88 fuses may at first sight appear marginally to fail to protect the protective conductors of PVC twin and earth cables; before rejecting this combination of fuse and cable the actual disconnecting time should be calculated. The circuit impedance is often limited by voltage drop, and a small reduction in impedance could reduce the disconnecting time to a value which enables a BS 88 fuse to protect such protective conductors.

With all fuses, shorter disconnecting times give lower I^2t let-through; in the case of Type 1 and Type 2 mcbs to BS 3871 whose characteristics are given in the IEE Regulations, this is not so; in the case of Type 1 and Type 2 there is no 5 s disconnecting time, the 'knee' in the characteristic occurring at 20 and 7 s, respect-

Table 20.17 Cross-sectional area of protective conductor protected by a BS 88 fuse.

Fuse rating (A)	Type of protective conductor*	
	PVC insulated single-core (mm²)	Protective conductor in PVC two-core + earth cable (mm²)
5 s disconnecting time		
6	1	1
10	1	1
16	1	1.5
20	1.5	?.5
25	2.5	2.5
32	2.5	2.5
40	4	4
50	4	6
63	6	6
0.4 s disconnecting time		
6	1	1
10	1	1
16	1	1
20	1	1
25	1	1
32	1	1.5
40	1.5	1.5
50	2.5	2.5
63	2.5	4

* The armour of PVC/SWA/PVC cables and the sheath of heavy-duty MI cables for the above circuits are adequate as protective conductors.

Table 20.18   Cross-sectional area of protective conductor protected by BS 3036 (semi-enclosed) fuses.

Fuse rating (A)	Type of protective conductor*	
	PVC insulated single-core (mm²)	Protective conductor in PVC two-core + earth cable (mm²)
5 s disconnecting time		
5	1	1
15	1	1
20	1	1.5
30	1.5	2.5
45	2.5	2.5
60	4	4
100	10	10
0.4 s disconnecting time		
5	1	1
15	1	1
20	1	1
30	1	1.5
45	1.5	2.5
60	2.5	4
100	4	6

* The armour of PVC/SWA/PVC cables and the sheath of heavy-duty MI cables for the above circuits are adequate as protective conductors.

ively. Thereafter the tripping time is anything between the 'knee' time and 0.1 s, so tables similar to those above cannot be prepared and the manufacturers should be consulted concerning the I^2t let-through of their devices.

Steel conduit has an adequate cross-sectional area for the size of fuse which will protect the cables drawn into it. The continued adequacy of steel conduit as a protective conductor is dependent upon its having been properly installed in the first place. Good threads must be cut on the conduit and the screwed joints must be tight. Damaged paint or galvanizing must be made good and, in those cases where corrosion as opposed to moisture is likely, special protective tapes must be used.

If, because of poor installation practice, it cannot be guaranteed that the joints in the conduit will remain sound, it is no answer to draw in a separate protective conductor into the conduit. Even if the protective conductor is connected to every box (which would not be possible) there is no way of bonding a length of conduit to it, and so, while the equipment supplied by the circuit in the conduit would be protected, lengths of conduit themselves may become and remain alive.

Attention is drawn to the minimum sizes required by IEE Regulations for protective conductors not enclosed in a wiring system and by Regulations for buried earthing conductors.

Voltage drop

It is usually found that the choice of a cable size for any particular circuit is influenced more by voltage drop than by any other factor and often the advantage of the higher current ratings of some cables is lost. Information on the voltage drop of cables is given in Appendix 4 of the Regulations in the form of millivolt drop per amp per metre (mV/A/m), taking into account all the conditions of a circuit.

For larger cables separate figures for the voltage drop arising from the resistive (mV/A/mr) and reactive (mV/A/mx) components are given together with that arising from their impedance (mV/A/mz). The use of these values without adjustment for power factor or operating temperature may lead to the selection of an unnecessarily large cable and advice on the effect of power factor on the voltage drop of larger cables and of temperature of all cables is given in the Regulations. While the savings to be had in the selection of the larger sizes of cable may be considerable, particularly where the run is long, for most final circuits the effort involved in making this adjustment may not be worthwhile.

Because voltage drop in most cases governs the choice of cable size it is advantageous to arrive at the cable size from voltage drop considerations first and then to ascertain whether the cable size arrived at will carry the design current. The method is shown by the following example.

Load current 20 A; circuit length 16 m.
Permitted voltage drop, 2½% of 240 V = 6000 mV.
Maximum mV/A/m is: $6000/20 \times 16 = 18.75$ mV/A/m.

The cables listed in Table 20.19 would therefore be suitable from the voltage drop point of view.

It is clear that the apparent advantage of the higher current-carrying capacities of certain cables is lost under voltage drop considerations. However, voltage drop is not subject to grouping or ambient temperature correction factors and for groups of cables the advantages of higher temperature cables once more become apparent.

Table 20.20 shows the maximum lengths of certain different cables which will give a 2½% voltage drop on a 240 V system with a 20 A load, single-phase, selected on their current-carrying capacities. Similar tables can be prepared from the IEE Regulations for other types of load and 4% voltage drop.

It can be seen that voltage drop is in many cases more limiting than current rating and that advantages gained in current rating are lost in voltage drop penalties.

Earth-loop impedance

For any voltage to earth, fuse and disconnecting time, there is a maximum earth-loop impedance. Some of that impedance is external to the circuit or installation concerned; a typical value for a TN-C-S (PME) service from the public supply network is 0.35 Ω. For a circuit at 240 V to earth a 20 A Type 3 mcb requires a

Table 20.19 Voltage drop for different cables.

Cable	Size (mm^2)	Voltage drop (mV/A/m)	Current-carrying capacity (A)
Single-core PVC conduit	2.5	18	24
PVC/PVC two-core + earth	2.5	18	24
PVC/SWA/PVC	2.5	18	28
Single-core 85°C rubber in conduit	2.5	18	30
Light duty MICC two-core PVC covered	2.5	17	31
Pirelli FP200*	2.5	17	29

* Based on BASEC Certificate No. 88100690.

Notes:

(1) Judged by current-carrying capacity only 1.5 mm^2 PVC/SWA/PVC or 1.5 mm^2 85°C rubber singles or 1.5 mm^2 two-core MICC would be adequate.

(2) In marginal cases the running of an oversized cable below its rated operating temperature might reduce the voltage drop slightly, allowing a smaller cable to be used than indicated by the above type of calculation. In sub-circuits the relationship between the calculated to installed length of cable is not usually that accurate.

Table 20.20 Maximum length of circuits.

Cable	Size (mm^2)	Voltage drop (mV/A/m)	Length for 2.5% voltage drop (m)
Single-core PVC in conduit	2.5	18	16.6
PVC/PVC two-core + earth	2.5	18	16.6
PVC/SWA/PVC two-core	1.5	29	10.3
Single-core 85°C rubber in conduit	1.5	31	9.6
Light duty MICC, PVC covered	1	45	6.7
Pirelli FP200	1	45	6.7

current of 200 A to disconnect either socket outlet or fixed apparatus circuits so that the maximum earth-loop impedance cannot exceed 1.2 Ω. If the external impedance is 0.35 Ω then the total impedance of the phase and protective conductors in the circuit must not exceed 0.85 Ω. Table 20.21 shows the length of various types of cable to give a phase and circuit protective conductor resistance of 0.85 Ω (assuming a circuit load of 20 A). It is based on the resistance of conductors at 115°C which is the temperature of the conductor of these cables when the fault is disconnected. Similar tables can be prepared for other types of protective devices and ratings.

Table 20.21   Lengths of circuit to give a total impedance of 0.85 Ω.

Cable	Size (mm^2)	Line + Protective conductor (Ω/m)	Length for 0.85 Ω (m)
PVC single-core + equal sized protective conductor	2.5	0.020	43
PVC/PVC two-core + earth	2.5	0.026	33
PVC/SWA/PVC two-core*	1.5	0.032	27
85°C rubber single-core + equal sized protective conductor	1.5	0.037	23
Light duty MICC[†]	1.5	0.019	45
Pirelli FP200[‡]	2.5	0.020	43

* Based on maximum values given in BS 6346.
[†] Based on maximum values given in BS 6207.
[‡] Based on values given in BASEC Certificate No. 88100690.

INSTALLATION METHODS

Guidance is given in the IEE Guidance Note No. 1, *Selection and erection*, on methods of installation including spacing of supports for conduit and trunking and current-carrying capacities of cables housed in them.

Conduit and trunking capacities

The cable capacity of conduit is dependent not only on the percentage cross-section of the conduit which is occupied by the cables, but on the number of cables which can be pulled in without excessive force. The length between draw boxes and the number of bends also affects the drawing-in tension required. Excessive force damages the insulation, or in extreme cases stretches the cable and so reduces its cross-sectional area. Hence the IEE Regulations provide two sets of cable and conduit factors, one for short straight runs (less than 3 m) and one for long runs or runs incorporating bends; the total of the cable factors must be less than the conduit factor.

For instance, 20 mm conduit, with a factor of 460 for short straight runs, can accommodate $16 \times 1.5\,\text{mm}^2$ PVC single cables. However, eight circuits must be derated by a group factor to 0.51 times their full rated current, so that 6 mm^2 cables are required to carry 20 A. These need 25 mm conduit. This is still more economical than the alternative of 2.5 mm^2 cables in three separate 20 mm diameter conduits. If a 3 m run with two bends of 20 mm conduit is used, only ten 1.5 mm^2 cables can be accommodated.

There are therefore major economic advantages in maintaining straight runs and in keeping between 3 m distances between draw boxes. This usually can only be a site decision and most drawings do not provide sufficient detail to be certain of the bends or sets required. Factors for trunking installations are also provided and the cable factors are different from those for use with conduit because no pulling in is required. The factors should be strictly adhered to even though it is possible to force a greater number of cables into trunking than the factors allow but at the cost of pressure on the cables. Although the IEE Regulations allow a 45% space factor, care should be taken during installation to ensure that the cables do not press excessively on the trunking, particularly at bends.

Cable supports

Unsheathed plastics and rubber cables need to be enclosed in conduit or trunking. Plastics or metal-sheathed cables may be supported on clips or saddles. The spacing of supports for conduit and trunking is based upon the loads and spans that can be supported without undue sag. The requirement for supports within 300 mm of bends acknowledges the additional stresses imposed by thermal expansion. Closer spacing of supports is called for where the installation may be expected to receive mechanical shocks. Fixings may be by way of screws through the back of boxes. Cables in vertical conduits and trunking must be supported at intervals to prevent undue pressure on the insulation at the top.

The internal edges in trunking at bends, etc., must be well rounded and the use of site constructed adaptations in place of factory-made fittings should be avoided unless care is exercised to avoid sharp edges. The top of vertical runs exceeding 5 m length should be fitted with a bend and not with an elbow or with a box. One of the most frequent causes of high mechanical pressure on wiring cable is the treatment it receives when pushed back into a box behind an accessory. All incoming holes in the box must be properly bushed and good workmanship exercised to arrange the cables as smoothly as possible.

Sheathed cables may be fixed directly to a surface or to tray or ladder supports. Spacings given in the IEE Guidance Note are those which will not impose undue stresses on the cables at the point of support. Pressure may arise when supporting the weight of a vertical run of cable; clamps should be distributed uniformly and sharp bends at the top of the run must be avoided. In order to get satisfactory dressing of cables, closer fixing centres may be required.

Fixing clips may be metal or plastics; when fixing MI cables, caution should be exercised if the cables are rated 'not exposed to touch' as their temperature may be sufficient to distort the clip.

Terminations and connections

Insulation should be stripped back only just far enough to permit complete insertion of the conductor into an appropriate terminal. Care must be exercised not to nick or damage the conductor and proper wire strippers are desirable; sharp knives are not recommended. If a knife is used then a glancing action, as though sharpening a pencil, is best. All the strands of a conductor should be securely fixed in or under the terminal and, where appropriate suitable tag washers should be used. Crimped connections should be made only with lugs or connectors of a size intended for the conductors concerned and with a tool recommended by the manufacturer of the connector.

Where cables enter into terminating boxes or equipment the sheath is taken through into the box and is stripped off inside. If the box or cover is made of metal the sheath must be protected from sharp edges by the use of a grommet, and for surface work or where foreign substances might enter the box the cable should be mechanically anchored and the entrance hole properly filled by using a suitably shaped gland which grips the cable.

If the cable is armoured, the gland must be of a type which provides adequate mechanical anchorage for the wires or strips. As the armour is often used as a protective conductor the gland must also provide good electrical contact and in turn be reliably connected to the enclosure or other earthed metal work. On installations out of doors or in wet environments the whole gland should be protected against corrosion by a watertight covering which goes well back over the outer sheath.

The ends of MI cables must be completely sealed against the entry of moisture. This is effected by screwing a small brass pot on to the end of the sheath and filling it with a special non-setting water-resistant compound, as shown in Fig. 20.6 or by using heat-shrink seals (see BS 6081). The conductors, where they emerge from the seal, are covered with insulating sleeving. The materials used for the seal and the sleeving must be selected to suit the expected working temperature of the cable.

Where MI cable enters terminal boxes and equipment it should be anchored by the use of a gland, which also provides electrical continuity for the copper sheath and, if required, watertightness. Here again it is often appropriate to cover the entire gland with a heat-shrink sleeve which goes well down over the outer PVC covering of the cable.

Terminating the sheath of MI cables calls for the same care to ensure adequate electrical continuity as with the armour of other cables, as both are generally used as protective conductors and must carry the earth-fault current of the circuit. Single-core MI cables involve a further consideration in that a small voltage is induced in them during normal operation which can be much higher during a short-circuit. It is normal practice to avoid danger from these voltages by bonding all the sheaths of each circuit together at each end and earthing them. As a result current circulates continually around the sheath even during normal operation. Providing care has been taken to ensure that bonding connections are sound and adequately robust no harm can arise.

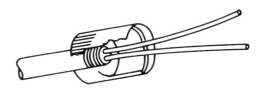

Fig. 20.6 MI cable termination (*BICC*).

The individual cables or conductors of single-core cables must not be separated by magnetic material; all the glands belonging to each circuit should be mounted on a non-magnetic plate (or an air gap must be formed in the plate) and glands should not be mounted on steel extensions formed by steel sockets or couplers, etc.

Busbar systems

Enclosed busbars, of copper or aluminium, mounted at intervals on suitable insulators form a flexible way of providing either main or final circuits. In the first case they are often seen in the form of vertical risers distributing supplies between floors and in the latter case popular forms are overhead busbar trunking systems supplying larger pieces of equipment and overhead lighting systems, including lighting tracks, in industrial and commercial installations.

Generally this type of system comes in manufactured lengths and is assembled on site. The manufacturers are responsible for providing electrical data such as current-carrying capacity, fault-current capacity and impedance.

In most types the bars are arranged in line across the enclosure, an arrangement which introduces unequal impedances and hence voltage drops between phases. The impedances are higher than for corresponding cables. However, for short distances and numerous tapping off points they offer good current-carrying capacities and flexibility of tap-off position which could not be economically obtained by any other means.

There are special systems, chiefly for rising mains where distances between tap-off points are fixed, where special forms of conductor result in bars which are much more compact for a given current rating. These offer a lower impedance and take up less space. The George Ellison Lambar system consists of interleaved and bonded flat sheets of copper and insulation, available with ratings up to 1000 A. Its form of construction provides a continuous fire barrier. Another system by Coaxial Risers and Switchgear Ltd is a concentric arrangement of bonded sheets of copper and insulation around a central copper rod.

Overhead busbars provide a ready means of changing or extending a distribution system in industrial premises. Such bars are normally arranged so that fused tap-offs or fused switches, up to 400 A rating, can be plugged into them at intervals of about 1 m upwards. Important points to note for plug-in tapping devices, see Fig. 20.7, are:

(1) A means of polarization must be provided so that insertion can only connect the outgoing conductors in the correct sequence.
(2) The plug-in unit must be firmly attached to the busbar enclosure.
(3) The fuses must be capable of clearing the busbar fault current and the plug-in device must be constructed so as to withstand such currents without distortion or ejection.
(4) A fifth bar can be added to provide a protective conductor connection.
(5) The tolerances on the means of location and the bar contact clips must be of a standard which ensures reliable contact between bars and tee-off conductors.

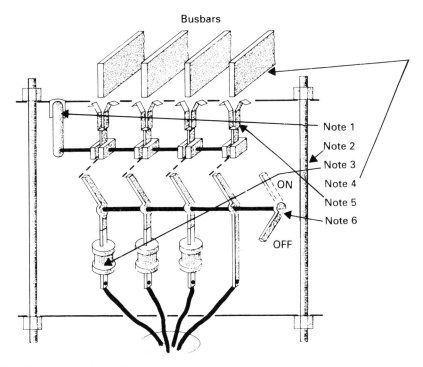

Busbars

Note 1
Note 2
Note 3
Note 4
Note 5
Note 6

ON

OFF

Fig. 20.7 Busbar plug-in and isolating unit.

(6) There must be an interlock to prevent insertion or removal of the plug-in device when the isolator contacts are closed. The position of the isolator contacts must be clearly indicated on the outside of the device.

Busbars can be used as main feeders. As these are often assembled on site, care must be taken to ensure that phases and neutral are not reversed after bars have negotiated bends, Fig. 20.8. It is prudent to allow for changing the order of the bars at a panel, although some manufacturers include a length of busbar which does this.

A reduction in the size of busbar is possible where the feed is at or near to the centre of the distributed load, in which case the bar can be about half the cross-section of an end feed unit, Fig. 20.9.

Long runs of bar should include expansion joints to absorb their change in length due to temperature variations. Where bars pass through fire-rated walls or floors fire barriers must be included to local authority requirements.

Cable management

The increasing and widening use of data-processing equipment has led to the installation of substantial quantities of data cables. This chapter is not concerned with the installation of such cables, which are the subject of cable management

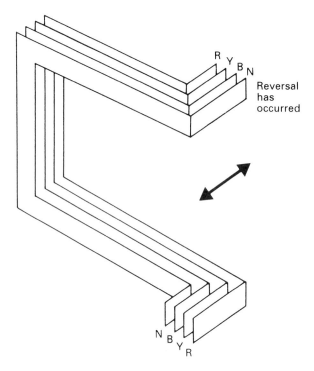

Fig. 20.8 Sketch showing how busbar phases can become reversed during installation.

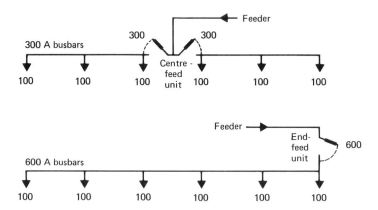

Fig. 20.9 Busbar systems with either centre- or end-feed units.

programmes. However, the performance of mains voltage cables may be adversely affected if they are covered by banks of data cables. Similarly, there may be electromagnetic interference between the power cables and the data cables. The latter commonly lie on the bottom of cable voids and power cables which are to be installed beneath a computer room floor or in a cable way carrying data cables should be fixed as high as possible.

Fire spread

The PVC sheath of cables can contribute to the spread of fire and potentially to the production of poisonous smoke. This is a particularly serious problem in designated escape routes and in cable routes where there are substantial numbers of cables. Where the installation of cables in escape routes or in places from which smoke and fumes can travel into those routes is unavoidable, low smoke and fume cables should be used.

Problems

Properly selected and installed cables have few problems but the most common are low insulation resistance and breakdown under overvoltage.

Persistent low insulation resistance in MI cables is usually due to defective seals at terminations and can only be cured by driving the moisture out of the cable with heat and re-sealing. However, low insulation resistance immediately after sealing an MICC cable is not uncommon and usually disappears within 24 hours. MICC cables should therefore not be tested for 24 hours after sealing; cables then having a low insulation resistance should be re-sealed. Low insulation resistance can occur because of surface tracking over the insulation at terminations due to damp or dust or in the case of braided cables to the trapping of the braid in the connection. This can also affect PVC and rubber-insulated cables and is cured by cleaning the cables. Low resistance also appears to be caused by excessive pressure of PVC cables on earthed metal and is cured by relieving the pressure.

Any cable will break down under overvoltage if the voltage is sufficiently high. In this respect transient voltages in l.v. networks can be as high as 4 kV. This can be withstood by most types of cable, but MI cable can be adversely affected, particularly at sharp bends. Such overvoltages either occur by design (such as lamp ignitor circuits) or result from the rapid collapse of magnetic fields such as those in control equipment coils or fluorescent lighting chokes when the coil circuit is interrupted. This situation is often made worse by chattering contacts.

In the case of ignitor circuits, MI cables should not be used; to do so is wrongly selecting the cable.

In the case of surges from equipment, the first action should be to ensure that there are no intermittent contacts and secondly, to fit surge diverters, available from the manufacturers, either across the terminals of the offending equipment or across the terminals to which the cable is connected. It is often only one length of cable in an installation which is affected.

CHAPTER 21

Control and Protection of Low Voltage Installations

H.R. Lovegrove, IEng, MIEIE
(Consultant)

ISOLATION AND SWITCHING

General

An essential part of every electrical installation is the means of control, both of the circuits and the current-using equipment.

There are four requirements for switching, three of which are safety requirements:

(1) isolation;
(2) switching-off for mechanical maintenance;
(3) emergency switching;
(4) functional switching.

One of the fundamental requirements for safety listed in BS 7671 (The IEE Wiring Regulations) calls for effective means to be provided in a suitable position, to cut off all voltages from every installation and every part of an installation in order to prevent or remove danger.

More importantly, the law in the form of Regulation 12 of the Electricity at Work Regulations 1989 requires suitable means to be provided to cut off the supply to equipment, and to isolate equipment from the supply in order to prevent danger.

The purpose of these requirements is to ensure that any person working on parts of an electrical system or electrical equipment that is normally live, or may become live, is protected from the effects of electric shock or injury from mechanical movement caused by equipment being unintentionally or accidentally energized. Also, persons operating equipment must be protected from injury in the event of the equipment malfunctioning.

Isolation

The international definition of 'isolation' is 'A function intended to cut off for reasons of safety the supply from all, or a discrete section of, the installation by separating the installation or section from every source of electrical energy.'

An isolator, also known as a disconnecter, is a device which disconnects the

572

supply to the system, or parts of the system, in the off-load condition. Generally, this should be accessible to skilled persons only.

An isolating switch, or switch disconnecter, should be capable of disconnecting the supply to the system under all possible conditions, i.e. operating, overload and fault conditions. This is intended for use by unskilled persons.

Isolators are often confused with isolating switches so when selecting isolation devices it is essential that the correct application be known and the appropriate device be selected. Isolators should be situated in a readily accessible position having regard to the use of the installation and premises, the capabilities of personnel and the nature of the activity.

An isolating device should comply with BS EN 60947−3 which sets out the creepage and clearance requirements when in the open condition. It should have a positive indication of the 'off' or 'open' position which should not be displayed until the gap between the contacts has been achieved. The purpose of the device, and if necessary the parts of the installation it controls, needs clearly marking on or adjacent to it so that there is no possibility of confusion. Access to the contacts and terminals should not be possible whilst the device contacts are closed. Provision for securing in the open position is required wherever it is likely that inadvertent or unintentional operation could cause danger.

Switching off for mechanical maintenance

This term is used to indicate the need to be able to disconnect equipment from the supply to enable non-electrical maintenance, repairs or replacements to equipment, to be carried out safely by persons who may not necessarily be electrically skilled. The intention is to provide a means of local control for the person doing maintenance work so as to ensure that the equipment being worked on cannot start up whilst this work is in progress.

The switch does not necessarily have to have the same characteristics required of the isolator and does not have to be connected to the loaded conductors. However, if connected into the load conductors it must be able to switch off the load current under full operating conditions. Particular regard should be given to inductive and capacitive loads. Where the switch is connected to a control circuit it should be so arranged that a fault on the circuit will not have the effect of bipassing the control switch and thus allow the equipment to operate.

Where switches provide for mechanical maintenance work these should be positioned so as to be under the complete control of the person(s) carrying out the work. They should be sited away from the equipment but must be positioned so that they can either be secured in the 'off' or 'open' position, or be in a lockable enclosure. In the latter situation, the equipment must be clearly labelled to indicate the position of the disconnecting switch.

Emergency switching

The definition of emergency switching is 'Rapid cutting-off of electrical energy to remove any unexpected hazard to persons, livestock or property. The hazard may be either electrical or mechanical'.

Do not confuse emergency switching with emergency stopping which provides a means of stopping any dangerous movement.

Emergency switching devices should act directly on loaded conductors to disconnect the supply as quickly as possible. This device may be connected to the load conductors although it is more commonly connected to the control circuit. The switch again should be positioned as close to the equipment operating position or within easy reach and line-of-sight of the operator. Where this is not possible or practical the position of the device should be clearly indicated and in sight of the operator.

Some installations, such as machine workshops, may require emergency switching to disconnect supplies to all equipment simultaneously. In this case the device(s) should be sited in a prominent position near to walkways and access points. BS 7671 states the particular requirements for fireman's emergency switching for high voltage discharge lighting.

Functional switching

A functional switch is a device that controls the operation of the equipment in its normal working mode. The requirements for the device are usually contained in the British Standards relating to the equipment.

Applications

A single switching device complying with BS EN60947−3 may be able to perform all of the required switching functions provided it is capable of interrupting the supply on load. Other devices may be used for individual or multiple functions but may not necessarily be able to perform all functions, in which event more devices should be used. Table 21.1 indicates the applications for some devices; the list is not exhaustive.

Detailed requirements

The detailed requirements for isolation and switching are contained in BS 7671 (the IEE Wiring Regulations) and the associated IEE Guidance Note No. 2. The installation designer needs to consider all of the relevant Regulations in formulating its design.

In particular consideration should be given to:

• the type of building, i.e. domestic, commercial, industrial, etc.;
• the type of user, i.e. adult, children, elderly, disabled;
• the environment, i.e. inside, outside, wet, dry, explosive, etc.

Isolation

It is necessary to provide for the isolation of all circuits from all supplies at the origin of an installation. The means of isolation must be such that it interrupts all live conductors of the system, except in three-phase supplies connected to TN

Table 21.1 Applications for switching devices.

Item	Isolation	Mechanical maintenance	Emergency switching	Functional switching
Isolator (disconnecter) to BS EN60947–3	×			
Switch-disconnecters isolating switches	×	×	×	
Circuit-breakers	×	×		×
Plugs and sockets	×	×		
Fuse links	×			
Links	×			
Fuse switches	×	×*	×*	×*
RCDs	×			

* Provided it is capable of breaking the full load current.

earthing systems where the neutral conductor need not be isolated if it can be assured that it is connected to earth.

Not withstanding the above condition, provision must be made on all installations for disconnection of the neutral conductor from earth for testing purposes. So, in practice, it is normally necessary to provide facilities for isolating all live conductors regardless of the earthing system.

Installations with supplies from more than one source must have isolating facilities provided for all supply sources. This may be by means of separate, linked or interlocked devices. Isolating devices must be clearly marked indicating the supply source so that there can be no possibility of confusion.

Domestic types of installation, or installations controlled by a single distribution board, may only require a single isolating device incorporated either in the consumer unit or distribution board. In larger and more complex installations a single isolating device will prove to be very inconvenient for the user. It is, therefore, necessary to provide a number of isolators to control different sections of the premises and different operating systems. In such installations the designer should consider the various aspects of control, use and system maintenance; and, in addition to the main isolator, provide for isolation of small circuit groups and even individual circuits. The minimum standard should be for every distributor and submain to be controlled by an isolator. If the isolation requirements are not properly considered and thus not properly planned, switching off live parts for maintenance or repair is likely to cause considerable disruption and inconvenience and can result in persons taking chances and working on live equipment without first isolating the supply – a dangerous practice which is virtually prohibited under the Electricity at Work Regulations.

Where an earthing system is either TN-S or TN-C-S, isolators fitted downstream from the main isolator need only interrupt phase conductors. Where the earthing system is TT or IT, all isolators must interrupt all live conductors.

Switching off for mechanical maintenance

This function of switching off for mechanical maintenance is restricted to maintenance which does not involve access to live parts.

Where mechanical maintenance may involve a risk of burns or injury from mechanical movement, provision must be made for the equipment to be switched off whilst the maintenance is carried out. This type of maintenance may be done by a person who is not electrically skilled and who has no understanding of the control systems for the equipment. It is, therefore, essential that the switch is positioned within easy access of and as close as possible to the equipment.

Care must be taken to ensure that the switch cannot be unintentionally or inadvertently operated and it must be under the control or supervision of the person doing the work, at all times. If it is not possible for the switch to be under such control then facilities must be provided for securing the switch in the 'off' or 'open' position. This is normally effected by a locking device fitted to the switch but may also be effected by siting the switch in a locked room where access is restricted to authorized personnel only.

The switch should directly interrupt the load conductors if and where practical. When this is not practical, as in the situation of a motor controlled by a star/delta starter, the switch may operate on the control circuit of the starter unit. In this case the switch must be of the stop/lock type and the control circuit wiring must be designed and arranged to ensure that the switch cannot be over-ridden, either unintentionally or by a fault occurring in the control circuit.

When designing an installation it may be advantageous to the user to select a single device that will perform the functions of both isolation and switching off for mechanical maintenance.

Emergency switching

Emergency switching is necessary where a possible danger may arise from equipment whilst it is performing its normal function. Such equipment may be heating, cooling or radiation machinery and machinery where there may be of necessity exposed live parts constituting a shock risk.

The emergency switch must act directly on the supply conductors by a single operation of the device. It must be readily accessible, clearly visible, have its purpose marked on or adjacent to it and, for additional identification, be coloured red.

Switching devices may be:

- inserted directly into the supply circuit;
- a push switch in the control circuit;
- an electro-sensitive safety control system incorporated into a machine.

Where a push switch in the control circuit is used it should be capable of being latched or locked in the 'open' position. A plug and socket arrangement should not be selected as an emergency switching device.

Emergency switching devices should disconnect all conductors of single-phase circuits, d.c. circuits, and three-phase circuits of TT and IT earthing systems, but only phase conductors of TN-S and TN-C-S systems need to be disconnected. A switching device or system must act as quickly as is possible on the supply conductors but, at the same time, must not introduce further dangers.

The designer of the installation should have an understanding of the working and operating controls of the equipment prior to selecting the means of emergency switching. For example, in an industrial plant a single operation on the supply to one section may cause danger in other parts of the plant, and similarly, switching-off the whole plant may give rise to further danger; in which case the emergency switching may have to start a sequential or partial shut-down process rather than simply switch it off completely.

Some equipment may continue to run on its own inertia after the supply has been switched off and the dangerous condition may continue until the machine stops. This may necessitate the use of emergency stopping devices and procedures in addition to the switching devices.

In workshops where there are number of machines being used simultaneously, i.e. training establishments, it is more likely to be necessary to have the emergency switching acting on the machines collectively rather than individually. This can be effected by siting, at strategic positions in the area, a number of emergency stop switches which will release a contactor wired into the supply conductors to the controlling distribution board.

Fireman's switches

The application for a fireman's switch is for both isolation and emergency switching so these devices need to meet the requirements for both functions. Their main use is in connection with high voltage discharge lighting. The positioning of these devices is important and it is, therefore, advisable to consult the local fire authority at the design stage of the proposed installation to ascertain the most suitable and convenient positions.

Fireman's switches should be coloured red, marked 'FIREMAN'S SWITCH', have the 'on' and 'off' positions clearly indicated (with the top being the 'off' position), and have the facility to prevent being inadvertently switched to the 'on' position. These requirements are fully detailed in BS 7671.

PROTECTION

This section is concerned with devices that provide excess current, earth leakage and electric shock protection for circuits and equipment of low voltage installations. Other sections deal with the requirements and applications of other methods for providing protection for safety in respect of electric shock, thermal effects, overcurrent and mechanical damage.

The Electricity at Work Regulations state 'effective means, suitably located, shall be provided for protecting from excess current every part of a system as may be necessary to prevent danger'.

Excess current can be in the form of either overload current or fault current. An overcurrent occurs when a conductor's load current is greater than its installed current rating. A fault current occurs when there is a short-circuit between live conductors or when there is an insulation breakdown or failure between a phase conductor and an earthed conductor.

Generally, all circuits and equipment have to be protected against the effects of overload and fault currents, earth leakage and conditions that may cause dangerous electric shocks. Under certain conditions, it is possible to dispense with overload protection if the equipment connected to the circuit is unable to create an overload condition.

In many installations overcurrent devices can provide protection against all of the conditions. With other installations, essentially those with TT or IT earthing systems, residual current devices (rcds) have to be used for earth leakage and shock protection.

The effectiveness of the protective device is measured by the time it takes to operate under overload or fault conditions. The operating time is dependent on the impedances of the system's circuit conductors and protective conductors.

Circuit protective devices may be either fuses or circuit-breakers. A fuse is designed to be destroyed by the element melting at a current in excess of its rated current, without given time parameters. A circuit-breaker is a mechanical device with switching contacts designed to open at a current in excess of its rated current, or under certain conditions, when there is imbalance between live conductors, again within given time parameters. For these devices to be effective they must operate rapidly to cut off the current before danger can arise.

Types of protective devices

The following types of protective devices can be used:

- overcurrent and earth leakage: fuses and circuit-breakers;
- earth leakage only: residual current circuit-breakers (rcds);
- combination units: combined overcurrent and residual current circuit-breakers.

All protective devices have to comply with the relevant British Standards.

British Standards set out the requirements to which a supplier must design, build and test the equipment. This includes:

- conditions for in-service operation
- classification
- characteristics:
    rated voltage, current and frequency, rated power dissipation, rated power acceptance of fuse holders
- limits of time−current characteristics:
    breaking range and breaking capacity, cut-off current and I^2t characteristics
- markings

- construction
- testing.

Fuses

There are two types of fuse in general use:

(1) enclosed fuses in the form of cartridges that have the fuse element sealed within, and
(2) semi-enclosed (rewirable) which have a copper wire element bridging the fuse holder contacts.

The relevant British Standards for fuses are:

- BS 88 and BS 1361 for enclosed fuses;
- BS 3036 for semi-enclosed fuses.

BS 88 fuses have a far greater range of current ratings, connecting arrangements, and dimensions. They will break higher fault currents than both BS 1361 and 3036 fuses. BS 88 fuses provide for two categories: G and M, and a number of different types. Category G fuses are for general circuit protection, typically non-inductive loads. Category M are slower operating devices which withstand higher in-rush currents, as in motor starting.

BS 1361 fuses are, again, slower operating devices and have a lower fault rating than BS 88 fuses. The physical dimensions of the 5–30 A range are such that they are non-interchangeable which is a safety consideration where they are likely to be replaced by non-electrical persons.

BS 3036 (rewirable fuses) are slower to operate and have a much lower fault current rating than the cartridge type fuse. They should not be used in situations where there is a prospective fault current greater than 2 kA, without additional back-up protection. Being considerably less expensive to buy and replace, the main advantage of the device is cost. Also, because it is slower in operation it will handle higher in-rush currents. The disadvantage of the device is its susceptibility to abuse. The element can be easily replaced with one of a higher rating than is suitable for the circuit wiring. When rewirable fuses are used the circuit cable's current-carrying capacity has to be reduced using a multiplying factor of 0.727. This may mean increasing the cable size which will, in turn, increase the cost of the installation.

Circuit breakers

There are two standards which apply to the manufacture of circuit breakers:

(1) BS EN60898: 1991, *Specification for overcurrent protection for household and similar installations*;
(2) BS EN60947: 1992 *Low voltage switchgear and control gear* and *Part 2, Circuit breakers*.

These standards have replaced BS 3871 and BS 4752 which applied to miniature circuit-breakers (mcbs), and moulded case circuit-breakers (mccbs).

BS EN60898: 1991 is applicable to circuit-breakers with a nominal current rating range of 6−125 A and with a fault current rating range up to 25 kA.

BS EN60947: 1992 applies to circuit breakers with nominal current rating range of 100−2500 A. The fault current rating range is not specified but the devices have to be tested at 50 kA. This type of circuit-breaker is intended for industrial and commercial applications.

As with fuses, their characteristics vary considerably. It is, therefore, important that the circuit-breaker is matched to the characteristics of the load and the impedances of the earthing system.

Circuit-breakers that are manufactured to BS EN60898 are available in three types: B, C, and D. The type is determined by reference to multiples of nominal current at which the device operates instantaneously (within 0.1 s). See Table II of BS EN60898.

All of the types may be produced with a fault current handling capacity as indicated in Table 21.2.

Circuit-breakers manufactured to BS EN60947−2 are in two categories:

(1) Category A: are not intended for discrimination with devices on the load side;
(2) Category B: are intended to be selective with load-side devices and normally incorporate a time delay facility.

The characteristics of the devices are not dictated by the Standard and are obtained by reference to the manufacturer's technical data publications.

Table 21.2 Fault current capacities included in BS EN60947.

1500 A
3000 A
4500 A
6000 A
10 000 A

Table 21.3 Ranges of instantaneous tripping characteristics as shown in BS EN60947.

Type	Range
B	between $3 \times I_a$ and $5 \times I_a$
C	between $5 \times I_a$ and $10 \times I_a$
D	between $10 \times I_a$ and $20 \times I_a$

Residual current circuit breakers (rccbs)

An rccb is a particular type of residual current device (rcd). Not all rcds are rccbs. The current British Standard for rcds is BS 4293, therefore rccbs have to be manufactured to meet the requirements of this Standard. *Note*: The Standard is likely to change in the near future in order to harmonize with IEC 1008. Many rcds currently in use are already being manufactured to the IEC Standard.)

Rcds are identified by the out-of-balance sensitivity of the tripping components. The tripping current is referred to as the $I_{\Delta n}$ of the device. The important feature of an RCD is its ability to operate rapidly when it senses a leakage current.

The British Standard requires that an RCD operates as follows:

- at the rated $I_{\Delta n}$ trips within 200 ms;
- at five times the rated $I_{\Delta n}$ trips within 40 ms;
- at half the rated $I_{\Delta n}$ it must not trip.

The main use of rccbs is to provide for protection against electric shock. It is, therefore, essential that the device maintains its sensitivity and speed of operation throughout the life of an installation. To facilitate this maintenance of the rcd sensing circuit it must incorporate a test switch circuit for the user and it is important that this test switch is operated every 3 months.

Combination units

Combination units are composed of a combined overcurrent and residual current circuit-breaker. These are currently manufactured to the relevant requirements of both IEC 1008 and BS EN60898. These devices are a simple and compact means of providing additional protection where there is a high shock risk.

Applications

There are a number of factors affecting the selection of the method and type of protective device to be used in a given situation. These are important and need to be given careful consideration.

Environment and utilization
These factors include:

- type of installation: industrial, commercial, domestic;
- Is it subject to the EAW Regulations?
- the occupancy: are the protective devices going to be replaced or reset by electrically skilled persons?
- the cost of the devices.

Electrical
These factors include:

- the maximum disconnection time allowed for the circuit, the type of utilization of the circuit and the load characteristics, including starting or inrush currents;
- the circuit earth loop impedance;
- the prospective fault current at the point of installation.

The main advantage fuses have over circuit-breakers is:

- cost;
- BS 88 fuses can be used where there is a high prospective fault current.

The main disadvantages are:

- they are costly to replace after a fault, and
- because fuses can readily be replaced and easily interchanged a person may, by fitting the wrong type of fuse, compromise the safety of the installation.

The main advantage of circuit-breakers is they are easy to reset after a fault and cannot be tampered with.

Circuit-breaker types

Type B circuit-breakers are mainly suitable for domestic and commercial install-ations where there is little or no switching surges, and where multi-core cables with reduced size protective conductors are used.

Type C circuit-breakers are for use in commercial and industrial applications where there are inductive lighting motor loads with switching surges. For these much lower circuit earth fault loop impedances are needed to achieve the required disconnection time and all insulated wiring systems may not be able to achieve the disconnection time without an rccb.

Type D circuit breakers are for use in applications with abnormally high inrush currents, such as main frame computers, welding equipment, transformers and X-ray equipment. Again, all insulated wiring systems may not be able to achieve the disconnection times.

Allowing for the constraints of prospective fault current at the point of installation and circuit impedances, fuses and circuit-breakers offer the same quality of protection and are, therefore, equally safe. As a rule, circuit-breakers are prefer-able to fuses where operation, maintenance and replacement is likely to be by unskilled persons. It is important that the correct circuit-breaker type is selected for the load and installation conditions but the choice is very often a matter of personal preference.

Rccbs

These devices should be used where the circuit earth loop impedances are too high for the required disconnection times, in places where there is an increased risk of electric shock (damp/wet situations) and where there are TT or IT earthing systems. Rccbs are often required in addition to fuses or circuit-breakers to provide shock protection and where high surge currents require slow-acting over-current protection.

When selecting rccbs consideration has to be given to the standing earth leakage current in the circuit or equipment being protected. Any inherent leakage current will effectively increase the sensitivity of the device which may be detrimental to its functioning and cause nuisance tripping.

CHAPTER 22

Protective Systems

B. Dakers, BSc, CEng, MIEE

D. Robertson, CEng, MIEE

Revised by J.R. Murray
(Rolls-Royce Industrial Power Group, Reyrolle Protection)

The term 'protective systems' refers to the electrical protection of plant associated with industrial systems and power systems such as alternators, motors, transformers and cables. Although this chapter is predominantly concerned with protective systems applicable to industrial systems and distribution systems up to 11 kV the basic principles and requirements apply equally to power systems operating at higher voltages.

A power system is designed to deliver electrical energy without interruption to the points where it is utilized. By definition, therefore, a fundamental requirement in the design of a system is flexibility. If one part of the system becomes faulty it should be disconnected quickly and ideally without affecting other parts of the system. Switchgear and protective gear provide this isolation flexibility and thus can be considered in the context of an insurance against loss of supply. The switchgear must be designed to be capable of interrupting the current resulting from a fault while the protective gear must be capable of recognizing a fault condition and initiating the switchgear to disconnect the faulty part of the power system with minimum disturbance to the rest of the system.

PROTECTIVE SYSTEM REQUIREMENTS

Protective gear is the collective name given to all the components necessary for recognizing, locating and initiating the removal of a faulty part of the power system. The most important qualities of a protection scheme are reliability and selectivity, with other important qualities being speed, stability and sensitivity, none of which however, is acceptable unless they are provided in such a way that reliability and selectivity are maintained, and the equipment is within reasonable cost.

Although total reliability is impossible, duplicate schemes, alarm supervision circuits and regular maintenance, have resulted in the number of incorrectly cleared faults being less than 5%.

Selectivity, or discrimination as it is more commonly known, may be defined as the quality of a protective system whereby it detects and responds to a fault condition so that only the faulty part of the circuit is disconnected. To improve overall reliability, protective schemes are arranged with back-up and overlapping zones. Because of this, selectivity is of the utmost importance to ensure that only the correct 'zone' is disconnected.

To avoid unnecessary damage to plant, protection must operate quickly, thus, speed of recognition and disconnection of a fault are very important. This is becoming more important with the increased size of plant and the resultant increase in fault current where widespread damage to plant can occur much more quickly.

Stability may be defined as the quality of the protective system whereby it remains unaffected by all power system conditions, faulty or otherwise, other than those for which it is designed to operate. The stability limit of protection can be quoted as the ratio of the maximum through-fault current which may cause maloperation against the relay setting current. If a protection system is to be completely satisfactory, the relay setting must ensure operation under minimum fault conditions. This, however, is not just a question of setting a relay as low as possible, since the resultant relay burden may cause saturation of the current transformers (CTs) under external fault conditions, resulting in instability of the protection. A compromise must therefore be found between stability and sensitivity.

For the most part, the problems existing can be solved by the application of standard equipment. When considering the choice of a protective system for a particular application, two questions arise: how much protection is required and which type of protection system to use.

In deciding how much protection should be used, a number of facts need to be considered including system fault statistics; type, nature and importance of plant to be protected; and degree of complexity compared with level of maintenance staff and risk of possible mal-operation.

The greatest assistance one can obtain in reaching a final decision on how much protection should be applied is to draw on experience. If experience dictates, for example, that the number of faults on a pilot cable over a few years has been negligible, there is little reason for applying supervisory equipment.

There are a number of possible solutions to known problems and the choice of which system is determined by assessing the required performance, and matching these protection requirements to that of the available protective systems in relation to cost, maintenance and installation.

Having chosen the type of protection system based on the factors above, it must be confirmed that the chosen system is suitable for a particular application. This involves investigating a set of facts which broadly can be divided into two groups:

(1) The behaviour of the components of the power system and the limits within which they may be operated, for example, permissible fault clearance time and available fault current during internal and external fault conditions.
(2) The characteristics of the protection system and the limits between which the protection system will perform correctly, for example, operating time, fault settings and stability.

UNIT AND NON-UNIT PROTECTION

Protection systems can be broadly defined in two categories, unit and non-unit protection. Economics dictate that protection systems cannot always be fully discriminative and the simpler forms of protection tend to disconnect more of the power system than just the faulted unit. This characterizes the fundamental difference between unit protection, which is concerned only with determining whether the fault is within or external to the power system unit it is protecting, and non-unit protection which disconnects the minimum amount of plant by tripping only the circuit-breakers nearest to the fault.

Because power system faults are fairly rare events, the concept of reliability in protection is very much related to stability, (i.e. mechanical or electrical interference must not cause mal-operation). On the other hand, when a fault does occur it is vital that operation is ensured and protection is always applied with at least one back-up system. The arrangement of non-unit protection, Fig. 22.1, allows primary protection for a power system unit and back-up protection for an adjacent unit to be obtained from one set of relay equipment, i.e. a fault between relays 2 and 1 has relay 2 as primary protection and relay 3 as back-up protection. Figure 22.2 shows that the fundamental principles of unit protection dictate that it will not provide back-up protection for adjacent units so that separate back-up protection must be provided. Unit protection is not always universally applied throughout a power system so that some power system units rely on the back-up protection as their primary protection. Busbars are typically in this category because the low incidence of busbar faults indicates that this is a reasonable risk. However, the catastrophic nature of a busbar fault means that the clearance times of back-up protection should be reviewed as a power system develops so that the rare event of a busbar fault and its predictable serious effects are constantly appreciated.

Discrimination

Discrimination with load current is perhaps the most fundamental requirement and thus phase-fault settings in simple protection have to be set above maximum load current which therefore limits the protection obtainable for phase faults. Fortunately, with earth faults, a residual connection of the CTs, Fig. 22.3, can be used which eliminates all current except earth-fault current from the relay so that low earth-fault settings are possible and this enables coverage to be given to power systems which have earth-fault current deliberately limited by earthing impedances. Settings must always be related to the required stability level because errors in CTs increase with energization level and the out-of-balance currents

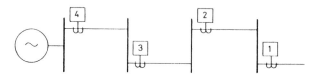

Fig. 22.1 Non-unit protection.

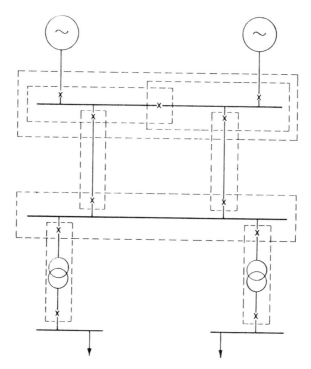

Fig. 22.2 Unit protection in overlapped zones.

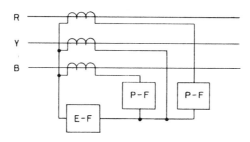

Fig. 22.3 Residual connection of earth-fault (E-F) relay with phase-fault (P-F) relays.

between CTs having different errors will be fed to residually connected earth-fault relays thus affecting discrimination if the earth-fault relay settings are not above the expected steady out-of-balance current.

To achieve the requirement that only the minimum of plant is disconnected during a fault, it is necessary that protection schemes discriminate with each other accurately in the presence of fault conditions. Two fundamental types of discrimination exist, first, in the case of non-unit schemes, Fig. 22.1, discrimination is obtained by ensuring that the circuit-breaker nearest to the fault operates while those further from the fault remain stable. This is achieved by current or time grading, or more generally, a combination of both. Thus in this case discrimination is dependent upon the intrinsic accuracy of the CTs, the relays and the circuit-breaker clearance times.

The second type of discrimination is obtained by the through-fault stability of unit protection. In Fig. 22.2, each unit protection involved in a fault must measure end-to-end and decide to trip or stabilize depending on whether the fault is found to be external or internal to the zone bounded by the CTs feeding that protection. With unit protection, the accuracy requirements are therefore to ensure correct comparison and in this case, intrinsic accuracy is not important but comparative accuracy is vital. This has required the establishment of a special category of CTs (BS 7626, Chapter 3, Clause 33) for unit protection where the turns ratio is specified to close limits and the output capability is specified in terms directly related to the requirements of unit protection (i.e. knee point voltage and winding resistance).

In general, protection should be as fast as possible to avoid too much damage and/or disturbance to the power system. However, in non-unit schemes time is often used, at least partially, to provide discrimination so that longer clearance times than may be desirable are involved. Also, it is fundamental to the nature of graded time protection that longer times of operation will be at the source end of radial or ring system, i.e. in Fig. 22.1 relay 4 must be the slowest relay, where the fastest time of clearance is preferable and this is the basic objection to simple graded protection schemes.

With unit protection, each unit has the same basic speed of operation so that a fault anywhere within the power system is cleared in the very fast times obtained by unit protection.

Distance protection is perhaps the most complex form of non-unit protection but the most widely used non-unit protection in distribution power systems is graded protection which is used as primary and back-up protection in radial power systems and as back-up to unit protection in interconnected power systems. The relays used to provide graded time protection are generally current-dependent time-lag relays with an inverse definite minimum time-lag characteristics, (idmtl relays).

Inverse time relays

An inverse time characteristic is an attractive feature because high levels of fault are cleared quickly and low levels of fault cleared relatively slowly allowing differentiation between faults and discrimination to be achieved. The relay characteristic has developed with a definite minimum time at high levels where the circuit-breaker operating time dominates the clearance time and if all the relays in a power system are subject to high levels of fault current, discrimination is obtained virtually on a definite time basis.

Some power systems use definite time-lag relays, controlled by overcurrent starting elements, to provide discrimination and this can have advantages because definite-time-lag (dtl) relays can be made much more accurate than idmtl relays. In the general case, however, the advantage of variable time with current level can be exploited because the magnitude of fault current varies with fault position and also with the amount of generation plant connected. Thus the idmtl relay is a combination of the requirement for definite-time-lag discrimination when the circuit-breaker time is dominant and inverse time lag at low levels where fault position can influence the current and hence the time of operation.

The concept of fault position giving different fault levels can be exploited by current grading in a radial system where maximum possible fault current for each relay position is easily calculated. Thus a fast overcurrent relay may be set to operate at high levels and when this is combined with idmtl relays closer grading can be obtained because the high current end of the overall characteristic is current graded and the idmtl curve is only required to ensure discrimination below this level. However, fast overcurrent relays may operate at much lower currents than their setting when an offset transient is present in the energizing current. This is referred to as transient overreach because they are operating for a fault current magnitude which is the calculated level for a fault position further away from the source than the relay setting indicates. Specially designed fast overcurrent relays are available which maintain their setting (within declared limits) in the presence of the offset transient; they are referred to as transient free. Microprocessor overcurrent and earth-fault faults generally include transient free high set elements as standard.

Because continuity of supply is one of the prime requirements in power system design, the radial power system develops naturally into a closed-ring type of circuit so that any feeder can be switched out without loss of supply to any busbar. Protection of this type of power system requires that discrimination must be obtained along both paths to the fault. This is achieved by directional control of the idmtl relays so that only the relays nearest to and either side of the fault operate. A typical arrangement is shown in Fig. 22.4 where it can be seen that the first circuit-breaker to trip opens the ring which then becomes a radial system allowing the fault to be cleared by the circuit-breaker nearest the fault.

The directional relay determines the direction of the fault current in relation to the voltage at the relay point, and is chosen with a maximum torque angle

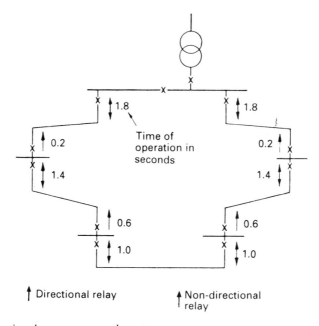

Fig. 22.4 Directional overcurrent relays.

appropriate to the angle between fault current and energizing voltages. In general, the faulted phase voltages are not used but the relay is nevertheless compensated to give correct directional operation at low voltages because all voltages collapse during three-phase faults.

Earth-fault directional relays are fed from residual current and residual voltage.

The directional elements are in effect reverse current relays which are generally arranged to have a maximum torque angle appropriate to the phase angle of the fault current. Overcurrent directional elements have a relay characteristic angle of 0 degrees or 45 degrees lead and obtain an overall lagging maximum torque angle by using appropriate phase-to-phase or phase-to-neutral voltages. Earth-fault directional elements are generally connected to residual current and residual voltage and may have a characteristic angle between 0 degrees and about 75 degrees lag.

Directional relays should not be confused with reverse power relays which are not voltage compensated and are used to detect unusual power flow conditions subsequent to power system disturbances. The power relay is not normally required to respond to fault conditions and its output contacts are therefore time lagged.

Unit protection is characterized by its sensitivity and speed of operation both of which make it susceptible to transient out-of-balance between the input and output signals which should theoretically sum to zero during through-fault conditions. Each type of power system unit has characteristics which affect the design of its unit protection. These are discussed later.

GRADED PROTECTION

In order to apply time-graded protection the relays must be in a radial configuration as shown in Fig. 22.1. Consideration of Fig. 22.1 illustrates how the principles of time-graded protection depend upon the fault current in each relay being uniquely identifiable so that the relay response can be determined from its chosen settings. When a common source feeds several radial feeders it is worth-while to separate and identify each set of relays clearly and the impedances between them. This will show how the relay at the source must be co-ordinated to suit the feeder giving the longest times.

Where the circuit between two relay positions has parallel paths it is necessary to identify the worst conditions for grading each pair of relays. Worst-case load and fault conditions could be determined by computer studies of the full spectrum of power system running and outage conditions. Engineering judgement is generally used to limit the power system studies to critical conditions. Thus grading must include power system analysis and sometimes requires compromise decisions.

The current in each relay is determined by its CT ratio and the relay response is determined by its current setting (plug setting) and its time setting (time multiplier). Discrimination is ensured by the inverse relationship between current and relay time of operation and the applied specified limits. BS 142 specifies the standard inverse curve for idmtl relays by the check points as illustrated in Fig. 22.5. The allowable errors are also specified in BS 142 and these have to be assessed when deciding the time that will be used as the grading step between each adjacent pair of relays.

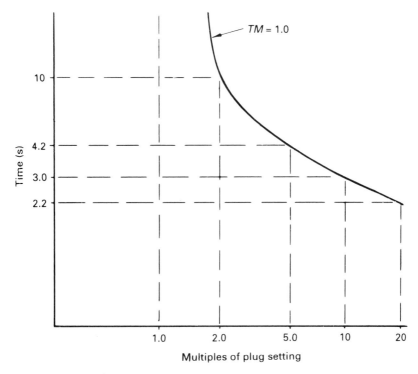

Fig. 22.5 Standard idmtl curve.

Microprocessor overcurrent and earth-fault relays generally have less error and provide choice of characteristics. This is an additional setting which complicates the grading procedure but is a great advantage when relays have to be provided in advance of finalization of the power system data.

Effect of impedance

The magnitude of the fault current is determined by the impedances of the power system feeding the fault and in the simplest form can be represented by source impedance (Z_s) and the impedance between the two relays being considered (Z_L) as shown in Fig. 22.6. Assuming that Z_s is much greater than Z_L, the fault current at each relay position is approximately the same, i.e. $I_F = V/Z_s = V/(Z_s + Z_L)$. If, in addition, the load current and CT ratio are the same for each relay position, discrimination could be obtained by time multiplier setting only.

The consideration of load current is important because the plug setting of overcurrent relays, which have a plug setting range of typically 50–200% of rating, determine the basic pick-up and reset levels of the relay. Thus for discrimination with load conditions, the relay plug setting must be related to the CT ratio and the maximum possible load current to ensure that the relay resets when clearance of a fault may leave the circuit with increased loading due to an outage.

When Z_L is significant in relation to Z_s, the fault current at the two relay points is different and the plug setting adjustment and/or CT ratio can be used in combination with the time multiplier setting to give discrimination.

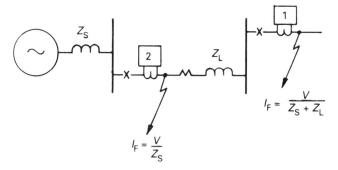

Fig. 22.6 Basic circuit for grading.

In the general case, Z_s is determined by the plant running condition and therefore is variable between minimum and maximum plant limits, hence the reason why idmtl relays are used. At very high levels of fault current the CT output may be limited by steady-state saturation of its core material. This limits the effective output of the CT to approximately a current level whose average value is that given by the knee point voltage of the CT's excitation curve divided by the CT winding resistance plus the relay winding resistance. Thus the combined performance of the CT and relay gives the inverse time curve up to the point of CT saturation and a definite time above the current at which the CT saturates. If the standard idmtl curve is used, CT saturation above twenty times the plug setting does not change the overall characteristic. However, very inverse and extremely inverse relay characteristics are modified by CT saturation so that this should be taken into account if high levels of energization are expected.

The effective burden of some types of relay varies with setting and thus when low settings are used, the possibility of CT saturation is increased. Thus with earth-fault relays and high multiples of setting the guidance on application of CTs as detailed in BS 7626: 1993 should be taken into account.

Microprocessor overcurrent and earth-fault relays have extremely low burden. This allows low settings to be used without increasing the possibility of CT saturation. However the time–current curves of these relays to not follow their theoretical characteristic at high multiples of setting so that their use with low settings may require this discontinuity in their curves to be included in the grading procedure.

Grading procedure

The procedure for grading a set of relays is first to establish a common base upon which to compare the curve of each relay with each adjacent relay. This common base is generally primary current referred to the base voltage of the power system. Settings have to be chosen for each relay to maintain the grading step at the highest possible fault current and also to ensure operation at the lowest possible fault current. Thus power system data must be available for maximum and minimum plant condition.

The grading step must take into account maximum circuit-breaker clearance time, CT errors, relay errors and relay overshoot. Generally, 0.4s has been used as a minimum but lower values are feasible, e.g. 0.25s, with fast circuit-breakers and accurate relays.

Relay overshoot is related to the inertia of an electromechanical relay and is the time difference between the operate time for a specified level of current and the time the relay has to be energized at this level of current to obtain operation. In other words, the relay will operate (and its time of operation is unimportant in this case) if it is energized for a time less than its operate time because the movement has stored energy which will cause it to continue to move after the energizing current ceases. It is tempting to believe that because semiconductor designs have no mechanical inertia in their measuring circuits, static relays will not have any overshoot. However, trapped charge in smoothing and timing circuits give much the same effect and although special circuits can obviously be designed to minimize this, economics may dictate that a static relay has as much overshoot as an electromechanical relay.

Having established plug settings for each overcurrent relay to ensure non-operation during maximum load conditions, the current multiple of the plug setting of the relay furthest from the source (relay 1 in Fig. 22.1) for maximum through-fault conditions will determine the required time multiplier setting for this relay. Often relay 1 can be set at 0.1 time multiplier because it is known that other protective systems, fuses, motor protective relays, etc., have such lower relative current settings that co-ordination is not required. Sometimes a minimum time multiplier is specified by the authority controlling the protection in the circuits being fed by relay 1.

If relay 1 has to be co-ordinated with other protection the time curve of this protection must be included in the grading exercise.

With the plug settings of each relay set to suit load currents the calculated plug setting multiple for the fault level at each relay is determined. The relay curve is only specified up to twenty times the plug setting and it is possible that CT saturation will affect time of operation at levels higher than this so that the plug setting should be increased if necessary until the fault current is below twenty times the plug setting. If a plug setting above the 200% limit is called for, the plug setting multiple and relay burden must be assessed against CT output.

Having decided the setting of relay 1, by one means or another, its time of operation for maximun fault level at its location can be established; this is the point of grading with relay 2. The multiple of relay 2 plug setting for this fault condition and the time of operation of relay 1 plus the grading step give the required time multiplier for relay 2. Figure 22.7 shows grading between two relays illustrating the way in which the two curves obtained from plug settings and time multipliers are drawn to a common base to show the overall protection obtained. The alternative settings for relay 2 which give the same grading step between the two curves show the improved fault coverage obtained for relay 2 if plug settings are held to the lowest value needed to ensure discrimination against maximum load conditions.

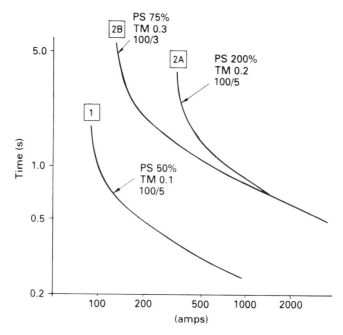

Fig. 22.7 Grading between two relays.

High-set relays

If high-set relays are used they must be set so that they can be included in the coordination of the relays. Setting of the high-set relays must take account of the stability requirements for a fault at the next relay point with due allowance made for the 50% overreach of simple high-set elements. If twice the setting requirement for operate current is also introduced, it is often difficult to achieve viable settings for the high-set elements.

Having selected the high-set settings, the operation point of the relays for grading is obtained using the lowest value of current multiple as determined by the high-set setting or the fault current. If the value is less than twice the plug setting the relay current is too low to ensure operation and a change in CT rating should be considered.

The high-set setting of relay 1 or the fault level at relay 1 is then chosen to determine the grading on the basis of which the lowest multiple of the plug setting at relay 2 is obtained. With the lowest multiple of plug setting for relay 2 given by the high-set setting of relay 1 or a fault at relay 1, the time multiplier or relay 2 is determined. If a time multiplier setting of greater than 1 is called for the plug setting must be increased.

Grading by computer

The grading of relays is a relatively simple task which can easily be done by a computer, however, the problem of collecting data and authenticating the data remains. Often data are only questioned when engineering judgement indicates that the final result is unusual. This means that the computer should be used to assist and not replace the engineer.

Graphics workstations provide interactive means of grading by displaying the relay co-ordination curves and thus allowing the engineer to study options and compromises. If the workstation is interfaced with a CAD/CAM system, hard copy of the final results may be produced for protection audits.

UNIT PROTECTION

Unit protection operates for faults which it measures as being within the protected unit and is stable for faults which it measures as being external to the protected unit. Differential protection, which forms the majority of unit protection, can be classified by the means used to obtain stability. In the current balance system, current is circulated in the secondaries between the sets of CTs at each end of the protected zone and the relay is fed by the summation of the currents which is theoretically zero under external fault conditions. The voltage balance system has voltages derived from and proportional to the currents which are balanced under external fault conditions, so that theoretically no current flows in the relays which are connected in series between ends. The basic arrangement of both systems is shown in Figs 22.8 and 22.9.

Each type of balance has limitations and the means of overcoming these limitations characterise relay design. Originally, the use of distributed air-gap CTs favoured the voltage balance scheme because this type of CT naturally produced a voltage output proportional to current. However, as can be seen in Fig. 22.9, significant pilot capacitance causes spill current in the relays of a voltage balance pilot wire differential scheme and thus provision must be made for this in the relay design. Similarly, in a current balance scheme, Fig. 22.8, the voltage drop due to the circulating current dictates that the relays must be connected at the electrical centre point.

In pilot wire differential schemes it is not practical to connect relays at the

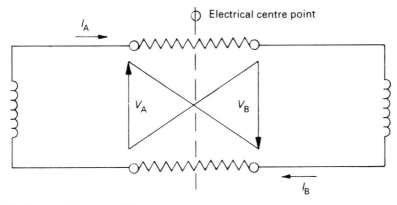

Fig. 22.8 Current balance stability.

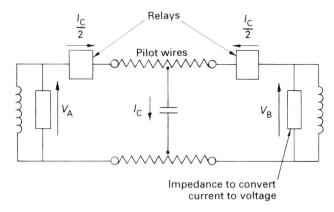

Fig. 22.9 Voltage balance stability.

electrical centre point because this would be geographically in the centre of the feeder. Thus the design of a feeder protective scheme must arrange for tripping contacts at each end and the relay equipment is therefore duplicated at each end, with both ends and the communication channel between ends forming the measuring circuit.

Steady-state out-of-balance spill currents are overshadowed by the severe out-of-balance during unequal saturation of the CTs caused by the offset transient in the primary current and the remanence left in the CTs from previous fault conditions. Considering that these effects are transient in nature, they are related to the dynamic performance of a protection which must therefore be established in relation to the primary transient and its effect upon the combination of relays, CTs and connections between them which make up the protective scheme.

These effects are best understood and analysed by computer simulation. Comprehensive dynamic representation of the power system, CTs and relay circuit can be set up using standard software packages (e.g. EMTP) which will allow step-by-step calculation and graphical display of the currents, voltages and fluxes in the circuits. This is essential for design and development but also allows special applications and unexplained field experiences to be analysed.

Low and high impedance schemes

Current balance schemes are classified as low or high impedance by the relative impedance of the relay used. The high impedance differential circulating current scheme allows for transient unbalance more definitely than any other type of protection because it assumes complete saturation of the CTs at one end with the CTs at the other end producing full output. With correct relay design in a high impedance protection system it should be possible to remove the CT at one end and replace it by its winding resistance only and still obtain stability.

Low impedance current balance differential schemes can be considered basically as two CTs feeding an effective short-circuit which is the low impedance relay. Thus assuming the relay has zero impedance, the excitation current of each CT is determined by the through-fault current, the CT winding resistance and the lead burden resistance. The excitation currents are not equal if different CTs or lead burden resistance are used and the difference between the excitation currents flows in the relay giving a tendency to operate during the through-fault condition.

In contrast to the low impedance differential scheme, the high impedance current balance scheme can be regarded as two CTs connected back-to-back so that the current from one secondary is absorbed by the current in the other secondary. Assuming a voltage-operated relay of infinite impedance there is no other path for current to flow and the excitation currents at each end must be equal. Thus, in this case, with unbalanced CTs or lead burdens, the total voltage produced by both CTs has a magnitude determined by the total loop burden of both CT winding resistance and both lead burden resistance and this total voltage is shared between the two CTs in proportion to the appropriate points on their respective excitation curve, with each CT having the same excitation current.

Practical high impedance differential schemes have relays with significant impedance values with respect to the excitation impedance of CTs so that spill current flows in the relay due to unbalance between CTs and this tends to reduce the fundamental voltage unbalance. The steady-state unbalance spill currents must be considered in relation to the sensitivity of the relay which has an operation level of typically 30 mA so that small levels of ratio error in the CTs can be significant, particularly when 5 A CTs are used and the prospective circulating secondary current will therefore be 250 A for a fifty times through-fault. Ratio error in CTs introduces additional error to that produced by unequal excitation currents and thus must be minimized by correct choice of CTs. CTs to BS 7626 which have special requirements as detailed in Chapter 3, Clause 33, should be used for differential protection but other classes of protection CTs have been successfully used.

The practice of using class 5P CTs (IEC 185) in differential protection has developed due to the fact that some CT manufacturers make them physically to be the same as special class CTs and thus the only difference is in the specified limits. However, the CT specification allows a broader approach and in general terms the only suitable CTs for differential protection remain the special class. As stated in BS 7626, class 5P CTs may be suitable for differential protection but this depends upon the difference between the CTs at the two ends rather than the protective scheme and the crucial factors are stated in Appendix A of BS 7626. These steady-state considerations must be remembered when settings are being chosen even though the basic stability requirements of high impedance protection are calculated for transient unbalance conditions. Thus steady-state stability requirements may require settings to be higher than transient stability requirements.

High impedance protection is limited to those applications where the lead length between CTs is less than about 1000 m and where CTs can be low reactance and have the same ratio. Pilot wire differential protection and transformer differential protection are therefore based upon low impedance principles.

Differential protection using low impedance relays requires some means to overcome the unbalance between ends during through-faults. Errors in CTs, both steady state and transient, produce differential current which appears to the relay as an internal fault and consequently there is always a limit to the sensitivity that can be used.

Transient unbalance between ends is generally much higher than steady-state unbalance and early low impedance differential schemes used slow relays to withstand these transient conditions. This technique is still used in very sensitive earth-fault protection and in simple differential schemes which may use idmtl relays as measuring elements.

However, speed of operation is one of the basic advantages of differential protection so that some means other than time delay is preferable to allow for transient unbalance and in low impedance differential schemes this has been generally achieved by biasing the relays.

The concept of bias is best understood by observing the change in relay setting with through-fault current in relation to spill current as shown in the idealized composite characteristics of Fig. 22.10.

The step in the current required to operate the relay curves is deliberately introduced into the design so that the operating current setting level during internal faults is not affected by the bias current. This gives a reduction in the ratio between spill current and current required to operate causing a potentially unstable condition at a relatively low level of energization; hence the requirement to type test low impedance biased schemes stability at levels of through-fault current below maximum. The different slopes of the 'current required to operate' curves for internal and external fault are generally due to only half the bias circuit being energized by an internal fault fed from one end.

Power system units take many different forms which dictate the design of the protection that will be applied. Thus each type of unit has unique features related to the particular problems of that type of unit and are each subjects in their own right. However, the salient points are summarized below.

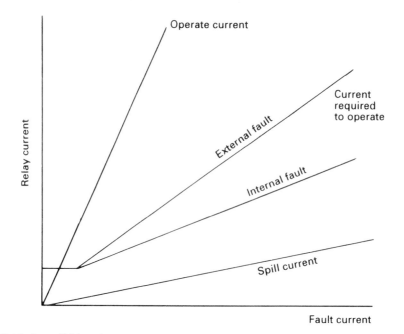

Fig. 22.10 Overall bias characteristic.

Feeder differential protection

The basic unit protection for feeders is differential protection using pilot wires between ends to effect the end-to-end comparison. The limitations imposed by the pilot wires dictate that sensitive measuring relays are required and the steady-state and transient unbalance between ends is generally catered for by biasing the relay.

For economy of pilots, the CT outputs are combined in such a way that all faults can be measured by comparison across one pair of pilots. British practice is to use a simple summation transformer to combine the fault currents but the alternative of sequence networks is used in some areas.

In common with all circuits which run in parallel with the power system, the pilot wires are subject to induced voltages when zero sequence current is flowing in the parallel path. Thus earth-fault conditions can cause relatively high induced voltages depending upon the length of common path between the pilots and zero sequence fault current, the coupling coefficient between the two circuits and the magnitude of the zero sequence current. Pilot wire protection must therefore be designed to cater for this induced voltage. Unfortunately, the simple approach used in telecommunication circuits of by-passing the induced energy to earth by surge suppressors cannot be used with pilot wire protection because the pilot wires are required to be active during the fault condition to stabilize the protection or to enable tripping. The approach has always been to float the pilots so that they cannot be affected by the flow of longitudinal zero sequence current and thus very high voltage withstand levels are required on the pilot circuits and these have been standardized as 5 kV of 15 kV. The 5 kV level is applicable to distribution cable networks where the effective coupling coefficient between pilots and zero sequence current path is relatively low. The 15 kV level is required where the coupling coefficient between the two paths is relatively high or at transmission voltage levels where induced voltages are generally higher. The development of differential protection using, for example, rented telephone circuits also means that surge suppressors cannot be used, and again, 15 kV isolation transformers must always be connected between the relay and the telephone system.

Solkor R system

A pilot wire protection is characterized by the way in which the comparison of the two ends is done and the way in which the transient unbalance between ends is catered for. A typical scheme is shown in Fig. 22.11. This is Solkor R protection and uses the principle of current balance to obtain stability during through-fault conditions. The basic current balance scheme requires connection of the relay at the electrical centre point of the pilots. This theoretically ensures that the secondary currents will cancel because the CT excitation currents will be equal. In practice, dissimilar CTs and CT transient saturation mean that the excitation currents can be different during steady-state and considerably different during transient conditions leading to substantial spill currents.

The physical siting of the relay in the middle of the pilots is not practical and the scheme requires a relay at each end. Diode switching is arranged so that the electrical centre of the pilots is moved end-to-end in response to the polarity of the summation transformer output thus keeping both relays stable.

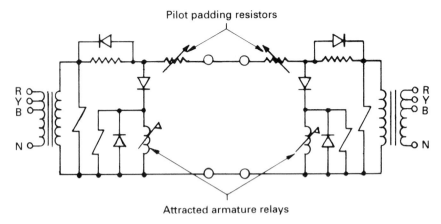

Fig. 22.11 Solkor R pilot wire protection.

Although Solkor R is in essence a current balance scheme, a current-to-voltage conversion is made by the summation transformer and pilot voltage limiting non-linear resistance. This voltage is then used to drive current around the pilot circuit which is fixed in ohmic value by having pilot padding resistors. The diodes in this pilot loop ensure that the relays are always at the negative end of the voltage divider created by the fixed resistor and the pilot resistance. By making the fixed resistor 1750 Ω and padding the pilots to 1000 Ω the relays have a negative bias which is effective because they are positively polarized by diodes connected in series with them. The important factor to note is the absolute and instant application of the bias. In biased relays generally the smoothing sometimes required (e.g. when biasing is arranged within magnetic circuits) causes delay in the establishment of the bias and hence the requirement to use a relatively slow measuring relay to avoid transient instability.

Solkor M system
This protection is a development of the Solkor range of relays using microprocessors, see Figs 22.12 and 22.13, the algorithms being modelled upon the same Solkor Rf summation techniques. The primary difference being that the end-to-end communication method uses modems operating at voice frequency. Thus there are no directly connected pilot wires and no circulating current. The end-to-end communication circuit is inherently monitored and data storage, fault recording and interfaces to computer terminals are provided to meet modern protection philosophy brought about by advances in technology.

Microphase FM system
This protection is a further development of the Solkor M using sophisticated sequence component algorithms which enhance the fault detection performance and meets requirements for the highest transmission system voltages. An additional feature incorporated in this system (and in Solkor M) is the ability to measure the propagation delay of the telecommunications link and compensate for the phase changes which can occur due to switching of the link route.

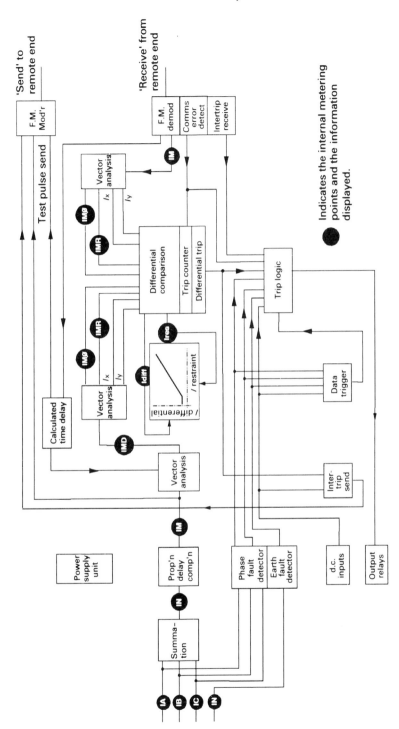

Fig. 22.12 Block diagram of Solkor M relay points.

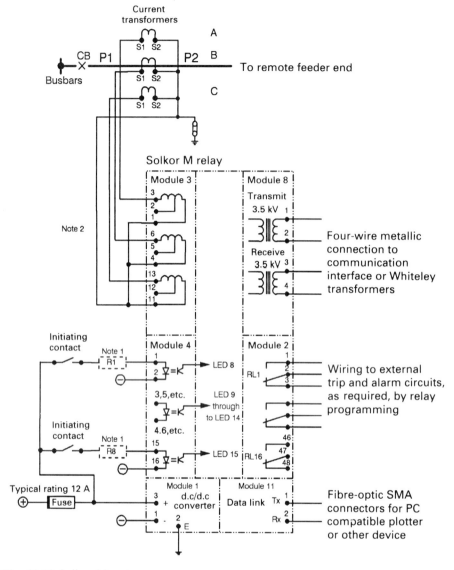

Fig. 22.13 Solkor M system.

Transformer differential protection

Transformer differential protection compares the primary and secondary currents of a transformer and is stable when these indicate that a fault is external to the transformer. The problem is complicated by the variable ratio of the power transformer (due to tap change) and the transformer winding configuration and earthing which can cause different fault current distribution each side of the transformer. In general, the transformer winding configuration is allowed for by using star-connected CTs with power transformer delta windings, and delta-connected CTs with power transformer star windings. This also excludes zero sequence current from the differential relay which avoids the problems caused by different earthing arrangements of the two power systems linked by the power transformer. Transformer protection is dealt with in detail later.

Restricted earth-fault protection

Restricted earth-fault protection uses the high impedance principle which gives extremely good stability with low settings because the settings are determined by characteristics of the CTs which are not directly related to the characteristics of the CTs which determine stability, see Fig. 22.14. Thus low settings with high stability levels can be obtained and this gives very good earth-fault coverage.

The stability criterion of a high impedance system is that the relay setting voltage is higher than that calculated by assuming that the maximum current is flowing in the saturated CT and thus producing a voltage equal to the maximum secondary current times the CT winding resistance plus lead burden resistance. The setting is determined by the secondary current required to energize the CTs and the relay at the relay setting voltage, plus the current taken by a setting resistor which is generally connected in parallel with the relay in order to increase the setting to a value reasonably above the steady-state unbalance that may be expected from the CTs. Details of restricted earth-fault protection are discussed later.

Busbar zone protection

The requirements of busbar zone protection are dominated by the stability requirements because the incidence of busbar faults is so low and the incorrect operation of busbar zone protection can have such wide ranging effects.

With fully enclosed metal-clad switchgear, simple busbar protection is possible by lightly insulating the switchgear enclosing frame and connecting it to earth through a CT feeding a simple relay. This system of frame-leakage protection only detects earth fault so that the possibility of phase-to-phase faults free from earth is an important factor in deciding whether this scheme is suitable.

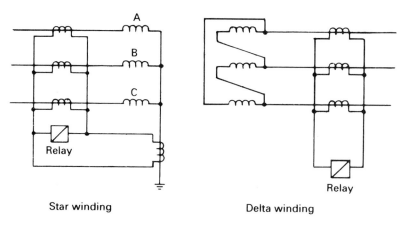

Star winding   Delta winding

Fig. 22.14 Restricted earth-fault protection.

With important busbars, the advent of a busbar fault, although an unlikely occurrence, generally causes so much disruption to the power system that high speed differential protection is applied.

The problems of differential busbar protection are: the through-fault current has the same magnitude as internal faults because the busbars have virtually no impedance; the through-fault current may be shared by several circuits feeding into the busbar and leave the busbar by only one circuit thus creating conditions for unbalance in the combination of the CT outputs; and complex busbar arrangements require circuits to be switched from one busbar to another which creates the potentially hazardous requirement of switching CT secondary circuits. This latter problem is overcome by using an overall check zone and applying sensitive alarm relays which detect the unbalance caused by load current when a CT is not connected.

The problem of high through-fault currents coupled with the sharing of the input current is overcome by using the high impedance circulating current differential scheme, i.e. busbar zone protection uses the same basic scheme as that used in restricted earth-fault protection.

Circulating current protection

There is a tendency to use mesh and breaker and a half substation arrangements and in these the circuits are not switched, so that check zones are not required and the high impedance scheme is used as a plain circulating current scheme. It is used whenever possible because of its good performance and simplicity. Reactors, interconnectors, auto-transformers and generator stators may be protected by high impedance circulating current protection, the only limitation being the capability of the CTs (which must be low reactance type) to support the lead burden.

TRANSFORMER PROTECTION

Protective schemes applied to transformers play a vital role in the economics and operation of a power system. Their percentage cost is extremely small, making it totally uneconomic to apply anything less than a complete scheme of protection especially to large transformer units. On smaller transformers where their loss may not be so important to system operation the protection applied must be balanced against economic considerations.

To design a protective scheme for a transformer it is necessary to have a knowledge of faults that have to be detected.

Figure 22.15 shows the types of fault that can be experienced, on a transformer. These are earth-fault on h.v. external connections; phase-to-phase on h.v. external connections; internal earth-fault on h.v. windings; internal phase-to-phase fault on h.v. windings; short-circuit between turns h.v. windings; earth-fault on l.v. external connections; phase-to-phase fault on l.v. external connections; internal earth-fault on l.v. windings; internal phase-to-phase fault on l.v. windings; short-circuit between turns l.v. windings; earth-fault on tertiary winding; short-circuit between turns tertiary winding; sustained system earth-fault; and sustained system phase-to-phase fault.

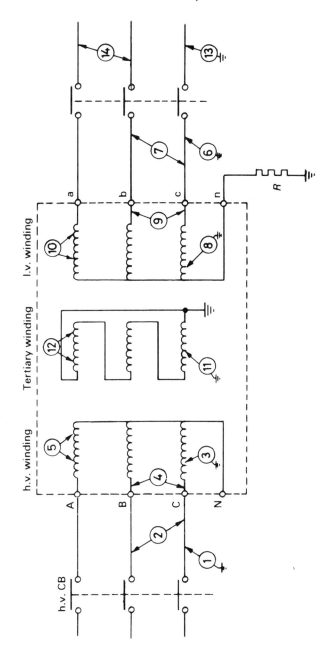

Fig. 22.15 Types of transformer fault.

Under fault conditions, current in the transformer windings is distributed in different ways and is of varying quantity depending upon the transformer winding connections and, in the case of earth-faults, the method of earthing. An understanding of these is essential for the design of protection schemes, particularly balanced differential schemes, the performance of directional relays and setting of overcurrent and earth-fault relays.

A detailed analysis is outside the scope of this chapter. A full treatise is given in *Fault Calculations* by C.H.W. Lackey, published by Oliver & Boyd Ltd.

Current balance schemes

Most protective schemes applied to transformers are based on the current balance principle of magnitude comparison of currents flowing into and out of the transformer, Fig. 22.10. This principle can be used to protect the transformer winding separately or as an overall unit. However, in the latter case certain refinements are necessary.

Separate winding protection
Relating the current balance principle to each winding of the transformer separately it can be seen, Fig. 22.12 that under magnetizing, normal load and through-fault conditions the current circulates between the CTs which, provided they have similar characteristics, results in no current flowing in the relay. Stability is therefore ensured under all external conditions. If however an earth fault occurs within the zone of the CTs an out-of-balance condition exists whereby current flows through the relay.

The main problem experienced in first applying a current balance scheme was retaining stability on through-faults accompanied by unequal saturation of the CTs during the first few cycles after the fault zero. This was overcome by using a relay of high impedance which has a large value stabilizing resistor connected in the relay circuit.

This scheme using high impedance relays has now been in use for many years, the simplicity in application being that the performance of the protection on both stability and fault setting can be calculated with certainty. For a given through-fault current (I_a) the maximum voltage that can occur across the relay circuits is given by:

$$V_R = \frac{I_a}{N} (R + X)$$

where V_R = maximum voltage across relay circuit
         R = maximum lead resistance
         X = CT secondary resistance
         N = CT ratio.

This is based on the assumption of the worst condition when one CT completely saturates and ceases to transform any part of the primary fault current, whilst the other CTs continue to transform accurately.

If the setting voltage of the relay is made equal to or greater than this voltage the protection will be stable for currents above the maximum through-fault current level. In practice, the knee point voltage of the CTs is designed to be at least twice this value.

The fault setting of the protection is calculated from:

$$I_{FS} = N(I_R + I_A + I_B)$$

where I_{FS} = fault setting
  I_R = relay circuit current at relay setting voltage
  I_A, I_B = CT magnetizing currents at relay setting voltage
  N = CT ratio.

The primary fault setting should be adjusted to the level required by the addition of resistors connected across the relay circuit to increase the relay circuit current I_R.

The circuit of a typical high impedance restricted earth-fault relay is shown in Fig. 22.16. The operating element is an attracted-armature relay energized from a full-wave rectifier. The capacitor C in conjunction with the resistors form a low-pass filter circuit. The function of this is to increase the setting at harmonic frequencies, thus retaining stability when high frequency currents are produced in certain installations during switching.

The links enable the voltage setting to be adjusted and non-linear resistor M1 limits the peak voltage output from the CTs during internal faults and so protects the relay and secondary wiring.

Summarizing, separate winding current balance schemes are: unaffected by load current, external fault or magnetizing inrush currents; and unaffected by the ratio of transformer. The complete winding can be protected with solidly earthed neutral but not when resistance earthed, and it will not detect phase faults (three-phase protection), shorted turns or open-circuits.

Overall differential protection
The current balance principle can also be applied to cover both primary and

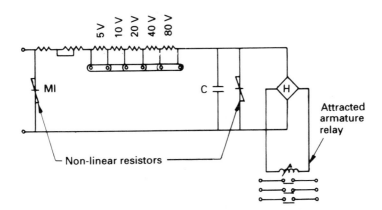

Fig. 22.16 High impedance relay.

secondary windings. However an overall scheme is affected by magnetizing current, although it remains balanced under normal load or through-faults providing CT ratios are matched; mismatch in CTs causes spill current to flow in the relay circuit. As most transformers are equipped with tap-changing the design of an overall scheme for three-phase transformers must also take account of this mismatch under through-fault conditions.

Thus the application of an overall scheme to three-phase transformers requires a biased relay with characteristics as illustrated in Fig. 22.10 to maintain stability on tap-changing and during magnetizing inrush currents.

In both cases the out-of-balance current flowing through the relay circuit may be several times the basic fault setting. A bias winding ensures that the relay remains stable under these conditions, as explained on p. 598. Usual practice is to arrange the bias characteristic with a slope of at least twice the slope of the expected steady-state spill current characteristic.

During internal faults the whole of the available CT secondary current passes through the relay operating circuit giving the rapid rise in operating current.

The second reason, already mentioned, for using a biased relay for overall transformer protection is that spill current may flow during a magnetizing surge. This spill current contains a large percentage of second and higher harmonics and it is convenient to convert these harmonics into bias current, thereby preventing the relay from operating during magnetizing inrush conditions.

One thing to be considered with 'harmonic bias' is that harmonics are also present during internal faults due to CT saturation. To ensure that the relay will operate under all internal fault conditions the harmonic bias unit should preferably be designed to use only second harmonic which predominates in a magnetizing surge.

Buchholz relay

On all but the smallest transfomers it is usual practice to install a Buchholz gas relay. This device detects severe faults from the resultant surge in the oil and low level faults by the measurement of accumulation of gas. It consists of two pivoted floats carrying mercury switches contained in a chamber which is connected in the pipe between the top of the transformer tank and the oil conservator. Under normal conditions the Buchholz relay is full of oil, the floats are fully raised and the mercury switches are open.

An electrical fault inside the transformer tank is accompanied by the generation of gas and, if the fault current is high enough, by a surge of oil from the tank to the conservator. Gas bubbles due to a core fault are generated slowly and collect in the top of the relay. As they collect, the oil level drops and the upper float turns on its pivot until the mercury switch closes. This generates an alarm. Similarly, incipient windings insulation faults and interturn faults which produce gas by decomposition of insulation material and oil will be detected. Such faults are of very low current magnitude and the Buchholz relay is the only satisfactory method of detection. As these faults are not serious, operation of the relay generates an alarm but does not trip out the transformer.

Serious electrical faults, such as a flashover between connections inside the main tank generate gas rapidly and produce a surge of oil which forces the lower float to rotate about its pivot, causing the lower mercury switch to close. This is arranged to trip both h.v. and l.v. circuit-breakers. In addition to the above, serious oil leakage is detected initially by the upper float which gives an alarm and finally, by the lower float, which disconnects the transformer before dangerous electrical faults result.

Application of two-winding transformer

It is common practice to apply both overall differential relays and restricted earth-fault relays to transformers over about 5 MVA. The amount of protection must be assessed against the importance to the system and economic considerations. A typical scheme of protection for a star/delta transformer is shown in Fig. 22.17. The reason for applying both types of relays is to obtain maximum coverage of earth-faults and phase faults.

Consider a transformer that is resistance-earthed. The current available on an internal earth-fault for operation of a differential form of protection could be

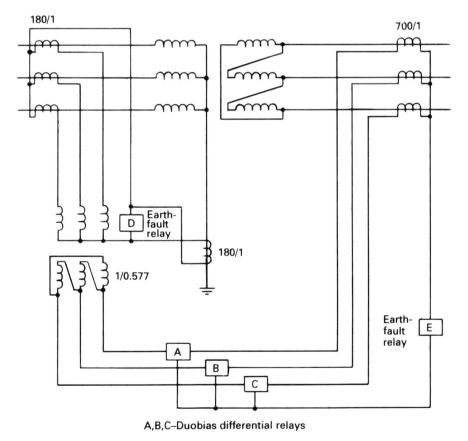

A,B,C–Duobias differential relays

Fig. 22.17 Differential protection with supplementary earth-fault protection.

inadequate because of the transformer action. This is the reason why a separate form of earth-fault protection is added even when solid earthing is employed.

Some care must also be exercised in choosing the CT ratios and connections. The CT ratio must compensate for the difference in primary and secondary currents of the transformer and their connections must compensate for the phase difference.

The restricted earth-fault relay can be operated from a completely separate set of line CTs or it can be combined as shown in Fig. 22.17 with the overall protection by incorporating it into the interposing CT circuit. A CT is required, of course, in the neutral-to-earth connection. The advantage of the restricted earth-fault relay is that it is energized from a CT which 'sees' the whole of the fault current and not just the primary side equivalent of it. Where the system is solidly earthed an overall transformer protection with a setting of about 30% gives complete phase-to-earth fault protection of the delta winding and about 80% of the star winding. In that case additional restricted earth-fault protection is not required for the delta winding, but if it is fitted to the start winding it will detect faults much nearer to the neutral end of the winding.

In addition to overall protection it is usual practice to protect all but the smallest transformers against interturn faults using a Buchholz relay. Back-up protection is normally provided by idmtl overcurrent relays. In applying restricted earth-fault (REF) protection to distribution networks such as 415 V, three-phase, four-wire systems care must be taken in the primary connection of the neutral CT, which is dependent upon the earthing position. Two neutral transformers may be required in order to detect the total zero sequence current flowing as a result of a fault.

Table 22.1 gives a general guide as to the protection suitable for different transformer ratings, although it should be noted that modern relays such as Duobias M allow economical protection over the complete range.

Table 22.1 Transformer protection.

	Type of transformer	Type of protection
1	Distribution Rating < 5 MVA	idmtl o/c ⎫on each REF ⎬winding
2	Distribution Rating > 5 MVA	Overall differential REF each winding
3	Two winding power transmission	Overall differential REF each winding idmtl o/c back-up SBEF*
4	Generator/transformer	Overall differential REF each winding h.v. idmtl o/c l.v. idmtl o/c SBEF*
5	Auto-transformer	Overall circulating current

* Standby earth fault.

Duobias M system

Although Duobias protection, using transductor type relays described above, is used throughout the world in great numbers, advances in technology have allowed the development of a microprocessor version known as Duobias M, see Figs 22.18 and 22.19. This protection incorporates within software two restricted earth-fault

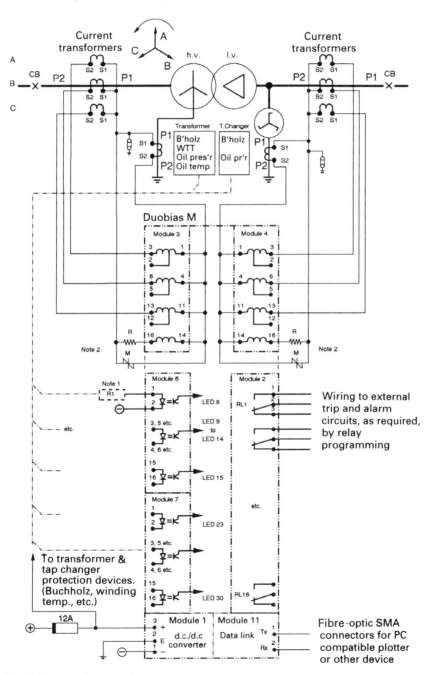

Fig. 22.18 Duobias M metering points.

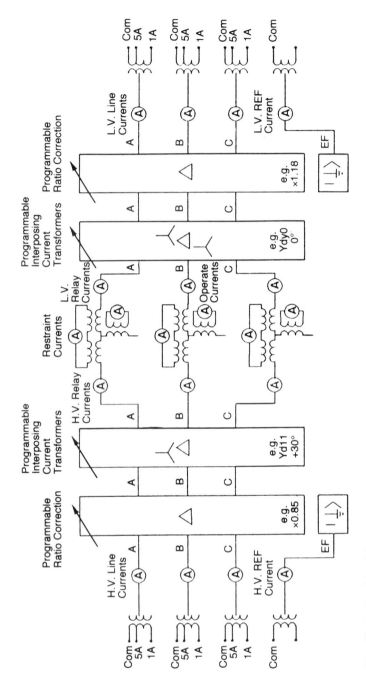

Fig. 22.19 Duobias M system.

elements, models of interposing CTs for almost every primary transformer configuration and inherent phase angle change compensation. The algorithms also include operation and indication for additional inputs such as Buchholz and winding temperature protection and the data storage and fault recording is included. This development allows an extremely economic but very comprehensive system to be applied to a full range of power transformers.

ROTATING PLANT

The operation of rotating machines can be affected by faults within or by external disturbances on the network. The protection of a machine must therefore be designed to be highly discriminative to react efficiently on internal faults and external disturbances.

The number and type of protective relays applied to a machine is a function of the machine characteristics such as size, driving device, single or parallel operation, short-circuit withstand current (and that of the network), and type and protection of network earthing. It is also related to economics such as cost of protection in relation to that of machine and the consequences resulting from a breakdown.

Alternators

On occurrence of a short-circuit at the terminals of an alternator the short-circuit current is initially between five and ten times that of full load of the machine because the initial stator current is limited only by the sub-transient leakage reactance. This is related to the flux set up by the stator mmf which fails to cross the air gap. The increase in stator current causes a demagnetizing effect by opposing the air gap flux but it is an appreciable time before a major change in the air gap flux can be effected.

The net effect is a gradual decrease in the short-circuit current over a period of seconds to a value which may be considerably below full load current of the machine.

Possible faults on alternators

An alternator may experience various types of faults the most common being stator faults which consist of earth-faults, phase-to-phase short-circuits, short-circuits between turns, open-circuits in windings and over-heating. These failures are usually caused by overvoltage and/or deterioration of the insulation.

Protection against external faults

This type of alternator fault can be detected by an impedance relay or by an overcurrent relay capable of responding to the fall-off in fault current caused by the impedance change described above, while having under normal operating conditions a pick-up above the nominal current and a time-delay characteristic that will co-ordinate with other relays.

This is achieved by using the voltage on the machine terminals to determine the time–current characteristics of the relay. Under normal or near normal voltage conditions such as might occur on overload the relay has a long inverse time characteristic. However, under short-circuit conditions when the terminal voltage falls the time−current characteristic is automatically selected to a normal time−current characteristic.

Unbalanced loads

Alternators can usually only support a small percentage of unbalanced loading permanently and must be disconnected from the system before it reaches too high a level. This condition is detected by a negative phase sequence relay with a time−current characteristic of the form $I_2^2 t$ where I_2 = negative phase sequence current in terms of full load rating, and t = time.

This form of protection is usually only applied to the higher rated machines.

Overloads

Overloads causing heating of the stator windings must be eliminated before a temperature dangerous for the machine is reached. Depending on the rating of the machine the overload protection may take the form of an overcurrent relay, a thermal relay or temperature sensors. For small machines, (i.e. hundreds of VA) there are relays which simultaneously provide overload protection by means of low-set elements and external phase-to-phase fault protection by means of high-set elements. For machines above 2 MVA platinum temperature sensors are generally provided. These are embedded in the stator windings and a decision to use these must be made before the machine is manufactured.

Reverse power conditions

It is usual practice to apply protection against failure of the prime power on back-pressure turbine sets and this usually takes the form of a sensitive reverse power relay, having a setting of 0.5%. This type of relay can also be used as a low forward power interlock relay to prevent circuit breaker tripping causing turbine runaway.

As a general rule all alternators, if they may be operating in parallel with other sources, should also be provided with reverse power protection which, depending on type of machine, may not have to be extremely sensitive, for example, on diesel generators a setting of 5−10% is quite adequate.

Protection against internal faults

There are a number of different types of internal faults requiring specific protective systems.

Stator sensitive earth-fault protection

The type of earth-fault protection applied depends on the type of earthing of the machine. There are two forms generally employed − resistor earthing or an earthing transformer.

An earth-fault occurring at the terminals of the alternator with resistor earthing causes full load current to flow. However, should the fault occur closer to the neutral point the voltage available to drive fault current is reduced and therefore its magnitude is reduced. Eventually a point in the winding is reached where the voltage is just sufficient to drive current equal to the fault setting of the protection. The remainder of the winding between this point and the neutral is thus not protected against earth-faults. Care must therefore be taken when selecting the value of the neutral resistor and the setting of the relay to ensure that this unprotected section is as small as practical. The protection consists of a current relay fed from a ring CT in the neutral connection to earth.

Where a distribution or earthing transformer is used a voltage operated relay is connected across the secondary side of the transformer associated with the earthing impedance.

Stator, phase and earth-fault
The standard scheme of protection for the stator windings is the simple overall current balance scheme using a high impedance relay. This employs similar CTs at the line and neutral end of each phase of the alternator, as explained earlier in this chapter. It is possible, although not strictly necessary, to apply an overall biased differential relay for this purpose such as that described for transformers.

Rotor earth-fault
Modern alternators operate with their field winding system unearthed but it is still necessary to protect against breakdown of insulation. A number of schemes are available to provide this protection the most common being a relay which applies a d.c. voltage between the circuit rotor and earth to detect any circulating d.c. current.

A similar scheme using a.c. injection is also available.

Failure of field system
Protection against loss of field is provided by an impedance relay with an offset circular characteristic (called 'offset mho'). Care must be taken in the setting of the relay to ensure that it is not affected by power swings. This protection is usually only applied to larger turbo-alternators.

Application of protection to alternators
In considering which of the foregoing types of protection to apply one must, as already stated, look at rating, importance and cost.

As a general guide, if considering the protection of a diesel alternator with rating up to hundreds of VA one would recommend the following schemes as minimum: overall phase and earth-fault; stator earth-fault; voltage restrained overcurrent; and reverse power.

For a turbo-alternator rated at tens of MVA all of the schemes described would be applied.

Motors

Motors, both synchronous and asynchronous, form an important part of every industrial plant or power system network. The usual faults experienced by motors are: sustained overloads; single-phasing; and phase faults and earth-faults on motor windings and connections.

Overloads and single-phasing

The most generally applied motor protection relay is a thermal relay. To be effective the relay must have a setting slightly in excess of the motor full load current but also remain stable under motor starting conditions which can result, depending upon method of starting, in currents of many times full load for several seconds.

Several types of thermal relays are available but generally they are based either on a bimetallic strip principle or on a static thermal image principle.

It is usual to have, in the same relay, an unbalanced load-detecting circuit to prevent operation of the motor under single-phasing conditions. This is necessary for although a motor will continue to run on only two phases it may overheat.

Phase and earth-faults

All motors above 75 kW should be provided with instantaneous relays to detect the above faults. These relay elements can also be incorporated in the thermal relay. In addition an REF system should be provided for earth-faults, using a similar arrangement to that for transformers.

The use of instantaneous elements depends also on the controlling circuitry of the motor. If a motor is controlled by a contactor which incorporates fuses then it is normal practice not to include instantaneous elements, allowing the fuses to cover multi-phase faults. However some form of earth-fault protection is still recommended.

As far as relays performing other functions are concerned no definite rule can be given to decide above what rating they should be applied. Economic considerations, principally the cost of protection including the necessary CTs, must be compared with the cost of the motor as well as the importance of the motor in the operation of an industrial process, and the consequences of its being out of service.

The use of a starting device incorporating either resistors, an auto-transformer or inductance in the stator circuit does not modify the protection requirements. However, the relay characteristics and settings must be defined in relation to the starting current and time corresponding to the use of the starting device.

Other forms of essential protection which may be used depending upon the type of motor, in addition to those described are: undervoltage; loss of field; negative sequence; locked rotor; and undercurrent.

The use of static components has made a big impact on the protection of machines. It is now possible to incorporate many of the protective relays described into an overall protection module using the international 483 mm rack mounted principle.

RELAY ACCOMMODATION

As well as changes in the technology of relay circuits with the introduction of microprocessors, there has been a significant change in the types of cases housing the relay elements.

For a number of years relay elements have been housed in separate cases for mounting on sheet steel panels or switchgear top plates.

Relays are now smaller and a modular form of case has been developed which is suitable for rack mounting or traditional panel mounting. The case is of uniform height (international standard 4U) and is available in a number of case widths, sizes 2, 3, 4, 6, and 8. They have the facility of being grouped horizontally or vertically in tiers giving great flexibility to layout design within a rack.

This enables several relays of different types to be grouped together to provide a system of protection for a particular application. The modular case system is shown in Fig. 22.20.

For front of panel testing, a test block is available housed in a size 2 case which provides the link between the test supplies and the relay as scheme wiring. Removal of the test block cover can be made to isolate the auxiliary d.c. supply to all connected relays automatically. Insertion of a test plug into the test block can be arranged to short-circuit the CT circuits and enable injection testing by external test equipment through the test sockets on the plug front face and also link the d.c. supply as required. The test block and plug are shown in Fig. 21.21.

The basic construction of a relay should not affect commissioning because relay specifications ensure that relays are constructed to allow for these procedures. However if for some reason the covers of a relay need to be removed suitable precautions will be needed dependent upon the relay construction.

Relays with sensitive moving parts and magnetic circuits will need to be kept clean with particular avoidance of ferrous metal particles.

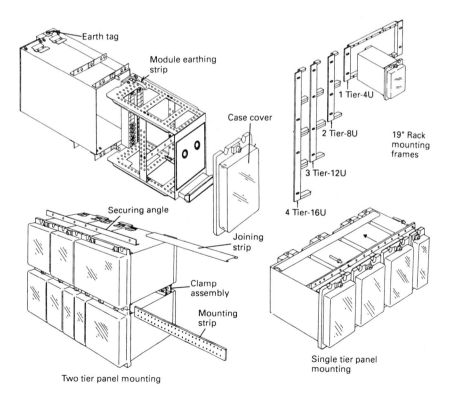

Fig. 22.20 Modular case system.

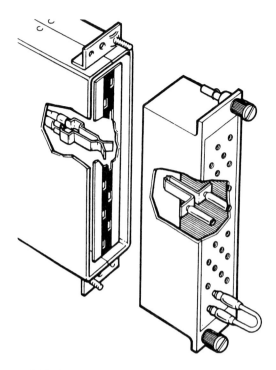

Fig. 22.21 Test plug and block (simplified).

Microprocessor relays will require special precautions against static discharge as outlined in the following section.

Handling of static sensitive devices/sub-assemblies

All components can be said to be static sensitive, some being more sensitive than others. The degree of sensitivity is much greater in individual devices than when these same devices are mounted on a pcb. However, static awareness and pre-cautionary handling procedures are essential to ensure that even pcb sub-assemblies carrying static sensitive devices (SSD) are not stressed by exposure to static discharge.

Static discharge
The charge which can build up on the human body due to normal movement is more than enough to cause irreparable damage to SSDs. This static charge can be made much worse, and build up much more quickly, if the person is wearing, for example, clothes made from man-made fibres or plastics-soled shoes. To ensure that the danger of damage to SSDs is kept to a minimum, they should only be handled by persons who are aware of the nature of static discharge, and only in a static protective workstation.

Static protective workstation

This is an area which has been specially prepared to reduce the risk of damage to SSDs and it is constructed as follows: clean the surface of the bench to be used; place on this a metal plate; cover the plate with a sheet of conducting plastics; and on top place an area of anti-static polyethylene.

Firmly fix the last mentioned item to the worktop and connect it to a good earth via a resistor in the range 200 K to 500 K. This is now the anti-static work surface.

Use of the workstation

The operator should not wear nylon or synthetic overgowns and should endeavour to ensure that his clothing does not come into contact with any SSD, i.e. roll-up sleeves. The operator must make sure that the personal grounding system, such as grounding wrist straps, are attached electrically to the anti-static work surface. Note that the operator should be connected to the grounding system for at least 10 s before handling the SSD.

Handling of pcbs

Do not remove or insert pcbs with SSD on them with the power on as transient voltages can damage the devices. Switch off all supplies, remove/insert the pcb with the minimum of handling and on pcb edges only, and only place the pcb on the anti-static work surface.

COMMISSIONING TESTS

Commissioning tests are required to check that the relays are correctly connected to the appropriate instrument transformers and that their operation is within reasonable limits. Generally, commissioning tests are limited by the need to transport the necessary equipment to site and the site conditions which may include temporary power supplies; thus the commissioning test instructions have to take this into account.

In preparing a test instruction, there is a dilemma in deciding the level of basic knowledge of the engineer receiving it. In general, information more appropriate to national standards and text books is not included in a particular test instruction, because these are available to supplement the manufacturer's test instructions. Because protective relays are generally measuring current, the concept of the current source is very important. A voltage source is characterized by the fact that a load applied to it does not cause the voltage to change in any way. Conversely, a current source is characterized by the fact that a load introduced into the current circuit does not change the current in any way. The character of a CT fed by a power system dictates that protective relays in service are energized by very good current sources.

Test equipment based upon mains supply is obviously fundamentally more a voltage source than a current source, so unless precautions are taken, the inservice condition is not correctly represented by simple test arrangements and incorrect results will be obtained.

For example, when timing an overcurrent relay, the inductance and resistance of the coil dictate the time constant of current build-up within the coil if it is fed from a voltage source. Thus a slower time of operation can be obtained to the true in-service condition where with the current source of power system feeding a CT, the current is established in the coil at full magnitude without any dealy.

However, the most common problem with using a poor current source is due to the non-linear nature of relay impedance. An induction disc idmtl relay obtains its current–time characteristic by saturation of the core material of the induction disc motor. This means that at an energizing level above the point at which the definite minimum time is obtained (twenty times the plug setting) the relay impedance is virtually the winding resistance of the coil, whereas at lower levels the inductance of the coil has significant value. Typically, a 5 A relay impedance varies from 0.12 to 0.5 Ω and a 1 A relay from 3.2 to 1.5 Ω at 100% tap over its working range. Other taps give higher or lower impedance and generally at the higher percentage taps the variation between impedance at low currents and high currents is not so marked. Of course, the impedance of the relay varies with instantaneous values of current so that the impedances quoted are somewhat fictitious because they represent a combination of a sinusoidal current and a non-linear voltage. Also, the voltage is processed by an instrument giving a value appropriate to the type of instrument and its calibration, generally the r.m.s. equivalent of the average value, i.e. a moving coil type of a.c. instrument.

Once the non-linearity of the relay impedance is understood the requirement for a suitable current source is evident. Obviously it is not practical to produce a perfect current source and it is therefore necessary to establish the rules for determining the suitability of the various degrees of perfection in a current source. In the laboratory, it is possible to establish very good current sources by the use of heavy current rigs feeding CTs. The next level below this is to use phase-to-phase mains voltage with resistance to limit the current and below that to use phase-to-neutral mains voltage with resistance to limit the current. Tests on site are generally done with a fixed resistor fed by a constantly variable autotrans-former which is used for fine control. If the same value of resistor has to be used for all currents, the auto-transformer may be at a very low voltage for some current levels which immediately questions the suitability of the current source. The first check on the suitability of a current source is to observe the ammeter feeding the relay from the current source while short-circuiting the relay. Obviously with a perfect current source the ammeter should show no change whether the relay is in circuit or not. If significant change is detected in the current, then the linearity of the relay impedance should be checked before settings and timings obtained with that particular test rig are agreed.

The requirement for fundamental accuracy in non-unit protection leads to a necessity to prove that the current source is suitable to determine the true setting or timing of a relay. This can be achieved by repeating the test with an improved current source until no change is apparent in the result between say, maximum resistance fed by phase-to-neutral voltage and the equivalent maximum resistance fed by phase-to-phase voltage. In some cases, or when absolute assurance is required, the heavy current rig is the only recourse but, even in this case, care must be taken that saturation of the various current limiting reactors or impedance changing transformers does not distort the current feeding the primary of the CT. This is particularly important in high impedance schemes where, for an internal fault, the CT is virtually unloaded and thus the impedance reflected into the primary, although extremely small, may still be significant if the heavy current machine is not carefully matched to the CT and the relay combination. It should also be appreciated that with a distorted waveshape the ammeter is unlikely to have the same relative response as the relay, so that the only true measurement is when the current is completely sinusoidal because relay and ammeter are both calibrated using sine-wave current.

In some critical relays, the harmonic content in the driving voltage can cause incorrect measurement and this must then be reduced by using inductance to provide a current source simulating the power system conditions more accurately than the more commonly used resistance current limiting.

Another common problem is the determination of direction of current. This is important in a relative sense for establishing the residual connection of CTs and in a fundamental sense, for setting up the operate direction of directional relays.

In a REF arrangement, using four CTs, each phase CT may be connected to give minimum spill current with the neutral CT as shown in Fig. 22.22 but with only three CTs the phasing-out must be done by phase-to-phase primary injection as shown in Fig. 22.23 and observation of the out-of-balance current obtained between each pair of CTs. In each case, ammeters should be available as shown for the energized CTs so that the CT ratios can be checked coincidentally.

Because conventions are not universally standardized for windings, it is not possible to be precise whether connecting start to start and finish to finish will give a reversal of polarity or not. Thus the direction of a directional element should be confirmed in relation to an independent quantity. This quantity is generally load current, for which the direction, in relation to the power system, is clearly known. In some cases the load may have a low power factor and thus the phase angle between load current and voltage at the relay terminals may not be decisive in defining the direction. If the relay has a variable characteristic angle, study of the vectors may allow direction to be checked at a different relay angle before finally setting the relay back to the setting required for fault current conditions.

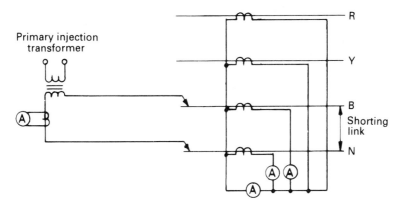

Fig. 22.22 Ratio and directional tests with neutral CT.

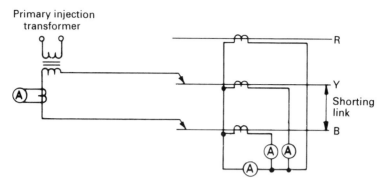

Fig. 22.23 Ratio and directional tests without neutral CT.

Directional earth-fault relays are particularly difficult to check for polarity because they are fed from residual voltage which gives a reversal of the effective voltage signal which can be confusing. Again, load current gives the best overall check on direction and a red phase-to-earth fault is represented by disconnecting and short-circuiting the yellow and blue CTs to provide only red load current as a residual current to the relay, and disconnecting the red phase winding of the VT to represent the total collapse of the red phase voltage which occurs due to a close-up red earth-fault. The red phase main secondary winding should be short-circuited to provide the correct in-fault loading on the directional element.

An important aspect in all testing is to take account of the heat generated in a relay when subject to relatively long energization times at currents in excess of rating. Apart from the possibility of damage to the relay by excessive heat, lower levels of heat could affect the performance if the relay is susceptible to self-heating.

Finally, it is worthwhile considering the accuracy of basic test equipment and simple cross-checks that can be done if some doubt exists with results. With multi-range meters it is simple to change their role or cross-check them against each other. Fundamental instrument accuracy (which is related to full scale) can be of the same order as relay accuracy, so that some thought is necessary in establishing the true accuracy of settings even when supplies, waveshape, temperature, frequency are within reference conditions. British protective relays are generally made to BS 142 which prescribes the reference conditions for which the various aspects of relay performance shall be declared.

CHAPTER 23

Power Factor Correction and Tariffs

T. Longland, CEng, MIEE, AMEME
(Technical Manager, Johnson & Phillips (Capacitors) Ltd)

Power capacitors have been employed in many and varied ways in industry over the past 60 years, but their increasing use is often limited by the apparent lack of practical information on their application. This chapter provides information and guidance on the selection, and use, of capacitors for power factor correction.

With increasing electricity charges, and the need to save energy, it is of paramount importance to both industrial and commercial users of electricity to ensure that their plant operates at maximum efficiency. This implies that the plant power factor must be at an economic level.

IMPORTANCE OF POWER FACTOR

Most a.c. electrical machines draw from the supply apparent power in terms of kilovolt amperes (kVA) which is in excess of the useful power, measured in kilowatts (kW), required by the machine. The ratio of these quantities is known as the power factor of the load, and is dependent upon the type of machine in use. Assuming a constant supply voltage this implies that more current is drawn from the electricity authority than is actually required.

$$Power\ factor = \frac{true\ power}{apparent\ power} = \frac{kW}{kVa}$$

A large proportion of the electrical machinery used in industry has an inherently low power factor, which means that the supply authorities have to generate much more current than is theoretically required. This excess current flows through generators, cables, and transformers in the same manner as the useful current. It is understood that there are resistive loads such as lighting and heating but these are generally outweighed by the motive power requirements.

If steps are not taken to improve the power factor of the load all the equipment from the power station to the factory sub-circuit wiring, has to be larger than necessary. This results in increased capital expenditure and higher transmission and distribution losses throughout the whole supply network.

To overcome this problem, and at the same time ensure that generators and cables are not overloaded with wattless current (as this excess current is termed), the supply authorities often offer reduced terms to consumers whose power factor is high, or impose penalties for those with low power factor.

THEORY OF POWER FACTOR CORRECTION

The kVA in an a.c. circuit can be resolved into two components, the in-phase component which supplies the useful power (kW), and the wattless component (kvar) which does no useful work. The phasor sum of the two is the kVA drawn from the supply. The cosine of the phase angle ϕ_1 between the kVA and the kW components represents the power factor of the load.

The phasor diagram for this is shown in Fig. 23.1. The load current is in phase with the kVA so that it lags the supply voltage by the same phase angle.

To improve the power factor, equipment drawing kvar of approximately the same magnitude as the load kvar, but in phase opposition (leading) is connected in parallel with the load. The resultant kVA is now smaller and the new power factor, cos ϕ_2 is increased. Thus any value of cos ϕ_2 can be obtained by controlling the magnitude of the leading kvar added. This is shown in Fig. 23.2.

POWER FACTOR IMPROVEMENT

In practice two types of equipment are available to produce leading kvar:

(1) *Rotary equipment*   Phase advancers, synchronous motors and synchronous condensers. Where auto-synchronous motors are employed the power factor correction may be a secondary function.
(2) *Static equipment*   Capacitors.

When installing equipment, the following points are normally considered: reliability of the equipment to be installed, probable life of such equipment, capital cost, maintenance cost, running costs and space required, and ease of installation.

Generally, the capital cost of rotating machinery, both synchronous and phase

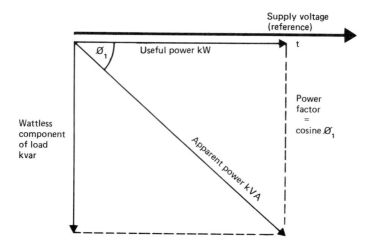

Fig. 23.1 Phasor diagram of plant operating at a lagging power factor.

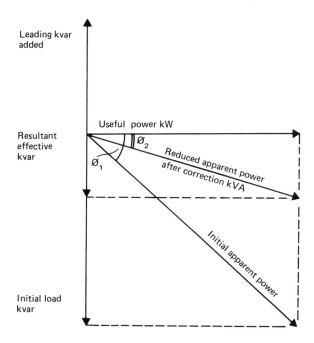

Fig. 23.2 Power factor improvement by adding leading kvar to a lagging power factor.

advancing, makes its use uneconomical, except where one is using rotating plant for a dual function-drive and power factor correction. In addition the wear and tear inherent in all rotary machines involves additional expense for upkeep and maintenance.

Capacitors have none of these disadvantages. Compared with other forms of correction, the initial cost is very low, upkeep costs are minimal and they can be used with the same high efficiency on all sizes of installation. They are compact, reliable, highly efficient, convenient to install and lend themselves to individual, group or automatic methods of correction. These facts indicate that, generally speaking, power factor correction by means of capacitors is the most satisfactory and economical method.

The static capacitor, owing to its low losses, simplicity and high efficiency, is now used almost universally for power factor correction.

ECONOMIC CONSIDERATIONS

When considering the economics of power factor correction it is important to remember that any plant used for this purpose does, in general, compensate for losses and lower the loadings on supply equipment such as cables, transformers, switchgear and generators.

The rating of the capacitor required to improve the power factor and the saving achieved depend largely on the electricity tariff. Charges can be related to kVAh, kvarh, or to md kVA, all of which can be reduced by installing power factor correction capacitors.

In the UK the Electricity Act 1989 came into force on 31 March 1990. As a result of this act independence was given to the 12 area electricity boards in England and Wales. The new companies are now generally referred to as 'regional electricity companies' (RECs).

The main change in Scotland was the transfer of the SSEB's nuclear generation to a new company, Scottish Nuclear. The remaining generation and distribution activities of the SSEB and those of the NSHEB were vested in

- Scottish Power
- Scottish Hydro Electric.

Each of the above companies has its own tariff structure. Now, however, the regional electricity companies are free to offer to supply any consumer anywhere in the country whose demand is greater than 100 kW. This has meant that each REC has a number of 'contract tariffs' designed for individual consumers some of which do and some of which do not impose penalties for poor power factor although all expect their consumers to have a power factor better than 0.85.

The published tariff, however, for each REC have not changed since privatization in their structure except that Eastern Electricity now incorporates an additional reactive unit charge. These published tariffs fall into two categories as follows:

(1) kVA maximum demand (kVA md) charge plus a unit charge dependent on this maximum demand. Each charge has a sliding tariff scale for the different ranges of kVA md, and number of units per kVA md respectively. Such tariffs are levied by the following regional companies: Southern Electricity; Eastern Electricity; East Midlands Electricity; Midlands Electricity and the Yorkshire Electricity.
(2) kW maximum demand charge plus a unit charge dependent on this md, the demand charge being related to the power factor. Both charges have sliding scales. Such tariffs are further subdivided, into two types, depending upon the basis of the 'demand charge'.
  (i) Demand charge increased according to the amount of average lagging power factor below a set base value. Such tariffs are found in the following regional companies: North Western Electricity; South Western Electricity and North Eastern Electricity.
  (ii) Demand charges varied according to the ratio of base power factor to the average power factor, or when the power factor is outside set limits. Such tariffs are found in the following: South Eastern Electricity regional companies; London Electricity; South Wales Electricity; Merseyside and North Wales Electricity and Scottish Hydro-Electric.

Commercial tariffs also include penalty clauses for low power factor and the most economic power factor is assessed on the same basis as for industrial tariffs. Domestic tariffs are not affected because the power factor is of the order of unity.

Power factor correction should always be regarded as an investment with two opposing considerations. First, the expenditure incurred in the overheads charged against the capacitor installation, and, second, the income brought about by the saving in the cost of electricity, together with the reduction of losses in the electrical system. The main capacitor overheads are depreciation, interest on capital, electrical losses and maintenance costs. The last two items, in most cases, are covered by the savings on the losses in the electrical system.

The efficiency of a capacitor installation remains almost constant throughout the 10–12 year life of the capacitor. It is, therefore, usually estimated that an overhead allowance of approximately 8% to 10% of the installed cost of the equipment be used.

Where tariffs are based upon a standing charge per kVA, plus a charge for each unit (kWh) supplied, the most economical degree of correction is found when the final power factor is approximately 0.97/0.98. It is never economic to attempt to improve the power factor to unity, as the nearer the approach to unity the more is the kvar that must be installed for a given improvement. This can be seen from Fig. 23.3.

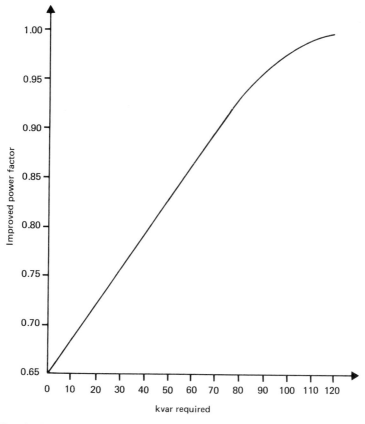

Fig. 23.3 Graph showing how the kvar to be connected varies with the power factor to be achieved. The curve is based on an initial load at 100 kW at a power factor of 0.65.

In the average power factor tariff, where there is a bonus or penalty clause if the power factor is above, or below, some datum figure, the most economic power factor is the datum one.

Table 23.1 indicates how the most economic power factor depends on the type of tariff. The table has been confined to tariffs applied by the UK electricity companies.

Three worked examples of savings to be made are shown in Table 23.2.

CALCULATION OF CAPACITOR SIZE

There are a number of methods of calculating the capacitor size required. Figures 23.4 and 23.5 together with Table 23.3 have been prepared with the object of providing speedy and simple methods of ascertaining if any benefit can be derived from improving the power factor of an industrial or commercial a.c. load.

PRACTICAL POWER FACTOR IMPROVEMENT

Figures 23.4 and 23.5 and Table 23.3 enable the capacitor requirements to be calculated knowing the initial power factor. In practice the problem is to determine this initial power factor.

The type of tariff that a consumer is charged not only determines the level of correction, but also the method by which the capacitor size is determined.

Maximum demand tariffs

Under an md tariff the md value is recorded on an appropriate meter which measures it either in kW or kVA. This meter usually measures twice the largest number of kVA (or kW) of md supplied during any half hour period, in any month.

With these tariffs, therefore, it is only necessary to ensure that the capacitors are in circuit at the time the md is being reached, if the prime function of the capacitor installation is to save money by reduced md charges.

Average power factor tariffs

Average power factor tariffs incorporate a power factor penalty clause based on the average power factor determined from the kWh and kVAh recorded in any metering period.

With this type of tariff, therefore, it is necessary to reduce the number of kvar hours recorded during the metering period. To do this the capacitor must be in circuit whenever reactive units are being recorded.

A refinement on this type of tariff, usually found on the continent, penalizes all reactive units, be they leading or lagging. With this type of tariff small steps of capacitors are required to ensure that there is no reactive power recorded at any loading condition.

Table 23.1 The most economic power factor required by the various UK electricity companies.

REC	Maximum demand charge based on	Power factor should be improved to	Type of power factor penalized
Eastern	kVA	0.97/0.98	Maximum demand
East Midlands	kVA	0.97/0.98	Maximum demand
London	kW	0.95	Average
Merseyside and North Wales	kW	0.93	Maximum demand
Midlands	kVA	0.97/0.98	Maximum demand
North Eastern	kW	0.90	Average
Scottish Hydro Electric Hydro	kW	0.90	Average
North Western	kW	0.90	Average
South Eastern	kW	0.90	Average
Southern	kVA	0.97/0.98	Maximum demand
Scottish Power	kVA	0.97/0.98	Average
South Wales	kW	0.95	Average
South Western	kW	0.90	Average
Yorkshire	kVA	0.97/0.98	Maximum demand

Table 23.2 Examples of savings to be made by power factor improvement.

Example	(£)
Example 1	£
Initial load conditions: 300 kVA, 0.67 power factor, 201 kW	
Tariff. Maximum demand charge each month based on:	
each kVA of the first 200 kVA of maximum demand	0.86
each kVA of the next 300 kVA of maximum demand	0.83
Total charges, before correction:	
200 kVA per month, at the above rate	172.00
100 kVA per month, at the above rate	83.00
Total charge, per month	255.00
Total charge, per annum	3060.00
Correction: 182 kvar of capacitors is required to improve the power factor to	
a level of 0.98, at an approximate cost of:	940.00
Improved load conditions: 205 kVA, 0.98 power factor, 201 kW	
New total charge per annum after correction	2113.80
Savings in electricity charges per annum	946.20
Note: Cost of the capacitors is recovered in 12 months.	

Table 23.2 *(contd)*

Example	(£)

Example 2
Initial load conditions: 85 kVA, 0.60 power factor, 51 kW
Tariff. Maximum demand charge each month based on:

the first 10 kW of maximum demand	12.75
each kW of the next 10 kW of maximum demand	1.27
each kW of the remaining kW of maximum demand	1.17

In addition, if the power factor in the month is less 0.9 the demand charge, for that month, is increased by 1% for each 1% by which the power factor is below that figure
Total charges, before correction:

51 kW per month, at the above rate	61.72
Power factor penalty clause, 30% of the above	18.52
Total charge, per month	80.24
Total charge, per annum	962.88
Correction: 51 kvar improves the power factor to a level of 0.95, thus allowing for future load increased, at a cost of:	280.00

Improved load conditions: 53.6 kVA, 0.95 power factor, 51 kW

New total charge per annum, after correction	740.64
Savings in electricity charges per annum	222.24

Note: Cost of the capacitors recovered in 15 months.

Example 3
Initial load conditions: 517 kVA, 0.58 power factor, 300 kW
Tariff. Maximum demand charge each month based on:

each kW of the first 200 kW of maximum demand	1.25
each kW of the remaining kW of maximum demand	1.22

In addition, if the power factor is lagging, the kW of maximum demand recorded is increased by dividing it by the average lagging power factor and then multiplying the figure so obtained by 0.95.

Total charges, before correction:
The chargeable kW of maximum demand is increased in line with the power factor penalty clause as follows:

$$\text{Chargeable demand} = \frac{300}{0.58} \times 0.95 = 490\,\text{kW}$$

490 kW per month, at the above rate	603.80
Charge per annum	7245.60
Correction: 330 kvar of capacitors is required to improve the power factor to a level of 0.95, at an approximate cost of:	2010.00

Improved load conditions: 316 kVA, 0.95 power factor, 300 kW
New total charge:

$$\text{Chargeable demand} = \frac{300}{0.95} \times 0.95 = 300\,\text{kW}$$

Total charge per annum	4464.00
Savings in electricity charges per annum	2781.60

Note: Cost of the capacitors is recovered in 12 months.

Table 23.3 Table of factors for calculating size of capacitor for power factor improvement.

Initial power factor	Factor for improving power factor to:								
	Unity	0.99	0.98	0.97	0.96	0.95	0.90	0.85	0.80
0.40	2.291	2.148	2.088	2.040	1.999	1.962	1.807	1.617	1.541
0.41	2.225	2.082	2.022	1.974	1.933	1.896	1.741	1.605	1.475
0.42	2.161	2.018	1.958	1.910	1.869	1.832	1.677	1.541	1.411
0.43	2.100	1.957	1.897	1.849	1.808	1.771	1.616	1.480	1.350
0.44	2.041	1.898	1.838	1.790	1.749	1.712	1.557	1.421	1.291
0.45	1.984	1.841	1.781	1.733	1.692	1.655	1.500	1.364	1.234
0.46	1.930	1.787	1.727	1.679	1.638	1.601	1.446	1.310	1.180
0.47	1.878	1.735	1.675	1.627	1.586	1.549	1.394	1.258	1.128
0.48	1.828	1.685	1.625	1.577	1.536	1.499	1.344	1.208	1.078
0.49	1.779	1.636	1.576	1.528	1.487	1.450	1.295	1.159	1.029
0.50	1.732	1.589	1.529	1.481	1.440	1.403	1.248	1.112	0.982
0.51	1.686	1.543	1.483	1.435	1.394	1.357	1.202	1.066	0.936
0.52	1.643	1.500	1.440	1.392	1.351	1.314	1.159	1.023	0.893
0.53	1.600	1.457	1.397	1.349	1.308	1.271	1.116	1.980	0.850
0.54	1.559	1.416	1.356	1.303	1.267	1.230	1.075	1.939	0.809
0.55	1.519	1.376	1.316	1.268	1.227	1.190	1.035	0.899	0.769
0.56	1.480	1.337	1.277	1.229	1.188	1.151	0.996	0.860	0.730
0.57	1.442	1.299	1.239	1.191	1.150	1.113	0.958	0.822	0.692
0.58	1.405	1.262	1.202	1.154	1.113	1.076	0.921	0.785	0.655
0.59	1.369	1.226	1.166	1.118	1.077	1.040	0.885	0.749	0.619
0.60	1.333	1.190	1.130	1.082	1.041	1.004	0.849	0.713	0.583
0.61	1.229	1.156	1.096	1.048	1.007	0.970	0.815	0.679	0.549
0.62	1.265	1.122	1.062	1.014	0.973	0.936	0.781	0.645	0.515
0.63	1.233	1.090	1.030	0.982	0.941	0.904	0.749	0.613	0.483
0.64	1.201	1.058	0.998	0.950	0.909	0.872	0.717	0.581	0.451
0.65	1.169	1.026	0.966	0.918	0.877	0.840	0.685	0.549	0.419
0.66	1.138	0.995	0.935	0.887	0.846	0.809	0.654	0.518	0.388
0.67	1.108	0.965	0.905	0.857	0.816	0.779	0.624	0.488	0.358
0.68	1.078	0.935	0.875	0.827	0.786	0.749	0.594	0.458	0.328
0.69	1.049	0.906	0.846	0.798	0.757	0.720	0.565	0.429	0.299
0.70	1.020	0.877	0.817	0.769	0.728	0.691	0.536	0.400	0.270
0.71	0.992	0.849	0.789	0.741	0.700	0.663	0.508	0.372	0.242
0.72	0.964	0.821	0.761	0.713	0.672	0.635	0.480	0.344	0.214
0.73	0.936	0.793	0.733	0.685	0.644	0.607	0.452	0.316	0.186
0.74	0.909	0.766	0.706	0.658	0.617	0.580	0.425	0.289	0.159
0.75	0.882	0.739	0.679	0.631	0.590	0.553	0.398	0.262	0.132
0.76	0.855	0.712	0.652	0.604	0.563	0.526	0.371	0.235	0.105
0.77	0.829	0.686	0.626	0.578	0.537	0.500	0.345	0.209	0.079
0.78	0.802	0.659	0.599	0.551	0.510	0.473	0.318	0.182	0.052
0.79	0.776	0.633	0.573	0.525	0.484	0.447	0.292	0.156	0.026

Table 23.3   *(contd).*

Initial power factor	Factor for improving power factor to:								
	Unity	0.99	0.98	0.97	0.96	0.95	0.90	0.85	0.80
0.80	0.750	0.607	0.547	0.499	0.458	0.421	0.226	0.130	–
0.81	0.724	0.581	0.521	0.473	0.432	0.395	0.240	0.104	–
0.82	0.698	0.555	0.495	0.447	0.406	0.369	0.214	0.078	–
0.83	0.672	0.529	0.469	0.421	0.380	0.343	0.188	0.052	–
0.84	0.646	0.503	0.443	0.395	0.354	0.317	0.162	0.026	–
0.85	0.620	0.477	0.417	0.369	0.328	0.291	0.136	–	–
0.86	0.593	0.450	0.390	0.243	0.301	0.264	0.109	–	–
0.87	0.567	0.424	0.364	0.316	0.275	0.238	0.083	–	–
0.88	0.540	0.397	0.337	0.289	0.248	0.211	0.056	–	–
0.89	0.512	0.369	0.309	0.261	0.220	0.183	0.028	–	–
0.90	0.484	0.341	0.281	0.233	0.192	0.155	–	–	–
0.91	0.456	0.313	0.253	0.205	0.164	0.127	–	–	–
0.92	0.426	0.283	0.223	0.175	0.134	0.097	–	–	–
0.93	0.395	0.252	0.192	0.144	0.103	0.066	–	–	–
0.94	0.363	0.220	0.160	0.112	0.071	0.034	–	–	–
0.95	0.329	0.186	0.126	0.078	0.037	–	–	–	–
0.96	0.292	0.149	0.089	0.041	–	–	–	–	–
0.97	0.251	0.108	0.048	–	–	–	–	–	–
0.98	0.203	0.060	–	–	–	–	–	–	–
0.99	0.143	–	–	–	–	–	–	–	–

This table gives values of tan ϕ for various of power factor, cos ϕ, and therefore provides a simple method of calculating values of kvar, and the size of capacitor required to improve the power factor.
Example:
Given 100 kW load to be improved from 0.77 to 0.95 power factor. Factor from table is 0.5.

Capacitor (kvar) = load (kW) × factor to improve from existing to proposed power factor
= 100 × 0.5 = 50 kvar

CAPACITOR SIZE RELATED TO TARIFF

A different calculation is necessary for ascertaining the correct capacitor rating depending on the particular tariff applicable.

Maximum demand tariff

The first step is to obtain past md records over a period of several years if possible. These normally indicate either md kVA, or md kVA and kW. In large systems recorders as well as instruments and meters are usually installed close to the point of metering, thus simplifying the calculations. If the meter records indicate both kVA and kW it is easy to obtain the md power factor from the relationship between kVA and kW outlined earlier. If, however, the md records only indicate kVA it is then necessary to estimate a power factor.

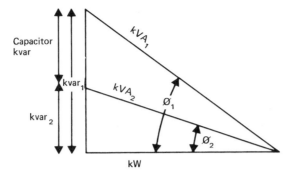

Initial conditions:

$$\text{Power factor} = \cos \phi_1 = \frac{kW}{kVA_1}$$

$$\tan \phi_1 = \frac{kvar_1}{kW}$$

$$kvar_1 = kW \times \tan \phi_1$$

Improved conditions:

$$\text{Power factor} = \cos \phi_2 = \frac{kW}{kVA_2}$$

$$\tan \phi_2 = \frac{kvar_2}{kW}$$

$$kvar_2 = kW \times \tan \phi_2$$

Capacitor kvar required to improve power factor from $\cos \phi_1$ to $\cos \phi_2$

$$= (kvar_1 - kvar_2)$$
$$= kW (\tan \phi_1 - \tan \phi_2)$$

This value of capacitor kvar can be determined either by drawing the vector diagram to scale, or by calculation using values from trigonometrical tables.

Fig. 23.4 Reduction of kVA loading for constant kW loading by improvement of power factor.

Prior to this, however, the md records should be studied to see if there is any pattern to the demands, or if there is any marked variation in demand from month to month. If such a pattern, or variation, is found the reason for it should be established.

If only the kVA demand is known the power factor has to be obtained from site measurements. The kWh and kVAh meters are read at the beginning, and end, of a known time period. The difference between the readings at the start and end of the period enables the relevant kW, kVA, and kvar components of the load during the test period to be calculated.

Ideally these tests should be performed at the same time as the md occurs, although in practice this very rarely happens. It is necessary, therefore, to determine

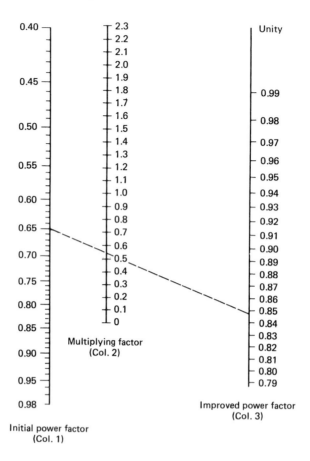

Fig. 23.5 Nomogram for determining size of capacitor for power factor improvement.

Instructions
Place a ruler to join the value of the initial power factor (column 1) to that of the improved power factor required (column 3).
Read the multiplying factor where the ruler crosses column 2. For example, a 100 kW load is to be improved from 0.65 to 0.85. power factor.
Using the chart, the multiplying factor is 0.55.

∴ capacitor (kvar) = 100 kW × 0.55 = 55 kvar

what plant is in use at periods of md which was not in use during the test period and make allowance for this in the calculations.

Example
A study of a consumer's electricity bill indicated an average md of 288 kVA. A test carried out at the plant provided the following load figures: 240 kVA, 168 kW, 0.7 power factor, 171 kvar.

It was noted, however, that at the time of the test there was 20 kW of resistive load not in circuit, together with some 31 kW of fluorescent lighting. Both items

Table 23.4 Calculations of md from known operating conditions.

	kVA	kW	Power factor	kvar
Test load	240	168	0.7	171
Resistive load		20	1.0	0
Lighting load		31	0.9	15
Calculated maximum demand	288	219	0.76	186

could be in use at times of md. The md power factor can be calculated as shown in Table 23.4.

Utilizing the tables already referred to it will be seen that to improve the md power factor of 0.76 to an economic level of 0.97 requires 132 kvar of capacitor correcting equipment.

Average power factor tariff

The following information is required to enable a capacitor size to be arrived at: md records; kWh consumed during the month; kvarh consumed during the month; and working hours of the plant during the month.

The example below shows how this information is used to calculate a capacitor size.

Example

A consumer is charged on a tariff which imposes a penalty charge when the average lagging power factor falls below 0.9. From a study of the electricity accounts, and a knowledge of the plant the following information was obtained: kWh consumed during the month 48 000; kvarh used during the month 56 000; average power factor (calculated from the above) 0.65; working hours 160 per month.

To improve the average lagging power factor from a level of 0.65 to an economic level of 0.90 some 33 300 kvarh must be removed from circuit by means of capacitors. Therefore:

capacitor size required = 33 300/160 = 206 kvar

In practice this capacitor size would probably be increased by 5 or 10% to ensure that the average power factor was kept above 0.90. The capacitor size arrived at in the above example is the minimum possible size. It does not necessarily follow that the installation of this capacitor would, in fact, give the required value of average power factor, as a number of other factors have to be considered. If load variations occur so that the capacitor gives a leading power at times, then the full kvar rating of the capacitor is not available for the reduction of kvarh during these periods. It will prevent, during such periods, the recording of any kvarh units, so that the difference between the reactive component of the uncorrected load and the capacitor kvar is lost. This is based on the assumption that the kvarh meters are fitted with devices to prevent 'unwinding' under leading power factor conditions.

It, therefore, follows that the actual capacity required must be increased over and above the minimum value to allow for this.

During periods when the factory is not in production, kWh, and kvarh, may still be recorded due to small items of plant that run continuously, or to transformer magnetizing currents (this fact is discussed later). It may, therefore, be necessary to provide for a portion of the total capacity to be left in circuit during light load periods. It is reasonable to assume that the supply authority will object to the whole capacity being in circuit 24 hours a day.

DETERMINATION OF LOAD CONDITIONS

The first step in designing any practical power factor correction scheme must be to obtain accurate details of the load conditions with values of kW, kVA and power factor at light, average and full load, together with type and details of the loads.

This may be achieved in one of the following ways: measurement of kW and kvar; measurement of voltage, current and kW; measurement of kVA and kvar; or use of a power factor indicating instrument, voltmeter and ammeter.

Use of tariff metering

In many instances it is possible to use the supply authority's meter to arrive at a plant loading condition. On the disc of the meter will be found a small mark, usually a red or black band, which can be watched. Count the number of revolutions of the disc for about one minute, note the number of revolutions made, and the time in seconds to make the revolutions, then:

$$X = 3600 \ N/Rt$$

where X = instantaneous reading in kWh (kvarh)
$\quad\quad N$ = number of revolutions of the disc in t seconds
$\quad\quad t$ = time in seconds for N revolutions
$\quad\quad R$ = meter constant in revolutions per kWh (kvarh).

The meter constant is stamped on the rating plate of the meter and is in revolutions per kWh or kvarh.

Where the meter constant is shown as units per revolution U, then the formula becomes:

$$X = 3600 \ NU/1000\,t$$

The meter constant in this case is in either Wh or var per revolution.

It should be appreciated that the readings obtained by this method are instantaneous so that they should be taken when load conditions are normal.

METHODS OF CORRECTION

Each power factor correction scheme requires individual consideration and, as the successful operation of a scheme depends largely on the correct positioning of the capacitors in the network, the importance of studying all relevant factors is emphasized. The relevant factors are: tariff in force; metering point; details of light, average, and full load kVA, kW and power factor; position of motors, welding equipment, transformers or other large plant causing bad power factor; and supply system problems such as harmonics.

Capacitors themselves do not generate harmonics, but they can either reduce or increase them, depending upon particular circumstances. The major sources of harmonics are such things as thyristors, rectifiers and arc furnaces.

The siting of the capacitors does, to some extent, depend on whether each piece of equipment, e.g. a motor, or a transformer, is being individually corrected or the plant as a whole, or part, is being corrected as a block (generally known as bulk or group correction). In the first case the capacitor and motor, or capacitor and transformer, are as close together as possible, in the second case the capacitor is located at some convenient point in the system, such as a substation.

On small installations individual correction can be applied to motors which are constantly in operation or, in the case of kVA md tariffs, on certain motors known to be in operation at the time of md. This method reduces the current loading on the distribution system with consequent improvement in voltage regulation and, generally speaking, is more economic. No additional switchgear is required as the capacitor is connected directly to the piece of equipment it is correcting, and can, therefore, be switched with that piece of plant.

Where a capacitor is connected to a motor it is connected directly across the motor terminals, and is switched with the motor starter, resulting in complete automatic control. The balance of the correction required can then be connected to the main busbars of the supply system, and controlled by a fuse-switch. It should be noted, however, that some supply authorities stipulate the maximum amount of kvar which may be switched in this manner.

Automatic switching of capacitors is recognized as an ideal method of obtaining the full electrical and financial benefits from a capacitor installation, the resulting economics and convenience far outweighing the initial cost. Optimum power factor is achieved under all conditions and there is no possibility of the equipment being inadvertently left out of commission. A bank of capacitors with the required total capacitor kvar controlled in equal stages by a multi-step relay and air break contactors connected to the main busbars is used in many applications.

Large industrial sites involving different kinds of manufacturing processes often require a combination of bulk and individual correction to provide the most economic means of power factor correction.

In providing for power factor correction it should be remembered that distribution boards and circuits can carry a greater useful load if the capacitors are installed as near as possible to the source of low power factor. For this reason either bulk or individual correction, rather than correction at the intake point, can almost invariably be justified.

Individual correction of motors

The practice of connecting a capacitor across the starter of an induction motor, and switching the motor and capacitor as one unit, is now universally established, and is to be recommended where there are not objections on technical and economic grounds. One size of capacitor gives an almost constant value of power factor over the normal load range since variations in motor kvar are comparatively small.

Care should be taken in deciding the kvar rating of the capacitor in relation to the magnetizing kVA of the machine. If the rating is too high damage may result to both motor and capacitor, because the motor, while still revolving after disconnection from the supply, may act as a generator by self-excitation and produce a voltage higher than the supply voltage. If the motor is switched on again before the speed has fallen to about 80% of the normal running speed, the high voltage will again be superimposed on the supply circuit and there may be a risk of damage to other types of equipment. As a general rule the correct size of capacitor for individual correction of a motor should have a kvar rating not exceeding 90% of the no-load magnetizing kVA of the machine.

Connecting a capacitor direct to a motor results in a lower load current under all load conditions and, therefore, the overload settings on the starter must be reduced in order to obtain the same degree of protection.

Correction of individual motors is to be recommended where they are used for group drives, or where they are used continuously during maximum load conditions, but it should not be applied where the motors are used for haulage, cranes, colliery winders, or where 'inching' or 'plugging' and direct reversal takes place. Individual correction of tandem, or two-speed motors should be avoided. If correction of a two-speed machine is necessary the capacitor should never be connected directly to the low speed component but a contactor arrangement installed using one capacitor for both windings. Care should also be taken when offering capacitors for direct connection to motors whose braking system is intended to be operated by loss of voltage, as the voltage remaining across the capacitor, when the main supply is removed, may be such as to prevent operation of the braking system.

Where capacitors are connected direct to motors it is not usual to provide the capacitor with any protection or isolating gear other than that afforded by the control gear of the machine. Separate protection of the capacitor is usually only provided when the drive is of such importance that it is undesirable for a failure of the capacitor to put the motor out of service. For the individual correction of h.v. motors, hbc fuses should be placed in the circuit between the motor and the capacitor.

Where star-delta starting is used a standard three-terminal delta-connected capacitor should be employed, which gives maximum power factor correction at the start when the power factor of the motor is low. The capacitor is connected as shown in Fig. 23.6.

Typical capacitor size for individual connection of standard motors is given in Table 23.5.

Individual correction of transformers

In any electrical distribution system the one item of plant in circuit continuously is the transformer. It is often convenient, therefore, to connect a capacitor directly across the transformer terminals. The benefit of this to a consumer charged on an

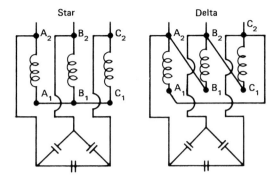

Fig. 23.6 Use of a three-terminal capacitor with a motor fitted with a star-delta starter.

Table 23.5 Standard capacitor kvar ratings for direct connection to 415 V three-phase 50 Hz induction motors.

Motor (kW)	Capacitor size related to motor speed					Motor (h.p.)
	two-pole 3000 r.p.m.	four-pole 1500 r.p.m.	six-pole 1000 r.p.m.	eight-pole 750 r.p.m.	twelve-pole 500 r.p.m.	
0.75	0.5	0.5	0.5	0.5	0.9	1.0
1.1	0.5	0.5	0.5	0.9	1.6	1.5
1.5	0.5	0.5	0.9	0.9	1.6	2.0
2.2	0.9	0.9	1.6	1.6	2.6	3.0
3.0	0.9	0.9	1.6	1.6	2.6	4.0
4.0	0.9	1.6	1.6	2.6	2.6	5.5
5.5	1.6	1.6	2.6	2.6	4.2	7.5
7.5	2.6	2.6	2.6	4.2	4.2	10.0
11.0	2.6	4.2	4.2	4.2	6.2	15.0
15.0	4.2	4.2	4.2	6.2	9.0	20.0
18.5	4.2	4.2	6.2	9.0	10.5	25.0
22.0	6.2	6.2	6.2	9.0	12.5	30.0
30.0	6.2	9.0	10.5	12.5	15.0	40.0
37.0	9.0	10.5	12.5	15.0	17.5	50.0
45.0	10.5	12.5	15.0	15.0	21.5	60.0
55.0	12.5	12.5	15.0	17.5	21.5	75.0
75.0	15.0	19.5	21.5	25.0	30.0	100.0
90.0	19.5	25.0	25.0	27.0	35.0	125.0
105.0	25.0	27.0	27.0	30.0	37.5	140.0
110.0	25.0	27.0	30.0	30.0	40.0	150.0
115.0	27.0	27.0	30.0	30.0	40.0	155.0
125.0	27.0	30.0	30.0	32.0	45.0	167.0
132.0	30.0	32.0	32.0	35.0	50.0	175.0
140.0	30.0	32.0	35.0	37.5	50.0	190.0
150.0	32.0	35.0	35.0	43.0	54.0	200.0
168.0	32.0	36.0	37.5	45.0	60.0	225.0
185.0	40.0	40.0	40.0	54.0	65.0	250.0

average power factor tariff is that the size of capacitor required to correct all the other plant can be reduced considerably. Where a tariff penalizes lagging kVA, transformer correction is a necessity. If the site has numerous transformers, and a high diversity of l.v. load, improving the power factor of the transformers often dispenses with the need for a central automatically controlled bank of capacitors.

The size of capacitor to be connected to the transformer depends on its no-load kVA. As a general rule up to 10% of the transformer rating, in kVA, is acceptable for the capacitor rating, e.g. the maximum size of capacitor for use with a 1000 kVA transformer is 100 kvar.

Bulk power factor correction

It is sometimes impossible to connect capacitors in the desired location owing to high temperature, restricted space, or the presence of explosive gases. Alternatively, it may not be economic to correct individually each piece of electrical equipment because they are small items in terms of electrical load. For these reasons, or when an installation operates with a high diversity factor, bulk correction may be employed.

Capacitors used for bulk power factor correction may be controlled manually or automatically. The manual type of scheme utilized fuse-switches, circuit-breakers or other switching devices to control the capacitors.

Manually controlled capacitors are normally employed for factory loads which are too small to warrant splitting the total capacitance required, or where high diversity of motor load makes individual correction uneconomic. Manual control can only be justified technically for continuous process work where minimum switching is required and where there is sufficient reactive kVA available in the circuit continuously to warrant the capacitor being connected all the time.

In the UK manual control of capacitors is limited to capacitors less than 40 kvar.

Automatic power factor correction

In larger installations, automatic power factor correction is being increasingly recognized as the ideal method of obtaining the full electrical and financial benefits from a capacitor installation; the resulting economies and convenience far overweigh the initial cost. Optimum power factor is achieved under all conditions and there is no possibility of the equipment being inadvertently left out of commission.

The equipment normally consists of a capacitor bank sub-divided into a number of equal steps, each step being controlled by a contactor. The contactors, in turn, are controlled by a relay responding to reactive kVA, of which there are several reliable makes available. The relay consists of a voltage coil connected across two phases of the load supply system, and the current coil, normally rates at 5 A, connected, through a current transformer, to the remaining phase.

The number of stages installed is usually a compromise between the technical requirements and cost. The aim of the automatic system is to have each contactor switching its maximum rated capacitance and, at the same time, have the capacitor bank divided into the most economic sub-sections, so that all meaningful variations in load can be corrected.

LOCATION OF CAPACITORS

Mention has, so far, only been made of the point of connection of the capacitor to the system to achieve the required, corrected power factor. In addition care must be taken as to the physical location of the capacitor to avoid problems that could lead to malfunction of the equipment.

As the maximum working temperature of capacitors is lower than that of other electrical equipment any conditions which give rise to unacceptable over-loads must be avoided. This, therefore, stipulates ambient temperature, maximum voltage and overload current.

Care should be taken in choosing a site for the capacitor to minimize the ambient temperature. If the site is outdoors, the direct rays of the sun should be avoided, especially in tropical climates. Rooms housing capacitor assemblies must be adequately ventilated.

If there are harmonic producers on site, such as thyristor-controlled machines, then the capacitors need to be located well away from the source of such harmonics, or designed to cope with any overloads produced in the capacitor by harmonic currents or voltages.

Because the ambient temperature is an important consideration in the life of any capacitor installation it is important that it is correctly specified by the purchaser to the supplier. Three temperature categories are available, as will be seen from Table 23.6.

Capacitors can be supplied for either indoor or outdoor use. It should be appreciated, however, that whilst an outdoor capacitor can be installed indoors, the reverse is not true.

Table 23.6 Limits of ambient temperature for capacitor installations (taken from IEC 871−1, *Shunt capacitors for a.c. power systems having a rated voltage above 1000 V* and IEC 831-1, *Shunt capacitors of the self healing type for a.c. systems having a rated voltage up to and including 1000 V*).

Symbol	Ambient air temperature (°C)		
	Max.	Highest mean over any period	
		24 h	1 year
A	40	30	20
B	45	35	25
C	50	40	30
D	55	45	35

INSTALLATION OF CAPACITORS

Installation and maintenance of capacitors should be carried out generally as specified in BS 1650: 1971, IEC 831 and IEC 871.

The cable supplying either an automatic capacitor bank, or a permanently connected capacitor should be fitted with some form of protection such as hbc fuses for capacitors up to 660 V, and thermal overloads and earth-fault protection for capacitors up to 11 kV and a means of isolating both capacitor and supply cable.

In selecting suitable switchgear, or fusegear, for capacitor duty it must be appreciated that the duty imposed on such equipment is more onerous then when used with other equipment of equivalent kVA loadings.

The reasons for this can be summarized as follows:

(1) At the instant of switching a large transient current flows.
(2) High overvoltage transients can occur when capacitors are disconnected by switching devices which allow restriking of the arc.
(3) The switchgear has to carry continuously the full rated current of the capacitor whenever it is in circuit, i.e. there is no allowance for diversity.
(4) At light loads, when the voltage may be higher than normal, the capacitor current is increased.
(5) If harmonics are present in the supply voltage the capacitor current is increased.

In view of these factors the following limitations are imposed on the capacitor by the manufacturer:

(1) The capacitor must be suitable for operation under abnormal conditions to an overvoltage of 1.1 times rated voltage.
(2) The capacitor must be suitable for continuous operation with a current of 1.3 times normal current.

In addition to these factors the capacitor manufacturer is permitted a manufacturing tolerance on output of $-0 + 10\%$.

In view of these conditions it is normal practice to derate switchgear and fuses used with capacitors. Tables 23.7 to 23.9 give recommended hbc fuse and switch sizes, together with cable sizes for low voltage capacitor applications. For systems not covered by the tables the advice of the capacitor manufacturer should be sought. This is specially important in the case of capacitors used above 660 V where large derating factors are required. At voltages of 3.3 kV and 11 kV it is not unusual to find fuses rated at three or four times capacitor current. With such large fuses it becomes important to ensure that the energy required for correct operation of the fuse is considerably less than the energy required to cause the capacitor tank to burst. It is also important to ensure that there is adequate discrimination between fuses used for capacitor protection and fuses further back in the system.

In addition to precautions when connecting the capacitor bank it is also necessary,

Table 23.7 Recommended sizes of fuses and cables for use with capacitors; PVC insulated cables to BS 6004/6346 having copper conductors (ambient temperature 30°C, conductor operating temperature 70°C).

Nominal capacitor current (A)	hbc fuse rating (A)	PVC armoured three-core (mm²)		PVC non-armoured single-core (mm²)	
		clipped direct	on perforated horizontal cable tray	clipped direct	on perforated horizontal cable tray
19.5	32	4	4	4	–
33	50	10	10	10	–
40	63	16	16	16	–
67	100	25	25	25	25
100	160	70	50	50	50
139	200	95	70	70	70
208	300	150	150	120	120
278	400	240	240	185	185
347	500	2 × 150	2 × 120	300	240
417	600	2 × 240	2 × 185	400	400
556	800	2 × 300	2 × 300	630	630

Table 23.8 Recommended sizes of fuses and cables for use with capacitors; cables with copper conductors having thermosetting insulation to BS 5467 (ambient temperature 30°C, conductor operating temperature 90°C).

Nominal capacitor current (A)	hbc fuse rating (A)	PVC armoured three-core (mm²)		PVC non-armoured single-core (mm²)	
		clipped direct	on perforated horizontal cable tray	clipped direct	on perforated horizontal cable tray
19.5	32				
33	50				
40	63	16	16		
67	100	25	25		
100	160	50	35		
139	200	70	70	50	50
208	300	120	95	95	95
278	400	185	150	150	120
347	500	240	240	240	185
417	600	2 × 150	300	300	240
556	800	2 × 240	2 × 185	630	500

Table 23.9 Recommended fuse-switch ratings for use with capacitors at various voltages.

Fuse-switch rating (A)	Nominal capacitor ratings (A)	Nominal capacitor kvar ratings at various voltages (V)				
		380	400	415	440	500
32	19.5	13	13.5	14	15	17
60	33	21.5	23	24	25	28.5
60	40	26	27.5	28.5	30	34.5
100	67	43	46	48	50	57
200	100	65	69	72	76	86
200	139	91	96	100	105	120
300	208	136	144	150	157	180
400	278	182	192	200	210	240
600	347	227	240	250	262	300
600	417	273	288	300	315	360
800	556	364	384	400	420	480
1000	695	455	480	500	525	600

in the case of automatic equipment, to supply a current and voltage signal to the reactive relay.

In the past reactive sensing relays were supplied as separate items and it was, therefore, necessary to provide both voltage and current signals. Modern practice, however, is to mount the relay within the automatic capacitor bank, thus only a current signal has to be provided. This is normally obtained by means of a current transformer having a 5 A secondary current. It is important, however, to ensure that the CT monitors the total load, including the capacitors, if the equipment is to work correctly. It is also important to note that there has to be a phase displacement between voltage and current signals to the relay. This phase displacement is dependent on the type of relay used.

After completing a capacitor installation the insulation resistance should be tested. This is done by shorting together all the capacitor terminals and applying a voltage between terminals and container. The value of this voltage, which must be maintained for 10 s is given in Table 23.10 which is taken from BS 1650: 1971.

Table 23.10 Routine test voltages between capacitor terminals and container.

Rated insulation level (kV r.m.s.)	Test voltage (kV r.m.s.)
0.6	3
1.2	6
2.4	11
3.6	16
7.2	22
12.0	28

It is important to note that such a test cannot be made between capacitor terminals as this will only result in the capacitor being charged and will not give an indication of insulation resistance.

CAPACITOR MAINTENANCE

Capacitors, being static apparatus, do not need the same attention as rotating machinery, but, nevertheless, require regular maintenance. Normally a power factor correction capacitor should be inspected at least every twelve months and preferably every six months. The time interval between inspections is, however, governed mainly by the conditions on site. Where capacitors are installed in a humid atmosphere, or are subjected to chemical fumes, or exposed to dirt and dust, more frequent attention should be given.

Before any examination always ensure that the capacitor has been disconnected from the supply, and then wait for at least one minute (capacitors up to 660 V) or five minutes (capacitors operative at voltages higher than 660 V) and then ensure discharge is complete by measuring the voltage between terminals. Finally, short all terminals together before testing. Shorting the terminals to earth is not effective for adequate discharge.

The following points should be observed when carrying out any capacitor maintanance.

Physical examination

(1) Examine externally that there is no damage or leakage of impregnant.
(2) Check that all cables are securely fixed and that all earth bonds are tight.
(3) Measure, if possible, the running temperature of the capacitor.

Testing

(1) Examine all insulators for signs of tracking, clean terminals and check for tightness.
(2) Ensure that discharge resistors, where fitted, are in order.
(3) Check all connections for tightness.
(4) Measure insulation resistance of the terminals to case.
(5) If a capacitance bridge is available measure the capacitance prior to energizing. If such a device is not available the capacitor should be energized and the line current measured by means of a 'clip-on' ammeter. The current measured should be compared with the current obtained from the following equation:

$$I = \frac{\text{kvar}}{\sqrt{3}V}$$

where V is line-to-line voltage.

In considering the question of maintenance it is important to know that most capacitors lose output during the course of their life. Some capacitors fail completely after a number of years of operation. The failure mode of most capacitors is a gradual loss of output without any noticeable signs of defect. Some capacitors, however, especially of the bulk oil filled variety, can fail with disastrous consequences.

Many capacitors, manufactured between the early 1950s and the late 1970s, are filled with a liquid whose chemical name is polychlorinated biphenyl generally called askarel. It has a variety of trade names, such as, Aroclor, Biclor, Pyraclor, etc. The use and disposal of this fluid is covered by stringent regulations, including the 'Disposal of Poisonous Wastes Act'. Legislation was introduced under section 100 of the Control of Pollution Act, 1974, to implement the sixth amendment (85/467 EEC) to directive 76/769 EEC, such that from 30 June 1986 it was illegal to manufacture, buy, sell or install pcb capacitors. Existing pcb filled capacitors may continue to remain in use only while they remain in a serviceable condition without leakage.

It is expected, however, following the Third North Sea Conference in 1990 and a consultation document issued by the Department of the Environment that all pcb equipment will need to be registered by the end of 1995 and must be destroyed by the end of 1999.

It is important, therefore, both from an economic point of view, and in order to comply with the Health and Safety at Work Act, that capacitor installations are checked periodically.

British Standards

Reference lists of applicable standards are given in the following chapters:

Chapter		Table
9	Lightning Protection	9.6
18	Lighting	18.4
19	Mains Cables	19.1
20	Selection of Wiring Systems	20.1

Reference is made in most chapters to BS 7671, Requirements for Electrical Installations (the IEE Wiring Regulations). A comprehensive list of applicable British Standards is given in Appendix 1 of the Wiring Regulations.

There are references to the following standards throughout this book.

British Standard no.	Subject	Page no. or chapter	British Standard EN no.
88	Low voltage fuses	Ch. 16, 579	60269-1
142	Characteristics for IDMTL relays	590	
148	Insulating oil for transformers and switchgear	148	
171	Power transformers	Ch. 13	
1010	Loading guide for transformers	325	
1013	Earthing code	65, 205	
1361	Cartridge fuses – domestic	579	
1362	Fuses – domestic and plugs	436	
1650	Installation and maintenance of capacitors	642	
2757	Thermal classification of electrical machines	406	
2914	Surge diverters	238	
3036	Fuses – semi enclosed up to 100 A	436, 545, 579	
3871	Circuit breakers – miniature	102, 543, 561, 579, 580, 581	60898
4099	Indicator lamps	47	
4343	Socket outlets – industrial	102, 246, 254	60309-2
4363	Distribution units on building sites	97, 249	
4553	PVC insulated split concentric cables	504	
4941	Contactors and motor starters	433	60947-4

Index

649